D1029602

HANDBOOK OF APPLIED MATHEMATICS FOR ENGINEERS AND SCIENTISTS

Other McGraw-Hill Books of Interest

Avallone & Baumeister • MARKS' STANDARD HANDBOOK FOR MECHANICAL ENGINEERS
Bhushan & Gupta • HANDBOOK OF TRIBOLOGY
Brady & Clauser • MATERIALS HANDBOOK
Bralla • HANDBOOK OF PRODUCT DESIGN FOR MANUFACTURING
Brunner • HANDBOOK OF INCINERATION SYSTEMS
Corbitt • STANDARD HANDBOOK OF ENVIRONMENTAL ENGINEERING
Dudley • GEAR HANDBOOK
Ehrich • HANDBOOK OF ROTORDYNAMICS
Elliot • STANDARD HANDBOOK OF POWERPLANT ENGINEERING
Freeman • STANDARD HANDBOOK OF HAZARDOUS WASTE TREATMENT AND DISPOSAL
Ganić & Hicks • THE McGRAW-HILL HANDBOOK OF ESSENTIAL ENGINEERING INFORMATION AND DATA
Gieck • ENGINEERING FORMULAS
Grimm & Rosaler • HANDBOOK OF HVAC DESIGN
Harris • HANDBOOK OF ACOUSTICAL MEASUREMENTS AND NOISE CONTROL
Harris & Crede • SHOCK AND VIBRATION HANDBOOK
Hicks • STANDARD HANDBOOK OF ENGINEERING CALCULATIONS
Jones • DIESEL PLANT OPERATIONS HANDBOOK
Juran and Gryna • JURAN'S QUALITY CONTROL HANDBOOK
Karassik et al. • PUMP HANDBOOK
Maynard • MAYNARD'S INDUSTRIAL ENGINEERING HANDBOOK
Parmley • STANDARD HANDBOOK OF FASTENING AND JOINING
Rohsenow, Hartnett, & Ganić • HANDBOOK OF HEAT TRANSFER FUNDAMENTALS
Rohsenow, Hartnett, & Ganić • HANDBOOK OF HEAT TRANSFER APPLICATIONS
Rosaler & Rice • STANDARD HANDBOOK OF PLANT ENGINEERING
Rothbart • MECHANICAL DESIGN AND SYSTEMS HANDBOOK
Shigley & Mischke • STANDARD HANDBOOK OF MACHINE DESIGN
Tuma • HANDBOOK OF NUMERICAL CALCULATIONS IN ENGINEERING
Wadsworth • HANDBOOK OF STATISTICAL METHODS FOR ENGINEERS AND SCIENTISTS
Young • ROARK'S FORMULAS FOR STRESS AND STRAIN

HANDBOOK OF APPLIED MATHEMATICS FOR ENGINEERS AND SCIENTISTS

Max Kurtz, P.E.

Consulting Engineer and Educator; Life Member, National Society of Professional Engineers; Author, Handbook of Engineering Economics, Structural Engineering for Professional Engineers' Examinations, Engineering Economics for Professional Engineers' Examinations, Comprehensive Structural Design Guide, Steel Framing of Hip and Valley Rafters; *Project Editor,* Civil Engineering Reference Guide; *Contributing Author,* Standard Handbook of Engineering Calculations

McGRAW-HILL, INC.

New York St. Louis San Francisco Auckland Bogotá
Caracas Hamburg Lisbon London Madrid
Mexico Milan Montreal New Delhi Paris
San Juan São Paulo Singapore
Sydney Tokyo Toronto

Library of Congress Cataloging-in-Publication Data

Kurtz, Max, date.
Handbook of applied mathematics for engineers and scientists / Max Kurtz.
p. cm.
Includes bibliographical references (p.) and index.
ISBN 0-07-035685-8
1. Engineering mathematics—Handbooks, manuals, etc. I. Title.
TA332.K87 1991
620'.001'51—dc20 91-711
CIP

1 2 3 4 5 6 7 8 9 0 DOC/DOC 9 7 6 5 4 3 2 1

ISBN 0-07-035685-8

The sponsoring editor for this book was Robert Hauserman, the editing supervisor was Peggy Lamb, and the production supervisor was Suzanne W. Babeuf. This book was set in Times Roman by Datapage.

Printed and bound by R. R. Donnelley & Sons Company.

CONTENTS

PREFACE

This handbook is intended as both a reference work and a guide in problem solving for engineers and scientists. It presents the definitions, basic equations, principles, and techniques of mathematics, and it illustrates their application by means of 313 fully solved examples. In addition to these numerical examples, the handbook explains in complete detail the procedures for solving other types of problems. Thus, this handbook presents both the structure and the dynamics of mathematics.

The material in this handbook ranges from elementary algebra and trigonometry to advanced calculus. While thoroughly covering the classical areas of mathematics, this handbook also covers those areas that have acquired major importance in modern engineering and science, such as boolean algebra, cybernetics, computer graphics, celestial mechanics, vector analysis, matrix algebra, Markov probability, and nondecimal numeral systems.

Moreover, this handbook stresses the computational techniques that are particularly suited for use with computers and calculators. For example, the sum of a power series can be found in minimum time by constructing the series in nested form, and Art. 1.18 demonstrates the procedure. Similarly, the values of a polynomial fraction corresponding to numerous values of the independent variable can be obtained in minimum time by transforming the fraction to an equivalent continued fraction, and Art. 1.21 demonstrates the procedure. Article 12.3.6 presents the Newton-Raphson method of finding the roots of an equation by successive approximation, which is now being widely applied.

This handbook uses a *practical* approach to mathematics wherever possible in order to make the material meaningful, interesting, and readily understandable. For example, the basic principles of boolean algebra are developed by referring to a simple illustrative situation. This device makes these principles instantly clear, and the entire subject thus becomes easily comprehensible rather than esoteric. Similarly, the laws of probability are developed by simple logic, using illustrative situations that keenly arouse the reader's interest. Example 9.12, in which we calculate the probability of winning a contrived state lottery, is typical in this respect. In Example 9.44, we demonstrate how the steady-state probabilities associated with a Markov process can be found instantly by use of a recurrent chain, which is easy to construct by applying the transition probabilities. Some individuals encounter difficulty grasping the terminology of statistical inference, and Sec. 10 solves highly simplified examples to help them surmount this difficulty.

This handbook has been designed to meet the specific mathematical needs of the engineer and scientist. As a result, it covers extensively certain areas that receive little or no attention in books in pure mathematics. For example, Art. 3.2.3 presents in complete detail those properties of the parabola that make this curve (and the paraboloid) extremely useful in the construction of numerous objects. Similarly, Sec. 12 explores the geometric properties of areas and solid bodies in a practical manner from the perspective of the engineer, for these properties arise in structural analysis, machine design, and fluid mechanics. With reference to quadric surfaces, the engineer wishes to know which surfaces can be constructed by the use of straight-line elements and which surfaces are developable, and Sec. 3 supplies this information. Article 3.4.10 demonstrates how the rectangular hyperbolic paraboloid is generated. Most problems in three-dimensional trigonometry that arise in practice involve simply the intersection of three planes; no sphere is present. Example 2.18 shows how the relationships of spherical trigonometry can be applied in solving these problems.

This handbook is self-contained, for the mathematical principles and techniques that are required for understanding one part of the book are present in another part. For example, in the study of computer graphics in Sec. 7 we apply matrix algebra, and the latter subject is presented in Sec. 5. Similarly, in the study of cybernetics in Sec. 11 we apply Markov probability, and the latter subject is presented in Sec. 9.

A problem in algebra that arises frequently in practice is that of transforming a second-degree equation with two unknowns to some simpler form, such as one that is devoid of the xy term. Article 3.1.7 offers a *visual* approach to this problem by means of Fig. 3.8. This diagram shows how the coefficients in the equation vary as the coordinate axes are rotated, and it thus makes the problem vivid, interesting, and simple to solve. Figure 3.8*a* is similar to Mohr's circle of stress and to Mohr's circle of moment of inertia, which appears in Fig. 12.47.

I am extremely grateful to my wife, Ruth Ingraham Kurtz, B.E. in E.E., project manager in computer systems development, for her invaluable assistance in the preparation of the manuscript. One of the many services she rendered was writing the computer program in Fig. 1.4. I can only thank her by means of this note. I am also grateful to the McGraw-Hill staff, and particularly Peggy Lamb, for encouragement and assistance in bringing this project to fruition.

Max Kurtz

HANDBOOK OF APPLIED MATHEMATICS FOR ENGINEERS AND SCIENTISTS

SECTION 1
ALGEBRA

1.1 CLASSIFICATION AND PROPERTIES OF NUMBERS

An *integer* is a whole number, and it can be positive, negative, or 0. A number that can be expressed in the form p/q, where p and q are integers and $q \neq 0$, is said to be *rational*; a number that cannot be expressed in this form is called *irrational.* For example, it can be demonstrated that the square root of any positive integer that is not a perfect square is irrational. Since we may set $q = 1$, all integers are rational.

The square root of a negative number is described as *imaginary*; for contradistinction, all other numbers are described as *real.* A number that is formed by combining a real and an imaginary number is called a *complex number.* Thus, $5 + \sqrt{-7}$ and $8 - \sqrt{-2}$ are complex numbers.

Let x denote a real number. The *absolute* (or *numerical*) *value* of x is denoted by $|x|$. The absolute value of a positive number equals the number itself; the absolute value of a negative number equals the corresponding positive number. Thus,

$$|9| = 9 \qquad |0| = 0 \qquad |-3| = 3$$

Let a, b, and c denote integers such that $a = bc$. The numbers b and c are called *factors* or *divisors* of a, and a is said to be a *multiple* of b and of c. For example, 8 is a factor of 32, and 32 is a multiple of 8. A positive integer that has factors in addition to 1 and itself is described as a *composite number*, and one that lacks such factors is called a *prime number.* By convention, 1 is not considered to be a prime number, and the first 10 prime numbers are 2, 3, 5, 7, 11, 13, 17, 19, 23, and 29. The *prime factors* of a number N are numbers that are primes and factors of N. For example, the prime factors of 20 are 2 and 5, and the prime factors of 36 are 2 and 3. Two positive integers are *relatively prime* if they have no common factors other than 1. Thus, 21 and 25 are relatively prime, but 21 and 24 are not.

Consider an equation of the form

$$a_0x^n + a_1x^{n-1} + a_2x^{n-2} + \cdots + a_n = 0$$

where the a's are all integers and n is a positive integer. A number that satisfies such an equation is called an *algebraic number*, and one that fails to do so is called a *transcendental number.* An algebraic number can be real or complex. For example, $3 + \sqrt{-7}$ is an algebraic number because it satisfies the equation $x^2 - 6x + 16 = 0$. Similarly, $\sqrt{5}$ and $\sqrt[3]{14}$ are algebraic numbers because they satisfy the equations $x^2 - 5 = 0$ and $x^3 - 14 = 0$, respectively. On the other hand, it is known that π is a transcendental number. It follows at once that every rational number is an algebraic number. For example, 9/17 is an algebraic number because it satisfies the equation $17x - 9 = 0$.

1.2 DEFINITIONS PERTAINING TO SUBTRACTION AND DIVISION

Let a and b denote real numbers. In the operation $a - b$, the quantities a and b are called the *minuend* and *subtrahend*, respectively.

Now let c and d denote two positive integers having the characteristics that $c > d$ and c is not a multiple of d. Let qd denote the largest multiple of d that is less than c. We may write $c = qd + r$, where $r < d$. If we divide c by d, the quantities q and r are the *quotient* and *remainder*, respectively. For example, for dividing 33 by 7, we may write $33 = 7 \times 4 + 5$. Therefore, when we divide 33 by 7, the quotient is 4 and the remainder is 5.

1.3 DEFINITIONS PERTAINING TO EXPRESSIONS

When numbers (or letters representing numbers) are connected by the signs for arithmetic operations, they constitute an *expression*. If an expression involves solely multiplication, division, and exponentiation, either singly or in combination, it is called a *term*. Thus, $8x^3/y^2$ is a term. When terms are combined by addition or subtraction (or both), they constitute a *multinomial*. Thus, $2xy^4 + 7x^{1.8} - 6y^2 + 5z$ is a multinomial. (A single number or letter may be regarded as a special term for defining a multinomial.) A *binomial* and *trinomial* are multinomials consisting of two and three terms, respectively.

If all exponents in a multinomial are positive integers, the expression is a *polynomial*. For example, $7xy^5 + 9x^3 - 3y^2$ is a polynomial. Assume that a polynomial contains the term $ax^py^qz^r$, where x, y, and z are variables and a is a constant. The *degree* of this term is the sum of the exponents of the variables, or $p + q + r$. The degree of a polynomial equals the highest degree of its terms. As an illustration, consider the polynomial $8x^2y - 13xy^3 - 2y^2z^5$. The third term is of highest degree, namely, 7; therefore, the polynomial is of degree 7.

1.4 LAWS OF ALGEBRAIC OPERATIONS

The basic laws pertaining to addition and multiplication are as follows:

1. Addition and multiplication are *commutative*. Thus,

$$a + b = b + a \qquad \text{and} \qquad ab = ba$$

2. Addition and multiplication are *associative*. Thus,

$$(a + b) + c = a + (b + c) \qquad \text{and} \qquad (ab)c = a(bc)$$

3. Multiplication is *distributive* with respect to addition and subtraction. Thus,

$$a(b+c)=ab+ac \qquad \text{and} \qquad a(b-c)=ab-ac$$

By an extension of the distributive law, we have

$$(a+b)(c+d)=ac+ad+bc+bd$$

Let a and b denote positive numbers; then $-a$ and $-b$ are negative numbers. The laws governing operations with negative numbers are as follows:

$$a+(-b)=a-b \qquad a-(-b)=a+b$$

$$a(-b)=-ab \qquad (-a)(-b)=ab$$

$$\frac{a}{-b}=\frac{-a}{b}=-\frac{a}{b} \qquad \frac{-a}{-b}=\frac{a}{b}$$

Now let c and d denote any real numbers. The number c/d is positive if c and d are both positive or both negative; otherwise, c/d is negative.

The laws governing the addition, multiplication, and division of fractions are as follows:

$$\frac{a}{b}+\frac{c}{b}=\frac{a+c}{b}$$

$$\frac{a}{b}\frac{c}{d}=\frac{ac}{bd} \qquad \frac{a/b}{c/d}=\frac{a}{b}\frac{d}{c}=\frac{ad}{bc}$$

The numerator and denominator of a fraction can be multiplied or divided by the same number without changing the value of the fraction. The fraction is then said to be converted to an *equivalent* fraction. Expressed symbolically,

$$\frac{a}{b}=\frac{ac}{bc}=\frac{a/c}{b/c}$$

The plus and minus signs are referred to as *algebraic signs*. Where an algebraic sign is omitted, the plus sign is understood. If an expression in parentheses is to be added to or subtracted from some quantity, the parentheses may be removed in accordance with this rule: Keep the algebraic signs within the expression if the expression is to be added; change the algebraic signs within the expression if the expression is to be subtracted. Thus,

$$a+(b-c-d+e-f)=a+b-c-d+e-f$$

$$a-(b-c-d+e-f)=a-b+c+d-e+f$$

1.5 EXPONENTS

Let m and n denote positive integers. The notation a^n denotes a number having a as a factor n times. Thus,

$$3^5=3\cdot 3\cdot 3\cdot 3\cdot 3$$

In the expression a^n, the number n is called the *exponent* or *power of a*. From the definition of an exponent, it follows that

$$a^m a^n = a^{m+n} \tag{1.1}$$

$$(a^m)^n = (a^n)^m = a^{mn} \tag{1.2}$$

$$(ab)^n = a^n b^n \tag{1.3}$$

$$\left(\frac{a}{b}\right)^n = \frac{a^n}{b^n} \tag{1.4}$$

Assume that $m > n$. Then

$$\frac{a^m}{a^n} = a^{m-n} \tag{1.5}$$

By extension, the foregoing laws are assumed to be valid for *all* values of m and n. Therefore,

$$a^0 = \frac{a^n}{a^n} = 1 \tag{1.6}$$

$$a^{-n} = \frac{a^0}{a^n} = \frac{1}{a^n} \tag{1.7a}$$

$$a^n = \frac{1}{a^{-n}} \tag{1.7b}$$

As an illustration, we have the following:

$$\frac{a^{2.5}(a^{1.2})^3}{a^{0.7}} = \frac{a^{2.5}a^{3.6}}{a^{0.7}} = a^{2.5}a^{3.6}a^{-0.7} = a^{5.4}$$

From Eq. (1.2), we have

$$(a^{1/n})^n = (a^n)^{1/n} = a^1 = a$$

If $a = b^n$, then $a^{1/n} = b$, and we say that b is the *nth root of a*. Alternatively, the nth root of a can be denoted by $\sqrt[n]{a}$. In this notation, the number a is called the *radicand* and n is called the *index*. Thus, $\sqrt[n]{a} = a^{1/n}$. We also have

$$a^{m/n} = (a^{1/n})^m = (a^m)^{1/n}$$

Consider the quantity e^A, where e is the number defined in Art. 1.28. If the expression for A is rather involved, it is convenient to replace the symbol e^A with the expression "exp A." For example,

$$8 + 3 \exp x^2 = 8 + 3e^{x^2}$$

The quantity $(-1)^n$ has the value $+1$ if n is an even integer and -1 if n is an odd integer.

1.6 LOGARITHMS

Let a denote a positive number other than 1, and let $a^x = N$. Then x is the *logarithm of N* to the base a. Expressed symbolically, $x = \log_a N$. Thus, the logarithm of a number N is the *power* to which the base must be raised to obtain N. For example, since $6^3 = 216$, then $\log_6 216 = 3$. Since a is positive, N is positive, and the definition of a logarithm applies solely to positive numbers.

The following laws of logarithms stem from the laws of exponents:

$$\log_a PQ = \log_a P + \log_a Q \tag{1.8}$$

$$\log_a \frac{P}{Q} = \log_a P - \log_a Q \tag{1.9}$$

$$\log_a N^u = u \log_a N \tag{1.10}$$

$$\log_a b = \frac{1}{\log_b a} \tag{1.11}$$

Let $N = P/Q$. From Eq. (1.9), we deduce the following relationships, which are independent of the base: $\log N > 0$ if $N > 1$; $\log N = 0$ if $N = 1$; $\log N < 0$ if $N < 1$. From the definition of a logarithm, we have $\log_a a = 1$.

Figure 1.1 is the graph of $\log_a N$ versus N where $a > 1$.

Two systems of logarithms are widely used. The *common* or *Briggs system* uses 10 as base; the *natural* or *napierian system* uses as base the number e, which is defined in Art. 1.28. The notation $\ln N$ is often used in place of $\log_e N$. Both common and natural logarithms are directly obtainable by use of the calculator. If the logarithm of a number in some other system is required, the value can be obtained by a simple procedure that we shall now illustrate.

EXAMPLE 1.1 Evaluate $\log_3 17$.

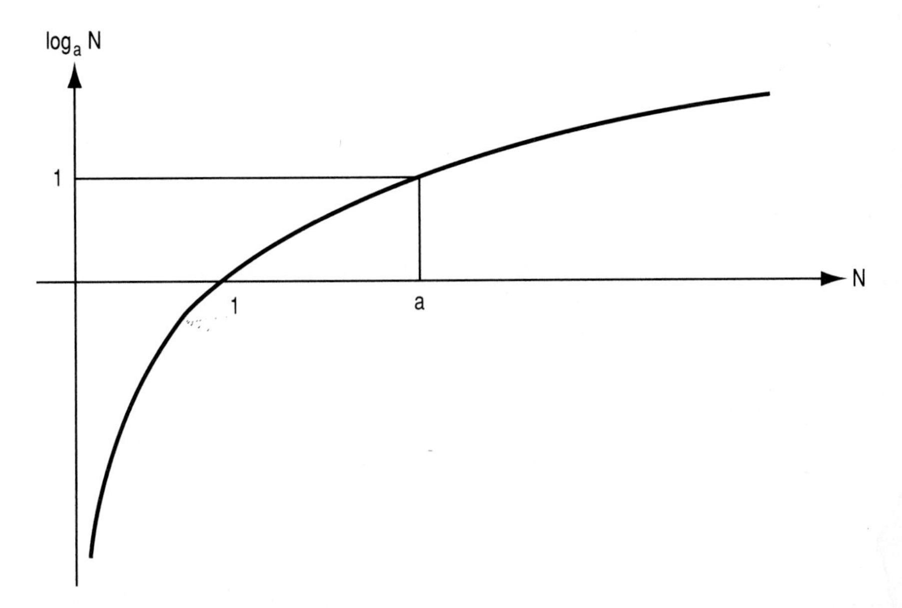

FIGURE 1.1 Logarithmic curve.

SOLUTION Let x denote the logarithm. Then $3^x = 17$. Applying Eq. (1.10), we have

$$x \log_{10} 3 = \log_{10} 17$$

$$x = \frac{\log_{10} 17}{\log_{10} 3} = \frac{1.230449}{0.477121} = 2.57890$$

As proof of this result, we obtain the following by calculator: $3^{2.57890} = 17$.

Generalizing on the basis of Example 1.1, we obtain

$$\log_b N = \frac{\log_a N}{\log_a b} \tag{1.12}$$

This equation provides a means of converting the logarithm of a number from one system to another.

1.7 DECIMAL FRACTIONS

Let a and b denote two positive integers, with $a < b$. The fraction a/b is in its *lowest terms* if a and b are relatively prime. Now assume that a/b is to be converted to an equivalent decimal fraction; i.e., to one having the form $c/10^m$, where c and m are positive integers. This conversion is possible if b has no prime factors other than 2 and 5. Thus, b can have the value 2, 5, 10, 20, 40, 50, 80, 100, etc. On the other hand, b cannot have the value 60, since this number has 3 as a prime factor. Thus, we have $4/5 = 8/10 = 0.8$, and $31/80 = 3875/10{,}000 = 0.3875$. The decimal fractions 0.8 and 0.3875 are described as *terminating*.

A fraction that fails to meet the stipulated requirement can be converted to an equivalent *infinite* decimal fraction. Where a decimal fraction is infinite, a digit or group of digits eventually repeats itself indefinitely. For example, $3/11 = 0.272727\ldots$, $69/110 = 0.6272727\ldots$, and $169/220 = 0.76818181\ldots$. By convention, an infinite decimal fraction is displayed by recording the repeating part only once and placing a bar above it. Thus, we write $1/3 = 0.\overline{3}$ and $169/220 = 0.76\overline{81}$.

1.8 OPERATIONS WITH COMPLEX NUMBERS

Let a denote a number that is real and positive. By Eq. (1.3), we have $\sqrt{-a} = \sqrt{a(-1)} = \sqrt{-1}\,\sqrt{a}$. The number $\sqrt{-1}$ is taken as the unit of imaginary numbers, and it is denoted by i. (In electrical engineering, the designation j is used.) We may therefore write

$$\sqrt{-a} = i\sqrt{a}$$

For example,

$$\sqrt{-10} = i\sqrt{10} \qquad \sqrt{-16} = i\sqrt{16} = 4i$$

Starting with i^0 and multiplying successively by i, we obtain the following:

$$i^0 = 1 \qquad i^1 = i \qquad i^2 = -1 \qquad i^3 = -i$$

This set of values then repeats itself cyclically. Therefore, if n is any integer, positive or negative, we have

$$i^{4n} = 1 \qquad i^{4n+1} = i \qquad i^{4n+2} = -1 \qquad i^{4n+3} = -i$$

For example, $i^{12} = 1$, $i^{19} = -i$, $i^{25} = i$, and $i^{-17} = i^{4(-5)+3} = -i$.

In the following material, all lowercase letters except i denote real numbers. A complex number, which is the sum of a real and an imaginary number, can be represented by $a + bi$. Let $A = a + bi$ and $A' = a' + b'i$. The numbers A and A' are equal if and only if $a = a'$ and $b = b'$. Now let $C = c + di$. We have the following:

$$A + C = (a + bi) + (c + di) = (a + c) + (b + d)i$$

$$A - C = (a + bi) - (c + di) = (a - c) + (b - d)i$$

$$AC = (a + bi)(c + di) = (ac - bd) + (ad + bc)i$$

$$\frac{A}{C} = \frac{a + bi}{c + di} = \frac{(ac + bd) + (bc - ad)i}{c^2 + d^2}$$

(Operations with complex fractions are discussed in Art. 1.9.)

Again let $A = a + bi$ and let $A^* = a - bi$. The numbers A and A^* are called *conjugates* of one another. Then

$$A + A^* = 2a \qquad A - A^* = 2bi \qquad AA^* = a^2 + b^2$$

$$\frac{A}{A^*} = \frac{(a^2 - b^2) + 2abi}{a^2 + b^2}$$

The reciprocal of a complex number is

$$\frac{1}{a + bi} = \frac{a - bi}{a^2 + b^2}$$

Operations with complex numbers can also be performed vectorially, in the manner discussed in Art. 2.1.9.

1.9 SIMPLIFYING AN EXPRESSION

It is often advantageous to replace a given algebraic expression with an equivalent expression of simpler form. There are three types of expressions that lend themselves to simplification, and they are as follows:

1. *Expressions with irrational numbers:* It is sometimes possible to transform the given expression to one where the radicand (the number under the radical sign) is of lesser value. As an illustration, consider the number $\sqrt[3]{448}$. The radicand 448 has 64 as a factor, and 64 is a perfect cube. By Eq. (1.3), we have

$$\sqrt[3]{448} = \sqrt[3]{64 \times 7} = \sqrt[3]{64}\,\sqrt[3]{7} = 4\,\sqrt[3]{7}$$

2. *Fractions with irrational denominators:* It is sometimes possible to replace the fraction with one in which the denominator is rational. This process is known as *rationalizing the denominator*. For example,

$$\frac{a}{b+\sqrt{c}} = \frac{a}{b+\sqrt{c}}\frac{b-\sqrt{c}}{b-\sqrt{c}} = \frac{a(b-\sqrt{c})}{b^2-c}$$

Following a similar procedure, we obtain the following:

$$\frac{a}{b^{1/2}+c^{1/2}} = \frac{a(b^{1/2}-c^{1/2})}{b-c}$$

$$\frac{a}{b^{1/2}-c^{1/2}} = \frac{a(b^{1/2}+c^{1/2})}{b-c}$$

$$\frac{a}{b^{1/3}+c^{1/3}} = \frac{a(b^{2/3}-b^{1/3}c^{1/3}+c^{2/3})}{b+c}$$

$$\frac{a}{b^{1/3}-c^{1/3}} = \frac{a(b^{2/3}+b^{1/3}c^{1/3}+c^{2/3})}{b-c}$$

3. *Fractions in which the denominator is a complex number:* The fraction can be replaced with one in which the denominator is a real number, in this manner:

$$\frac{a}{b+ci} = \frac{a}{b+ci}\frac{b-ci}{b-ci} = \frac{a(b-ci)}{b^2+c^2}$$

1.10 PROPORTIONS AND PROPORTIONALITIES

The notation $a : b = c : d$ is read "a is to b as c is to d," and it means that $a/b = c/d$. Thus, a and b have the same proportion (or ratio) as c and d.

The notation $y \propto x$ means that y is directly proportional to x, or y varies directly as x. If k is a constant, the relationship can be expressed as $y = kx$.

Similarly, the notation $y \propto 1/x$ means that y is inversely proportional to x, or y varies inversely as x. The relationship can be expressed as $y = k/x$.

Numerical problems pertaining to variation can be solved by expressing the given information in the form of an equation by use of the constant k.

EXAMPLE 1.2 A simply supported steel shaft of diameter D and length L is subjected to a uniformly distributed load W. The deflection d of the shaft varies directly as W, directly as L^3, and inversely as D^4. The deflection is 3.88 cm when $D = 10$ cm, $L = 6$ m, and $W = 14$ kN. Find the deflection when $D = 11.5$ cm, $L = 7$ m, and $W = 13.2$ kN.

SOLUTION The equation for d becomes

$$d = \frac{kWL^3}{D^4}$$

Applying the subscripts 1 and 2 to denote the original and revised conditions, respectively, we have

$$\frac{d_2}{d_1} = \frac{W_2}{W_1}\left(\frac{L_2}{L_1}\right)^3\left(\frac{D_1}{D_2}\right)^4$$

$$d_2 = 3.88\left(\frac{13.2}{14}\right)\left(\frac{7}{6}\right)^3\left(\frac{10}{11.5}\right)^4 = 3.32 \text{ cm}$$

1.11 DEFINITIONS PERTAINING TO EQUATIONS

An equation that contains letters representing numbers is described as *literal.* A literal equation is termed an *identity* if arbitrary values can be assigned to the letters, and a *conditional equation* if such is not the case. For example, the equations $(x + y)(x - y) = x^2 - y^2$ and $\sin^2 x + \cos^2 x = 1$ are universally true, and therefore they are identities. On the other hand, the equation $x^2 + 2x = 15$ holds only if x is 3 or -5, and therefore it is a conditional equation. Some authors restrict the symbol $=$ to a conditional equation and use the symbol $\equiv$ for an identity.

The letters in a conditional equation are called *unknowns*, and a set of values of the unknowns that satisfies the equation is termed a *solution.* Thus, $x = 3$, $y = -7$ is a solution of the equation $x^2 + 2y + 5 = 0$. If the equation contains a single unknown, a solution is also called a *root* of the equation.

An equation with multiple unknowns is described as *symmetrical* if any interchange of the unknowns leaves the equation unchanged. Thus, the equation $A(x^2 + y^2 + z^2) + Bxyz + C(x + y + z) + D = 0$ is symmetrical.

A *polynomial equation* is formed when a polynomial is set equal to 0. The degree of a polynomial equation equals that of the polynomial itself. For example, the equation $x^2y + 5x - 7y = 0$ is of the third degree. Equations of the first, second, third, and fourth degrees are known, respectively, as *linear*, *quadratic*, *cubic*, and *quartic* (or *biquadratic*) equations.

1.12 SIMULTANEOUS LINEAR EQUATIONS IN TWO UNKNOWNS

Consider the following system of simultaneous linear equations containing the unknowns x and y:

$$3x + 4y = 19$$

$$7x - 6y = -63$$

The coefficients of y have 12 as their lowest common multiple. Therefore, we can obtain an equivalent system of equations in which the coefficients of y are identical if we multiply the first equation by 3 and the second by 2. The result is

$$9x + 12y = 57$$

$$14x - 12y = -126$$

Adding the two equations, we obtain $23x = -69$, and $x = -3$. Substituting this value of x in either equation, we obtain $y = 7$.

The general form of this system of equations is

$$a_1 x + b_1 y = c_1$$

$$a_2 x + b_2 y = c_2$$

The foregoing method of solution yields the following:

$$x = \frac{b_2 c_1 - b_1 c_2}{a_1 b_2 - a_2 b_1} \qquad y = \frac{a_1 c_2 - a_2 c_1}{a_1 b_2 - a_2 b_1} \tag{1.13}$$

If $a_1/b_1 \neq a_2/b_2$, the denominator has a nonzero value, and the system has a definitive solution.

A system of simultaneous linear equations containing more than two unknowns can be solved by the methods presented in Arts. 5.7.4 and 5.7.5.

1.13 INEQUALITIES

In the following discussion, it is understood that all quantities are real. If $a - b$ is positive, $a > b$; if $c - d$ is negative, $c < d$. Two inequalities have the *same sense* or *opposite sense* according to whether their inequality signs point in the same or opposite directions, respectively.

An *absolute* inequality is one that is not subject to restrictions; a *conditional* inequality is one that contains variables and is valid for a restricted range of values of the variables. For example, $4 > -9$ and $x^2 + y^2 > 0$ are absolute inequalities. On the other hand, $x^2 + 7 > 16$ is a conditional inequality because it applies only if $|x| > 3$.

The laws governing operations with inequalities are as follows:

1. The sense of an inequality is not changed if both sides are increased or decreased by the same number.

   For example, if $3x > 8$, then $3x - 2 > 6$. Similarly, if $a + b > c$, then $a > c - b$. Thus, a term can be transposed from one side of the inequality to the other if its algebraic sign is changed.

2. The sense of an inequality is not changed if both sides are multiplied or divided by the same *positive* number.

   For example, if $4x + 7y > 23$, then $8x + 14y > 46$.

3. The sense of an inequality is reversed if both sides are multiplied or divided by the same *negative* number.

   For example, if $3x - 9y < 15z$, then $-x + 3y > -5z$.

Many inequalities stem from the fact that a^n is positive if n is an even integer, irrespective of the algebraic sign of a.

EXAMPLE 1.3 Let a and b denote two numbers other than 0, with $a \neq b$. Prove that

$$\frac{a}{b} + \frac{b}{a} > 2 \tag{a}$$

if a and b are both positive or both negative, and that

$$\frac{a}{b}+\frac{b}{a}<-2 \qquad (b)$$

if one number is positive and the other negative.

SOLUTION We start with $(a-b)^2>0$, expand the expression, and transpose, obtaining the following:

$$a^2-2ab+b^2>0 \qquad a^2+b^2>2ab$$

First consider that a and b have the same algebraic sign. If we divide both sides of the last inequality by ab, the sense of the inequality remains, and Inequality (a) results.

If we now multiply both sides of Inequality (a) by -1, we obtain

$$-\frac{a}{b}-\frac{b}{a}<-2 \qquad (c)$$

Now set $a'=-a$. Then $a'/b=-(a/b)$ and $b/a'=-(b/a)$. Inequality (c) now becomes

$$\frac{a'}{b}+\frac{b}{a'}<-2$$

Inequality (b) is thus established.

As an illustration, let $a=-4$ and $b=-5$. Then

$$\frac{-4}{-5}+\frac{-5}{-4}=0.8+1.25=2.05>2$$

Now let $a=4$ and $b=-5$. Then

$$\frac{4}{-5}+\frac{-5}{4}=-0.8-1.25=-2.05<-2$$

1.14 NOTATION FOR FUNCTIONS, SUMS, AND PRODUCTS

The notation $f(x)$ denotes a function of x, and $f(r)$ denotes the value of $f(x)$ when $x=r$. Thus, if

$$f(x)=4x^3-7x^2+9$$

$$f(5)=4(5^3)-7(5^2)+9=334$$

Where several functions are present, they can be denoted by $f(x)$, $g(x)$, $h(x)$, etc.

The Greek letters Σ and Π are used to denote the sum and product, respectively, of a group of homologous expressions in which the variable increases successively by 1. Entries placed above and below the symbol identify the variable and indicate its range of values. For example,

$$\sum_{x=3}^{7} \frac{x+1}{x^2} = \frac{4}{3^2} + \frac{5}{4^2} + \frac{6}{5^2} + \frac{7}{6^2} + \frac{8}{7^2}$$

$$\prod_{u=2}^{5} (x+u) = (x+2)(x+3)(x+4)(x+5)$$

The variable in the sigma or pi notation may be a subscript, as in the following case:

$$\sum_{j=2}^{7} a_j = a_2 + a_3 + a_4 + a_5 + a_6 + a_7$$

Where a double-subscript system is employed, it is necessary to use a separate letter for each subscript. As an illustration,

$$\sum_{i=4}^{7} \sum_{j=1}^{3} a_{ij} = a_{41} + a_{42} + a_{43} + a_{51} + a_{52} + a_{53} + a_{61} + a_{62} + a_{63} + a_{71} + a_{72} + a_{73}$$

1.15 MATHEMATICAL INDUCTION

Where a proposition is believed to be true but a direct proof is difficult to secure, an indirect proof is sometimes obtainable by a process of reasoning called *mathematical induction*. To illustrate the process, we shall prove the following:

$$1 \cdot 2 + 2 \cdot 3 + 3 \cdot 4 + \cdots + n(n+1) = \frac{n(n+1)(n+2)}{3} \qquad (d)$$

Let S_n denote the sum of the terms at the left, and set $n = 1$. The left and right sides of Eq. (d) yield the following values, respectively:

$$S_1 = 1 \cdot 2 = 2 \qquad S_1 = \frac{1 \cdot 2 \cdot 3}{3} = 2$$

Thus, Eq. (d) is valid for $n = 1$. Now assume it is valid for $n = k$, where k is an arbitrary positive integer. Adding the $(k+1)$th term in the series and expressing it as $3(k+1)(k+2)/3$, we obtain

$$S_{k+1} = \frac{k(k+1)(k+2)}{3} + \frac{3(k+1)(k+2)}{3}$$

$$= \frac{(k+1)(k+2)}{3}(k+3) = \frac{(k+1)(k+2)(k+3)}{3}$$

Comparing the expression for S_{k+1} with the expression in Eq. (d), we arrive at this conclusion: If Eq. (d) is valid for $n = k$, it is also valid for $n = k + 1$. We have already demonstrated that Eq. (d) is valid for $n = 1$. Therefore, it is valid for $n = 2$. Since it is valid for $n = 2$, it is also valid for $n = 3$; etc. It follows that Eq. (d) is valid for all positive integral values of n.

As a second illustration, we shall demonstrate that $x - y$ is a factor of the expression $x^n - y^n$ for all positive integral values of n. For this purpose, we write

$$x^{k+1} - y^{k+1} = x(x^k - y^k) + y^k(x - y)$$

Therefore, if the proposition is true for $n = k$, it is also true for $n = k + 1$. The proposition is true for $n = 1$, and the proof is now complete.

1.16 FACTORIAL NUMBERS

Let n denote a positive integer. The symbol $n!$ (read "n factorial") denotes the product of the first n integers. The integers are usually recorded in descending order of magnitude. For example,

$$5! = 5 \cdot 4 \cdot 3 \cdot 2 \cdot 1 = 120$$

Since $n! = (n+1)!/(n+1)$, we set

$$0! = \frac{1!}{1} = 1$$

The product of a group of consecutive integers can be expressed as the quotient of two factorial numbers. For example,

$$17 \cdot 16 \cdot 15 \cdot 14 \cdot 13 = \frac{17!}{12!}$$

In applying the factorial notation, it is imperative that parentheses be inserted where required. Thus, $3m!$ means 3 times $m!$, not $(3m)!$.

A double-factorial notation is used to denote the product of even or odd integers exclusively, in this manner:

$$(2n)!! = (2n)(2n-2)(2n-4) \cdots 6 \cdot 4 \cdot 2$$

$$(2n-1)!! = (2n-1)(2n-3)(2n-5) \cdots 5 \cdot 3 \cdot 1$$

For example,

$$10!! = 10 \cdot 8 \cdot 6 \cdot 4 \cdot 2 = 3840$$

$$9!! = 9 \cdot 7 \cdot 5 \cdot 3 \cdot 1 = 945$$

The relationships between single-factorial and double-factorial numbers are as follows:

$$(2n)!! = 2^n n! \tag{1.14}$$

$$(2n-1)!! = \frac{(2n)!}{(2n)!!} = \frac{(2n)!}{2^n n!} \tag{1.15}$$

The meaning of $n!$ is extended in Art. 12.13.1.

1.17 SERIES

A sequence of numbers (or expressions that represent numbers) that is generated in a prescribed manner is termed a *progression*. When these numbers are connected with plus or minus signs, they constitute a *series*. We shall let u_k denote the kth term in the series, n the number of terms, and S the algebraic sum of the series.

A series is described as *finite* or *infinite* according to whether n is finite or infinite, respectively. An infinite series is *convergent* if S approaches a limiting value as n increases beyond bound, and *divergent* if such is not the case. In the following material, all letters except x denote constants.

A *power series* is one having the form

$$a_0 + a_1 x + a_2 x^2 + \cdots + a_n x^n$$

However, some coefficients may be 0. An *alternating series* is one in which plus and minus signs alternate. For example,

$$x - 2x^3 + 3x^5 - \cdots + (-1)^{n-1} n x^{2n-1}$$

is an alternating series. The following types of series arise very frequently.

Arithmetic Series. This is one in which $u_{k+1} - u_k$ is constant. The general form is

$$a + (a + d) + (a + 2d) + \cdots + [a + (n - 1)d]$$

The sum is

$$S = \frac{n}{2}(u_1 + u_n) = \frac{n}{2}[2a + (n - 1)d] \qquad (1.16)$$

Geometric Series. This is one in which u_{k+1}/u_k is constant. Let r denote this constant. The general form is

$$a + ar + ar^2 + \cdots + ar^{n-1}$$

The sum is

$$S = \frac{a(r^n - 1)}{r - 1} \qquad (1.17)$$

An infinite geometric series is convergent if $|r| < 1$, and the limit of the sum is

$$S = \frac{a}{1 - r} \qquad (1.17a)$$

Harmonic Series. This is one in which the reciprocals of the terms form an arithmetic progression. The following series is illustrative:

$$\frac{1}{3} + \frac{1}{5} + \frac{1}{7} + \cdots + \frac{1}{2n + 1}$$

Arithmogeometric Series. This is a composite of an arithmetic and geometric series, the general form being

$$a + (a + d)r + (a + 2d)r^2 + \cdots + [a + (n - 1)d]r^{n-1}$$

The sum is

$$S = \frac{a(r^n - 1)}{r - 1} + \frac{dr}{(r - 1)^2}[(n - 1)r^n - nr^{n-1} + 1] \qquad (1.18)$$

An infinite arithmogeometric series is convergent if $|r| < 1$, and the limit of the sum is

$$S = \frac{a}{1-r} + \frac{dr}{(1-r)^2} \tag{1.18a}$$

EXAMPLE 1.4 Setting $n = 35$, find the sum of the following series:

$$-85 - 81 - 77 - \cdots + (4n - 89)$$

SOLUTION This is an arithmetic series, and $u_{35} = 4(35) - 89 = 51$. The first form of Eq. (1.16) yields $S = (35/2)(-85 + 51) = -595$.

EXAMPLE 1.5 Setting $n = 29$, find the sum of the following series:

$$4 - 7 + 10 - 13 + \cdots + (-1)^{n-1}(3n + 1)$$

SOLUTION Let k denote an even positive integer. Then

$$u_k = -3k - 1 \qquad u_{k+1} = 3k + 4$$

$$u_k + u_{k+1} = 3$$

Thus, we can pair all terms beyond the first, and each pair has an algebraic sum of 3. The number of such pairs is $(29 - 1)/2 = 14$. Then $S = 4 + 14 \times 3 = 46$.

EXAMPLE 1.6 Setting $n = 13$, find the sum of the following series, to two decimal places:

$$115 - 115\left(\frac{6}{7}\right) + 115\left(\frac{6}{7}\right)^2 - \cdots + (-1)^{n-1}115\left(\frac{6}{7}\right)^{n-1}$$

SOLUTION This is a geometric series in which $a = 115$ and $r = -6/7$. Equation (1.17) yields

$$S = \frac{115[(-6/7)^{13} - 1]}{-6/7 - 1} = \frac{115[(6/7)^{13} + 1]}{6/7 + 1} = 70.27$$

EXAMPLE 1.7 Setting $n = 9$, find the sum of the following arithmogeometric series, to four decimal places:

$$3 - 27(0.8) + 51(0.8)^2 - 75(0.8)^3 + \cdots$$

SOLUTION

$$a = 3 \qquad d = 24 \qquad r = -0.8$$

Substituting in Eq. (1.18), we obtain $S = 11.2752$.

The following series, in which m is a positive integer, is a speical type of geometric series:

$$a^m + a^{m-1}b + a^{m-2}b^2 + \cdots + b^m = \frac{a^{m+1} - b^{m+1}}{a - b} \tag{1.19a}$$

If we replace b with $-b$, we obtain

$$a^m - a^{m-1}b + a^{m-2}b^2 - \cdots + (-1)^m b^m = \frac{a^{m+1} - (-1)^{m+1}b^{m+1}}{a+b} \tag{1.19b}$$

Let $p = m + 1$. On the basis of Eq. (1.19), we arrive at these conclusions: The binomial $a - b$ is a factor of $a^p - b^p$; the binomial $a + b$ is a factor of $a^p - b^p$ if p is even, and it is a factor of $a^p + b^p$ if p is odd.

The following are special types of arithmogeometric series:

$$a + 2a^2 + 3a^3 + \cdots + na^n = \frac{na^{n+1}}{a-1} - \frac{a(a^n - 1)}{(a-1)^2} \tag{1.20a}$$

$$a - 2a^2 + 3a^3 - \cdots + (-1)^{n-1}na^n = (-1)^{n-1}\frac{na^{n+1}}{a+1} + \frac{a[(-1)^{n-1}a^n + 1]}{(a+1)^2} \tag{1.20b}$$

An important class of series is that in which $u_k = k^m$, where m is a positive integer. We have the following:

$$1 + 2 + 3 + \cdots + n = \frac{n(n+1)}{2} \tag{1.21}$$

$$1^2 + 2^2 + 3^2 + \cdots + n^2 = \frac{n(n+1)(2n+1)}{6} \tag{1.22}$$

$$1^3 + 2^3 + 3^3 + \cdots + n^3 = \left[\frac{n(n+1)}{2}\right]^2 \tag{1.23}$$

$$1^4 + 2^4 + 3^4 + \cdots + n^4 = \frac{n(n+1)(2n+1)(3n^2 + 3n - 1)}{30} \tag{1.24}$$

Many other series may be viewed as composites of these basic exponential series, and their sums are found on this basis. As an illustration, consider the following:

$$S = 1 \cdot 2^2 + 2 \cdot 3^2 + 3 \cdot 4^2 + \cdots + n(n+1)^2$$

Expressing S in sigma form and taking the limits of k as 1 and n, we obtain

$$\begin{aligned} S &= \Sigma\, k(k+1)^2 = \Sigma\, (k^3 + 2k^2 + k) \\ &= \Sigma\, k^3 + 2\,\Sigma\, k^2 + \Sigma\, k \\ &= \frac{n^2(n+1)^2}{4} + \frac{2n(n+1)(2n+1)}{6} + \frac{n(n+1)}{2} \\ &= \frac{n(n+1)(3n^2 + 11n + 10)}{12} = \frac{n(n+1)(n+2)(3n+5)}{12} \end{aligned}$$

Similarly, consider the series

$$S = 1 \cdot 4 + 2 \cdot 5 + 3 \cdot 6 + \cdots + n(n+3)$$

Proceeding as before, we obtain

$$\begin{aligned} S &= \Sigma\, k(k+3) = \Sigma\, k^2 + 3\,\Sigma\, k \\ &= \frac{n(n+1)(2n+1)}{6} + \frac{3n(n+1)}{2} = \frac{2n(n+1)(n+5)}{6} \end{aligned}$$

Another important class of series is one in which

$$u_k = k(k+1)(k+2)\cdots(k+m)$$

The sum is

$$S = \frac{n(n+1)(n+2)\cdots(n+m)(n+m+1)}{m+2} \tag{1.25}$$

For example, setting $m = 2$, we have

$$1\cdot 2\cdot 3 + 2\cdot 3\cdot 4 + 3\cdot 4\cdot 5 + \cdots + n(n+1)(n+2) = \frac{n(n+1)(n+2)(n+3)}{4}$$

1.18 NESTED FORM OF A SERIES

Assume that we must find the sum of a finite series and that no simple expression for this sum exists. In this situation, it is necessary to construct the series, term by term. We shall present an iterative procedure that simultaneously constructs the series and calculates the sum. In many instances, this procedure is very well suited for use by the calculator or computer.

We take the ratio u_n/u_{n-1} as the starting point. In cycle 1 of the calculations, we add 1 to this ratio and then multiply by the ratio u_{n-1}/u_{n-2}. The expression formed thus far is

$$\frac{u_{n-1}}{u_{n-2}}\left(1 + \frac{u_n}{u_{n-1}}\right) = \frac{1}{u_{n-2}}(u_{n-1} + u_n)$$

In cycle 2, we add 1 to this series and then multiply by the ratio u_{n-2}/u_{n-3}. The expression formed thus far is

$$\frac{1}{u_{n-3}}(u_{n-2} + u_{n-1} + u_n)$$

We continue in this manner, allowing the value of each subscript to diminish successively by 1. In the final cycle, we add 1 and multiply simply by u_1. Thus, the series

$$u_1 + u_2 + u_3 + \cdots + u_n$$

can be expressed in the alternative form

$$u_1\left(1 + \frac{u_2}{u_1}\left(1 + \frac{u_3}{u_2}\left(1 + \cdots + \frac{u_n}{u_{n-1}}\right)\right)\right)$$

The latter is termed the *nested form* of the series.

As an illustration, consider the following series:

$$1!x - 2!x^2 + 3!x^3 - 4!x^4 + \cdots + (-1)^{n-1}n!x^n$$

For this series,

$$\frac{u_k}{u_{k-1}} = -\frac{k!x^k}{(k-1)!x^{k-1}} = -kx$$

Setting $n = 5$ for brevity and substituting in the general expression, we obtain the following nested form of this series:

$$x(1 - 2x(1 - 3x(1 - 4x(1 - 5x))))$$

That this expression yields the required series can be demonstrated by starting with the expression $1 - 5x$ in the innermost parentheses and then proceeding outward.

As a second illustration, consider the following series:

$$\frac{1!}{x^2} + \frac{3!}{x^4} + \frac{5!}{x^6} + \cdots + \frac{(2n-1)!}{x^{2n}}$$

For this series,

$$\frac{u_k}{u_{k-1}} = \frac{(2k-2)(2k-1)}{x^2}$$

With $n = 5$, the nested form of this series is

$$\frac{1}{x^2}\left(1 + \frac{2 \cdot 3}{x^2}\left(1 + \frac{4 \cdot 5}{x^2}\left(1 + \frac{6 \cdot 7}{x^2}\left(1 + \frac{8 \cdot 9}{x^2}\right)\right)\right)\right)$$

For compactness, the standard form of a series can be represented by the sigma notation. Analogously, the nested form can be represented by a notation that uses the Greek letter lambda. This system of representation exhibits the binomial expression that appears between successive parentheses. The first binomial appearing at the left is taken as the first term of the nested series. For example,

$$1 - 3x(1 - 6x(1 - 9x(1 - \cdots - 3nx))) = \bigwedge_{k=1}^{n} (1 - 3kx)$$

$$\frac{1}{x}\left(1 + \frac{2}{x}\left(1 + \frac{3}{x}\left(1 + \cdots + \frac{n}{x}\right)\right)\right) = \frac{1}{x}\bigwedge_{k=2}^{n}\left(1 + \frac{k}{x}\right)$$

Now consider a series in which the exponents of x form an arithmetic progression, the general form being

$$a_1x^b + a_2x^{b+d} + a_3x^{b+2d} + \cdots + a_nx^{b+(n-1)d}$$

If the coefficients are not factorial numbers, it may be preferable to express the series in the following alternative form:

$$x^b(a_1 + x^d(a_2 + x^d(a_3 + \cdots + x^d(a_n))))$$

For an alternating series, the plus signs are replaced with minus signs.

As an illustration, consider the series

$$\frac{x^2}{2} + \frac{x^5}{4} + \frac{x^8}{6} + \cdots + \frac{x^{2+3(n-1)}}{2n}$$

With $n = 5$, the alternative nested form of the series is

$$x^2\left(\frac{1}{2} + x^3\left(\frac{1}{4} + x^3\left(\frac{1}{6} + x^3\left(\frac{1}{8} + x^3\left(\frac{1}{10}\right)\right)\right)\right)\right)$$

Similarly, the series

$$\frac{1}{x} - \frac{3}{x^2} + \frac{5}{x^3} - \frac{7}{x^4} + \frac{9}{x^5}$$

has the nested form

$$\frac{1}{x}\left(1 - \frac{1}{x}\left(3 - \frac{1}{x}\left(5 - \frac{1}{x}\left(7 - \frac{1}{x}(9)\right)\right)\right)\right)$$

The important advantage of the nested form of a series is that it yields a symmetric procedure for finding the sum of a series by calculator or computer, by use of a loop. We shall illustrate this procedure.

EXAMPLE 1.8 Setting $x = 12$ and $n = 6$, find the sum S of the following series, to four decimal places:

$$\frac{1}{1!} + \frac{x^{0.3}}{2!} + \frac{x^{0.6}}{3!} + \cdots + \frac{x^{0.3(n-1)}}{n!}$$

SOLUTION

$$\frac{u_k}{u_{k-1}} = \frac{x^{0.3}}{k}$$

Therefore, the nested form of the series is

$$1 + \frac{x^{0.3}}{2}\left(1 + \frac{x^{0.3}}{3}\left(1 + \frac{x^{0.3}}{4}\left(1 + \frac{x^{0.3}}{5}\left(1 + \frac{x^{0.3}}{6}\right)\right)\right)\right)$$

Although this expression provides running calculations, we shall resolve the calculations into discrete steps to exhibit intermediate values. Let A_1 denote the value of the expression in the innermost parentheses, A_2 the value of the expression in the second innermost parentheses, etc. With $x^{0.3} = 12^{0.3} = 2.107436$, we have the following:

$$A_1 = 1 + \frac{2.107436}{6} = 1.351239$$

$$A_2 = 1 + \frac{2.107436}{5}(1.351239) = 1.569530$$

$$A_3 = 1 + \frac{2.107436}{4}(1.569530) = 1.826921$$

$$A_4 = 1 + \frac{2.107436}{3}(1.826921) = 2.283373$$

$$A_5 = 1 + \frac{2.107436}{2}(2.283373) = 3.406031$$

Then $S = 3.4060$.

1.19 SYNTHETIC DIVISION

It is frequently necessary to divide a polynomial in x by a binomial of the form $x - b$, where b is a constant. The operation can be performed rapidly by the use of *synthetic division*, and we shall illustrate the procedure.

EXAMPLE 1.9 Divide $2x^7 - 9x^6 + 31x^4 - 18x^3 + 20x^2 + 3x - 15$ by $x - 3$.

SOLUTION The powers of x must be consecutive, and therefore we add a term $0x^5$. The procedure consists of placing numbers in three horizontal rows with a summation line between the second and third rows, as shown below. We place the coefficients of the given polynomial in the first row and then record the value of b (in this case, 3) at the extreme right, in the manner shown. We shall refer to b as the *modulus*. We bring the first number in the first row (which is 2) down to the third row. From this point on, the procedure is iterative. In cycle 1, we multiply the number 2 in the third row by the modulus to obtain $+6$, and we place the result in the second column of the second row. We now add the numbers -9 and $+6$ in the second column to obtain -3, and we record this number in the third row. In cycle 2, we multiply the number -3 in the third row by the modulus to obtain -9, and we place the result in the third column of the second row. We now add the numbers $+0$ and -9 in the third column to obtain -9, and we record this number in the third row. Continuing in this manner, we obtain the results indicated. The first seven numbers in the third row are the coefficients of the quotient, and the last number is the remainder. We now supply the powers of x to the numbers in the third row. Thus, when the given polynomial is divided by $x - 3$, the quotient is $2x^6 - 3x^5 - 9x^4 + 4x^3 - 6x^2 + 2x + 9$, and the remainder is 12.

$$\begin{array}{l|l} 2 - 9 + 0 + 31 - 18 + 20 + 3 - 15 & \underline{3} \\ \quad + 6 - 9 - 27 + 12 - 18 + 6 + 27 & \\ \hline 2 - 3 - 9 + \ 4 - \ 6 + \ 2 + 9 + 12 & \end{array}$$

EXAMPLE 1.10 With reference to Example 1.9, divide the given polynomial by $x + 2$.

SOLUTION The modulus is now -2, and the results obtained are shown below. The quotient is $2x^6 - 13x^5 + 26x^4 - 21x^3 + 24x^2 - 28x + 59$, and the remainder is -133.

$$\begin{array}{l|l} 2 - \ 9 + \ 0 + 31 - 18 + 20 + \ 3 - \ 15 & \underline{-2} \\ \quad - \ 4 + 26 - 52 + 42 - 48 + 56 - 118 & \\ \hline 2 - 13 + 26 - 21 + 24 - 28 + 59 - 133 & \end{array}$$

1.20 POLYNOMIALS IN SINGLE VARIABLE

In Art. 1.3, we defined the degree of a polynomial. Let $f(x)$ denote a polynomial of nth degree in a single variable x. Then

$$f(x) = a_0x^n + a_1x^{n-1} + a_2x^{n-2} + \cdots + a_n$$

where n is a positive integer, the a's are all constants, and $a_0 \neq 0$. The number a_n can be regarded as the coefficient of x^0.

The notation $f(r)$ denotes the value of $f(x)$ when $x = r$. Consider that we have a polynomial in x and wish to evaluate $f(r)$. Straightforward substitution of r for x in the given expression is usually a cumbersome process, and two alternative methods of evaluation are available. The first method consists of applying the following principle, which is known as the *remainder theorem*:

Theorem 1.1. If a polynomial in x is divided by $x - r$, where r is a constant, the remainder is $f(r)$.

As an illustration, consider the polynomial

$$f(x) = 2x^6 - 9x^5 + 3x^4 + 11x^3 - 23x + 10$$

When this polynomial is divided by $x - 4$ and by $x + 3$, the remainders are 366 and 3670, respectively. Therefore, $f(4) = 366$ and $f(-3) = 3670$. (The division should be performed by use of synthetic division, which is presented in Art. 1.19.)

Now consider the polynomial

$$f(x) = (3 + 2i)x^5 + (9 + 17i)x^4 + (-41 + 13i)x^3 + (156 - 34i)x^2 + (-208 + 91i)x - 337 + 18i$$

We wish to evaluate this polynomial when $x = -7$. We can segregate the real and imaginary constants to form two distinct polynomials. With i discarded, they are as follows:

$$3x^5 + 9x^4 - 41x^3 + 156x^2 - 208x - 337$$

$$2x^5 + 17x^4 + 13x^3 - 34x^2 + 91x + 18$$

Division by $x + 7$ yields a remainder of -5986 for the first polynomial and 459 for the second. Therefore, $f(-7) = -5986 + 459i$.

The second method of evaluating a polynomial, which is known as *Horner's rule*, consists of casting the polynomial in nested form. Applying the alternative form in Art. 1.18, we have

$$f(x) = a_n + x(a_{n-1} + x(a_{n-2} + x(a_{n-3} + \cdots + x(a_0))))$$

As an illustration, consider the polynomial

$$4x^7 + 5x^6 + 18x^5 - 31x^3 + 16x^2 - 15x + 71$$

Its nested form is

$$71 + x(-15 + x(16 + x(-31 + x(0 + x(18 + x(5 + x(4)))))))$$

We shall evaluate this polynomial for $x = -3$, and we shall exhibit intermediate values. Let A_1 denote the value of the expression in the innermost parentheses, A_2 the value of the expression in the second innermost parentheses, etc. Then

$$A_1 = 4 \qquad A_2 = 5 + (-3)4 = -7$$

$$A_3 = 18 + (-3)(-7) = 39 \qquad A_4 = 0 + (-3)39 = -117$$

$$A_5 = -31 + (-3)(-117) = 320 \qquad A_6 = 16 + (-3)320 = -944$$

$$A_7 = -15 + (-3)(-944) = 2817 \qquad A_8 = 71 + (-3)2817 = -8380$$

Then $f(-3) = -8380$.

1.21 TRANSFORMATION OF POLYNOMIAL FRACTIONS TO CONTINUED FRACTIONS

A *polynomial fraction* is one in which both numerator and denominator are polynomials in a single variable x. For the present purpose, the terms of each polynomial will be arranged in ascending powers of x. We shall define two operations pertaining to a polynomial fraction: decomposition and factor inversion.

In *decomposition*, the fraction is expressed as the sum of a constant and a new fraction having the same denominator as the given fraction. For example, consider the fraction

$$\frac{24 + 17x - 9x^2}{6 + 3x}$$

We perform the following calculations:

$$\frac{24}{6} = 4 \qquad 4(6 + 3x) = 24 + 12x$$

$$24 + 17x - 9x^2 - (24 + 12x) = 5x - 9x^2$$

Therefore, we have

$$\frac{24 + 17x - 9x^2}{6 + 3x} = 4 + \frac{5x - 9x^2}{6 + 3x}$$

Factor inversion is applied to a polynomial fraction in which the numerator lacks a constant term. The variable x is factored out of the numerator, and the other factor is placed in the denominator. The following operation illustrates the procedure:

$$\frac{9x + 3x^2}{6 - 2x} = \frac{x(9 + 3x)}{6 - 2x} = \frac{x}{\dfrac{6 - 2x}{9 + 3x}}$$

We shall now illustrate the procedure by which a polynomial fraction is transformed to an expression containing a continued fraction.

EXAMPLE 1.11 Transform the following fraction:

$$f_1(x) = \frac{42 - 229x + 250x^2}{28 - 176x + 80x^2}$$

SOLUTION The calculations proceed in cycles, and each cycle consists of two steps: a decomposition and a factor inversion. In the first step, we operate upon the given fraction; in each subsequent step, we operate upon the fraction formed in the preceding step. The calculations are as follows:

$$f_1(x) = 1.5 + A$$

$$A = \frac{35x + 130x^2}{28 - 176x + 80x^2} = \frac{x}{B}$$

$$B = \frac{28 - 176x + 80x^2}{35 + 130x} = 0.8 + C$$

$$C = \frac{-280x + 80x^2}{35 + 130x} = \frac{x}{D}$$

$$D = \frac{35 + 130x}{-280 + 80x} = -0.125 + E$$

$$E = \frac{140x}{-280 + 80x} = \frac{x}{F}$$

$$F = \frac{-280 + 80x}{140} = -2 + G$$

$$G = \frac{80x}{140} = \frac{x}{H}$$

$$H = \frac{140}{80} = 1.75$$

In cycle 1, we decompose $f_1(x)$ to form $1.5 + A$, where A is the fraction shown on the following line. We then apply factor inversion to A to obtain x/B, where B is the fraction shown on the following line. In cycle 2, we decompose B to form $0.8 + C$, and we then apply factor inversion to C to obtain x/D. Continuing in this manner, we obtain the results displayed. By replacing $A, B, C, \ldots, H$ with their expressions, we obtain the following:

$$f_1(x) = 1.5 + \cfrac{x}{0.8 + \cfrac{x}{-0.125 + \cfrac{x}{-2 + \cfrac{x}{1.75}}}}$$

EXAMPLE 1.12 Transform the following fraction:

$$f_2(x) = \frac{105 - 232x + 28x^2 + 8x^3}{30 - 62x}$$

SOLUTION Proceeding as before, we obtain the following:

$$f_2(x) = 3.5 + \cfrac{x}{-2.0 + \cfrac{x}{2.5 + \cfrac{x}{0.5 + \cfrac{x}{-1.0 + \cfrac{x}{1.5}}}}}$$

EXAMPLE 1.13 Transform the following fraction:

$$f_3(x) = \frac{15 + 4x}{30 - 7x + x^2}$$

SOLUTION Since the numerator is of lower degree than the denominator, it is advantageous to invert the given fraction, thus obtaining

$$f_3(x) = \cfrac{1}{\cfrac{30 - 7x + x^2}{15 + 4x}}$$

Proceeding as before, we obtain the following:

$$f_3(x) = \cfrac{1}{2 + \cfrac{x}{-1 + \cfrac{x}{-3 + \cfrac{x}{5}}}}$$

The benefit that accrues from the use of continued fractions is this: As we shall find in Art. 12.11.8, many power series in x can be approximated to the required degree of precision by means of polynomial fractions. If the sum of the power series is to be found for numerous values of x, the calculations can be performed more efficiently by computer if continued fractions are used.

1.22 BINOMIAL COEFFICIENTS AND THEIR RELATIONSHIPS

Let n and r denote nonnegative integers, with $n \geq r$. The expression $n!/[r!(n - r)!]$ is termed a *binomial coefficient*, and it is denoted by this notation:

$$\frac{n!}{r!(n - r)!} = \binom{n}{r}$$

The expression at the right is read "nCr." For example,

$$\binom{12}{7} = \frac{12!}{7!5!} = \frac{12 \cdot 11 \cdot 10 \cdot 9 \cdot 8}{5 \cdot 4 \cdot 3 \cdot 2} = 792$$

The following relationships exist:

$$\binom{n}{r} = \binom{n}{n - r} \tag{1.26}$$

$$r\binom{n}{r} = n\binom{n - 1}{r - 1} = (n - r + 1)\binom{n}{n - r + 1} \tag{1.27}$$

$$\binom{n}{r} + \binom{n}{r + 1} = \binom{n + 1}{r + 1} \tag{1.28}$$

$$\binom{n}{0}+\binom{n}{1}+\binom{n}{2}+\cdots+\binom{n}{n}=2^n \tag{1.29}$$

$$\binom{n}{0}-\binom{n}{1}+\binom{n}{2}-\cdots+(-1)^n\binom{n}{n}=0 \tag{1.30}$$

$$\binom{n}{r}+\binom{n-1}{r}+\binom{n-2}{r}+\binom{n-3}{r}+\cdots+\binom{r}{r}=\binom{n+1}{r+1} \tag{1.31}$$

$$\binom{n}{1}+2\binom{n}{2}+3\binom{n}{3}+\cdots+n\binom{n}{n}=2^{n-1}n \tag{1.32}$$

We have the following special cases:

$$\binom{n}{0}=\binom{n}{n}=1 \qquad \binom{n}{1}=\binom{n}{n-1}=n$$

From the definition, we have

$$\binom{n}{r}=\frac{n(n-1)(n-2)\cdots(n-r+1)}{r!}$$

We now define the following analogous binomial coefficient:

$$\binom{-n}{r}=\frac{-n(-n-1)(-n-2)\cdots(-n-r+1)}{r!}$$

The numerator contains r products. If we change the algebraic sign of each product, the expression at the right assumes this form:

$$(-1)^r\frac{(n+r-1)!}{r!(n-1)!}$$

It follows that

$$\binom{-n}{r}=(-1)^r\binom{n+r-1}{r} \tag{1.33}$$

1.23 BINOMIAL THEOREM

Let n denote a positive integer. The expansion of $(a+b)^n$ yields the following series:

$$\begin{aligned}(a+b)^n &= a^n+\frac{n}{1!}a^{n-1}b+\frac{n(n-1)}{2!}a^{n-2}b^2\\ &\quad+\frac{n(n-1)(n-2)}{3!}a^{n-3}b^3+\cdots\\ &\quad+\frac{n(n-1)(n-2)\cdots(n-k+1)}{k!}a^{n-k}b^k+\cdots+b^n\end{aligned} \tag{1.34}$$

This equation can be proved by mathematical induction. To expand $(a-b)^n$, it is simply necessary to replace b with $-b$ in the equation.

The expansion of $(a+b)^n$ can be expressed in terms of the binomial coefficients defined in Art. 1.22, and the series becomes

$$(a+b)^n = \binom{n}{0}a^n + \binom{n}{1}a^{n-1}b + \binom{n}{2}a^{n-2}b^2 + \binom{n}{3}a^{n-3}b^3 + \cdots + \binom{n}{k}a^{n-k}b^k + \cdots + \binom{n}{n}b^n$$

In sigma notation,

$$(a+b)^n = \sum_{k=0}^{n}\binom{n}{k}a^{n-k}b^k$$

The nested form of the binomial series is

$$a^n\left(1+\frac{n}{1}\frac{b}{a}\left(1+\frac{n-1}{2}\frac{b}{a}\left(1+\frac{n-2}{3}\frac{b}{a}\left(1+\cdots+\frac{1}{n}\frac{b}{a}\right)\right)\right)\right)$$

In lambda notation,

$$(a+b)^n = a^n \mathop{\Lambda}_{k=0}^{n-1}\left(1+\frac{n-k}{k+1}\frac{b}{a}\right)$$

For example

$$(a+b)^5 = a^5\left(1+\frac{5}{1}\frac{b}{a}\left(1+\frac{4}{2}\frac{b}{a}\left(1+\frac{3}{3}\frac{b}{a}\left(1+\frac{2}{4}\frac{b}{a}\left(1+\frac{1}{5}\frac{b}{a}\right)\right)\right)\right)\right)$$

Equation (1.34) is also valid if n is nonintegral or negative and $|b/a| < 1$. However, in this case the binomial expansion is an infinite series, and the term b^n never arises.

If we set $a = 1$ and replace b with x, Eq. (1.34) becomes

$$(1+x)^n = 1 + \frac{n}{1!}x + \frac{n(n-1)}{2!}x^2 + \frac{n(n-1)(n-2)}{3!}x^3 + \cdots + \frac{n(n-1)(n-2)\cdots(n-k+2)}{(k-1)!}x^{k-1} + \cdots \tag{1.35}$$

If $|x| \geq 1$, this equation is restricted to positive integral values of n. However, if $|x| < 1$, the foregoing equation is valid for all values of n. If n is not a positive integer, the series is infinite.

1.24 POLYNOMIAL EQUATIONS IN SINGLE UNKNOWN

A polynomial equation of nth degree in a single unknown x has the general form

$$a_0x^n + a_1x^{n-1} + a_2x^{n-2} + \cdots + a_n = 0$$

where n is a positive integer, the a's are all constants, and $a_0 \neq 0$. If both sides of the equation are divided by a_0, the equation assumes the following form, which is known as the *p form*:

$$x^n + p_1 x^{n-1} + p_2 x^{n-2} + \cdots + p_n = 0$$

The equation is described as real, rational, and integral if all the p's are real, rational, and integral.

The polynomial equation of nth degree has n roots, but some may be identical. Let $r_1, r_2, r_3, \ldots, r_n$ denote the roots. The remainder theorem in Art. 1.20 yields the following principle, which is known as the *factor theorem*:

Theorem 1.2. If r_i is a root of the equation $f(x) = 0$, then $x - r_i$ is a factor of $f(x)$.

The converse of this statement is also valid, and therefore the p form of the equation can be recast as

$$(x - r_1)(x - r_2)(x - r_3) \ldots (x - r_n) = 0 \tag{1.36}$$

When the indicated multiplication is performed, the following relationships ensue:

$$\begin{aligned} -p_1 &= r_1 + r_2 + r_3 + \cdots + r_n \\ p_2 &= r_1 r_2 + r_1 r_3 + \cdots + r_{n-1} r_n \\ -p_3 &= r_1 r_2 r_3 + r_1 r_2 r_4 + \cdots + r_{n-2} r_{n-1} r_n \\ &\cdots \\ (-1)^n p_n &= r_1 r_2 r_3 \cdots r_n \end{aligned} \tag{1.37}$$

For example, the equation

$$x^4 - 11x^3 + 17x^2 + 107x - 210 = 0$$

has the roots 2, -3, 5, and 7, as can readily be proved by substitution. We have the following:

$$\begin{aligned} -p_1 &= 11 = 2 + (-3) + 5 + 7 \\ p_2 &= 17 = 2(-3) + 2 \times 5 + 2 \times 7 + (-3)5 + (-3)7 + 5 \times 7 \\ -p_3 &= -107 = 2(-3)5 + 2(-3)7 + 2 \times 5 \times 7 + (-3)5 \times 7 \\ p_4 &= -210 = 2(-3)5 \times 7 \end{aligned}$$

When a root r_i of an equation becomes known, it is possible to divide the polynomial by $x - r_i$ and thereby depress the equation to one of $(n-1)$th degree.

The three theorems that follow yield information pertaining to the nature of the roots of a polynomial equation.

Theorem 1.3. If a real, rational, integral equation has a root r_i that is real and rational, then r_i is an integer and a factor of p_n.

Theorem 1.4. If the complex number $a + bi$ is a root of a real equation, the conjugate complex number $a - bi$ is also a root.

Theorem 1.5. If an irrational number of the form $a + \sqrt{b}$ is a root of a rational equation, the irrational number $a - \sqrt{b}$ is also a root.

These theorems are illustrated by the equation

$$x^5 - 6x^4 + 11x^3 + 110x^2 - 614x - 952 = 0$$

which has the roots $3 + 5i$, $3 - 5i$, $2 + \sqrt{11}$, $2 - \sqrt{11}$, and -4. The last root is a factor of -952.

When a real polynomial is arranged in descending powers of x, a *sign variation* occurs when two successive terms differ in algebraic sign. (The powers need not necessarily be consecutive.) As an illustration, consider the polynomial

$$2x^7 - 9x^6 + 12x^5 + 7x^3 - 13x + 5$$

With a plus sign before the first term, the sequence of algebraic signs is $+-++-+$, and four sign variations occur.

If the unknown x in a given equation is replaced with $-x$, the resulting equation is termed the *negative* of the given equation. With reference to a real polynomial equation, let P and N denote the number of positive and negative real roots, respectively, and let V and V' denote the number of sign variations in the given equation and in its negative, respectively. The principle that follows, which is known as *Descartes' rule of signs*, provides a relationship between P and V.

Theorem 1.6. The number P is either equal to V or it is less than V by an even integer.

The same relationship exists between N and V'. Thus, we obtain an upper limit to the values of P and N. A root that is neither a positive nor a negative real number is either imaginary, complex, or 0.

EXAMPLE 1.14 Determine the nature of the roots of the equation $x^6 - x - 4 = 0$.

SOLUTION The negative equation is $x^6 + x - 4 = 0$. Thus, $V = 1$ and $V' = 1$. Since P and N are restricted to positive values, it follows that $P = N = 1$. The equation has six roots. Therefore, the roots consist of one positive real number, one negative real number, and four complex numbers.

EXAMPLE 1.15 Determine the nature of the roots of the equation $x^5 - 8x^2 = 0$.

SOLUTION The negative equation is $x^5 + 8x^2 = 0$. Thus, $V = 1$ and $V' = 0$, and it follows that $P = 1$ and $N = 0$. By factoring, the equation is transformed to $x^2(x^3 - 8) = 0$. Therefore, two roots are 0, one root is 2, and two roots are complex numbers.

In many instances, Descartes' rule of signs leads to ambiguity concerning the nature of the roots. As an illustration, consider the equation

$$x^5 - 7x^4 + 9x^2 - 2 = 0$$

for which $V = 3$ and $V' = 2$. All we can conclude is that P is either 3 or 1 and N is either 2 or 0. Therefore, the number of complex roots is either 0, 2, or 4.

It is often desirable to establish a range of values within which the real roots of an equation are located. The principle that follows is helpful in this respect.

Theorem 1.7. Consider that the real polynomial $f(x)$ is divided by $x - b$ by synthetic division. If all numbers in the third row are positive, b is greater than all real roots of the equation $f(x) = 0$.

This principle imposes an upper limit on the real roots. A lower limit can be found by forming the negative of the given equation.

EXAMPLE 1.16 Demonstrate that the real roots of the equation

$$x^4 - 3x^3 - 5x^2 + 19x - 60 = 0$$

lie between -4 and 5.

SOLUTION We shall first prove that 5 is greater than all real roots. Dividing the polynomial by $x - 5$ in the manner discussed in Art. 1.19, we obtain the following results:

$$\begin{array}{l} 1 - 3 - \ \ 5 + 19 - \ \ 60 \quad \underline{|5} \\ \quad + 5 + 10 + 25 + 220 \\ \hline 1 + 2 + \ \ 5 + 44 + 160 \end{array}$$

Since all numbers in the third row are positive, the upper limit is established.

We shall now prove that -4 is less than all real roots. To do this, we must prove that 4 is greater than all real roots of the negative equation, which is

$$x^4 + 3x^3 - 5x^2 - 19x - 60 = 0$$

Dividing the negative equation by $x - 4$, we obtain the following results:

$$\begin{array}{l} 1 + 3 - \ \ 5 - 19 - \ \ 60 \quad \underline{|4} \\ \quad + 4 + 28 + 92 + 292 \\ \hline 1 + 7 + 23 + 73 + 232 \end{array}$$

Thus, 4 is greater than all real roots of the negative equation, and the lower limit of the given equation is established.

If $f(x) = 0$ is a real equation, a wealth of information pertaining to this equation can be obtained by constructing the graph of $f(x)$. Let

$$r_i = \text{real root of the equation}$$

$$m_i = \text{number of roots having the value } r_i$$

$$S_i = \text{slope of graph at } x = r_i$$

$$g(x) = \frac{f(x)}{x - r_i}$$

$$a + bi = \text{complex root of the equation}$$

The graph of $f(x)$ has the following properties:

1. Slope $S_i = g(r_i)$. Therefore, $S_i = 0$ if $m_i > 1$, and $S_i \neq 0$ if $m_i = 1$.
2. The graph intersects the x axis at $x = r_i$ if m_i is an odd number (including 1).
3. The graph is tangent to the x axis at $x = r_i$ if m_i is an even number.
4. The graph does not intersect the x axis at $x = a$ (unless a is a real root).

As an illustration, consider the equation

$$x^5 - 5x^4 - 12x^3 + 126x^2 - 280x + 200 = 0$$

which has the following roots: $r_1 = -5$; $r_2 = r_3 = 2$; $r_4 = 3 + i$; $r_5 = 3 - i$. Figure 1.2 is the graph of the polynomial. The graph intersects the x axis at $x = -5$, and it is tangent to the x axis at $x = 2$.

Now consider the equation

$$x^4 - 2x^3 - 12x^2 + 40x - 32 = 0$$

which has the following roots: $r_1 = -4$; $r_2 = r_3 = r_4 = 2$. Figure 1.3 is the graph of the polynomial. The graph intersects the x axis at $x = -4$ and at $x = 2$. and it has zero slope at $x = 2$. (The graph has a *point of inflection* at $x = 2$.)

Assume that a polynomial equation is real, rational, and integral and that the term p_n in the p form of the equation is not unduly large. The equation can be tested for real integral roots by applying Theorems 1.3 and 1.2, in that order. For example, consider the equation $x^4 + 8x^3 + 5x^2 - 74x - 120 = 0$. The factors of -120 are 1, -1, 2, -2, etc. Applying each factor in turn and using synthetic division, we find that $x + 2$ is a factor of the polynomial, and therefore -2 is

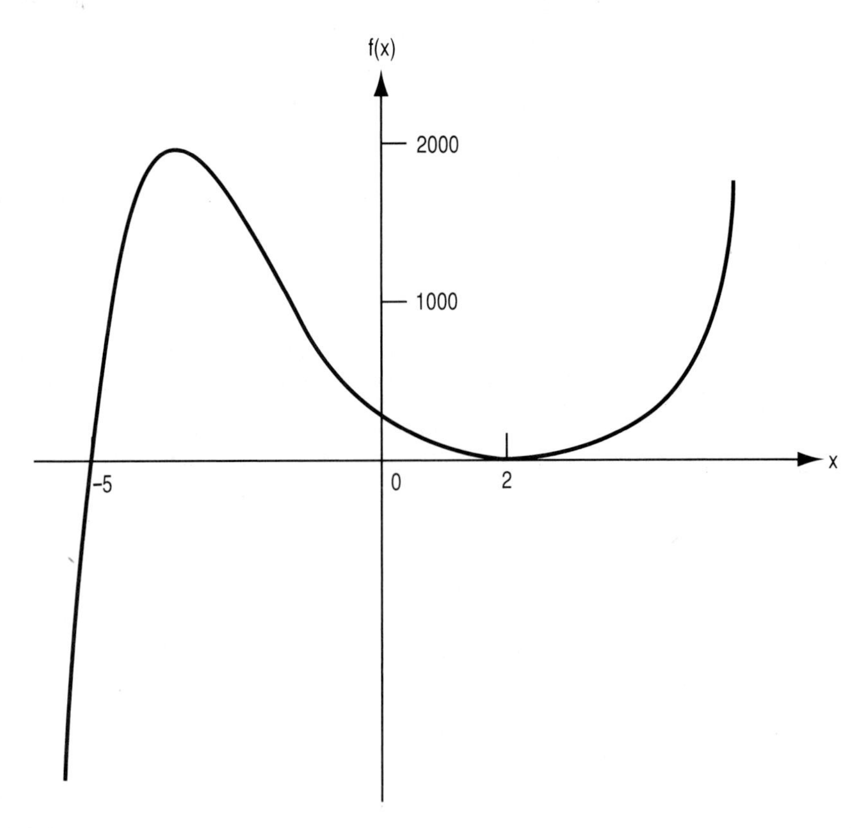

FIGURE 1.2 Graph of $f(x) = x^5 - 5x^4 - 12x^3 + 126x^2 - 280x + 200$.

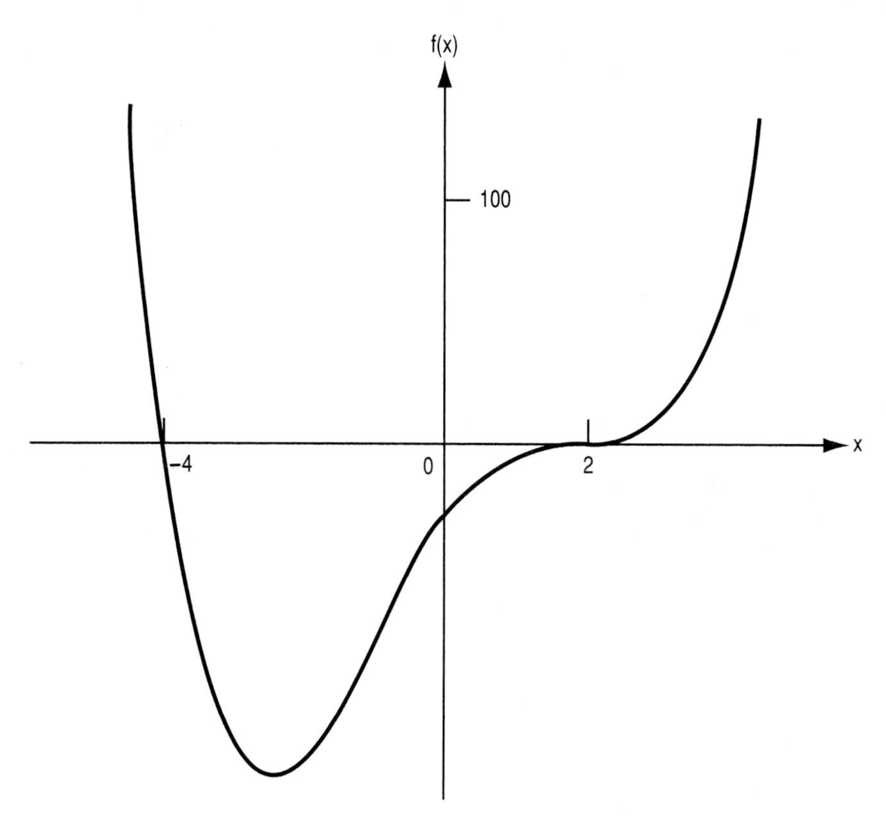

FIGURE 1.3 Graph of $f(x) = x^4 - 2x^3 - 12x^2 + 40x - 32$.

a root of this equation. After division by $x + 2$, the depressed equation is $x^3 + 6x^2 - 7x - 60 = 0$. Repeating the foregoing procedure, we find that 3 is a root of the depressed equation, and therefore of the original equation. Thus, all real integral roots of a given equation can be identified with relative ease. Article 12.3.6 presents a method of finding the real roots of an equation, to any required degree of precision, when these roots are nonintegral.

1.25 SOLUTION OF QUADRATIC EQUATION

A second-degree equation in a single unknown x has the general form

$$ax^2 + bx + c = 0$$

where a, b, and c are constants and $a \neq 0$. This equation has two roots, which we call x_1 and x_2. Let

$$f = b^2 - 4ac \tag{1.38}$$

The roots of the equation are as follows:

$$x_1 = \frac{-b + \sqrt{f}}{2a} \qquad x_2 = \frac{-b - \sqrt{f}}{2a} \tag{1.39}$$

In the present case, the relationships in Eq. (1.37) assume the following form:

$$x_1 + x_2 = -\frac{b}{a}$$
$$x_1 x_2 = \frac{c}{a} \tag{1.40}$$

EXAMPLE 1.17 Solve the equation $8x^2 + 10x - 63 = 0$.

SOLUTION

$$a = 8 \qquad b = 10 \qquad c = -63$$

$$f = 100 - 4(8)(-63) = 2116 \qquad \sqrt{f} = 46$$

$$x_1 = \frac{-10 + 46}{16} = 2.25 \qquad x_2 = \frac{-10 - 46}{16} = -3.5$$

These results are compatible with Eq. (1.40).

EXAMPLE 1.18 Solve the equation $x^2 - x\sqrt{3} - 6 = 0$.

SOLUTION

$$a = 1 \qquad b = -\sqrt{3} \qquad c = -6$$

$$f = 3 - 4(1)(-6) = 27 \qquad \sqrt{f} = 3\sqrt{3}$$

The roots are $2\sqrt{3}$ and $-\sqrt{3}$, and they are compatible with Eq. (1.40).

Assume that a, b, and c are real numbers. The character of the two roots is determined by the parameter f, and f is accordingly termed the *discriminant* of the equation. If $f > 0$, the two roots are real and unequal. If $f = 0$, the two roots are real and equal. If $f < 0$, the two roots are conjugate complex numbers.

An equation is said to be in *quadratic form* if it can be transformed to a quadratic equation by replacing one unknown with another. However, the transformed equation may contain more roots than the given equation.

EXAMPLE 1.19 Solve the equation $2x - \sqrt{x + 12} - 21 = 0$.

SOLUTION Set $z = \sqrt{x + 12}$. Then $x = z^2 - 12$. Substitution in the given equation yields $2z^2 - z - 45 = 0$. The roots of this transformed equation and the corresponding values of x are as follows:

$$z_1 = 5 \qquad x_1 = 13$$

$$z_2 = -4.5 \qquad x_2 = 8.25$$

Substituting these values of x in the original equation, we find that 13 is the only true root.

1.26 SOLUTION OF CUBIC EQUATION

A third-degree equation in a single unknown x can be expressed in the general form

$$x^3 + bx^2 + cx + d = 0$$

where b, c, and d are constants. We shall assume that they are real numbers. The cubic equation has three roots, which we call x_1, x_2, and x_3.

We introduce the following parameters:

$$f = \frac{b}{3} \qquad g = \frac{c}{3} - f^2 \qquad h = \frac{d}{2} + f^3 - \frac{fc}{2}$$

$$j = g^3 + h^2$$

The value of j determines the character of the roots, and we now classify cubic equations in the manner described in Table 1.1.

For the *type 1 equation* ($j = 0$), we introduce this additional parameter:

$$m = 2(-h)^{1/3}$$

The solution is as follows:

$$x_1 = m - f \qquad x_2 = x_3 = -\frac{m}{2} - f \tag{1.41}$$

For the *type 2 equation* ($j > 0$), we introduce the following additional parameters:

$$k = \sqrt{j}$$

$$l_1 = (-h + k)^{1/3} \qquad l_2 = (-h - k)^{1/3}$$

$$m = l_1 + l_2 \qquad n = (l_1 - l_2)\sqrt{0.75}$$

TABLE 1.1 Types of Cubic Equations

Type	Primary condition	Secondary condition
1	$j = 0$	
2	$j > 0$	
3*a*		$h = 0$
3*b*	$j < 0$	$h > 0$
3*c*		$h < 0$

The solution is as follows:

$$x_1 = m - f$$
$$x_2 = -\frac{m}{2} - f + ni \qquad x_3 = -\frac{m}{2} - f - ni \tag{1.42}$$

where i is the unit of imaginaries.

For the *type 3 equation* ($j < 0$), we set

$$p = 2\sqrt{-g}$$

For the type $3a$ equation ($h = 0$), the solution is as follows:

$$x_1 = -f$$
$$x_2 = p\sqrt{0.75} - f \qquad x_3 = -p\sqrt{0.75} - f \tag{1.43}$$

For the types $3b$ and $3c$ equations ($h \neq 0$), we set

$$k = \sqrt{-j} \qquad \alpha = \text{arc tan}\,\frac{k}{-h}$$

where α is an acute angle. For the type $3b$ ($h > 0$), we set $\beta = 180° + \alpha$, and for type $3c$ ($h < 0$), we set $\beta = \alpha$. Thus, β is obtuse if $h > 0$ and acute if $h < 0$. We now set $\gamma = \beta/3$, and the solution is as follows:

$$x_1 = p \cos \gamma - f$$
$$x_2 = p \cos (\gamma + 120°) - f \tag{1.44}$$
$$x_3 = p \cos (\gamma + 240°) - f$$

We now draw the following conclusions concerning the roots of a cubic equation:

Type 1: All three roots are real. The three roots are equal if $g = 0$, and only two roots are equal if $g \neq 0$.

Type 2: One root is real and two roots are conjugate complex numbers (or imaginary numbers of opposite sign).

Type 3: All three roots are real and unequal.

Equation (1.37) assumes the following form with respect to a cubic equation:

$$x_1 + x_2 + x_3 = -b$$
$$x_1x_2 + x_1x_3 + x_2x_3 = c \tag{1.45}$$
$$x_1x_2x_3 = -d$$

EXAMPLE 1.20 Solve the equation $x^3 - 6x^2 - 54x + 464 = 0$.

SOLUTION

$$b = -6 \qquad c = -54 \qquad d = 464$$

$$f = -2 \qquad g = -18 - 4 = -22$$

$$h = 232 - 8 - 54 = 170 \qquad j = (-22)^3 + 170^2 = 18{,}252$$

This is a type 2 equation. Continuing, we have

$$k = \sqrt{18{,}252} = 135.1$$

$$l_1 = (-170 + 135.1)^{1/3} = -3.2679$$

$$l_2 = (-170 - 135.1)^{1/3} = -6.7320$$

$$m = -3.2679 - 6.7320 = -10$$

$$n = (-3.2679 + 6.7320)\sqrt{0.75} = 3$$

The solution is as follows:

$$x_1 = -10 + 2 = -8 \qquad x_2 = 5 + 2 + 3i = 7 + 3i \qquad x_3 = 7 - 3i$$

The solution is compatible with Eq. (1.45).

EXAMPLE 1.21 Solve the equation $x^3 + 12x^2 - 123x + 190 = 0$.

SOLUTION

$$b = 12 \qquad c = -123 \qquad d = 190$$

$$f = 4 \qquad g = -57 \qquad h = 405 \qquad j = -21{,}168$$

This is a type $3b$ equation. Continuing, we have

$$p = 2\sqrt{57} = 15.0997 \qquad k = \sqrt{21{,}168} = 145.4923$$

$$\alpha = \text{arc tan}\left(\frac{145.4923}{-405}\right) = -19.7603^\circ$$

$$\beta = 180^\circ - 19.7603^\circ = 160.2397^\circ \qquad \gamma = 53.4132^\circ$$

The solution is as follows:

$$x_1 = 15.0997 \cos 53.4132^\circ - 4 = 5$$

$$x_2 = 15.0997 \cos 173.4132^\circ - 4 = -19$$

$$x_3 = 15.0997 \cos 293.4132^\circ - 4 = 2$$

The solution is compatible with Eq. (1.45).

Figure 1.4 presents a computer program for solving a cubic equation. In addition, it solves types $3a$ and $3c$ equations and the equation in Example 1.20. Angles are expressed in radians, with $180^\circ = \pi$ rad.

```
D Line# 1      7             IBM Personal Computer FORTRAN Compiler V2.0
      1 C
      2 C   SOLUTION OF CUBIC EQUATION
      3 C
      4         IMPLICIT REAL (A-H), REAL(O-Z)
      5         REAL J,M,K,L1,L2,N
      6         CHARACTER CHAR, CHARS(64)
      7         INTRINSIC SQRT, ATAN, COS
      8         DATA PI/3.1416/, CHARS/64*' '/
      9 C
     10     210 FORMAT (1X,'....SOLUTION OF CUBIC EQUATION....',/
     11       2     1X,'     X**3 + BX**2 + CX + D = 0',/)
     12     270 FORMAT (1X,' F= ',F10.4,' G= ',F10.4,' H= ',F10.4,//,
     13       2          1X,' J =',F14.4, /)
     14     290 FORMAT (1X,' X1=', F7.2,10X,'X2=',F7.2,' + ',F7.2,'i',
     15       2      /, 1X,       ,20X,  ' X3=', F7.2,' - ',F7.2, 'i')
     16     299 FORMAT (1X,'ANOTHER SOLUTION?(Y/N)...'\)
     17     300 FORMAT (A1)
     18 C
     19         WRITE (*,210)
     20 C
     21       5  CONTINUE
     22          CALL INPUT (CHARS,B,C,D)
     23 C
     24         F = B/3.
     25         G = C/3. - F**2
     26         H1=D/2.
     27         H2=F**3
     28         H3=F*C/2.
     29         H = H1+H2-H3
     30         J = G**3 + H**2
     31 C
     32         WRITE (*,270) F,G,H,J
     33 C
     34 C  IS J = 0?  (5)
     35         IF (J .EQ. 0.) THEN
     36            M = 2.* (-H)**.3333
     37            X1 = M - F
     38            X2 = -M/2. - F
     39            X3 = X2
     40            ICASE = 1
     41            GO TO 90
     42         ENDIF
     43 C J > 0 (9) - (11)
     44         IF (J .GT. 0.) THEN
     45            K = SQRT (J)
     46            CALL POWER (L1,-H,K,.3333)
     47            CALL POWER (L2,-H,-K,.3333)
     48            M = L1 + L2
     49            N = (L1 - L2) * SQRT(.75)
     50            X1 = M - F
     51            XR2 = -M/2. - F
     52            XR3 = XR2
     53            ICASE = 2
     54            GO TO 90
     55         ENDIF
```

FIGURE 1.4 Solution of cubic equation.

```
56 C   J < 0 (18), (19)
57        P = 2. * SQRT(-G)
58        IF (H .EQ. 0.) THEN
59           X1 = -F
60           X2 = P * SQRT(.75) - F
61           X3 = -X2 - F - F
62           ICASE = 3
63           GO TO 90
64        ENDIF
65        K = SQRT (-J)
66        FUNC = K / (-H)
67        ALPHA = ATAN (FUNC)
68 C H> 0 (25)
69        IF (H .GT. 0.) THEN
70           BETA = PI + ALPHA
71        ELSE
72           BETA = ALPHA
73        ENDIF
74        GAMMA = BETA/3.
75        X1 = P * COS(GAMMA) - F
76        X2 = P * COS(GAMMA + 2.*PI/3.) - F
77        X3 = P * COS (GAMMA + 4.*PI/3.) - F
78        ICASE = 4
79 C
80     90 CONTINUE
81 C
82        CALL PRINT (ICASE,M,K,N, X1,X2,X3)
83 C
84        IF (ICASE .NE. 2) GO TO 95
85 C
86        WRITE (*,290) X1, XR2,N, XR3,N
87 C
88     95 CONTINUE
89        WRITE (*,299)
90        READ (*,300) CHAR
91        IF (CHAR .EQ. 'Y') GO TO 5
92 C
93        STOP
94        END


 95         SUBROUTINE POWER (A,H,K,E)
 96         REAL K
 97 C   TO ASSURE BASE RAISED TO REAL POWER IS POSITIVE
 98 C
 99        IS = 1
100        Q = H+K
101        IF (Q .LT. 0.) THEN
102             IS = -1
103             Q = -Q
104        ENDIF
105        A = Q**E * IS
106        RETURN
107        END


108        SUBROUTINE PRINT (ICASE,M,K,N, X1,X2,X3)
109 C
110        REAL M,K,N
111 C
```

FIGURE 1.4 (*Continued*)

```
112    610 FORMAT (1X,'  M = ',F10.2,/)
113    611 FORMAT (1X,'  X1 = ',F10.2,/,
114      2        1X,'  X2 = ',F10.2,/,
115      3        1X,'  X3 = ',F10.2,/)
116    620 FORMAT (1X,'  K = ',F10.2,/)
117 C
118        GO TO (10,20,30,40), ICASE
119 C
120     10 WRITE (*,610) M
121        IF (ICASE .EQ. 2) RETURN
122 C
123     15 CONTINUE
124        WRITE (*,611) X1,X2,X3
125 C
126        RETURN
127 C
128     20 WRITE (*,620) K
129        IF (ICASE .EQ. 2) GO TO 10
130        GO TO 15
131 C
132     30 CONTINUE
133        GO TO 15
134 C
135     40 CONTINUE
136        GO TO 20
137 C
138        END

139        SUBROUTINE INPUT(CHARS,B,C,D)
140 C
141        CHARACTER CHARS(2)
142 C
143    200 FORMAT (1X,'ENTER TITLE OF EXAMPLE, IF DESIRED',/,
144      1  1X,\)
145    205 FORMAT (64A1)
146    230 FORMAT (/,
147      2        1X,'ENTER COEFFICIENTS WITH DECIMAL POINTS
148      3        1X,'                      B. '\)
149    235 FORMAT (F10.2)
150    240 FORMAT (1X,'                       C. '\)
151    250 FORMAT (1X,'                       D. '\)
152    260 FORMAT (1X,' B= ',F10.2,/,
153      2        1X,' C= ',F10.2,/,
154      3        1X,' D= ',F10.2,/)
155 C
156        WRITE (*,200)
157        READ (*,205) (CHARS(I),I=1,64)


158 C
159        WRITE (*,230)
160         READ (*,235) B
161         WRITE (*,240)
162        READ (*,235) C
163        WRITE (*,250)
164         READ (*,235) D
165 C
166        WRITE (*,260) B,C,D
167 C
168        RETURN
169        END
```

FIGURE 1.4 (*Continued*)

```
....SOLUTION OF CUBIC EQUATION....
    X**3 + BX**2 + CX + D = 0

ENTER TITLE OF EXAMPLE, IF DESIRED
    TYPE 2, J > 0

ENTER COEFFICIENTS WITH DECIMAL POINTS
                         B. -6.
                         C. -54.
                         D. 464.
 B=      -6.00
 C=     -54.00
 D=     464.00

 F=    -2.0000 G=   -22.0000 H=    170.0000

 J =    18252.0000

  K =     135.10

  M =     -10.00

 X1=  -8.00             X2=   7.00 +    3.00i
                        X3=   7.00 -    3.00i
ANOTHER SOLUTION?(Y/N)...

B>
B>CUBIC
....SOLUTION OF CUBIC EQUATION....
    X**3 + BX**2 + CX + D = 0

ENTER TITLE OF EXAMPLE, IF DESIRED
     TYPE 3A, J < 0 AND H = 0

ENTER COEFFICIENTS WITH DECIMAL POINTS
                         B. 9.
                         C. -39.
                         D. -171.
 B=       9.00
 C=     -39.00
 D=    -171.00

 F=     3.0000 G=   -22.0000 H=       .0000

 J =   -10648.0000

  X1 =      -3.00
  X2 =       5.12
  X3 =     -11.12

ANOTHER SOLUTION?(Y/N)...
```

FIGURE 1.4 (*Continued*)

```
B>                                                 CUBIC
....SOLUTION OF CUBIC EQUATION....
     X**3 + BX**2 + CX + D = 0

ENTER TITLE OF EXAMPLE, IF DESIRED
      TYPE 3C, J < 0 AND H < 0

ENTER COEFFICIENTS WITH DECIMAL POINTS
                         B. -  39, .
                         C. 414.
                         D. -1296.
 B=       -39.00
 C=       414.00
 D=     -1296.00

 F=    -13.0000 G=     -31.0000 H=    -154.0000

 J =      -6075.0000

  K =         77.94

  X1 =         24.00
  X2 =          6.00
  X3 =          9.00

ANOTHER SOLUTION?(Y/N)...N
Stop - Program terminated.
```

FIGURE 1.4 (*Continued*)

1.27 SOLUTION OF QUARTIC EQUATION

A fourth-degree equation in a single unknown x can be expressed in the general form

$$x^4 + bx^3 + cx^2 + dx + e = 0$$

where the coefficients b, c, d, and e are constants. We shall assume that they are real numbers.

The procedure for solving a quartic equation is as follows: Applying the values of the coefficients, set up the following cubic equation:

$$y^3 - cy^2 + (bd - 4e)y + (4c - b^2)e - d^2 = 0 \tag{1.46}$$

Applying any root of this equation, let

$$q^2 = \left(\frac{b}{2}\right)^2 - c + y \qquad s^2 = \left(\frac{y}{2}\right)^2 - e \tag{1.47}$$

If $(b/2)y - d$ is positive, consider both q and s to be positive; if this quantity is negative, consider q to be positive and s to be negative. Now set up these equations:

$$\begin{aligned} x^2 + \left(\frac{b}{2} + q\right)x + \frac{y}{2} + s = 0 \\ x^2 + \left(\frac{b}{2} - q\right)x + \frac{y}{2} - s = 0 \end{aligned} \tag{1.48}$$

The roots of this pair of quadratic equations are the roots of the quartic equation.

EXAMPLE 1.22 Solve the equation $x^4 - 9x^3 + 22x^2 + 28x - 120 = 0$.

SOLUTION

$$b = -9 \qquad c = 22 \qquad d = 28 \qquad e = -120$$

Equation (1.46) becomes

$$y^3 - 22y^2 + 228y - 1624 = 0$$

One root of this equation is 14, and Eq. (1.47) yields the following:

$$q^2 = \frac{81}{4} - 22 + 14 = \frac{49}{4} \qquad s^2 = 49 + 120 = 169$$

We have $(b/2)y - d = (-4.5)14 - 28 < 0$. Therefore, we set $q = 7/2$ and $s = -13$. Equation (1.48) assumes these forms:

$$x^2 - x - 6 = 0 \qquad x^2 - 8x + 20 = 0$$

The roots of the first equation are 3 and -2; the roots of the second equation are $4 + 2i$ and $4 - 2i$. These four numbers are the roots of the given quartic equation.

1.28 DEFINITION OF e

The quantity e has alternative definitions, and we select the following as the simplest one:

$$e = \lim_{n \to \infty} \left(1 + \frac{1}{n}\right)^n$$

By expanding the binomial in accordance with Eq. (1.34) and allowing n to become infinite, we obtain the following infinite series:

$$e = 1 + \frac{1}{1!} + \frac{1}{2!} + \frac{1}{3!} + \frac{1}{4!} + \cdots \tag{1.49}$$

The quantity e is an irrational number. To six significant figures, its value is 2.71828. Similarly, we have

$$e^x = \lim_{n \to \infty} \left(1 + \frac{1}{n}\right)^{nx}$$

Again expanding and allowing n to become infinite, we obtain

$$e^x = 1 + \frac{x}{1!} + \frac{x^2}{2!} + \frac{x^3}{3!} + \frac{x^4}{4!} + \cdots \tag{1.50}$$

1.29 NUMBERS IN FLOATING-DECIMAL-POINT SYSTEM

In the Arabic numeral system, the position occupied by a digit corresponds to a specific power of 10. For example,

$$4725.9 = 4(1000) + 7(100) + 2(10) + 5(1) + 9\left(\frac{1}{10}\right)$$

$$= 4(10^3) + 7(10^2) + 2(10^1) + 5(10^0) + 9(10^{-1})$$

Let n denote a positive integer. If a given number is multiplied by 10^n or 10^{-n}, the decimal point is displaced n places to the right or left, respectively. For example,

$$67.352(10^2) = 6735.2$$

$$\frac{5783.16}{10^3} = 5783.16(10^{-3}) = 5.78316$$

We may write the following:

$$3719.2 = [3719.2(10^{-3})]10^3 = 3.7192(10^3)$$

$$0.0815 = [0.0815(10^2)]10^{-2} = 8.15(10^{-2})$$

Thus, the decimal point in a number can be displaced n places to the right or left if we multiply the resulting number by 10^{-n} or 10^n, respectively.

Where numbers of widely varying magnitude are present, it is often desirable to record the numbers in a standard form in which the decimal point is placed at some assigned position in the sequence of digits. This procedure can be followed if each number is multiplied by the appropriate power of 10. Moreover, for simplicity, the notation (10^n) is replaced with the notation En. Thus,

$$389.16 = 0.38916\text{E}3 = 3891.6\text{E–}1$$

In the notation En, the number n is the exponent of 10 in the multiplier. However, we may interpret it as the number of places the decimal point must be displaced *to the right* to obtain the number in natural form. This method of expressing numbers is known as the *floating-decimal-point system.*

EXAMPLE 1.23 Record the following numbers in the floating-decimal-point system with the decimal point following the first digit: 47,281; 0.00296; 3.972.

SOLUTION The numbers are 4.7281E4, 2.96E–3, and 3.972E0.

If a given number is to be recast in such a manner that the exponent of 10 changes from n to m, the decimal point must be displaced $n - m$ places *to the right.* Thus, 5.89361E4 = 5893.61E1, and 326.79E2 = 3.2679E4. Where numbers are to be added or subtracted under the floating-decimal-point system, the exponents of 10 must first be made identical.

EXAMPLE 1.24 Perform the following operation:

$$29.7\text{E–}2 + 3.1356\text{E}1 + 0.148302\text{E}3$$

Express the sum in each of the following forms: first, with the decimal point immediately following the first digit and second, in natural form.

SOLUTION Let S denote the sum. We shall select the smallest exponent, which is -2, as the standard. Then

$$S = 29.7\text{E–}2 + 3135.6\text{E–}2 + 14{,}830.2\text{E–}2$$

$$= (29.7 + 3135.6 + 14{,}830.2)\text{E–}2 = 17{,}995.5\text{E–}2$$

In the specified forms, $S = 1.79955\text{E}2 = 179.955$.

1.30 BINARY AND OCTAL NUMERAL SYSTEMS

The decimal numeral system uses 10 digits (0 to 9, inclusive), but it is possible to devise a numeral system that uses any feasible number of digits. Specifically, the binary system uses 2 digits (0 and 1), and the octal system uses 8 digits (0 to 7, inclusive). The number of digits used in a system is called the *base* (or *radix*) of the system, and the base is displayed by means of a subscript appended to the number. For example, 5204_8 is a number in the octal system. Where the subscript is omitted, it is understood to be 10. We shall denote the base of a system by B.

In a *positional* numeral system, a number is expressed as the sum of powers of B, and the digits are so arranged that the corresponding powers of B diminish consecutively from left to right. If the number is nonintegral, the digits corresponding to the powers 0 and -1 are separated by a period, which is termed a *decimal, binary*, or *octal point*, depending on the system. Thus, if a digit d lies j places to the *left* of the period, it represents dB^{j-1}; if it lies k places to the *right* of the period, it represents dB^{-k}. For example, in the decimal system,

$$726.104 = 7(10^2) + 2(10^1) + 6(10^0) + 1(10^{-1}) + 0(10^{-2}) + 4(10^{-3})$$

It is frequently necessary to convert a number from one numeral system to another. The conversion of a fractional number requires special consideration. In Art. 1.7, we stated that a rational fraction a/b in its lowest terms can be converted to a terminating decimal fraction only if b has no prime factors other than 2 and 5. Correspondingly, this fraction can be converted to a terminating binary or octal fraction only if b has no prime factors other than 2. Table 1.2 exhibits these relationships, and the following conclusions emerge: A terminating binary or octal fraction always yields a terminating decimal fraction, and an infinite decimal fraction always yields an infinite binary or octal fraction. However, a terminating

TABLE 1.2

	Type of fraction	
Prime factors of denominator b	Binary or octal system	Decimal system
2 only	Terminating	Terminating
2 and 5 only or 5 only	Infinite	Terminating
2, 5, and another number	Infinite	Infinite

decimal fraction may yield a terminating or an infinite binary or octal fraction. Corresponding binary and octal fractions are either both terminating or both infinite.

For our purpose, the conversion of numbers from one system to another can be divided into four types. We shall study each type in turn.

Type 1. Binary or octal number to decimal number.

This form of conversion is straightforward. It is merely necessary to convert each digit in the given number to its equivalent decimal number and then sum the results. For example,

$$402.7_8 = 4(8^2) + 0(8^1) + 2(8^0) + 7(8^{-1})$$
$$= 256 + 0 + 2 + 0.875 = 258.875_{10}$$

$$11010.11_2 = 1(2^4) + 1(2^3) + 0(2^2) + 1(2^1) + 0(2^0) + 1(2^{-1}) + 1(2^{-2})$$
$$= 16 + 8 + 0 + 2 + 0 + 0.5 + 0.25 = 26.75$$

However, where numerous conversions are to be made, the calculations can be performed more expeditiously by expressing the binary or octal number in the form of a nested series. We apply the second form given in Art. 1.18 but replace x with the base B.

EXAMPLE 1.25 Express 5713.26_8 as a decimal number.

SOLUTION Let N denote the number. In standard form,

$$N = 5(8^3) + 7(8^2) + 1(8^1) + 3(8^0) + 2(8^{-1}) + 6(8^{-2})$$
$$= 3019.34375$$

Separating the integral and fractional parts of N and expressing them as nested series, we have

$$N = 3 + 8(1 + 8(7 + 8(5))) + 8^{-1}(2 + 8^{-1}(6))$$

The calculations for the first series are as follows:

$$8 \times 5 = 40 \qquad 7 + 40 = 47$$
$$8 \times 47 = 376 \qquad 1 + 376 = 377$$
$$8 \times 377 = 3016 \qquad 3 + 3016 = 3019$$

Thus, $5713_8 = 3019_{10}$. The calculations for the second series, with 8^{-1} replaced with 0.125, are as follows:

$$(0.125)6 = 0.75 \qquad 2 + 0.75 = 2.75$$
$$(0.125)(2.75) = 0.34375$$

Thus, $0.26_8 = 0.34375_{10}$, and $5713.26_8 = 3019.34375_{10}$.

Type 2. Decimal number to binary or octal number.

Let N denote a positive number, and assume it has the form $(d_3 d_2 d_1 d_0)_B$. Taking the digits in reverse order, we have

$$N = d_0 + d_1 B + d_2 B^2 + d_3 B^3$$

We shall divide N by B to obtain a quotient Q_1 and remainder R_1; then we shall divide Q_1 by B to obtain a quotient Q_2 and remainder R_2; etc. Then

$$R_1 = d_0 \qquad Q_1 = d_1 + d_2 B + d_3 B^2$$

$$R_2 = d_1 \qquad Q_2 = d_2 + d_3 B$$

$$R_3 = d_2 \qquad Q_3 = d_3$$

$$R_4 = d_3 \qquad Q_4 = 0$$

Thus, when an integral decimal number is given, successive division of this number by B yields the digits of the corresponding binary or octal number, but in reverse order.

EXAMPLE 1.26 Express 5917_{10} and 323_{10} as octal numbers, and express 367_{10} as a binary number.

SOLUTION The successive divisions and their results are recorded in Table 1.3. By reversing the order of the remainders, we obtain the following conversions:

$$5917_{10} = 13{,}435_8 \qquad 323_{10} = 503_8 \qquad 367_{10} = 101101111_2$$

This procedure for converting an integral decimal number to a binary or octal number is an inversion of that followed in Example 1.25, where we converted an octal number to its decimal equivalent.

Now let N denote a positive number that is less than 1, and assume it has the form $(0.d_1 d_2 d_3)_B$. Then

$$N = d_1 B^{-1} + d_2 B^{-2} + d_3 B^{-3}$$

TABLE 1.3

Conversion to octal numbers			Conversion to binary numbers		
Division	Q	R	Division	Q	R
5917/8	739	5	367/2	183	1
739/8	92	3	183/2	91	1
92/8	11	4	91/2	45	1
11/8	1	3	45/2	22	1
1/8	0	1	22/2	11	0
—	—	—	11/2	5	1
323/8	40	3	5/2	2	1
40/8	5	0	2/2	1	0
5/8	0	5	1/2	0	1

We form the following products:

$$P_1 = BN = d_1 + d_2B^{-1} + d_3B^{-2}$$

$$P_2 = B(P_1 - d_1) = d_2 + d_3B^{-1}$$

$$P_3 = B(P_2 - d_2) = d_3$$

Thus, when a fractional decimal number is given, successive multiplication of the type shown yields the digits of the corresponding binary or octal number. The digits emerge in their proper order.

EXAMPLE 1.27 Express 0.921875_{10} as an octal number and 0.59375_{10} as a binary number, and express 0.735_{10} as an octal number to five significant figures.

SOLUTION The calculations for the first number are as follows:

$$8(0.921875) = 7.375 \qquad 8(0.375) = 3.000$$

Taking the integral parts of the products, we have $0.921875_{10} = 0.73_8$. The reason why this fraction terminates in the octal system is that $0.921875 = 921{,}875/1{,}000{,}000 = 59/64$, and the only prime factor of 64 is 2.

The calculations for the second number are as follows:

$$2(0.59375) = 1.1875 \qquad 2(0.1875) = 0.375 \qquad 2(0.375) = 0.75$$

$$2(0.75) = 1.5 \qquad 2(0.5) = 1.0$$

Thus, $0.59375_{10} = 0.10011_2$. The reason why this fraction terminates in the binary system is that $0.59375 = 59{,}375/100{,}000 = 19/32$, and the only prime factor of 32 is 2.

The calculations for the third number are as follows:

$$8(0.735) = 5.88 \qquad 8(0.88) = 7.04 \qquad 8(0.04) = 0.32$$

$$8(0.32) = 2.56 \qquad 8(0.56) = 4.48 \qquad 8(0.48) = 3.84$$

Since $3 < 8/2$, we disregard the sixth digit. Then $0.735_{10} = 0.57024_8$, to five significant figures. The reason why this fraction is infinite in the octal system is that $0.735 = 735/1000 = 147/200$, and 200 has the prime factor 5 as well as 2.

Type 3. Octal number to binary number.

Since $8 = 2^3$, this form of conversion is extremely simple. We merely replace each digit in the octal number with its equivalent binary number. The binary number must be expressed in three digits, even if the first one is 0.

EXAMPLE 1.28 Express 365.24_8 as a binary number.

SOLUTION

$$3_8 = 011_2 \qquad 6_8 = 110_2 \qquad 5_8 = 101_2$$

$$2_8 = 010_2 \qquad 4_8 = 100_2$$

Discarding the 0's at the ends as superfluous, we obtain $365.24_8 = 11110101.0101_2$.

This result can be tested by converting the octal and binary numbers to their decimal equivalents. In both instances, the number is 245.3125.

Type 4. Binary number to octal number.

Inverting the procedure for the type 3 conversion, we divide the digits in the binary number into groups of 3, starting at the binary point. Each group of digits forms a number, and this number is the corresponding digit in the octal number.

EXAMPLE 1.29 Express 11010.1011_2 as an octal number.

SOLUTION To obtain groups of three digits, we add a 0 at the left end and two 0's at the right end. We then group the digits in this manner: 011 010.101 100.

$$011_2 = 3 \qquad 010_2 = 2 \qquad 101_2 = 5 \qquad 100_2 = 4$$

Thus, $11010.1011_2 = 32.54_8$. The corresponding decimal number is 26.6875.

It is frequently necessary to convert a binary, octal, or decimal fraction to an ordinary fraction, that is, one having the form a/b, where a and b are decimal integers. Assume that the given fraction is infinite. In accordance with the statement in Art. 1.7, this infinite fraction eventually assumes the form of a repeating digit or group of digits. Therefore, this infinite fraction embodies an infinite geometric series.

EXAMPLE 1.30 Convert the infinite octal fraction $0.6353535\cdots_8$ to an ordinary fraction.

SOLUTION Let N denote the number. The repeating group of digits is 35, and $35_8 = 29_{10}$. When this group of digits is displaced m places to the right, it is divided by 8^m. Therefore, in decimal form,

$$N = \frac{6}{8} + 29\left(\frac{1}{8^3} + \frac{1}{8^5} + \frac{1}{8^7} + \cdots\right)$$

The expression in parentheses is an infinite geometric series in which $a = 1/8^3$ and $r = 1/8^2$. Applying Eq. (1.17*a*), we obtain

$$N = \frac{6}{8} + 29\left(\frac{1}{504}\right) = \frac{407}{504}$$

1.31 HEXADECIMAL NUMERAL SYSTEM

This numeral system has 16 as the base. Uppercase letters are used to represent the digits beyond 9, in this manner: A for 10, B for 11, C for 12, D for 13, E for 14, and F for 15.

EXAMPLE 1.31 Express $D2A5_{16}$ and $4FB_{16}$ as decimal numbers.

SOLUTION Let N_1 and N_2 denote the corresponding decimal numbers. Then

$$N_1 = 13(16^3) + 2(16^2) + 10(16^1) + 5(16^0) = 53{,}925$$

$$N_2 = 4(16^2) + 15(16^1) + 11(16^0) = 1275$$

EXAMPLE 1.32 Express $2{,}538{,}169_{10}$ as a hexadecimal number.

SOLUTION The procedure is parallel to that in Example 1.26, and the results of the successive divisions are recorded in Table 1.4. Reversing the order of the remainders and expressing them in hexadecimal form, we obtain the hexadecimal number 26BAB9.

TABLE 1.4

Division	Q	R
2,538,169/16	158,635	9
158,635/16	9,914	11
9,914/16	619	10
619/16	38	11
38/16	2	6
2/16	0	2

1.32 MODULO ARITHMETIC

When an integer a is divided by another integer m, the result consists of a quotient q and a remainder r. (We include cases where $q = 0$ or $r = 0$.) In many instances, our interest centers about the remainder exclusively, the quotient being of no consequence. The branch of algebra that concerns itself with the study of remainders is known as *modulo arithmetic*. In the subsequent material, it is understood that all numbers are integers.

The divisor m is called the *modulus*, and we shall restrict it to positive values. The dividend a can be expressed as $a = qm + r$, where $|r| < m$, and a can be positive, negative, or 0. In modulo arithmetic, the remainder r is generally restricted to nonnegative values. Where a is negative, r can be made positive by decreasing q by 1. For example, if 17 is the modulus, we have $-81 = (-4)17 - 13$; thus, $q = -4$ and $r = -13$. However, we can express -81 in the alternative form $-81 = (-5)17 + 4$, making $q = -5$ and $r = 4$.

Let a and b denote two numbers that have the same remainder when they are divided by m. Then a and b are said to be *congruent* to each other with respect to m. In the notation devised by Gauss, we write

$$a \equiv b \qquad (\text{mod } m)$$

and this statement is read "a is congruent to b modulo m."

As an illustration, let $m = 7$, $r = 3$, and $a = 7q + 3$. Allowing q to vary from -2 to 2, we obtain the following set of congruences:

$$-11 \equiv -4 \equiv 3 \equiv 10 \equiv 17 \qquad (\text{mod } 7)$$

Congruences can be depicted geometrically by placing numbers on a circle, which is known as the *congruence circle*. To illustrate the procedure, we shall take 7 as the modulus. In Fig. 1.5, we divide a circle into seven equal parts and label the boundaries with uppercase letters. Starting at A and proceeding in a clockwise direction, we place the number 0 at A, 1 at B, 2 at C, etc., for two cycles. Now

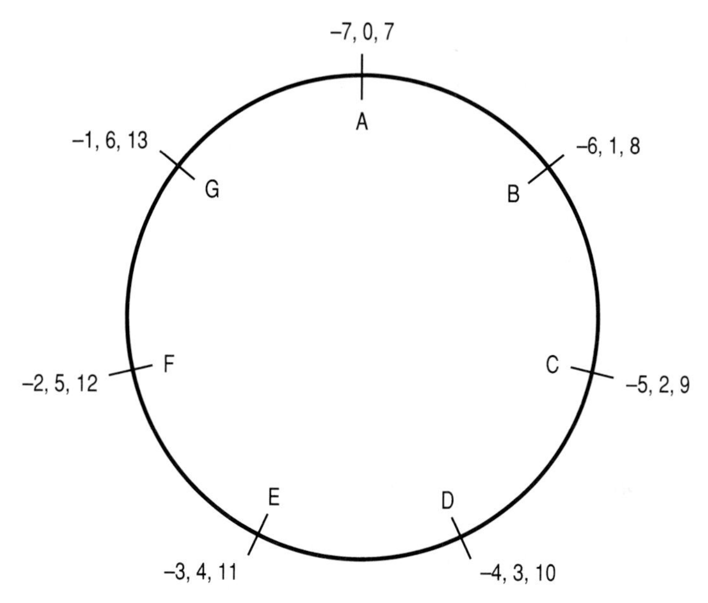

FIGURE 1.5 Congruence circle.

returning to A and proceeding in a counterclockwise direction, we place the number -1 at G, -2 at F, etc. Numbers that lie at the same point are congruent to one another. Thus, taking point E, we find that $-3 \equiv 4 \equiv 11 \pmod 7$.

The congruence circle confers a practical meaning upon the congruence relationship. Thus, consider that we spin a wheel in a game of chance. What is significant is merely the point at which the marker comes to rest; how many revolutions the wheel makes is irrelevant. Thus, with reference to Fig. 1.5, it would be immaterial whether the marker traversed 2, 9, or -5 spaces.

If $a \equiv b \pmod m$, then $a - b$ is a multiple of m, and we may write $a - b = pm$, or $a = b + pm$. Conversely, if $a - b$ is a multiple of m, then $a \equiv b \pmod m$. Now let $a \equiv a'$ and $b \equiv b' \pmod m$. Then

$$a = a' + cm \qquad b = b' + dm$$

and the following relationships exist:

$$a + b = a' + b' + (c + d)m$$

$$a - b = a' - b' + (c - d)m$$

$$ab = a'b' + (a'd + b'c)m + cdm^2$$

It follows that

$$a + b \equiv a' + b' \pmod m$$

$$a - b \equiv a' - b' \pmod m$$

$$ab \equiv a'b' \pmod m \qquad (e)$$

For example, since $12 \equiv 21$ and $5 \equiv 23 \pmod 9$, the foregoing equations yield the following congruences: $17 \equiv 44$, $7 \equiv -2$, and $60 \equiv 483 \pmod 9$.

Let m' be a factor of m. If $a \equiv b \pmod m$, then $a \equiv b \pmod{m'}$. The proof is as follows: Let $m = nm'$. From the foregoing discussion, we have $a - b = pm = pnm'$. Thus, $a - b$ is a multiple of m', and the conclusion follows. For example, since $23 \equiv 51 \pmod{14}$, then $23 \equiv 51 \pmod 7$. The converse of this statement is not necessarily true. For example, $105 \equiv 15 \pmod 6$, but $105 \equiv 9 \pmod{12}$ and $15 \equiv 3 \pmod{12}$.

The principle of transposition that pertains to equations is of course valid for congruences as well. For example, taking 11 as modulus, we have $34 - 19 \equiv 4$. Transposing the 19, we obtain $34 \equiv 23 \equiv 1$.

From Eq. (*e*), we deduce the following: If $a \equiv a' \pmod m$, then $a^n \equiv a'^n \pmod m$. This equality yields interesting relationships between a decimal number and its digits. We shall illustrate these relationships with 13 as modulus. We start with $10 \equiv -3$. Multiplying successively by -3, we obtain the following: $10^2 \equiv 9$, $10^3 \equiv -27 \equiv -1$, $10^4 \equiv 3$, $10^5 \equiv -9$, $10^6 \equiv 27 \equiv 1$. From this point on, the set of numbers -3, 9, -1, 3, -9, 1 recurs cyclically. By transposition, we obtain $10 + 3 \equiv 0$, $10^2 - 9 \equiv 0$, $10^3 + 1 \equiv 0$, etc.

Now consider that we have a seven-place decimal number $x = d_6 d_5 d_4 d_3 d_2 d_1 d_0$. Then

$$x = d_0 + 10d_1 + 10^2 d_2 + 10^3 d_3 + 10^4 d_4 + 10^5 d_5 + 10^6 d_6$$

We now form the corresponding number

$$y = d_0 - 3d_1 + 9d_2 - d_3 + 3d_4 - 9d_5 + d_6$$

Then

$$x - y = (10+3)d_1 + (10^2 - 9)d_2 + (10^3 + 1)d_3 + (10^4 - 3)d_4 + (10^5 + 9)d_5 + (10^6 - 1)d_6$$

Applying the previous results, we find that each expression in parentheses has a remainder of 0; that is, it is divisible by 13. Therefore, $x - y$ is divisible by 13. Then

$$x - y \equiv 0 \pmod{13} \qquad x \equiv y \pmod{13}$$

Thus, when x is divided by 13, it has the same remainder as y. In particular, if y is divisible by 13, x is also divisible by 13.

As an illustration, let $x = 5{,}182{,}741$. Then

$$y = 1 - 3 \times 4 + 9 \times 7 - 2 + 3 \times 8 - 9 \times 1 + 5 = 70$$

When x and y are divided by 13, the remainder is 5 in both instances.

By selecting 11 as the modulus and proceeding in the same manner, we discover that a decimal number is divisible by 11 if and only if the corresponding number $y = d_0 - d_1 + d_2 - d_3 + \cdots$ is divisible by 11. For example, the number 7,481,639 is divisible by 11 because the corresponding number $9 - 3 + 6 - 1 + 8 - 4 + 7 = 22$ is divisible by 11.

A *linear congruential sequence* (LCS) is a set of numbers that is generated by starting with an initial number (or *seed*) X_0 and then repeatedly applying the recursive linear equation

$$X_{n+1} \equiv aX_n + c \pmod m$$

where X_n is the nth number in the sequence, a and c are constants, all numbers are positive, and X_n, a, and c can range from 0 to $m - 1$. Since the number of possible values of X_n is m, every LCS eventually becomes a recurrent cycle of numbers.

To illustrate the procedure for generating an LCS, we shall set $X_0 = 2$ and apply the equation

$$X_{n+1} \equiv 4X_n + 7 \qquad (\text{mod } 9)$$

Then $X_1 \equiv 6$, $X_2 \equiv 4$, $X_3 \equiv 5$, $X_4 \equiv 0$, $X_5 \equiv 7$, $X_6 \equiv 8$, $X_7 \equiv 3$, $X_8 \equiv 1$, $X_9 \equiv 2$, and the set of numbers from X_0 to X_8 then recurs cyclically.

For our illustrative case, the following table records every possible value of X_n and the corresponding value of X_{n+1}:

X_n	0	1	2	3	4	5	6	7	8
X_{n+1}	7	2	6	1	5	0	4	8	3

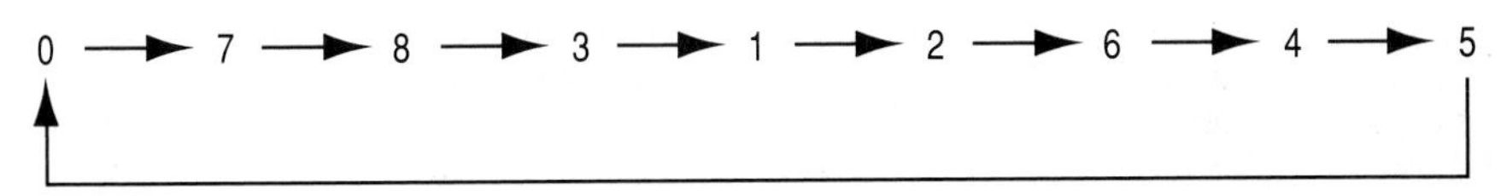

FIGURE 1.6 State diagram.

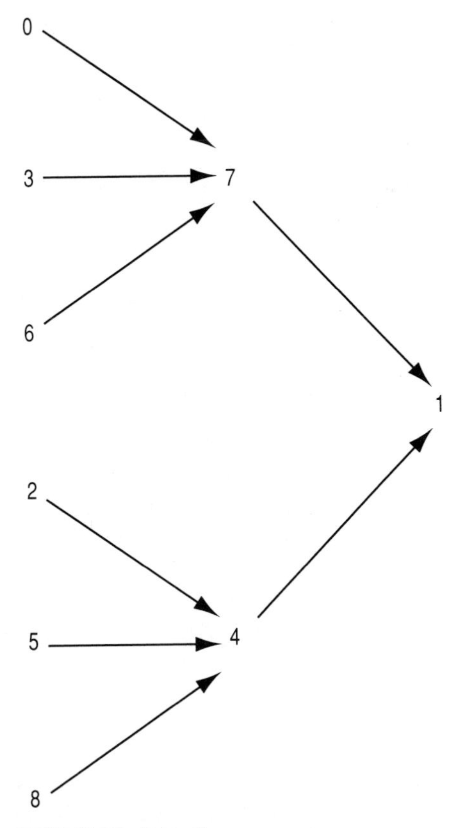

FIGURE 1.7 State diagram.

The results can be tested for consistency by constructing the congruence circle. When X_n increases by 1, the point corresponding to X_{n+1} advances four places in the clockwise direction.

The succession of numbers in our illustrative LCS can be depicted readily by the diagram in Fig. 1.6. This diagram contains every possible value of X_n, and an arrow leads from X_n to X_{n+1}. Thus, if we start with $X_0 = 8$, we obtain the recurrent sequence 8, 3, 1, 2, 6, 4, 5, 0, 7. Figure 1.6 is called a *state diagram*, and it corresponds to the *kinematic graphs* presented in Sec. 11.

If we now apply the equation

$$X_{n+1} \equiv 3X_n + 7 \qquad (\text{mod } 9)$$

we obtain the state diagram in Fig. 1.7. Whatever value X_0 may have, the sequence eventually culminates in 1, and this number then recurs indefinitely.

Linear congruential sequences are often applied to generate a set of pseudo-random numbers. The sequence must be designed to give the recurrent cycle of numbers the minimum desired length, and this is accomplished by assigning values to m, a, c, and X_0 in accordance with certain rules.

1.33 PYTHAGOREAN SETS

Let A, B, and C denote positive integers. If $A^2 + B^2 = C^2$, these integers constitute a *Pythagorean set*. The set is described as *primitive* if A, B, and C have no common factor.

Now let a, b, and c denote positive integers having the relationship

$$bc = \frac{a^2}{2}$$

We can construct a Pythagorean set by letting

$$A = a + b \qquad B = a + c \qquad C = a + b + c$$

TABLE 1.5

Generating set			Pythagorean set		
a	b	c	A	B	C
2	1	2	3	4	5
4	1	8	5	12	13
6	1	18	7	24	25
6	2	9	8	15	17
8	1	32	9	40	41
10	1	50	11	60	61
10	2	25	12	35	37
12	8	9	20	21	29
20	8	25	28	45	53

The numbers a, b, and c constitute the *generating set*. If a, b, and c have no common factor, the corresponding Pythagorean set is primitive.

Thus, by starting with $a = 2$, allowing a to assume successively higher even values, and selecting suitable values for b and c corresponding to each value of a, we can construct an endless group of Pythagorean sets. Table 1.5 presents specimen generating sets and their corresponding Pythagorean sets.

SECTION 2

PLANE AND SPHERICAL TRIGONOMETRY

2.1 PLANE TRIGONOMETRY

2.1.1 Characteristics of Angles

With reference to Fig. 2.1*a*, consider that line OA on the positive side of the x axis revolves about the origin O. The angle thus generated is positive or negative according to whether the direction of motion is counterclockwise or clockwise, respectively. Thus, $\angle AOB$ is positive and $\angle AOC$ is negative. In $\angle AOB$, OA and OB are the initial and terminal sides, respectively.

Three units of angular measure are used: the degree (°), the grad, and the radian (rad). The *radian* is the angle at the center of a circle subtended by an arc that has the same length as the radius. With reference to $\angle AOD$ in Fig. 2.1*b*, draw a circular arc having O as center and an arbitrary radius OE. The number of radians in the angle is the ratio $\widehat{EF}/OE$. The angle formed by a complete revolution is 360°, 400 grads, and 2π rad. Where the size of an angle is expressed without any unit, it is understood to be in radians.

Let d, g, and r denote the number of degrees, grads, and radians, respectively, in a given angle. Then

$$g = \frac{d}{0.9} \qquad r = \left(\frac{\pi}{180}\right)d \qquad r = \left(\frac{\pi}{200}\right)g$$

For example, if $d = 144$, $g = 160$, and $r = 2.5133$. In particular, we have

$$90^\circ = \frac{\pi}{2}\text{ rad} \qquad 180^\circ = \pi \text{ rad}$$

$$270^\circ = \frac{3\pi}{2}\text{ rad} \qquad 360^\circ = 2\pi \text{ rad}$$

The *degree* is divided into 60 min, and the minute is divided into 60 s. However, the calculator requires that the size of the angle be expressed solely in degrees. The conversion from mixed units to a single unit is achieved by dividing the number of seconds, and then the adjusted number of minutes, by 60. For example, $109^\circ 15'36'' = 109^\circ 15.6' = 109.26^\circ$.

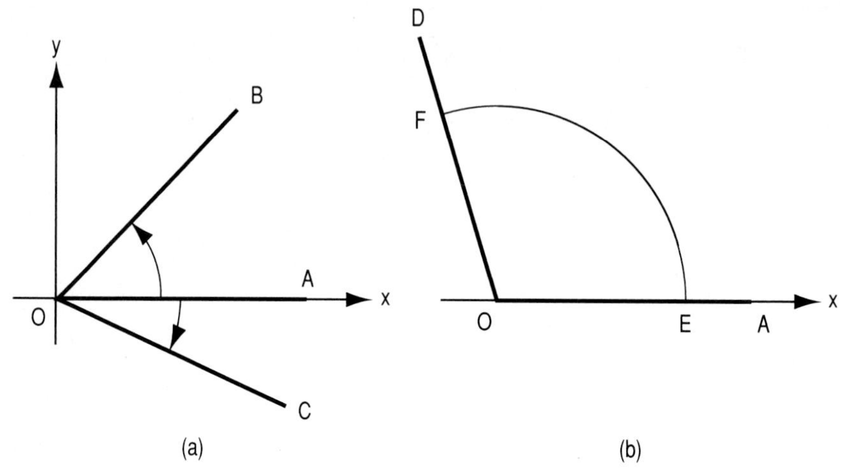

FIGURE 2.1 (*a*) Formation of an angle; (*b*) radian measure of an angle.

The appearance of an angle does not divulge how many complete revolutions (if any) occurred during its formation, but this number is often significant. Moreover, the appearance generally does not divulge the direction in which the generating line revolved. Therefore, it is often necessary to express the size of an angle in *general form* by adding to the apparent size k complete revolutions, where k is an arbitrary integer, positive, negative, or 0. Thus, if the apparent size of an angle is d°, the size is taken as $(360k + d)^\circ$; if the apparent size is r rad, the size is taken as

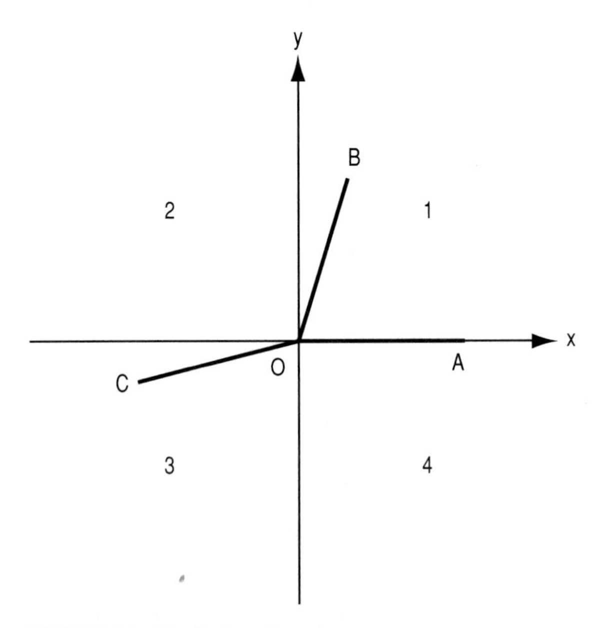

FIGURE 2.2 Numbering of quadrants.

$(2\pi k + r)$ rad. For example, if the apparent size of an angle is 65°, the angle may be considered to be 65°, 425°, 785°, −295°, etc. All these angles are *coterminal.*

The x and y axes divide the plane into quadrants, and these are numbered in the manner shown in Fig. 2.2. If OA is the initial side of an angle, the angle is said to lie in the quadrant in which its terminal side is located. Thus, in Fig. 2.2, $\angle AOB$ lies in the first quadrant and $\angle AOC$ lies in the third quadrant. An angle θ is described as *acute* if $0 < \theta < 90°$ and *obtuse* if $90° < \theta < 180°$.

If two positive angles have a sum of 90°, one angle is the *complement* of the other. If two positive angles have a sum of 180°, one angle is the *supplement* of the other.

2.1.2 Functions of an Angle

In Fig. 2.3, let θ denote an angle that has its initial side OA on the positive side of the x axis and its terminal side OB in any quadrant whatever. From an arbitrary point P on OB, drop a perpendicular QP to the x axis, and let $OQ = x$, $QP = y$, and $OP = r$. The functions of the angle are as follows:

$$\sin\theta = \frac{y}{r} \qquad \cos\theta = \frac{x}{r} \qquad \tan\theta = \frac{y}{x}$$

Corresponding to these basic functions are the following reciprocal functions:

$$\csc\theta = \frac{1}{\sin\theta} \qquad \sec\theta = \frac{1}{\cos\theta} \qquad \cot\theta = \frac{1}{\tan\theta}$$

The distance r is considered to be positive, and the algebraic sign of a function of θ depends on the quadrant in which θ lies. Table 2.1 exhibits the algebraic signs of the functions.

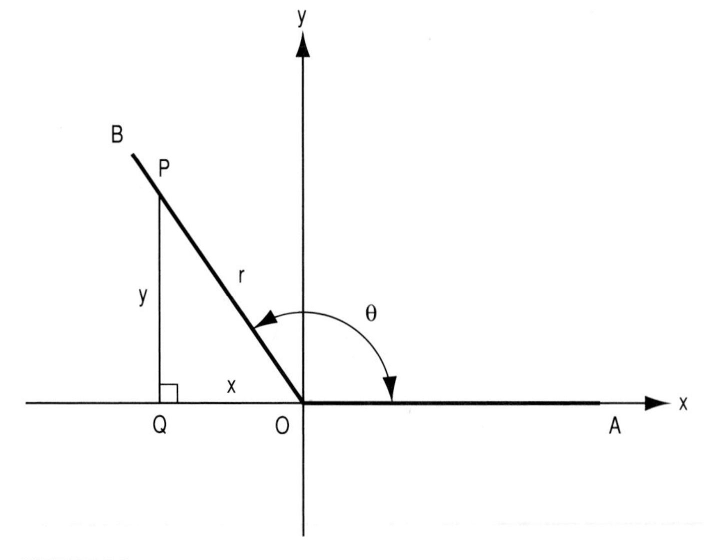

FIGURE 2.3

TABLE 2.1 Algebraic Signs of Trigonometric Functions

Function	Quadrant 1	2	3	4
$\sin\theta$	+	+	−	−
$\cos\theta$	+	−	−	+
$\tan\theta$	+	−	+	−

Assume that a function of an angle is to be raised to the mth power, where $m \neq -1$. The exponent is placed directly after the word. For example, $\sin^{3.5}\theta = (\sin\theta)^{3.5}$. However, if $m = -1$, parentheses must be used, as in $(\sin\theta)^{-1}$.

The functions of an angle have the following interrelationships:

$$\tan\theta = \frac{\sin\theta}{\cos\theta} = \frac{\sec\theta}{\csc\theta} \tag{2.1}$$

$$\sin^2\theta + \cos^2\theta = 1 \tag{2.2}$$

$$1 + \tan^2\theta = \sec^2\theta \tag{2.3}$$

$$1 + \cot^2\theta = \csc^2\theta \tag{2.4}$$

If θ is extremely small, $\sin\theta$ and $\tan\theta$ are very close to θ when θ is expressed in radians. Therefore, in practice these functions can often be replaced with θ, or vice versa, without introducing any appreciable error.

2.1.3 Functions of Sum and Difference of Two Angles

Let α and β denote two angles. The functions of their sum and difference are as follows:

$$\sin(\alpha + \beta) = \sin\alpha\cos\beta + \cos\alpha\sin\beta \tag{2.5}$$

$$\cos(\alpha + \beta) = \cos\alpha\cos\beta - \sin\alpha\sin\beta \tag{2.6}$$

$$\tan(\alpha + \beta) = \frac{\tan\alpha + \tan\beta}{1 - \tan\alpha\tan\beta} \tag{2.7}$$

$$\sin(\alpha - \beta) = \sin\alpha\cos\beta - \cos\alpha\sin\beta \tag{2.8}$$

$$\cos(\alpha - \beta) = \cos\alpha\cos\beta + \sin\alpha\sin\beta \tag{2.9}$$

$$\tan(\alpha - \beta) = \frac{\tan\alpha - \tan\beta}{1 + \tan\alpha\tan\beta} \tag{2.10}$$

Table 2.2 presents the relationships for the special cases where α is 0 or a multiple of 90°. In this table, β is replaced with θ. For example, $\tan(180° - \theta) = -\tan\theta$, and $\cos(270° + \theta) = \sin\theta$.

TABLE 2.2 Functions of Related Angles

	Function		
Value of α	$\sin\alpha$	$\cos\alpha$	$\tan\alpha$
$-\theta$	$-\sin\theta$	$\cos\theta$	$-\tan\theta$
$90^\circ-\theta$	$\cos\theta$	$\sin\theta$	$\cot\theta$
$90^\circ+\theta$	$\cos\theta$	$-\sin\theta$	$-\cot\theta$
$180^\circ-\theta$	$\sin\theta$	$-\cos\theta$	$-\tan\theta$
$180^\circ+\theta$	$-\sin\theta$	$-\cos\theta$	$\tan\theta$
$270^\circ-\theta$	$-\cos\theta$	$-\sin\theta$	$\cot\theta$
$270^\circ+\theta$	$-\cos\theta$	$\sin\theta$	$-\cot\theta$
$360^\circ-\theta$	$-\sin\theta$	$\cos\theta$	$-\tan\theta$

2.1.4 Functions of Multiples and Submultiples of an Angle

The equations of Art. 2.1.3 yield the following:

$$\sin 2\theta = 2\sin\theta\cos\theta \tag{2.11}$$

$$\cos 2\theta = \cos^2\theta - \sin^2\theta = 1 - 2\sin^2\theta = 2\cos^2\theta - 1 \tag{2.12}$$

$$\tan 2\theta = \frac{2\tan\theta}{1-\tan^2\theta} \tag{2.13}$$

The foregoing equations in turn yield the following:

$$\sin\frac{\theta}{2} = \pm\sqrt{\frac{1-\cos\theta}{2}} \tag{2.14}$$

$$\cos\frac{\theta}{2} = \pm\sqrt{\frac{1+\cos\theta}{2}} \tag{2.15}$$

$$\tan\frac{\theta}{2} = \pm\sqrt{\frac{1-\cos\theta}{1+\cos\theta}} = \frac{1-\cos\theta}{\sin\theta} = \frac{\sin\theta}{1+\cos\theta} \tag{2.16}$$

2.1.5 Transformation of Products of Functions

The equations of Art. 2.1.3 also yield the following:

$$\sin\alpha\sin\beta = -\tfrac{1}{2}\cos(\alpha+\beta) + \tfrac{1}{2}\cos(\alpha-\beta) \tag{2.17}$$

$$\sin\alpha\cos\beta = \tfrac{1}{2}\sin(\alpha+\beta) + \tfrac{1}{2}\sin(\alpha-\beta) \tag{2.18}$$

$$\cos\alpha\cos\beta = \tfrac{1}{2}\cos(\alpha+\beta) + \tfrac{1}{2}\cos(\alpha-\beta) \tag{2.19}$$

2.1.6 Inverse Functions

The statement $\theta = \sin^{-1} a$ means that θ is an angle whose sine is a. For example, since $\sin 30^\circ = 0.5$, then $30^\circ = \sin^{-1} 0.5$. The expression $\sin^{-1} a$ has the alternative forms arc sin a and invsin a. Analogous expressions apply with reference to the other functions.

Let θ and a denote an angle and a real number, respectively, that are related by a trigonometric function. There is a unique value of a corresponding to a given value of θ. On the other hand, there is an infinite number of values of θ corresponding to a given value of a, and the calculator provides only one value in this set. We shall call this value θ_c. Other values of θ may be obtained by applying the following relationships:

$$\sin(180° - \theta) = \sin\theta \qquad \cos(360° - \theta) = \cos\theta$$

$$\tan(180° + \theta) = \tan\theta$$

Moreover, a function of $360° + \theta$ equals the corresponding function of θ.

EXAMPLE 2.1 Applying the restriction $0 \le \theta \le 360°$, solve the following equations: (*a*) $\theta = \sin^{-1} 0.7$; (*b*) $\theta = \sin^{-1}(-0.7)$; (*c*) $\theta = \cos^{-1} 0.4$; (*d*) $\theta = \cos^{-1}(-0.4)$; (*e*) $\theta = \tan^{-1} 2.1$; (*f*) $\theta = \tan^{-1}(-2.1)$; (*g*) $\theta = \sec^{-1} 3.2$.

SOLUTION As Table 2.1 discloses, there are two values of θ that satisfy each equation. We shall call these values θ_1 and θ_2. With reference to Eq. (*g*) in Example 2.1, we have $\cos\theta = 1/(\sec\theta) = 1/3.2 = 0.3125$. The solutions are presented in Table 2.3.

TABLE 2.3

Equation (Example 2.1)	θ_c, °	θ_1, °	θ_2, °
a	44.43	44.43	$180 - \theta_c = 135.57$
b	−44.43	$180 - \theta_c = 224.43$	$360 + \theta_c = 315.57$
c	66.42	66.42	$360 - \theta_c = 293.58$
d	113.58	113.58	$360 - \theta_c = 246.42$
e	64.54	64.54	$180 + \theta_c = 244.54$
f	−64.54	$180 + \theta_c = 115.46$	$360 + \theta_c = 295.46$
g	71.79	71.79	$360 - \theta_c = 288.21$

2.1.7 Solution of Right Triangle

A triangle consists of six parts: three sides and three angles. If we are given three parts, at least one of which is a side, we can evaluate the three remaining parts, and this process is called *solving the triangle*. With reference to $\triangle ABC$ in Fig. 2.4*a*, the parts are labeled in the manner shown. The angles are related by

$$A + B + C = 180° \tag{2.20}$$

Now assume that $C = 90°$, as shown in Fig. 2.4*b*. In this right triangle, AB is the *hypotenuse*, and AC and BC are the *legs*. The following relationships exist:

$$a^2 + b^2 = c^2 \tag{2.21}$$

$$\sin A = \cos B = \frac{a}{c} \tag{2.22}$$

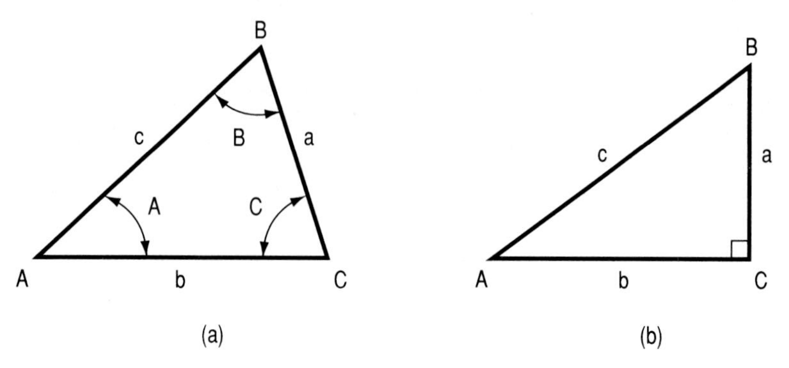

FIGURE 2.4 (*a*) Labeling of parts of a triangle; (*b*) right triangle.

$$\cos A = \sin B = \frac{b}{c} \tag{2.23}$$

$$\tan A = \cot B = \frac{a}{b} \tag{2.24}$$

Problems that require the solution of a right triangle can be classified on the basis of the given information, in this manner: angle and opposite leg; angle and adjacent leg; angle and hypotenuse; leg and hypotenuse; two legs.

EXAMPLE 2.2 Solve the triangle in which $a = 5$, $A = 36°$, and $C = 90°$.

SOLUTION Applying Eqs. (2.20), (2.22), and (2.24) in turn, we obtain the following:

$$B = 90° - A = 90° - 36° = 54°$$

$$c = \frac{a}{\sin A} = \frac{5}{\sin 36°} = 8.507$$

$$b = \frac{a}{\tan A} = \frac{5}{\tan 36°} = 6.882$$

Equation (2.23) provides a test of these results.

$$\sin B = \frac{b}{c} = \frac{6.882}{8.507} = 0.8090 \qquad \sin 54° = 0.8090$$

Equation (2.21) is also satisfied.

EXAMPLE 2.3 Solve the triangle in which $b = 12.5$, $A = 41°$, and $C = 90°$.

SOLUTION

$$B = 90° - A = 90° - 41° = 49°$$

$$c = \frac{b}{\cos A} = \frac{12.5}{\cos 41°} = 16.563$$

$$a = b \tan A = 12.5 \tan 41° = 10.866$$

Equation (2.22) provides a test.

$$\cos B = \frac{a}{c} = \frac{10.866}{16.563} = 0.6560 \qquad \cos 49^\circ = 0.6561$$

Equation (2.21) is also satisfied.

EXAMPLE 2.4 Solve the triangle in which $a = 10.3$, $c = 16.8$, and $C = 90°$.

SOLUTION

$$\sin A = \cos B = \frac{a}{c} = \frac{10.3}{16.8} = 0.6131$$

$$A = 37.81^\circ \qquad B = 52.19^\circ$$

$$b^2 = c^2 - a^2 = 16.8^2 - 10.3^2 = 176.15 \qquad b = 13.272$$

Equation (2.23) provides a test.

$$\cos A = \sin B = \frac{b}{c} = \frac{13.272}{16.8} = 0.7900$$

$$\cos 37.81^\circ = \sin 52.19^\circ = 0.7900$$

EXAMPLE 2.5 Solve the triangle in which $a = 8.2$, $b = 11.7$, and $C = 90°$.

SOLUTION

$$c^2 = a^2 + b^2 = 8.2^2 + 11.7^2 = 204.13 \qquad c = 14.287$$

$$\tan A = \frac{a}{b} = \frac{8.2}{11.7} = 0.7009$$

$$A = 35.03^\circ \qquad B = 54.97^\circ$$

Equation (2.22) provides a test.

$$\sin A = \cos B = \frac{a}{c} = \frac{8.2}{14.287} = 0.5739$$

$$\sin 35.03^\circ = \cos 54.97^\circ = 0.5740$$

2.1.8 Solution of Oblique Triangle

The oblique triangle in Fig. 2.4*a* has the following relationships:

$$\frac{a}{\sin A} = \frac{b}{\sin B} = \frac{c}{\sin C} \tag{2.25}$$

$$a^2 = b^2 + c^2 - 2bc \cos A \tag{2.26a}$$

$$b^2 = a^2 + c^2 - 2ac \cos B \tag{2.26b}$$

$$c^2 = a^2 + b^2 - 2ab \cos C \tag{2.26c}$$

Equation (2.25) is known as the *law of sines*, and Eq. (2.26) constitutes the *law of cosines*.

Several additional relationships have been formulated to convert the calculations to logarithmic form. However, the calculator has obviated the need for these relationships except for special purposes.

In general, an angle in an oblique triangle can be acute or obtuse. Therefore, when an angle is evaluated by the law of sines, two prospective values of the angle emerge, in accordance with the discussion in Art. 2.1.6. It is then necessary to identify the true value where only one value is possible.

Problems that require the solution of an oblique triangle can be classified on the basis of the given information, in the manner shown in Table 2.4. In solving a triangle, it is helpful to construct the triangle with ruler and compasses by applying the given information. This construction instantly clarifies the problem.

EXAMPLE 2.6 Solve the triangle in which $A = 42.9°$, $B = 31.6°$, and $a = 20.5$.

SOLUTION This is a type 1 problem.

$$C = 180° - (A + B) = 105.5°$$

$$b = \frac{a \sin B}{\sin A} = 15.780 \qquad c = \frac{a \sin C}{\sin A} = 29.020$$

These results satisfy Eq. (2.26*a*).

EXAMPLE 2.7 Solve the triangle in which $b = 13.4$, $c = 8.6$, and $C = 31.5°$.

SOLUTION This is a type 2 problem. Figure 2.5 reveals that this problem is

TABLE 2.4 Classification of Problems in Oblique Plane Triangles

Type	Given information
1	Two angles and one side
2	Two sides and the angle opposite one of them
3	Two sides and the included angle
4	The three sides

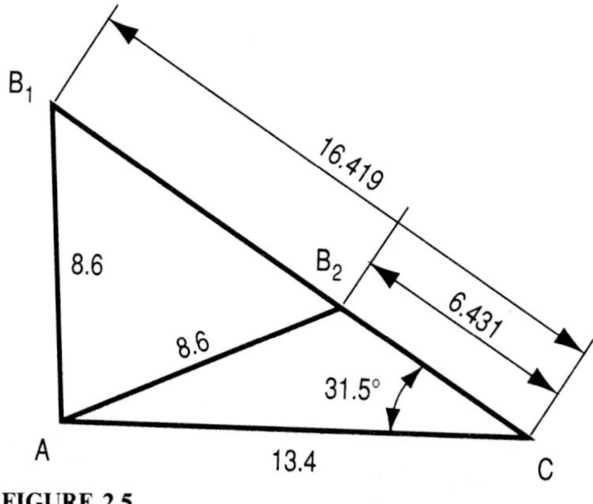

FIGURE 2.5

ambiguous because the construction yields two triangles rather than one. We shall apply the subscripts 1 and 2 to distinguish their values.

$$\sin B = \frac{b \sin C}{c} = 0.8141$$

$$B_1 = 54.5^\circ \qquad B_2 = 125.5^\circ$$

$$A_1 = 180^\circ - (B_1 + C) = 94^\circ \qquad A_2 = 180^\circ - (B_2 + C) = 23^\circ$$

$$a_1 = \frac{c \sin A_1}{\sin C} = 16.419 \qquad a_2 = \frac{c \sin A_2}{\sin C} = 6.431$$

These results satisfy Eq. (2.26*c*).

EXAMPLE 2.8 Solve the triangle in which $a = 18.3$, $b = 13.9$, and $C = 29^\circ$.

SOLUTION This is a type 3 problem. Refer to Fig. 2.6. By Eq. (2.26*c*), we obtain $c = 9.118$. Then

$$\sin A = \frac{a \sin C}{c} = 0.9730 \qquad \sin B = \frac{b \sin C}{c} = 0.7391$$

The prospective values of these angles are

$$A_1 = 76.7^\circ \qquad A_2 = 103.3^\circ \qquad B_1 = 47.7^\circ \qquad B_2 = 132.3^\circ$$

Figure 2.6 reveals that the acceptable values are $A = 103.3^\circ$ and $B = 47.7^\circ$. These values satisfy Eq. (2.20).

EXAMPLE 2.9 Solve the triangle in which $a = 9.0$, $b = 15.8$, and $c = 22.1$.

SOLUTION This is a type 4 problem. Refer to Fig. 2.7. By the law of cosines,

$$\cos A = \frac{b^2 + c^2 - a^2}{2bc} = 0.9408 \qquad A = 19.8^\circ$$

$$\cos B = \frac{a^2 + c^2 - b^2}{2ac} = 0.8038 \qquad B = 36.5^\circ$$

$$\cos C = \frac{a^2 + b^2 - c^2}{2ab} = -0.5547 \qquad C = 123.7^\circ$$

These results satisfy Eq. (2.20).

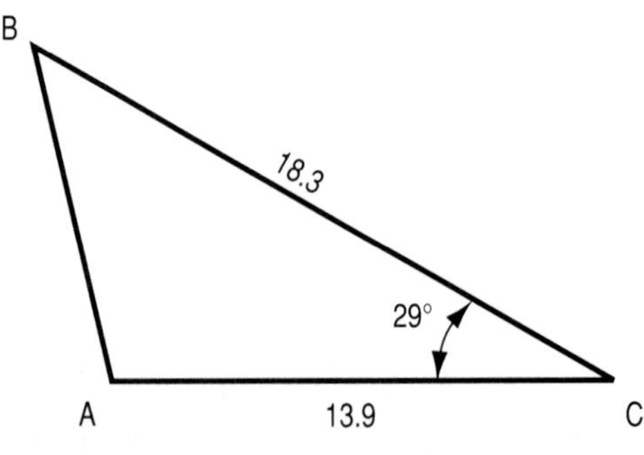

FIGURE 2.6

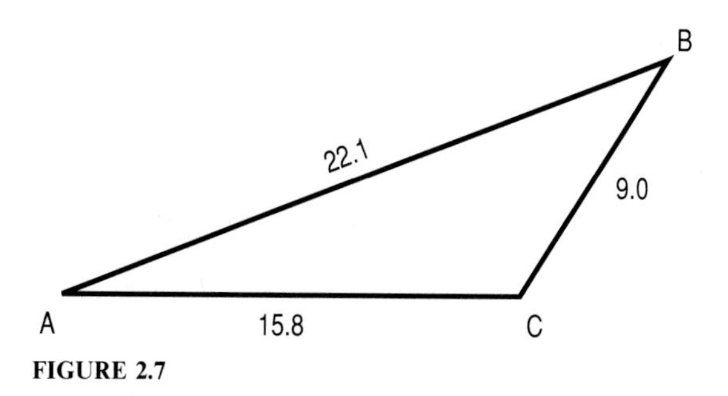

FIGURE 2.7

2.1.9 DeMoivre's Theorem and Applications

Let a and b denote real numbers, and let $z = a + bi$, where i is the unit of imaginaries. Thus, z is a complex number. By use of a simple scheme, it is possible to represent every number, real, imaginary, or complex, by a unique point in a plane. In Fig. 2.8*a*, real numbers are represented on the x axis, and imaginary numbers are represented on the y axis. Thus, a is represented by point M, and bi is represented by point N. The complex number z is represented by point P in Fig. 2.8*b*, which has the coordinates shown.

With reference to Fig. 2.8*b*, let $OP = r$ and $\angle MOP = \theta$. The distance r and angle θ are termed the *modulus* and *amplitude*, respectively, of the number z. Since $a = r \cos \theta$ and $b = r \sin \theta$, we may write

$$z = a + bi = r(\cos \theta + i \sin \theta)$$

The first expression for z is the *rectangular form*, and the second is the *polar form*.

Now let z_1 and z_2 denote complex numbers, where

$$z_1 = r_1(\cos \theta_1 + i \sin \theta_1) \qquad z_2 = r_2(\cos \theta_2 + i \sin \theta_2)$$

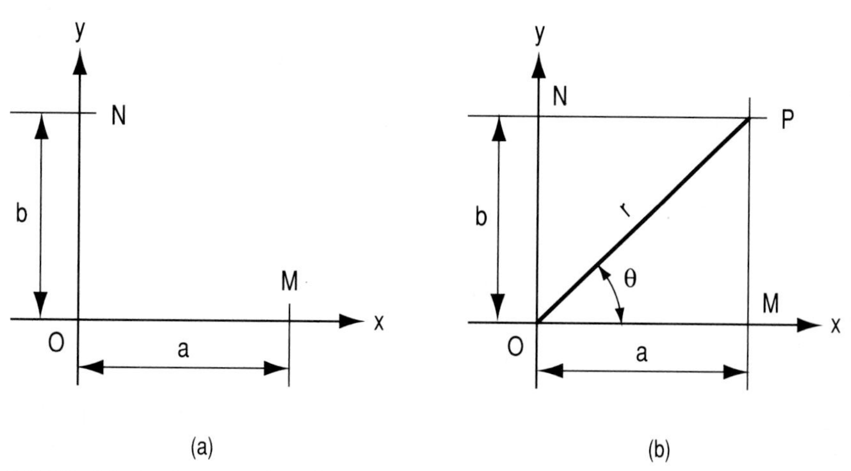

FIGURE 2.8 (*a*) Representation of real and imaginary numbers; (*b*) representation of complex numbers.

Performing the indicated operations, we obtain these results:

$$z_1 z_2 = r_1 r_2[\cos(\theta_1 + \theta_2) + i \sin(\theta_1 + \theta_2)] \tag{2.27a}$$

$$\frac{z_1}{z_2} = \frac{r_1}{r_2}[\cos(\theta_1 - \theta_2) + i \sin(\theta_1 - \theta_2)] \tag{2.27b}$$

Equation (2.27*a*) can be extended, and we arrive at the following principle:

Theorem 2.1. When several complex numbers are multiplied, the product in general is also a complex number. The modulus of the product equals the product of the moduli of the given numbers; the amplitude of the product equals the sum of the amplitudes of the given numbers.

Now consider that the complex number z is multiplied by itself repeatedly. Theorem 2.1 yields the following:

$$[r(\cos\theta + i \sin\theta)]^n = r^n(\cos n\theta + i \sin n\theta) \tag{2.28}$$

This relationship can be extended to include all rational values of n.

Equation (2.28) is known as *DeMoivre's theorem*, and it has numerous applications. We shall illustrate one application: finding the nth root of a complex number.

EXAMPLE 2.10 Find the cube roots of $-4 + 3i$.

SOLUTION Let r' and θ' denote the modulus and amplitude, respectively, of the given number, and refer to Fig. 2.9.

$$r' = \sqrt{(-4)^2 + 3^2} = 5$$

$$\tan\theta' = \frac{3}{-4} = -0.75 \qquad \theta' = 143.13°$$

Now let r and θ denote the modulus and amplitude, respectively, of a root. Then

$$[r(\cos\theta + i \sin\theta)]^3 = 5(\cos 143.13° + i \sin 143.13°)$$

Applying Eq. (2.28) and replacing 143.13° with its general expression as given in

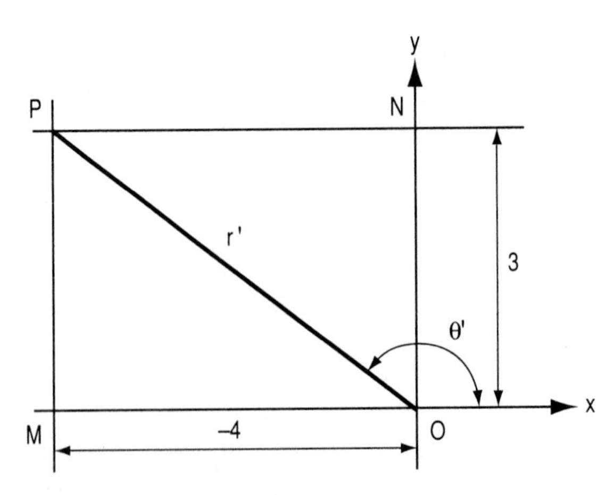

FIGURE 2.9

Art. 2.1.1, we obtain

$$r^3(\cos 3\theta + i \sin 3\theta) = 5[\cos(360k + 143.13)^\circ + i \sin(360k + 143.13)^\circ]$$

We now equate corresponding parts. Since r is a real positive number, we simply set $r = 5^{1/3} = 1.7100$. Setting k equal to 0, 1, and 2 in turn, and then dividing by 3, we obtain the values of θ recorded in Table 2.5. Setting $a = r \cos\theta$ and $b = r \sin\theta$, we obtain the values of a and b shown in the table. In rectangular form, the three cube roots of $-4 + 3i$ are as follows:

$$z_1 = 1.1506 + 1.2650i \qquad z_2 = -1.6708 + 0.3640i$$

$$z_3 = 0.5202 - 1.6290i$$

These cube roots are represented by points P_1, P_2, and P_3 in Fig. 2.10. The points are uniformly spaced on a circle of radius 1.7100.

TABLE 2.5

k	θ, °	a	b
0	$\frac{143.13}{3} = 47.71$	1.1506	1.2650
1	$\frac{503.13}{3} = 167.71$	−1.6708	0.3640
2	$\frac{863.13}{3} = 287.71$	0.5202	−1.6290

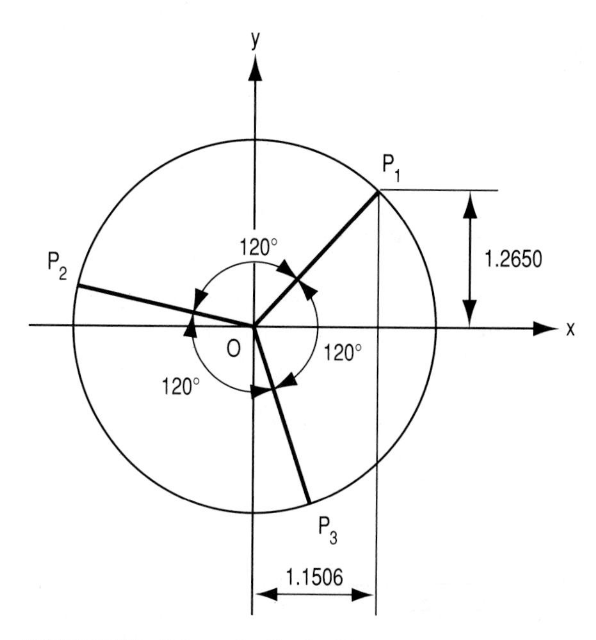

FIGURE 2.10 Cube roots of $-4 + 3i$.

2.2 SPHERICAL TRIGONOMETRY

2.2.1 Space Angles

When two planes intersect, a *dihedral angle* is formed. The two planes are the *faces* of the angle, and their line of intersection is the *edge* of the angle. In Fig. 2.11*a*, planes P and Q intersect along line AB, and the angle between these planes is θ. Lines CD and CE lie in planes P and Q, respectively, and they are both perpendicular to AB. Angle DCE is the *plane angle* corresponding to the dihedral angle, and its magnitude is also θ.

When three or more planes intersect at a point, a *polyhedral angle* is formed. We shall consider only *closed polyhedral angles*. This type of angle is illustrated in Fig. 2.11*b*, where five planes intersect at V. The point of intersection of the planes is the *vertex*, the lines of intersection of adjacent planes are the *edges*, the portions of the planes between the edges are the *faces*, and the plane angles between adjacent edges are the *face angles*. Thus, in Fig. 2.11*b*, V is the vertex, AV is an edge, plane AVB is a face, and $\angle AVB$ is a face angle. A polyhedral angle formed by three intersecting planes is called a *trihedral angle*, and it is illustrated in Fig. 2.11*c*.

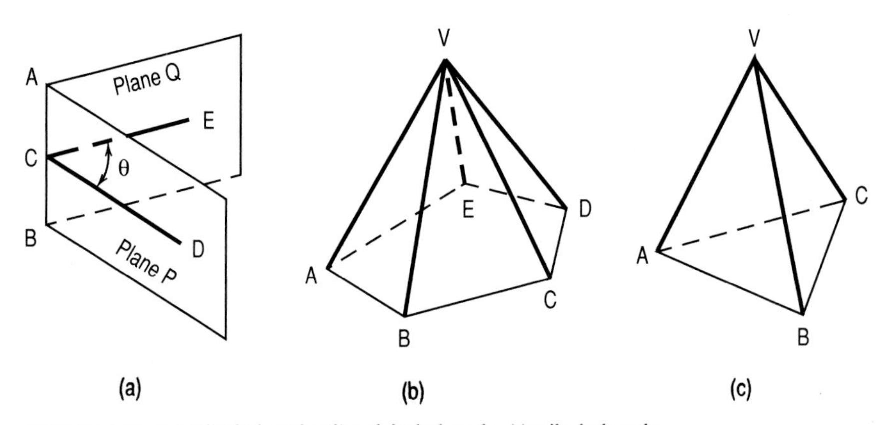

FIGURE 2.11 (*a*) Dihedral angle; (*b*) polyhedral angle; (*c*) trihedral angle.

2.2.2 Projected Angles

In Fig. 2.12, $ABCD$, $DCEF$, and $ABEF$ are rectangles in planes P, Q, and R, respectively. Planes P and Q are perpendicular to each other, and the angle between planes P and R is θ. Line AG is an arbitrary line in plane R, and GH is perpendicular to DC. Let $\angle FAG = \alpha$, $\angle GAB = \beta$, $\angle DAH = \alpha'$, and $\angle HAB = \beta'$. Angles α' and β' are the *projections* of α and β, respectively, on plane P. Angles α and β are complementary, and angles α' and β' are also complementary. We have the following:

$$\tan \alpha = \frac{FG}{AF} \qquad \tan \alpha' = \frac{DH}{AD}$$

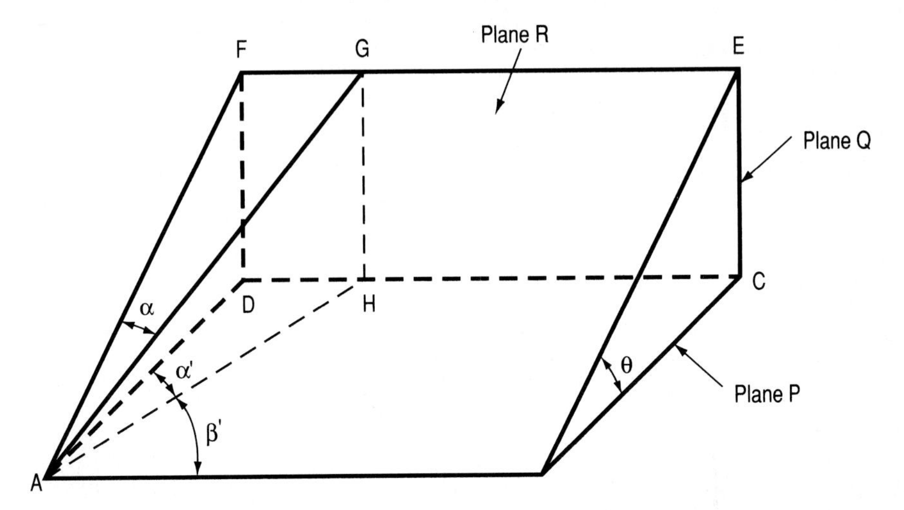

FIGURE 2.12 Projection of angles.

Setting $DH = FG$ and $AD = (AF) \cos \theta$ and substituting, we obtain

$$\tan \alpha' = \frac{\tan \alpha}{\cos \theta} \tag{2.29a}$$

Similarly,

$$\tan \beta' = \tan \beta \cos \theta \tag{2.29b}$$

If θ is obtuse, α' and β' are considered negative. In Fig. 2.12, $\alpha < \alpha'$ and $\beta > \beta'$.

Now consider that points A, B, F, and G remain stationary but plane P is displaced while remaining parallel to its original position. Angles α and β and their projected angles remain constant during this displacement. In general, therefore, when an angle on plane R is projected on plane P, Eq. (2.29*a*) applies if the angle has one side *perpendicular to* the line of intersection of the planes, and Eq. (2.29*b*) applies if the angle has one side *on or parallel to* the line of intersection.

In Fig. 2.12, let $\angle GAH = \phi$. Then

$$\sin \phi = \sin \theta \cos \alpha \tag{2.30}$$

If α and θ are both acute, $\phi < \theta$.

In Fig. 2.11*b*, *ABCDE* is a plane through the polyhedral angle. If each face angle of the polyhedral angle is projected on this plane, the sum of the projected angles is 360°. Applying the foregoing inequalities, we deduce the following:

Theorem 2.2. The sum of the face angles of a polyhedral angle is less than 360°.

In Fig. 2.11*c*, assume that $\angle AVC$ is the largest of the three face angles. If we project the two other face angles on plane *AVC*, the sum of the projected angles is $\angle AVC$. Applying the foregoing inequalities, we deduce the following:

Theorem 2.3. In a trihedral angle, the sum of any two face angles is greater than the third face angle.

2.2.3 Definitions Pertaining to a Sphere

Consider that a plane P intersects a sphere S. Plane P is a *central plane* if it contains the center of S. The line of intersection of P and S is a circle. This line is called a *great circle* if P is a central plane, as in Fig. 2.13*a*, and a *small circle* if P is a noncentral plane, as in Fig. 2.13*b*. The diameter of S that is perpendicular to P is the *axis* of the circle, and the extremities of the axis are the *poles* of the circle. Thus, in Fig. 2.13*a* and *b*, diameter EF is perpendicular to plane P. Therefore, EF is the axis of $\odot ABCD$, and E and F are the poles of the circle. One-fourth of a great circle is a *quadrant*.

The *spherical distance* between two points on a sphere is the length of the minor arc of the great circle that contains them. For example, in Fig. 2.13*a*, G and H lie on the great circle $ABCD$, and $\overset{\frown}{GH}$ is the spherical distance between these points. It is postulated that the spherical distance is the shortest distance between two points on a sphere.

The *polar distance* of a circle is the spherical distance from the nearer pole to any point on the circle. Thus, in Fig. 2.13*b*, AE is the polar distance of $\odot ABCD$. The polar distance of a great circle is a quadrant.

In Fig. 2.14, $C1$ and $C2$ are great circles of a sphere. These circles intersect to form a *spherical angle* at A (and at B). Let θ denote this angle. There are three ways of measuring θ. First, we draw lines AC and AD through A tangent to $C1$ and $C2$, respectively. The plane angle between these tangents equals θ. Second, θ equals the dihedral angle between the central planes of $C1$ and $C2$. Finally, we draw the great circle $C3$ that has A and B as its poles. This circle intersects $C1$ at E and $C2$ at F. Angle θ is measured by $\overset{\frown}{EF}$.

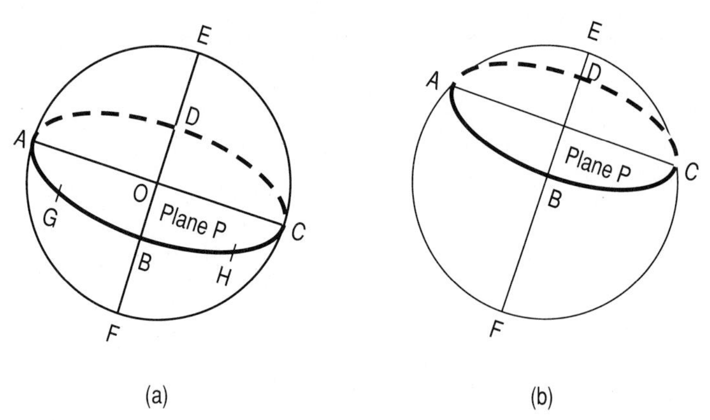

FIGURE 2.13 Intersection of plane and sphere. (*a*) Great circle of sphere; (*b*) small circle of sphere.

2.2.4 Basic Properties of a Spherical Triangle

Consider that three or more central planes are passed through a sphere. Since these planes all contain the center of the sphere, they form a polyhedral angle at the center. The planes also form great circles that intersect one another. A closed figure on the surface of the sphere bounded by arcs of these great circles is termed a *spherical polygon*. If the polygon contains three sides, it is called a *spherical triangle*.

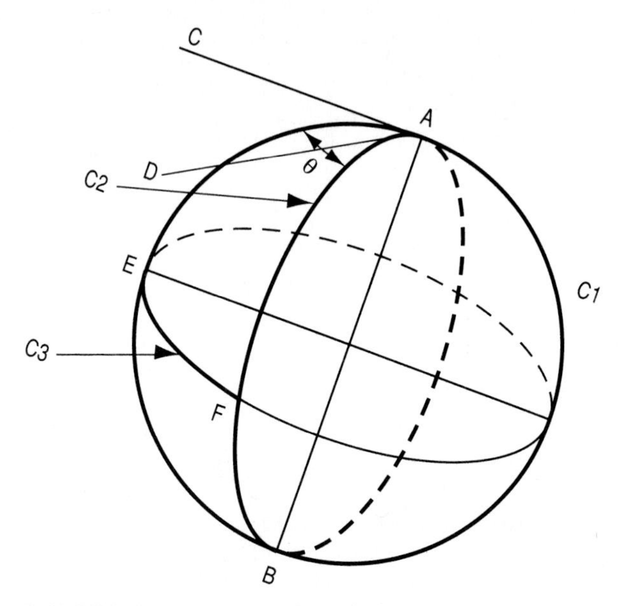

FIGURE 2.14 Measurement of spherical angle.

In Fig. 2.15, O is the center of a sphere, and AB, AC, and BC are arcs of great circles. Thus, ABC is a spherical triangle. The circular arcs are the *sides* of the triangle, and the spherical angles at the vertices A, B, and C are called simply the *angles* of the triangle. The parts of the triangle are labeled in a manner analogous to that of a plane triangle. Thus, the side opposite vertex A is labeled a, and the angle at vertex A is labeled A. Although the triangle is merely a spherical surface, it is helpful to visualize the entire solid bounded by the three central planes and the

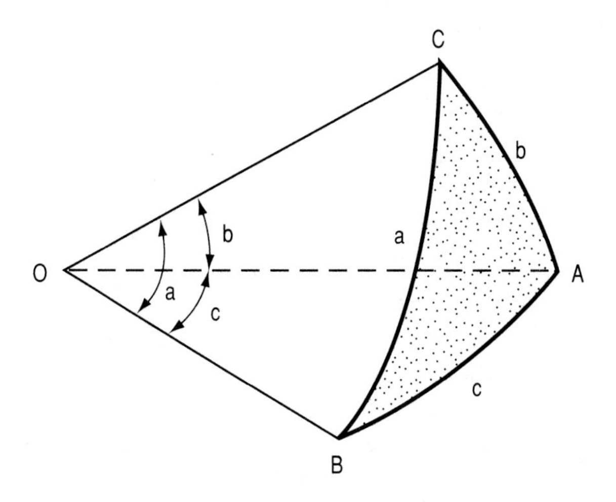

FIGURE 2.15 Spherical triangle.

sphere. The sides of the spherical triangle are measured in degrees, and they are equal in numerical value to the corresponding face angles of the trihedral angle at the center. For example, in Fig. 2.15, the sides a, b, and c are equal to $\angle$s BOC, AOC, and AOB, respectively. Similarly, $\angle$s A, B, and C are equal, respectively, to the dihedral angles between the following pairs of planes: BOA and COA; BOA and COB; BOC and AOC.

The definitions pertaining to a plane triangle apply to a spherical triangle as well. For example, a right spherical triangle is one in which one angle is 90°, and the side opposite this angle is the hypotenuse.

Theorems 2.2 and 2.3 yield the following:

Theorem 2.4. The sum of the sides of a spherical triangle is less than 360°.

Theorem 2.5. The sum of any two sides of a spherical triangle is greater than the third side.

The following relationship also applies:

Theorem 2.6. The sum of the angles of a spherical triangle is greater than 180° and less than 540°.

A spherical triangle consists of six parts: three sides and three angles. If three parts are given, the three remaining parts can be evaluated.

We shall demonstrate how the solution of a spherical triangle can be tested mathematically. However, the solution can also be tested empirically and with reasonable accuracy by constructing a simple model of the solid in Fig. 2.15, using either stiff cardboard or sheet metal. For example, selecting a convenient length for the radius of the sphere, we form face OBC in Fig. 2.16. In lieu of cutting this face along the circular arc BC, we cut it along lines BD and CD perpendicular to OB and OC, respectively. The mating edges of the faces are then joined by some suitable means. The dihedral angles between the faces of the solid, which equal the angles of the triangle, can then be measured by placing a protractor on the lines perpendicular to the edges. This simple model aids enormously in visualizing the

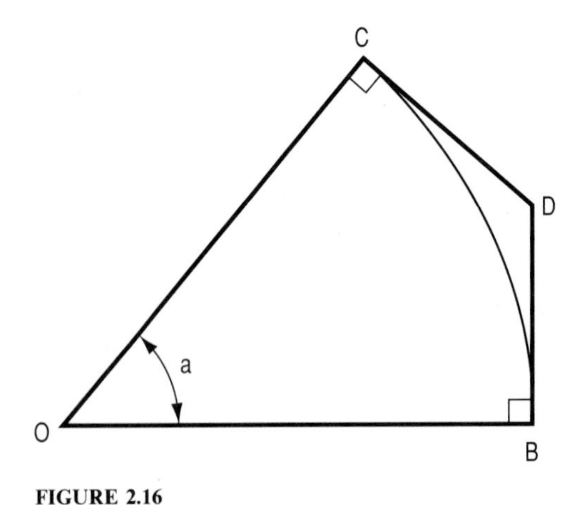

FIGURE 2.16

spherical triangle. It is to be observed, however, that the faces of the solid can be assembled in two different ways, thereby forming two distinct solids. One solid is the mirror image of the other.

2.2.5 Solution of Right Spherical Triangle

In a right spherical triangle, there are five variable parts, and two of these parts must be given. The number of combinations of five parts taken two at a time is 10, and consequently a right triangle is characterized by 10 relationships. Let C denote the right angle. The relationships are as follows:

$$\sin a = \sin c \sin A \tag{2.31a}$$

$$\sin b = \sin c \sin B \tag{2.31b}$$

$$\sin a = \tan b \cot B \tag{2.32a}$$

$$\sin b = \tan a \cot A \tag{2.32b}$$

$$\cos A = \cos a \sin B \tag{2.33a}$$

$$\cos B = \cos b \sin A \tag{2.33b}$$

$$\cos A = \tan b \cot c \tag{2.34a}$$

$$\cos B = \tan a \cot c \tag{2.34b}$$

$$\cos c = \cot A \cot B \tag{2.35}$$

$$\cos c = \cos a \cos b \tag{2.36}$$

Table 2.6 lists each possible combination of three variable parts of the triangle and the equation by which those parts are related.

After the sine of a part has been computed, it is necessary to ascertain whether that part is acute or obtuse. Two parts are said to be of the *same species* if they are both acute or both obtuse, and of *opposite species* if one is acute and the other obtuse. On the basis of Eqs. (2.33) and (2.35), we arrive at the following conclusions:

TABLE 2.6 Catalog of Equations for Right Spherical Triangle

Known parts	Equation
a, b, c	2.36
a, b, A	2.32b
a, b, B	2.32a
a, c, A	2.31a
a, c, B	2.34b
a, A, B	2.33a
b, c, A	2.34a
b, c, B	2.31b
b, A, B	2.33b
c, A, B	2.35

Theorem 2.7. An oblique angle and its opposite side are always of the same species.

Theorem 2.8a. If the hypotenuse is less than 90°, the two oblique angles are of the same species, and the two sides opposite these angles are of the same species.

Theorem 2.8b. If the hypotenuse is greater than 90°, the two oblique angles are of opposite species, and the two sides opposite these angles are of opposite species.

EXAMPLE 2.11 Solve the triangle in which $A = 43°$, $B = 61°$, and $C = 90°$.

SOLUTION Scanning Table 2.6 for the combinations containing A and B, we find that Eqs. (2.33) and (2.35) are applicable.

$$\cos a = \frac{\cos A}{\sin B} = 0.8362 \qquad a = 33.26°$$

$$\cos b = \frac{\cos B}{\sin A} = 0.7109 \qquad b = 44.69°$$

$$\cos c = \cot A \cot B = 0.5944 \qquad c = 53.53°$$

Equation (2.36) provides a simple test of these results, but Eq. (2.34) provides a more effective test because it involves functions that were not used in the calculations.

$$\cos A = \tan b \cot c = 0.7312 \qquad A = 43°$$

$$\cos B = \tan a \cot c = 0.4848 \qquad B = 61°$$

These results coincide with the given data.

EXAMPLE 2.12 Solve the triangle in which $a = 31°$, $A = 35°$, and $C = 90°$.

SOLUTION Scanning Table 2.6 for the combinations containing a and A, we find that Eqs. (2.32*b*), (2.31*a*), and (2.33*a*) are applicable.

$$\sin b = \tan a \cot A = 0.8581$$

$$b_1 = 59.11° \qquad b_2 = 180° - b_1 = 120.89°$$

$$\sin c = \frac{\sin a}{\sin A} = 0.8979$$

$$c_1 = 63.89° \qquad c_2 = 180° - c_1 = 116.11°$$

$$\sin B = \frac{\cos A}{\cos a} = 0.9557$$

$$B_1 = 72.87° \qquad B_2 = 180° - B_1 = 107.13°$$

To identify the correct combination of values, we apply Theorem 2.8 and the fact that a and A are both acute. If $c < 90°$, b and B are both acute. Therefore, b_1, c_1, and B_1 constitute one valid combination. If $c > 90°$, b and B are both obtuse. Therefore, b_2, c_2, and B_2 constitute another valid combination. This problem

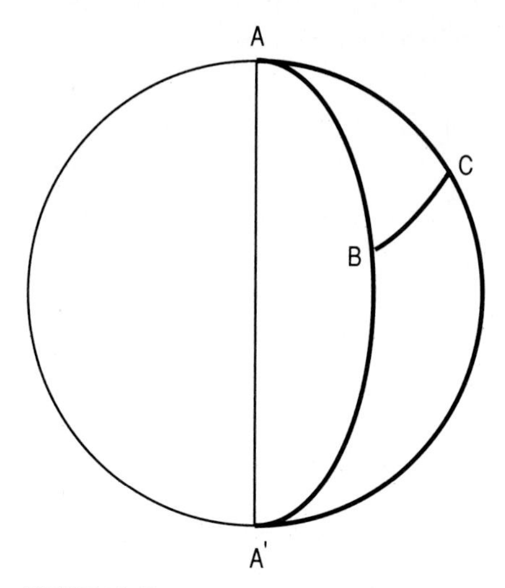

FIGURE 2.17

is therefore ambiguous, and the two possible triangles are ABC and $A'BC$ in Fig. 2.17.

To test these results, we apply Eq. (2.34), using either combination of values.

$$\cos A = \tan b \cot c = 0.8192 \qquad A = 35^\circ$$

$$\tan a = \tan c \cos B = 0.6010 \qquad a = 31^\circ$$

The results are thus confirmed.

2.2.6 Equations of Oblique Spherical Triangle

An oblique spherical triangle is characterized by a duality that enables us to transform a given equation to a corresponding equation by replacing each part of the triangle with the supplement of the opposite part. The second equation is called the *dual* of the first, and vice versa. The six parts of the triangle have the following relationships:

$$\frac{\sin A}{\sin a} = \frac{\sin B}{\sin b} = \frac{\sin C}{\sin c} \tag{2.37}$$

$$\cos a = \cos b \cos c + \sin b \sin c \cos A \tag{2.38a}$$

$$\cos b = \cos a \cos c + \sin a \sin c \cos B \tag{2.38b}$$

$$\cos c = \cos a \cos b + \sin a \sin b \cos C \tag{2.38c}$$

$$\cos A = -\cos B \cos C + \sin B \sin C \cos a \tag{2.39a}$$

$$\cos B = -\cos A \cos C + \sin A \sin C \cos b \tag{2.39b}$$

$$\cos C = -\cos A \cos B + \sin A \sin B \cos c \tag{2.39c}$$

$$\tan\left(\frac{a-b}{2}\right)=\frac{\sin[(A-B)/2]}{\sin[(A+B)/2]}\tan\left(\frac{c}{2}\right) \tag{2.40a}$$

$$\tan\left(\frac{a+b}{2}\right)=\frac{\cos[(A-B)/2]}{\cos[(A+B)/2]}\tan\left(\frac{c}{2}\right) \tag{2.40b}$$

$$\tan\left(\frac{A-B}{2}\right)=\frac{\sin[(a-b)/2]}{\sin[(a+b)/2]}\cot\left(\frac{C}{2}\right) \tag{2.41a}$$

$$\tan\left(\frac{A+B}{2}\right)=\frac{\cos[(a-b)/2]}{\cos[(a+b)/2]}\cot\left(\frac{C}{2}\right) \tag{2.41b}$$

Equations (2.40) and (2.41) are known as *Napier's equations*. In these equations, c and C can be substituted for a and A or b and B, respectively. We also have the following:

$$\cot a=\frac{\sin C}{\tan A\sin b}+\frac{\cos C}{\tan b} \tag{2.42}$$

$$\cot A=\frac{\sin c}{\tan a\sin B}-\frac{\cos c}{\tan B} \tag{2.43}$$

Again, appropriate substitutions of parts can be made. In the special case where $C=90°$, Eq. (2.42) reduces to Eq. (2.32*b*).

With reference to the trihedral angle at O in Fig. 2.15, consider that two face angles are projected on the plane of the third angle. The sum of the two projected angles equals the third angle. Therefore, applying Eq. (2.29*b*), we obtain the following:

$$a=\tan^{-1}(\tan b\cos C)+\tan^{-1}(\tan c\cos B) \tag{2.44a}$$

$$b=\tan^{-1}(\tan a\cos C)+\tan^{-1}(\tan c\cos A) \tag{2.44b}$$

$$c=\tan^{-1}(\tan a\cos B)+\tan^{-1}(\tan b\cos A) \tag{2.44c}$$

The principle of duality yields the following equations:

$$A=\cot^{-1}(\cos b\tan C)+\cot^{-1}(\cos c\tan B) \tag{2.45a}$$

$$B=\cot^{-1}(\cos a\tan C)+\cot^{-1}(\cos c\tan A) \tag{2.45b}$$

$$C=\cot^{-1}(\cos a\tan B)+\cot^{-1}(\cos b\tan A) \tag{2.45c}$$

In applying Eqs. (2.44) and (2.45), we may view each term at the right as an angle that lies in the first or second quadrant. If the sum of the two terms lies in the third or fourth quadrant, we adjust the result by subtracting 180° from the sum. One equation in this set of six equations may be applied when four parts of the triangle are known, and two equations in the set may be applied to test the solution of the triangle.

Where the species of a part is not readily apparent, the following principles may be applied:

Theorem 2.9. If a side differs from 90° in absolute value by a larger amount than another side in the triangle, it is of the same species as its opposite angle. An analogous statement applies with reference to an angle.

Theorem 2.10. Half the sum of two sides of a triangle is of the same species as half the sum of the two opposite angles.

2.2.7 Solution of Oblique Spherical Triangle

Problems involving the solution of an oblique spherical triangle can be classified in the manner shown in Table 2.7.

EXAMPLE 2.13 Solve the triangle in which $a = 62°$, $b = 54°$, and $c = 93°$.

SOLUTION This is a type 1 problem. Equation (2.38*a*) may be recast as

$$\cos A = \frac{\cos a - \cos b \cos c}{\sin b \sin c}$$

$$= \frac{0.46947 - (0.58779)(-0.05234)}{(0.80902)(0.99863)} = 0.61917$$

$$A = 51.74°$$

Similarly, we obtain $B = 46.01°$ and $C = 117.36°$. These results can be tested by applying Eq. (2.37).

A type 2 problem can be solved in an analogous manner by applying Eq. (2.39) to find the sides.

EXAMPLE 2.14 Solve the triangle in which $a = 83°$, $b = 65°$, and $C = 104°$.

SOLUTION This is a type 3 problem.

$$\frac{a+b}{2} = 74° \qquad \frac{a-b}{2} = 9° \qquad \frac{C}{2} = 52°$$

Equations (2.41) yield the following results:

$$\frac{A-B}{2} = 7.246° \qquad \frac{A+B}{2} = 70.343°$$

$$A = 70.343° + 7.246° = 77.59°$$

$$B = 70.343° - 7.246° = 63.10°$$

TABLE 2.7 Classification of Problems in Oblique Spherical Triangles

Type	Given information
1	The three sides
2	The three angles
3	Two sides and the included angle
4	Two angles and the included side
5	Two sides and the angle opposite one of them
6	Two angles and the side opposite one of them

Equation (2.38c) enables us to calculate c by applying solely the given data, and the result is $c = 99.56^\circ$.

Alternatively, A can be found by applying Eq. (2.43) but interchanging b and c and the corresponding angles. Similarly, B can be found by applying Eq. (2.43) but replacing a with b, b with c, and c with a, as well as the corresponding angles.

The results can be tested in numerous ways. We shall test them by applying Eq. (2.44).

$$a = \tan^{-1}(2.14451)(-0.24192) + \tan^{-1}(-5.93756)(0.45243)$$

$$a = 152.58^\circ + 110.42^\circ - 180^\circ = 83^\circ$$

$$b = \tan^{-1}(8.14435)(-0.24192) + \tan^{-1}(-5.93756)(0.21491)$$

$$b = 116.91^\circ + 128.09^\circ - 180^\circ = 65^\circ$$

The results are thus confirmed.

A type 4 problem is solved in an analogous manner.

EXAMPLE 2.15 Solve the triangle in which $a = 118^\circ$, $b = 80^\circ$, and $A = 126^\circ$.

SOLUTION This is a type 5 problem. By Eq. (2.37),

$$\sin B = \frac{\sin b \sin A}{\sin a} = 0.902348$$

$$B_1 = 64.468^\circ \qquad B_2 = 180^\circ - 64.468^\circ = 115.532^\circ$$

Theorems 2.9 and 2.10 fail to invalidate either value of B, for these reasons: First, $|90^\circ - b| < |90^\circ - a|$. Second, $(a + b)/2$ is obtuse, but both $(A + B_1)/2$ and $(A + B_2)/2$ are also obtuse. Therefore, the present problem is ambiguous, and we must retain both values of B.

We now apply Eqs. (2.44c) and (2.45c) with the computed values of B. Using subscripts for c and C corresponding to those for B, we have the following:

$$c_1 = \tan^{-1}(-1.88073)(0.43102) + \tan^{-1}(5.67128)(-0.58779)$$

$$c_1 = 140.97^\circ + 106.70^\circ - 180^\circ = 67.67^\circ$$

$$c_2 = 39.03^\circ + 106.70^\circ = 145.73^\circ$$

$$C_1 = \cot^{-1}(-0.46947)(2.09353) + \cot^{-1}(0.17365)(-1.37638)$$

$$C_1 = 134.50^\circ + 103.44^\circ - 180^\circ = 57.94^\circ$$

$$C_2 = 45.50^\circ + 103.44^\circ = 148.94^\circ$$

Alternatively, the values of c and C can be computed by applying Eqs. (2.40a) and (2.41a), respectively.

In summary, the two solutions are as follows:

$$B_1 = 64.47^\circ \qquad c_1 = 67.67^\circ \qquad C_1 = 57.94^\circ$$

$$B_2 = 115.53^\circ \qquad c_2 = 145.73^\circ \qquad C_2 = 148.94^\circ$$

The two triangles are shown in Fig. 2.18. The ambiguity of this problem stems from the fact that there are two central planes through C that intersect $\widehat{AB}$ in such

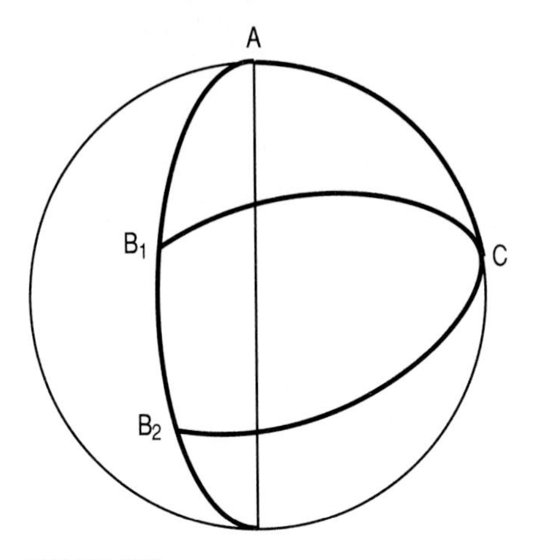

FIGURE 2.18

manner that $\widehat{BC} = 118°$. The two sets of results can be tested by substituting in Eq. (2.37) or any other equation that was not applied in the solution.

A type 6 problem is solved in an analogous manner.

2.2.8 Area of Spherical Triangle

The *spherical excess* of a triangle is the amount by which the sum of its angles exceeds 180°. The area of a triangle is directly proportional to its spherical excess.

Let E = spherical excess, degrees
K = area of triangle
r = radius of sphere

The relationship is

$$K = \frac{\pi r^2 E}{180°} \tag{2.46}$$

EXAMPLE 2.16 A triangle on a sphere of 65-cm radius has angles of 85°, 108°, and 133°. Compute the area of the triangle.

SOLUTION

$$E = 85° + 108° + 133° - 180° = 146°$$

$$K = \frac{\pi(65^2)146°}{180°} = 10{,}766 \text{ cm}^2 = 1.0766 \text{ m}^2$$

If all three sides of the triangle are given, the spherical excess can be found directly in this manner: Let $s = (a + b + c)/2$. Then

$$\tan\frac{E}{4} = \sqrt{\left(\tan\frac{s}{2}\right)\left(\tan\frac{s-a}{2}\right)\left(\tan\frac{s-b}{2}\right)\left(\tan\frac{s-c}{2}\right)} \tag{2.47}$$

This relationship is known as *L'Huilier's equation*.

2.2.9 Terrestrial Triangles

In calculations pertaining to the surface of the earth, we consider the earth to be a sphere with a diameter of 7916 mi (12,740 km). Therefore, the length of a 1° arc of a great circle is $\pi(7916)/360 = 69.080$ mi (111.17 km).

In Fig. 2.19*a*, *NS* is the axis of the earth, *N* being the north pole and *S* the south pole. Point *G* represents Greenwich, England. The great circle *ABCD* in the central plane perpendicular to *NS* is the equator. A *meridian* is a great semicircle that is

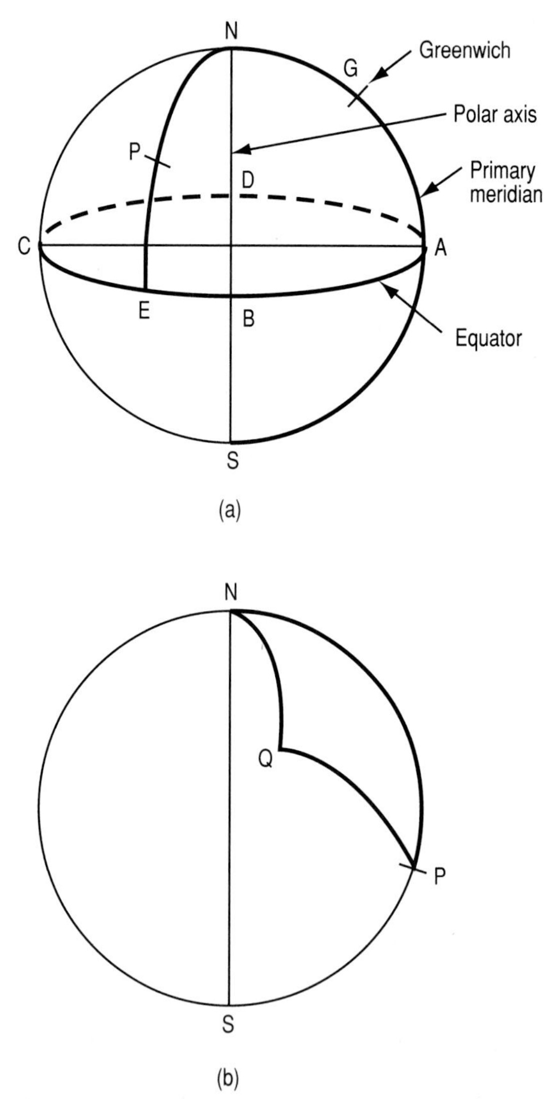

FIGURE 2.19 (*a*) Spherical coordinates of point on earth's surface; (*b*) terrestrial triangle.

bounded by N and S, and a *parallel* is a small circle in a plane parallel to the equatorial plane. The meridian through G is the *primary meridian.* A point P on the earth's surface is located by means of *spherical coordinates*, in this manner: The *latitude* of P is its angular distance from the equator, or $\widehat{PE}$. The *longitude* of P is the angle GNP between the primary meridian and the meridian through P; it is measured by $\widehat{AE}$ on the equator. The notation 28°N,37°E refers to a point that lies 28° north of the equator and 37° east of the primary meridian. In Fig. 2.19*b*, P and Q are points on the earth's surface. A spherical triangle is formed by the meridians through these points and the arc of the great circle through the points. When the coordinates of P and Q are specified, three parts of this triangle are known. The acute angle between the meridian through Q and $\widehat{PQ}$ is the *bearing* of P from Q. Its orientation must be specified.

EXAMPLE 2.17 Points P and Q on the earth's surface have the following locations: P, 18°S,15°W; Q, 41°N,92°W. Compute the geographical distance between these points and the bearing of one point from the other.

SOLUTION With reference to Fig. 2.19*b*,

$$\widehat{NP} = 90° + 18° = 108° \qquad \widehat{NQ} = 90° - 41° = 49°$$

$$\angle PNQ = 92° - 15° = 77°$$

In $\triangle PNQ$, two sides and the included angle are known. Thus, this is a type 3 problem. Proceeding as in Example 2.14, we obtain these results:

$$\angle PQN = 111°57.4' \qquad \angle QPN = 47°23.5'$$

$$\widehat{PQ} = 92.365° \qquad 180° - \angle PQN = 68°02.6'$$

$$\text{Distance} = (92.365)(69.080) = 6381 \text{ mi } (10{,}268 \text{ km})$$

$$\text{Bearing of } P \text{ from } Q = \text{S}68°02.6'\text{E}$$

$$\text{Bearing of } Q \text{ from } P = \text{N}47°23.5'\text{W}$$

2.2.10 General Problem of Intersecting Planes

Since the spherical angles of a spherical triangle are numerically equal to the dihedral angles formed by central planes, the equations of Art. 2.2.6 are applicable to any problem involving three intersecting planes. The point of intersection of the planes is considered to be the center of a sphere.

EXAMPLE 2.18 With reference to Fig. 2.20, planes P, Q, and R intersect at O. Line OA is the intersection of P and Q, line OB is the intersection of P and R, and line OC is the intersection of Q and R. The angle between P and Q is 57°, the angle between P and R is 43°, and $\angle AOB = 81°$, as shown. Compute the angle between Q and R, and locate OC on each of these planes.

SOLUTION The method of constructing orthographic drawings is discussed in Art. 3.3.2. Consider A, B, and C to be equidistant from O, and form the

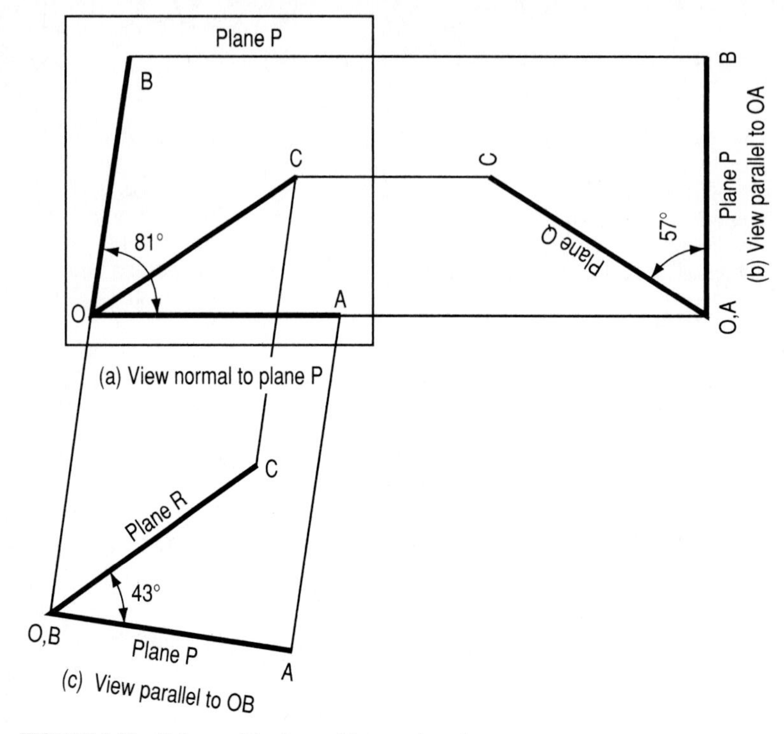

FIGURE 2.20 Orthographic views of intersecting planes.

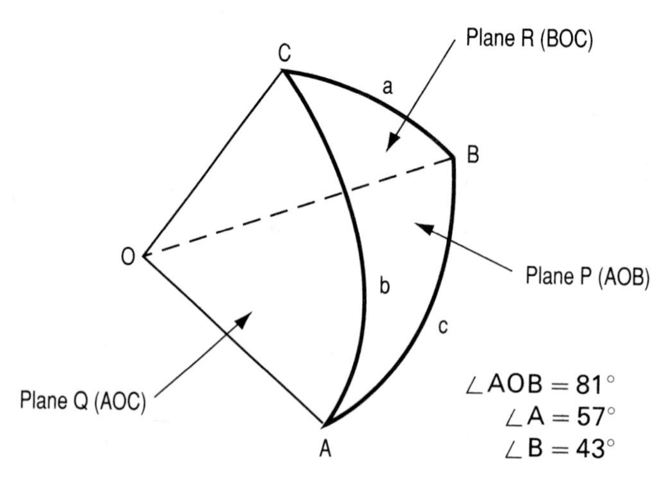

FIGURE 2.21

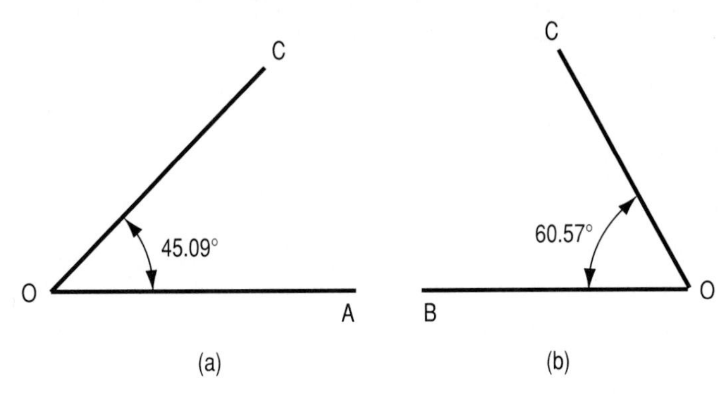

FIGURE 2.22 (*a*) View normal to plane Q; (*b*) view normal to plane R.

spherical triangle ABC in Fig. 2.21. The parts are labeled as indicated, and the known values are $A = 57°$, $B = 43°$, and $c = 81°$. This is a type 4 problem. Applying Eq. (2.40), we obtain $a = 60.57°$ and $b = 45.09°$. Applying Eq. (2.39*c*), we obtain $C = 108°$. Thus, the angle between planes Q and R is 108°, and the location of OC on planes Q and R is as shown in Fig. 2.22.

SECTION 3
PLANE AND SOLID ANALYTIC GEOMETRY

3.1 BASIC PLANE ANALYTIC GEOMETRY

3.1.1 Definitions and Notation

In Fig. 3.1*a*, P is an arbitrary point in the plane, and QP and RP are lines through P perpendicular to the x and y axes, respectively. The distance OQ is the x *coordinate* or *abscissa* of P, and the distance OR is the y *coordinate* or *ordinate* of P. The x and y coordinates are known collectively as the *rectangular* (or *cartesian*) *coordinates* of P. The designation $P(x_1, y_1)$ identifies a point P that has an abscissa of x_1 and ordinate of y_1.

If a displacement along a given line is considered to have an algebraic sign, the line is termed a *directed line*, and an arrowhead is used to identify the positive direction. A segment of such a line is a *directed segment*. Thus, in Fig. 3.1*a*, the x axis is a directed line. A displacement along this line is positive if it is to the right and negative if it is to the left. The directed segment ST is positive, and the directed segment TS is negative. An *undirected line* is one to which no positive direction is assigned.

In Fig. 3.1*b*, L is an arbitrary undirected straight line. The *inclination* of L is the angle α between the positive side of the x axis and L, as measured in the counterclockwise direction. The *slope* of L is defined as the tangent of α.

Straight lines or curves that have some common characteristic constitute a *family*. For example, circles that contain two given points constitute a family of circles, the size of the family being infinite.

If a curve approaches a straight line without intersecting it, the straight line is an *asymptote* of the curve. As we proceed along the curve, the undirected distance between the curve and the straight line becomes and remains smaller than any number we can specify.

Let A and B denote two points on a curve. The straight line connecting A and B is a *chord* of the curve. Now assume that the curve opens indefinitely. The *concavity* of the curve is defined by the position of the chord in relation to that of arc AB. Thus, the curve is *concave upward* if the chord lies above the arc, and *concave downward* if the reverse is true, as illustrated in Fig. 3.2.

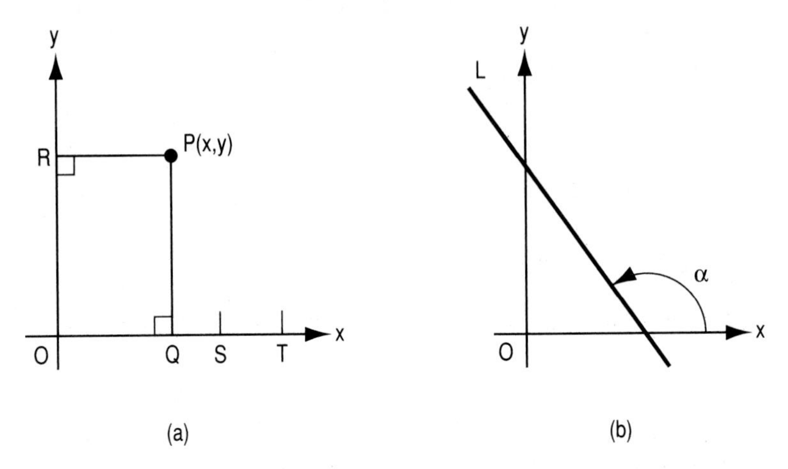

FIGURE 3.1 (*a*) Coordinates of a point and meaning of a directed line segment; (*b*) inclination of a straight line.

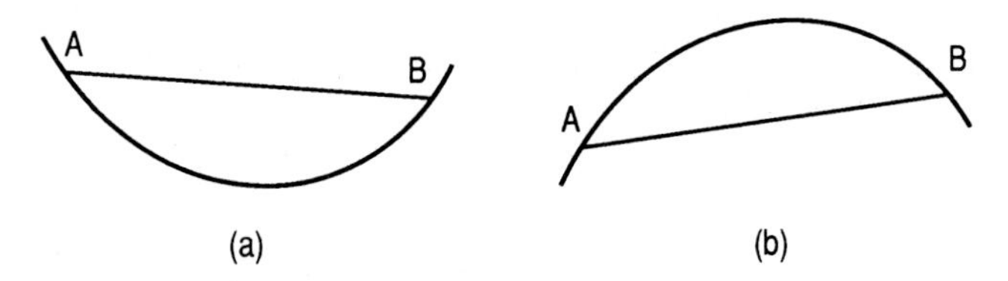

FIGURE 3.2 Concavity of a curve. (*a*) Concave-upward curve; (*b*) concave-downward curve.

3.1.2 Equation of the Straight Line

In Fig. 3.3*a*, L is the straight line through points P_1 and P_2, which have the coordinates shown. Let m denote the slope of L. Then

$$m = \tan \alpha = \frac{y_2 - y_1}{x_2 - x_1} \tag{3.1}$$

In Fig. 3.3*b*, the straight line L contains point P_1 and intersects the x and y axes at A and B, respectively. The distances a and b are called, respectively, the *x intercept* and *y intercept* of L. Again let m denote the slope of L. The equation for L can be given in any of the following forms:

$$y - y_1 = m(x - x_1) \tag{3.2}$$

$$y = mx + b \tag{3.3}$$

$$\frac{x}{a} + \frac{y}{b} = 1 \tag{3.4}$$

Equations (3.2), (3.3), and (3.4) are known as the *point-slope form*, *slope-intercept form*, and *intercept form*, respectively.

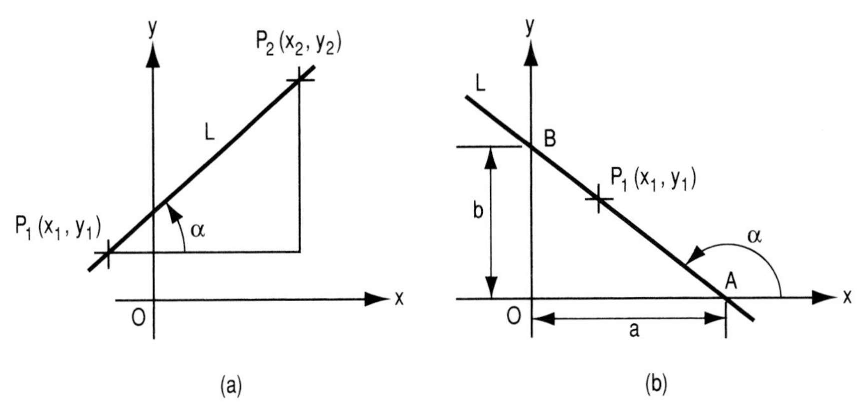

FIGURE 3.3 (*a*) Slope of the straight line through two given points; (*b*) intercepts of a straight line.

The *general form* of the equation of a straight line is

$$Ax + By + C = 0$$

where A, B, and C are real numbers. These coefficients are related to the parameters of the line by these equations:

$$m = \tan\alpha = -\frac{A}{B} \qquad a = -\frac{C}{A} \qquad b = -\frac{C}{B} \tag{3.5}$$

EXAMPLE 3.1 Write the equation of the straight line that contains points $P_1(-5, 7)$ and $P_2(3, 31)$.

SOLUTION Applying Eqs. (3.1) and (3.2) in turn, we have

$$m = \frac{31 - 7}{3 - (-5)} = 3 \qquad y - 7 = 3[x - (-5)] = 3x + 15$$

In general form, the equation is $3x - y + 22 = 0$. The coordinates of both P_1 and P_2 satisfy this equation, and it is thus confirmed.

3.1.3 Parallel and Perpendicular Straight Lines

The straight line L in Fig. 3.4 has the equation $Ax + By + C = 0$. In Fig. 3.4*a*, line L' is parallel to L, and the perpendicular distance between the lines is d. The equation of L' is

$$Ax + By + C \pm d\sqrt{A^2 + B^2} = 0 \tag{3.6}$$

Whether the plus or minus sign applies can be determined by computing the y intercepts of the lines.

In Fig. 3.4*b*, line L' is perpendicular to L, and it intersects L at $P(x_1, y_1)$. Let m and m' denote the slopes of L and L', respectively. Then

$$mm' = -1 \tag{3.7}$$

The equation of L' is

$$Bx - Ay - Bx_1 + Ay_1 = 0 \tag{3.8}$$

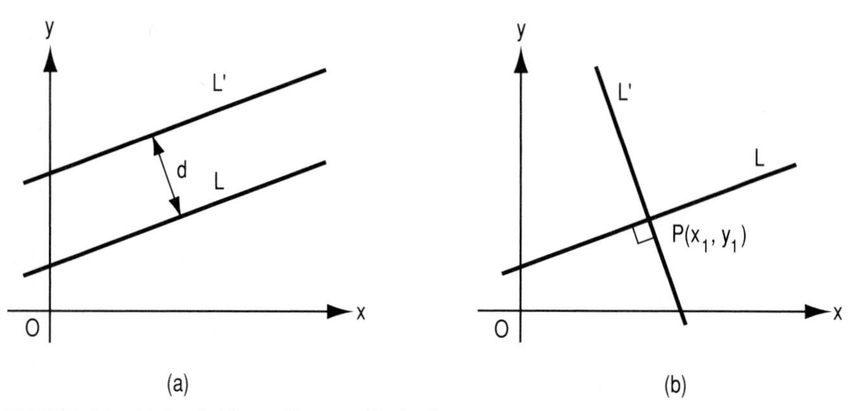

FIGURE 3.4 (*a*) Parallel lines; (*b*) perpendicular lines.

3.1.4 Angle between Two Straight Lines

Let ϕ denote the acute angle between two undirected straight lines. This angle can readily be found by computing the inclination of each line and then taking the difference between the results.

EXAMPLE 3.2 Lines L_1 and L_2 have the equations $4x + 9y - 36 = 0$ and $7x + 5y - 35 = 0$, respectively. Determine the angle ϕ between these lines.

SOLUTION The lines are plotted in Fig. 3.5. By Eq. (3.5),

$$\tan \alpha_1 = -\frac{4}{9} \qquad \tan \alpha_2 = -\frac{7}{5}$$

$$\alpha_1 = 156.04^\circ \qquad \alpha_2 = 125.54^\circ \qquad \phi = \alpha_1 - \alpha_2 = 30.50^\circ$$

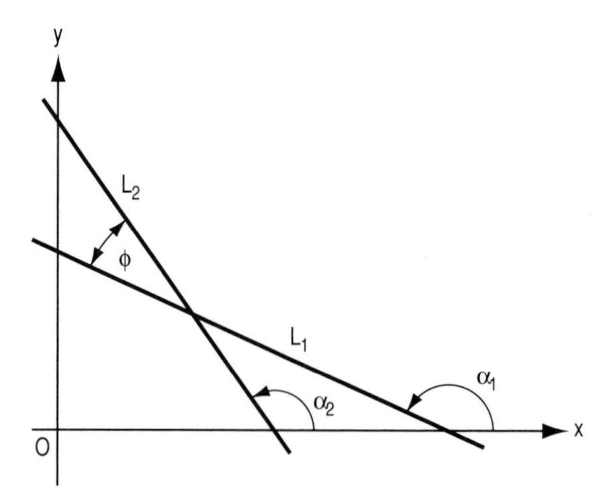

FIGURE 3.5 Angle between two straight lines.

In general, let the equations of L_1 and L_2 be $A_1x + B_1y + C_1 = 0$ and $A_2x + B_2y + C_2 = 0$, respectively. Then

$$\tan \phi = \left| \frac{-A_1B_2 + A_2B_1}{A_1A_2 + B_1B_2} \right| \tag{3.9}$$

3.1.5 Perpendicular from a Point to a Line

In Fig. 3.6, line L has the equation $Ax + By + C = 0$ and $P(x_1, y_1)$ is an arbitrary point in the plane. Point $Q(x_2, y_2)$ is the foot of the perpendicular from P to L. We wish to determine the perpendicular distance d and the coordinates of Q. Let

$$c = Ax_1 + By_1 + C \tag{3.10}$$

Then

$$d = \frac{|c|}{\sqrt{A^2 + B^2}} \tag{3.11}$$

$$x_2 = \frac{B^2x_1 - ABy_1 - AC}{A^2 + B^2} \tag{3.12a}$$

$$y_2 = \frac{-ABx_1 + A^2y_1 - BC}{A^2 + B^2} \tag{3.12b}$$

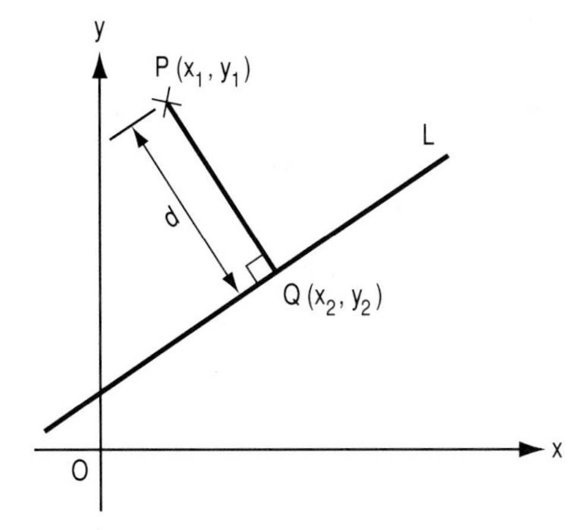

FIGURE 3.6 Perpendicular from a given point to a given line.

3.1.6 Transformation of Coordinates

It is often highly advantageous to move the coordinate axes to some new position, and this process is called a *transformation of coordinates*. Three types of transformation are available: a *translation of axes*, in which the x and y axes are displaced while remaining parallel to their original positions; a *rotation of axes*, in which the axes are rotated about the origin; and a composite of a translation and rotation. Let x and y denote the original coordinates of a point, and x' and y' denote the coordinates of the point following a single transformation. The equation of a curve in terms of x' and y' is known as the *transformed equation* of the curve.

In Fig. 3.7*a*, the x and y axes are displaced to the positions x' and y', respectively, thereby displacing the origin from O to O'. Then

$$x = x' + h \qquad y = y' + k \tag{3.13}$$

In Fig. 3.7*b*, the x and y axes are rotated about O in the counterclockwise direction through an angle θ. The relationships are as follows:

$$x' = x \cos\theta + y \sin\theta \tag{3.14a}$$

$$y' = -x \sin\theta + y \cos\theta \tag{3.14b}$$

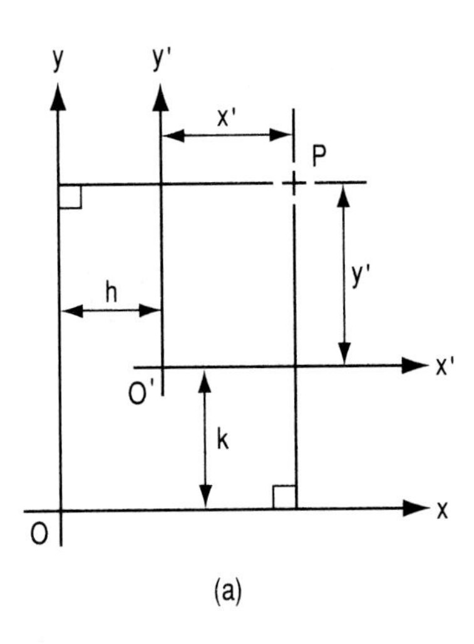

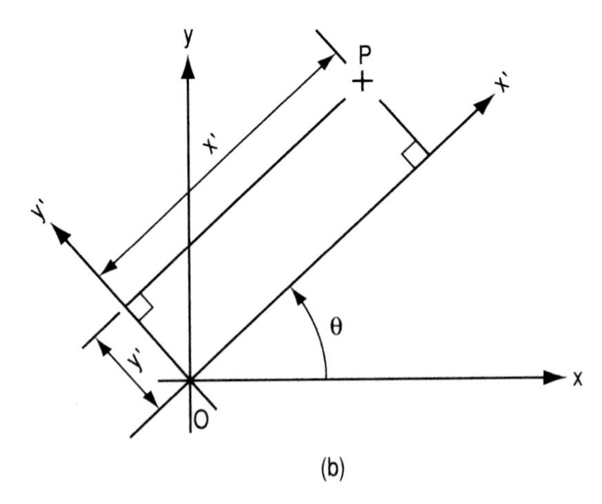

FIGURE 3.7 Methods of transforming axes. (*a*) Translation of axes; (*b*) rotation of axes.

$$x = x' \cos\theta - y' \sin\theta \tag{3.15a}$$

$$y = x' \sin\theta + y' \cos\theta \tag{3.15b}$$

The transformed equation of a curve can be obtained by replacing x and y in the original equation with their expressions in terms of x' and y'.

3.1.7 Transformation of Second-Degree Equation

The transformation of coordinates is particularly important with reference to a second-degree equation. The general second-degree equation in two unknowns can be expressed in the form

$$Ax^2 + Bxy + Cy^2 + Dx + Ey + F = 0$$

The first three terms are *quadratic*, and the next two terms are *linear*. The transformed equation can be expressed as

$$A'x'^2 + B'x'y' + C'y'^2 + D'x' + E'y' + F' = 0$$

The translation of axes shown in Fig. 3.7*a* produces the following results:

$$A' = A \qquad B' = B \qquad C' = C \tag{3.16a}$$

$$D' = 2Ah + Bk + D \qquad E' = Bh + 2Ck + E \tag{3.16b}$$

$$F' = Ah^2 + Bhk + Ck^2 + Dh + Ek + F \tag{3.16c}$$

We can obtain the condition $D' = E' = 0$ by setting

$$h = \frac{BE - 2CD}{4AC - B^2} \qquad k = \frac{BD - 2AE}{4AC - B^2} \tag{3.17a}$$

In this case,

$$F' = \frac{Dh + Ek}{2} + F \tag{3.17b}$$

EXAMPLE 3.3 A curve has the equation

$$3x^2 + 8xy - 14y^2 - 16x + 172y - 418 = 0$$

Translate the axes in a manner that causes the linear terms in the equation to vanish, and test the result.

SOLUTION

$$A = 3 \qquad B = 8 \qquad C = -14 \qquad D = -16 \qquad E = 172 \qquad F = -418$$

Equation (3.17*a*) yields $h = -4$, $k = 5$; Eq. (3.17*b*) yields $F' = 44$. Therefore, the transformed equation is $3x'^2 + 8x'y' - 14y'^2 + 44 = 0$.

This result can be tested by selecting a set of values that satisfies the transformed equation. Such a set is $x' = 6$, $y' = -2$. Equation (3.13) then yields $x = 2$, $y = 3$, and the new set of values satisfies the original equation.

When the coordinate axes are rotated in the manner shown in Fig. 3.7*b*, the coefficients in the transformed second-degree equation are as follows:

$$A' = \frac{A+C}{2} + \frac{A-C}{2}\cos 2\theta + \frac{B}{2}\sin 2\theta \tag{3.18a}$$

$$B' = -(A-C)\sin 2\theta + B\cos 2\theta \tag{3.18b}$$

$$C' = \frac{A+C}{2} - \frac{A-C}{2}\cos 2\theta - \frac{B}{2}\sin 2\theta \tag{3.18c}$$

$$D' = D\cos\theta + E\sin\theta \tag{3.18d}$$

$$E' = -D\sin\theta + E\cos\theta \tag{3.18e}$$

$$F' = F \tag{3.18f}$$

Figure 3.8 exhibits visually the manner in which the coefficients vary with θ, and the diagrams are accordingly termed *transformation diagrams*. In Fig. 3.8*a*, values of A' and C' appear on the horizontal axis; values of $B'/2$ appear on the vertical axis, with the downward direction positive. The steps in the construction are as follows: Plot the points $P(A, B/2)$ and $Q(C, -B/2)$, as shown. Draw the straight line PQ, intersecting the horizontal axis at M. Draw RP perpendicular to the horizontal axis. Then $OM = (A+C)/2$ and $MR = (A-C)/2$. Draw a circle having M as center and MP as radius. Now rotate the diameter PQ through a counterclockwise angle of 2θ, to the position $P'Q'$. The coordinates of P' are A' and $B'/2$, and the coordinates of Q' are C' and $-B'/2$.

In Fig. 3.8*b*, the values of D' and E' appear on the horizontal and vertical axes, respectively. The steps are as follows: Plot the point $P(D, E)$, as shown. Draw a circle having the origin as center and OP as radius. Now rotate the radius OP through a counterclockwise angle of θ, to the position OP'. The coordinates of P' are D' and E'.

Figure 3.8*a* yields a wealth of valuable information. For example, it exhibits the limits of A', B', and C', it reveals that A' and C' have identical limits but are 180° out of phase with respect to 2θ, and it reveals that B' has a limiting value when $A' = C' = (A+C)/2$. Let G denote the radius MP. Then

$$G^2 = \frac{(A-C)^2 + B^2}{4} \tag{3.19}$$

We can reduce B' to 0 by placing P' at S or at T. In either case, the requirement is

$$\tan 2\theta = \frac{B}{A-C} \tag{3.20}$$

This equation yields two values of 2θ that differ by 180°. When P' is at S,

$$A' = \frac{A+C}{2} + G \qquad C' = \frac{A+C}{2} - G \tag{3.21}$$

When P' is at T, the equations for A' and C' are interchanged.

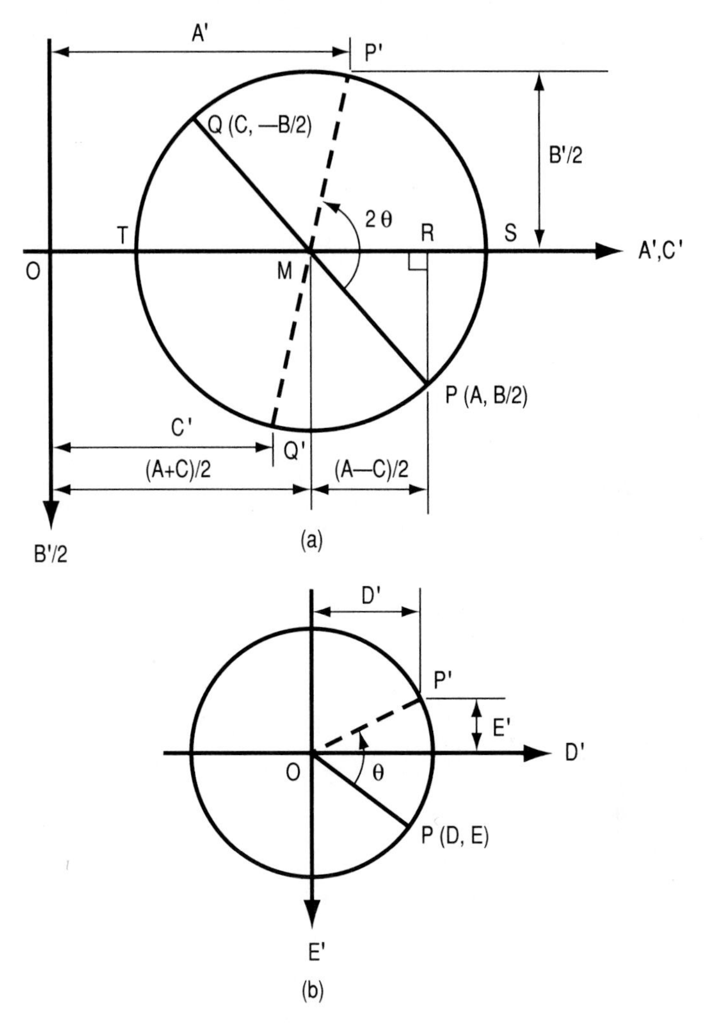

FIGURE 3.8 Transformation diagrams for rotation of axes. (*a*) Variation of A', B', and C'; (*b*) variation of D' and E'.

From the geometry of Fig. 3.8, we deduce the following:

$$A' + C' = 2(OM) = \text{constant} \tag{3.22a}$$

$$4A'C' - B'^2 = 4(OT)(OS) = \text{constant} \tag{3.22b}$$

$$D'^2 + E'^2 = \text{constant} \tag{3.22c}$$

Thus, the quantities $A + C$, $4AC - B^2$, and $D^2 + E^2$ remain *invariant* under a rotation of the axes.

EXAMPLE 3.4 A curve has the equation $2x^2 - 12xy + 7y^2 - 1 = 0$. Transform the equation of the curve to one that is devoid of an xy term, and find the angle through which the coordinate axes must be rotated to obtain this equation. Test the results.

SOLUTION

$$A = 2 \qquad B = -12 \qquad C = 7 \qquad D = E = 0 \qquad F = -1$$

$$\frac{A + C}{2} = 4.5 \qquad \frac{A - C}{2} = -2.5$$

Equation (3.19) yields $G = 6.5$, and Eq. (3.21) yields $A' = 11$ and $C' = -2$. Therefore, one transformed equation is $11x'^2 - 2y'^2 - 1 = 0$, and the other is $2x'^2 - 11y'^2 + 1 = 0$. By constructing the transformation diagram, we find that $\theta = -56.31°$ (or 123.69°) for the first equation and $\theta = 33.69°$ for the second equation.

To test the first transformed equation, we select the following set of values that satisfies this equation: $x' = 3, y' = 7$. Equation (3.15) then yields $x = 7.488$, $y = 1.387$, and the latter set of values satisfies the original equation. We also observe that $4A'C' = 4AC - B^2$, as Eq. (3.22*b*) requires.

3.1.8 Parametric Equations

In mechanics, a point moves in a prescribed manner and thereby generates a curve that is termed its *trajectory*. The instantaneous x and y coordinates of the moving point may both be functions of a third variable, such as elapsed time or angular displacement. This third variable is termed a *parameter*, and the two equations that relate the coordinates to the parameter are called *parametric equations*.

EXAMPLE 3.5 In the mechanism in Fig. 3.9*a*, the crank OA rotates about O, the rod AB is connected to OA with a smooth pin, and pin B is constrained to move along the x axis. The lengths are recorded in the drawing, and $b > a$. Express the coordinates of point C on AB in terms of the parameter α, and draw the trajectory of C.

SOLUTION Draw DA parallel to the y axis and EC parallel to the x axis. The rectangular coordinates of C are as follows:

$$x = OD + EC = OD + k(DB) = OD + k\sqrt{(AB)^2 - (AD)^2}$$

$$y = DA - k(DA) = (1 - k)(DA)$$

Expressing all distances in terms of a and b, we obtain

$$x = a \cos \alpha + k\sqrt{b^2 - a^2 \sin^2 \alpha}$$

$$y = a(1 - k) \sin \alpha$$

The trajectory of C is shown in Fig. 3.9*b* (to an enlarged scale). It is symmetric about the x axis.

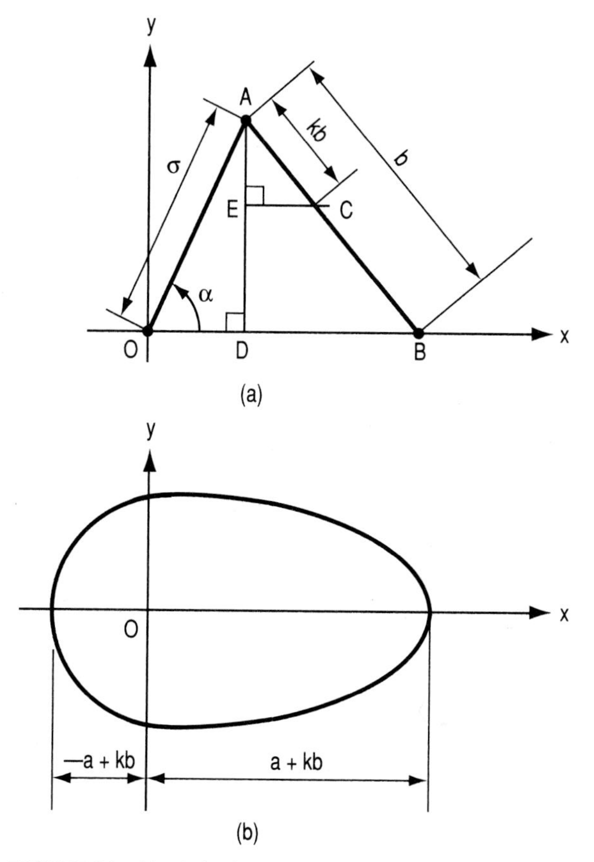

FIGURE 3.9 (*a*) Mechanism; (*b*) trajectory of point *C*.

3.1.9 Polar Coordinates

With reference to Fig. 3.10*a*, the position of point P in the plane can be expressed by specifying the distance of P from O and the angle between the line OP and the reference line OA. Let $r = OP$ and $\theta = \angle AOP$, a positive angle being measured in the counterclockwise direction. The quantities r and θ are called, respectively, the *radius vector* and *vectorial angle* of P, and they constitute the *polar coordinates* of that point. Point O is the *pole* of coordinates, and line OA is the *polar axis*. The notation $P(r_1, \theta_1)$ identifies a point P that has a radius vector r_1 and vectorial angle θ_1. It is often preferable to use polar coordinates in lieu of rectangular coordinates.

If r is negative, the given point lies on the extension of the terminal side of the vectorial angle. For example, the point $P(-7, 52°)$ is located in this manner: In Fig. 3.10*b*, draw OB at an angle of 52° with the polar axis. Now draw OC, the extension of OB. On OC, locate P at a distance of 7 units from O. Of course, P may also be considered to have the coordinates (7, 232°) or (7, −128°). Thus, a given point does not have a unique set of polar coordinates.

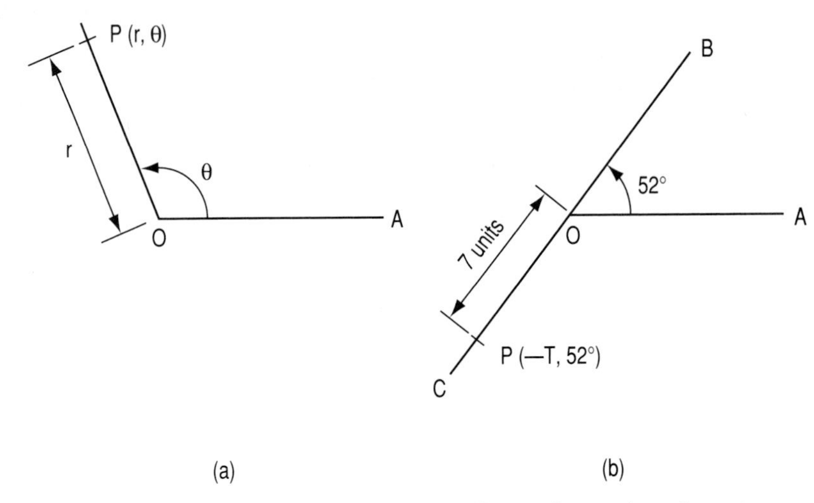

FIGURE 3.10 (*a*) Use of polar coordinates; (*b*) significance of a negative radius vector.

The conversion from rectangular to polar coordinates, or vice versa, is made by applying these relationships:

$$r^2 = x^2 + y^2 \qquad \theta = \tan^{-1}\frac{y}{x} \tag{3.23}$$

$$x = r\cos\theta \qquad y = r\sin\theta \tag{3.24}$$

EXAMPLE 3.6 A curve has the equation $10x^2 - 5xy + 3y^2 = 65$. Transform the equation to one having polar coordinates. Verify the result.

SOLUTION By replacing x and y with their expressions in Eq. (3.24), we obtain $r^2(10\cos^2\theta - 5\sin\theta\cos\theta + 3\sin^2\theta) = 65$. By applying the equations of Art. 2.1.4, we can now transform the second equation to $r^2(7\cos 2\theta - 5\sin 2\theta + 13) = 130$.

To verify the result, we select the following set of x and y values that satisfies the given equation: $x = 2$, $y = 5$. The corresponding polar coordinates as given by Eq. (3.23) are $r = \sqrt{29}$, $\theta = 68.20°$. Upon substitution, we find that these polar coordinates satisfy the polar equation.

3.2 *THE CONIC SECTIONS*

3.2.1 Introduction

The line of intersection of a plane and a right circular cone is termed a *conic section*, or simply a *conic*. However, despite the etymology of the term, a conic section can be defined and analyzed without reference to the cone, and the study of conic sections lies in the domain of plane analytic geometry. There are three types of conic sections, namely, the ellipse, parabola, and hyperbola, the circle being considered a degenerate form of the ellipse.

A conic section is generated by a point that moves in a plane in some prescribed manner. The distance from the moving point to a reference point or reference line is always considered to be positive.

When a conic section is positioned in such manner that its equation in rectangular coordinates is as simple as possible, the section is said to be in *standard position*, and the corresponding equation is termed its *standard form*. Since the x and y axes can be interchanged, there are two standard positions. We shall discuss each conic section in turn, and we shall place the section in standard position. If the conic is rotated 90° about the origin in the counterclockwise direction, the transformed equation is obtained by replacing x with y and y with $-x$ in the original equation.

3.2.2 General Definitions

There are several terms that are common to all three conic sections, and we shall now define them. For this purpose, let P denote the moving point that generates the curve.

With respect to the *ellipse* and *hyperbola*, our interest centers about the distance of P from two fixed points, F and F'. These fixed points are the *foci*, and the distances FP and $F'P$ are the *focal radii*. The line through F and F' is the *principal axis*, and the point on this axis that lies midway between F and F' is the *center* of the conic.

With respect to the *parabola*, our interest centers about the distance of P from a fixed point F and a fixed line L. Again, F is the focus and FP is the focal radius; line L is the *directrix*. The principal axis is the line through F that is perpendicular to L.

With respect to all three conics, the chord that passes through a focus and is perpendicular to the principal axis is the *latus rectum*. The point at which the principal axis intersects the conic is the *vertex*.

3.2.3 The Parabola

The moving point P generates a parabola if its distance from the focus F and its distance from the directrix L remain equal. (The terms are defined in Art. 3.2.2.) In Fig. 3.11, we place the principal axis on the x axis and the vertex at the origin. Let $AF = 2p$. Since O lies on the parabola, $AO = OF = p$. The equation of the parabola is

$$y^2 = 4px \tag{3.25}$$

The curve is symmetric about the x axis. The chord $K'K$ is the latus rectum, and its length is $4p$.

The parabola has four highly important properties, and we shall discuss them in turn.

Tangential Property. In Fig. 3.12, $P(x_1, y_1)$ is an arbitrary point on the curve, QP is parallel to the y axis, and T is the tangent to the curve at P. The tangent intersects the x and y axes at M and N, respectively. Then $MN = NP$, $MO = OQ = x_1$, and $ON = y_1/2$.

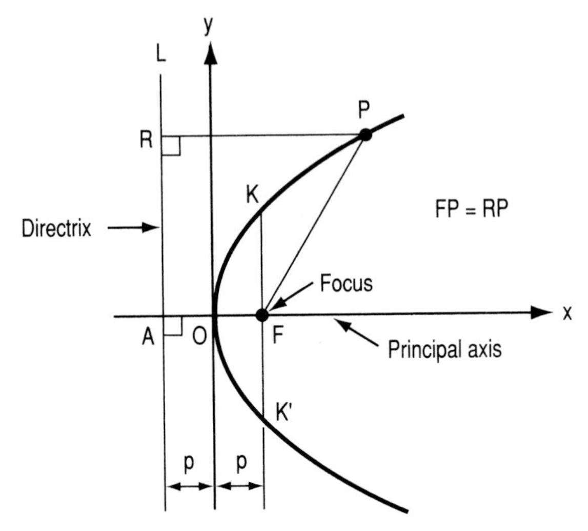

FIGURE 3.11 The parabola.

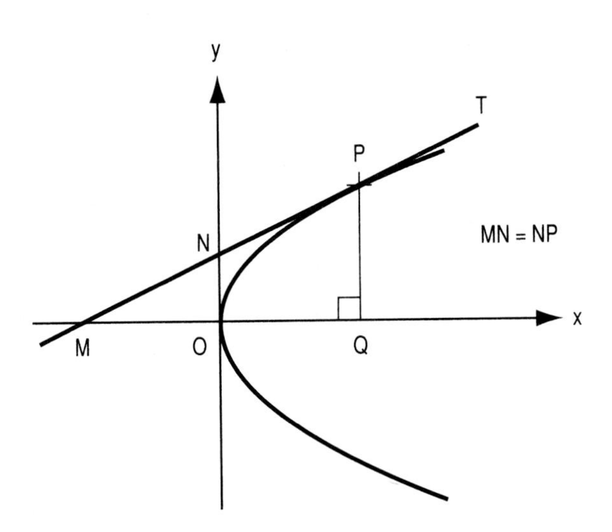

FIGURE 3.12 Tangential property of the parabola.

Focal Property. In Fig. 3.13, HP is the line through P parallel to the x axis, T is again the tangent to the parabola at P, and FP is the focal radius. Then $MF = FP$, and $\angle HPM = \angle FPM$. Therefore, in accordance with the law of reflection in optics, all rays of light that emanate from the focus of a parabolic mirror are reflected along lines parallel to the principal axis of the mirror.

Tangent-Offset Property. In Fig. 3.14, P_1 and P_2 are arbitrary points on the parabola, and T is the tangent to the curve at P_1. Let v denote the length of P_1P_2 as projected on the y axis, and let t denote the distance from T to P_2 as measured

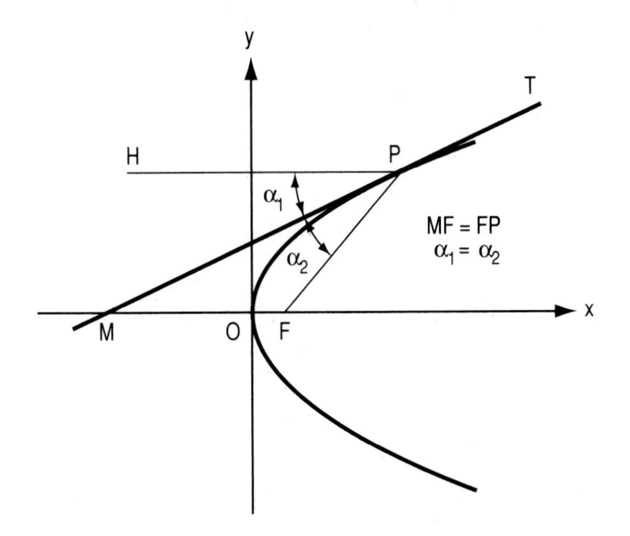

FIGURE 3.13 Focal property of the parabola.

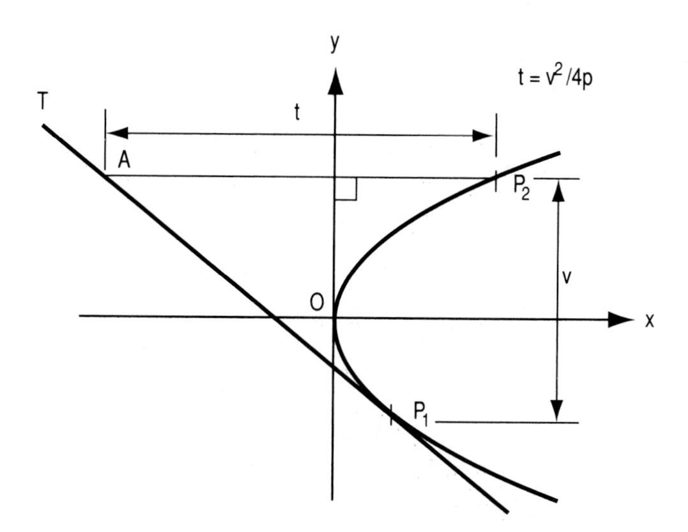

FIGURE 3.14 Tangent-offset property of the parabola.

parallel to the x axis. The distance t is called the *tangent-offset* of P_2 from P_1. Then $t = v^2/4p$; thus, t varies directly as v^2.

Tangent-Intersection Property. In Fig. 3.15, P_1 and P_2 are arbitrary points on the parabola, T_1 and T_2 are the tangents to the curve at P_1 and P_2, respectively, and AP_1 and BP_2 are parallel to the x axis. By the tangent-offset property, $AP_1 = BP_2$, and it follows that triangles AP_1Q and P_2BQ are congruent. Therefore, $AQ = QP_2$, $BQ = QP_1$, and $w_1 = w_2$.

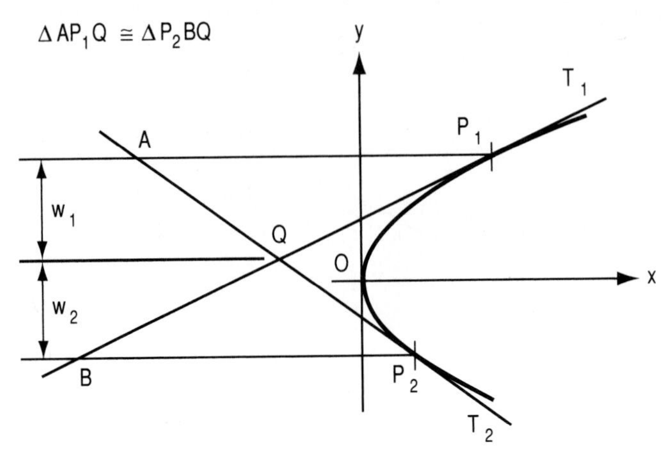

FIGURE 3.15 Tangent-intersection property of the parabola.

3.2.4 The Ellipse

The moving point P generates an ellipse if the sum of its distances from the foci remains constant. (The terms are defined in Art. 3.2.2.) In Fig. 3.16, we place the principal axis on the x axis and the center of the ellipse at the origin. Let

$$F'F = 2c \qquad F'P + FP = 2a \qquad b^2 = a^2 - c^2$$

From the definitions, $a > c$ and $a > b$. The equation of the ellipse is

$$\frac{x^2}{a^2} + \frac{y^2}{b^2} = 1 \tag{3.26}$$

The curve is symmetric about both the x and y axes.

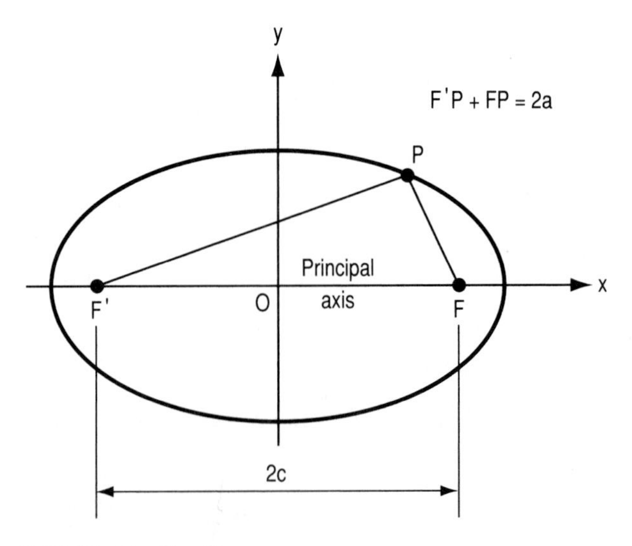

FIGURE 3.16 The ellipse.

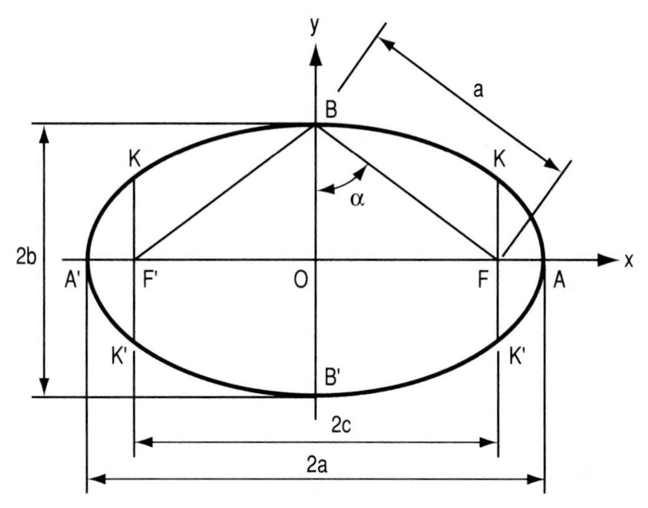

FIGURE 3.17 Dimensions of the ellipse.

In Fig. 3.17, points A and A' are the vertices of the ellipse, and the chords $A'A$ and $B'B$ are the *major axis* and *minor axis*, respectively. Each chord marked $K'K$ is a latus rectum. The lengths are as follows:

$$A'A = 2a \qquad B'B = 2b \qquad K'K = \frac{2b^2}{a} \tag{3.27}$$

By symmetry, $FB = F'B = a$.

In Fig. 3.18, C is a circle that circumscribes the ellipse E. Line Q is an arbitrary

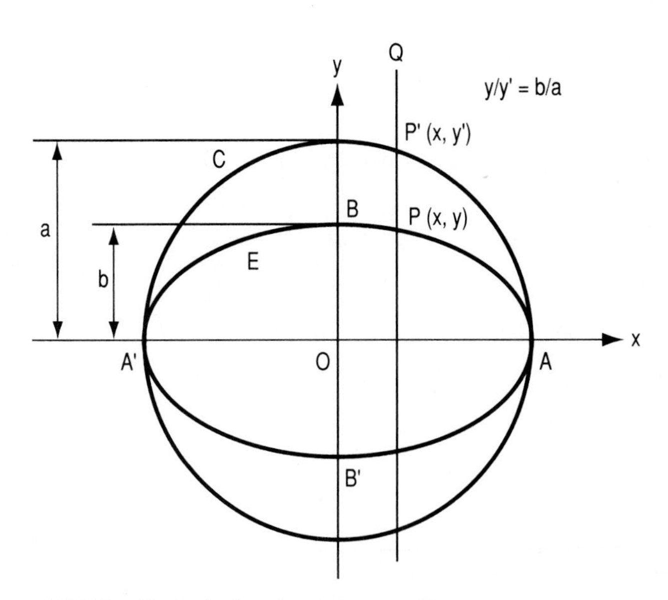

FIGURE 3.18 Reduction of a circle to an ellipse.

line parallel to the y axis, and it intersects the circle at $P'(x, y')$ and the ellipse at $P(x, y)$. Then

$$\frac{y}{y'} = \frac{b}{a} = \text{constant} \tag{3.28}$$

Thus, we may consider that the ellipse in Fig. 3.18 is derived from its circumscribed circle by reducing all ordinates of the circle by a constant proportion.

The *eccentricity* e of an ellipse is defined as $e = c/a$. Therefore, with reference to Fig. 3.17, $e = \sin\alpha$. Consider that b remains constant while c varies. When e approaches 0, the foci approach the origin, and the ellipse approaches a circle as its limiting form. As e increases toward 1, the foci diverge and the ellipse becomes increasingly flatter. Consequently, e is an index of the shape of the ellipse. The focal radii in Fig. 3.16 may be expressed in this manner:

$$F'P = a + ex \qquad FP = a - ex \tag{3.29}$$

$$\therefore\ F'P - FP = 2ex \tag{3.30}$$

In Fig. 3.19, line L is parallel to the y axis and lies at the indicated location. Line PR is parallel to the x axis. From Eq. (3.29), we obtain

$$\frac{FP}{PR} = e \tag{3.31}$$

Therefore, we may define the ellipse in this alternative manner: The moving point P generates an ellipse if the ratio of its distance from the focus F to its distance from the fixed line L remains constant and is less than 1. For this reason, L is called a *directrix of the ellipse*. The ellipse also has a directrix L' to the left of the y axis, and L' corresponds to the focus F'.

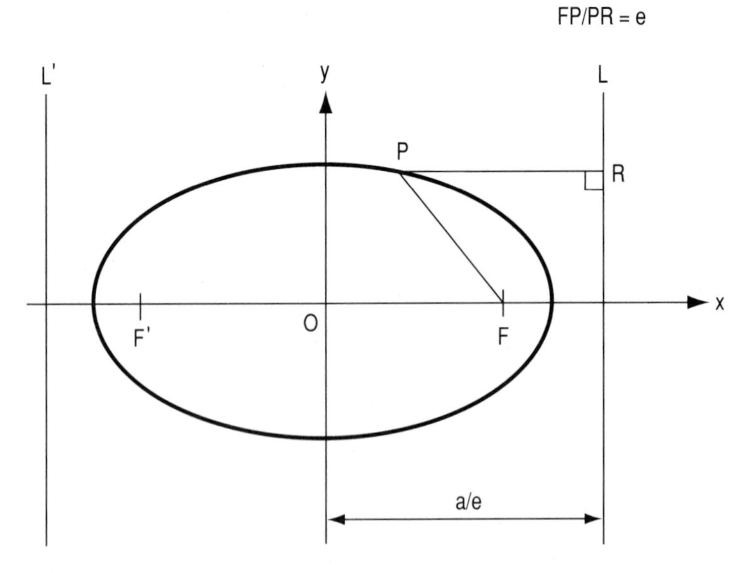

FIGURE 3.19 Directrices of the ellipse.

3.2.5 The Circle

The circle may be regarded as a degenerate form of ellipse in which the foci coincide. In the standard position, the center of the circle is at the origin. Let r denote the radius of the circle. The equation of the curve is

$$x^2 + y^2 = r^2 \tag{3.32}$$

3.2.6 The Hyperbola

The moving point P generates a hyperbola if the difference between its distances from the foci remains constant. (The terms are defined in Art 3.2.2.) In Fig. 3.20, we place the principal axis on the x axis and the center of the hyperbola at the origin. Let

$$F'F = 2c \qquad F'P - FP = \pm 2a$$

We apply the plus or minus sign, respectively, according to whether P lies to the right or to the left of the y axis. It follows that $c > a$. Now let

$$b^2 = c^2 - a^2$$

The ratio a/b can have any positive value whatever.

The equation of the hyperbola is

$$\frac{x^2}{a^2} - \frac{y^2}{b^2} = 1 \tag{3.33}$$

The curve has two branches, and it is symmetric about both the x and y axes.

In Fig. 3.21, points A and A' are the vertices of the hyperbola. Lines M and M' have the equations $y = bx/a$ and $y = -bx/a$, respectively, and they are the asymptotes of the hyperbola. Line $Q'Q$ is parallel to the y axis, and it contains A. We construct the rectangle shown. Lines $A'A$ and $B'B$ are the *transverse axis* and

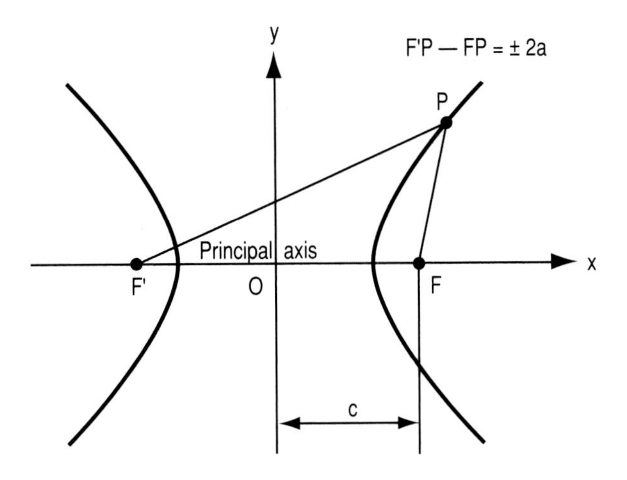

FIGURE 3.20 The hyperbola.

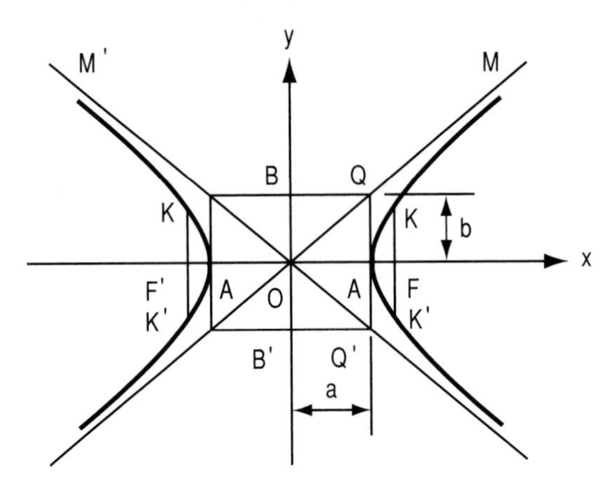

FIGURE 3.21 Axes and asymptotes of the hyperbola.

conjugate axis, respectively, of the hyperbola. Each chord marked $K'K$ is a latus rectum. The lengths are as follows:

$$A'A = 2a \qquad B'B = 2b \qquad K'K = \frac{2b^2}{a} \tag{3.34}$$

It follows that $OQ = c$.

The *eccentricity* e of a hyperbola is defined as $e = c/a$. Therefore, with reference to Fig. 3.22, $e = \sec \alpha$. The greater the eccentricity, the wider the hyperbola. The focal radii in Fig. 3.20 may be expressed in this manner:

$$F'P = ex + a \qquad FP = ex - a \tag{3.35}$$

$$\therefore\ F'P + FP = 2ex \tag{3.36}$$

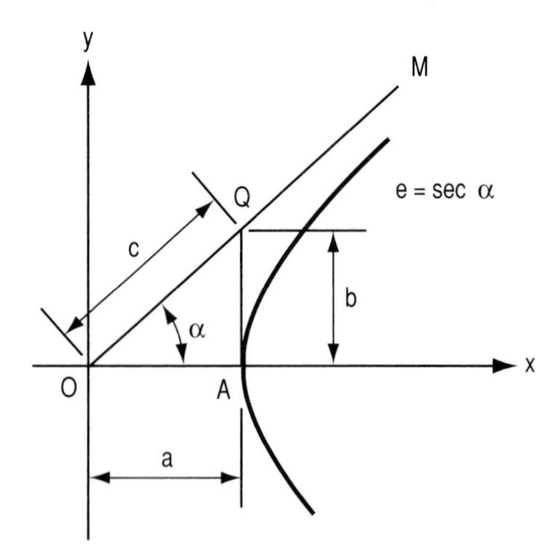

FIGURE 3.22 Interpretation of the eccentricity of the hyperbola.

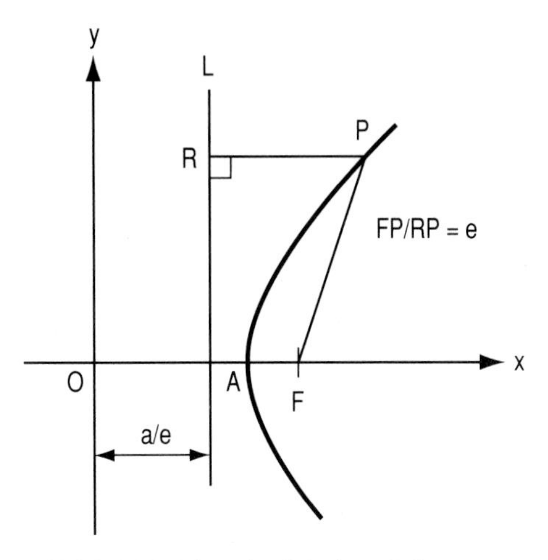

FIGURE 3.23 Directrix of the hyperbola.

In Fig. 3.23, line L is parallel to the y axis and lies at the indicated location, and line PR is parallel to the x axis. From Eq. (3.35), we obtain

$$\frac{FP}{RP} = e \tag{3.37}$$

Therefore, we may define the hyperbola in this alternative manner: The moving point P generates a hyperbola if the ratio of its distance from the focus F to its distance from the fixed line L remains constant and is greater than 1. For this reason, L is called a *directrix of the hyperbola.* The hyperbola also has a directrix that lies to the left of the y axis, and this directrix corresponds to the focus F'.

An *equilateral hyperbola* is one having the property $a = b$. Its asymptotes lie midway between the x and y axes, and its eccentricity is $\sqrt{2}$. If the curve is rotated about the origin through an angle of 45° in the counterclockwise direction, the equation of the curve becomes

$$xy = \frac{a^2}{2} \tag{3.33a}$$

The x and y axes are the asymptotes.

3.2.7 General Definition of a Conic Section

It is now possible to formulate an all-inclusive definition of a conic section. It is as follows:

The moving point P generates a conic section if the ratio of its distance from a fixed point F (the focus) to its distance from a fixed line L (the directrix) remains constant. Let e denote this ratio. The conic is an ellipse if $e < 1$, a parabola if $e = 1$, and a hyperbola if $e > 1$. The circle may be viewed as the limiting form approached by the ellipse if the directrix recedes indefinitely and consequently e approaches 0.

3.2.8 Criterion for Identifying a Conic Section

If a conic section is in nonstandard position, its equation has the general form $Ax^2 + Bxy + Cy^2 + Dx + Ey + F = 0$. The conic can be brought into standard position by a suitable transformation of the axes. However, the form of the conic can be discerned directly from the original equation by computing the quantity $J = 4AC - B^2$. Equation (3.22*b*) states that J is invariant under a rotation of the axes, and this quantity is called the *discriminant* of the conic. The form of the conic is as follows:

A hyperbola if $J < 0$

A parabola if $J = 0$

An ellipse if $J > 0$

The circle may be regarded as a special type of ellipse in which $A = C$ and $B = 0$.

Geometrically, the form of the conic can be discerned by means of the transformation diagram in Fig. 3.8*a*, the criterion being the position of the origin relative to the circle in that diagram. The form of the conic is as follows:

A hyperbola if O lies within the circle

A parabola if O lies on the circle

An ellipse if O lies outside the circle

The circular conic may be regarded as a special type of ellipse in which the circle in Fig. 3.8*a* has zero radius.

In general, every second-degree equation is the equation of a conic section.

EXAMPLE 3.7 Identify the conic section in Example 3.4.

SOLUTION

$$J = 4AC - B^2 = 4 \times 2 \times 7 - (-12)^2 = -88$$

Therefore, the conic section is a hyperbola.

The first transformed equation in Example 3.4 may be written as $11x^2 - 2y^2 = 1$. This is a particular form of Eq. (3.33) with $a = \sqrt{1/11}$ and $b = \sqrt{1/2}$. Our conclusion that the section is a hyperbola is thus confirmed.

3.3 BASIC SOLID ANALYTIC GEOMETRY

3.3.1 Methods of Representing Objects

In representing objects in three-dimensional space, we shall construct two types of drawings: oblique and orthographic. An *oblique drawing* is a pictorial representation of the object, and it distorts many features of the object. An *orthographic drawing*, which is the conventional type of drawing used in engineering, is formed by considering that an observer views the object from a specific position. However, rays of light from the object to the observer's eye are assumed to be parallel. An orthographic drawing thus consists of a series of *views*, each view corresponding to a particular direction. Orthographic drawings are the most effective means of conveying information and establishing relationships, and they enable us to test calculated results graphically.

3.3.2 Orthographic Drawings

We shall now present definitions and principles pertaining to orthographic drawings. The direction along which an observer views the object is the *line of sight* (LS). A view in which the LS is perpendicular to a given line or plane is a *normal view* of that line or plane; a view in which the LS is parallel to a given line or plane is an *edge view* of that line or plane.

In an edge view, a line appears as a point and a plane appears as a line. An edge view of a line is an edge view of every plane that contains that line. Let L and Q denote a line and plane, respectively, that are perpendicular to each other. An edge view of L is a normal view of Q, and vice versa. A normal view of L is an edge view of Q, and vice versa.

A line appears in its true length in a normal view of that line. The angle between two nonperpendicular lines appears in its true size only in a view that is a normal view of both lines. The angle between a line and a plane appears in its true size in a view that is a normal view of the line and an edge view of the plane.

3.3.3 Rectangular Coordinates of a Point

The x, y, and z coordinate axes in Fig. 3.24 are mutually perpendicular, they intersect at O (the *origin*), and they have the indicated positive directions. A plane that contains two coordinate axes is a *coordinate plane*, and it is labeled to indicate which axes it contains. For example, the xz plane contains the x and z axes. The coordinate axes divide space into eight regions, called *octants*.

Let P denote a point in space. The *rectangular* (or *cartesian*) *coordinates* of P are found by passing planes through P parallel to the coordinate planes, thereby forming the rectangular block in Fig. 3.24. Then $x = OA$, $y = OB$, $z = OC$. Thus,

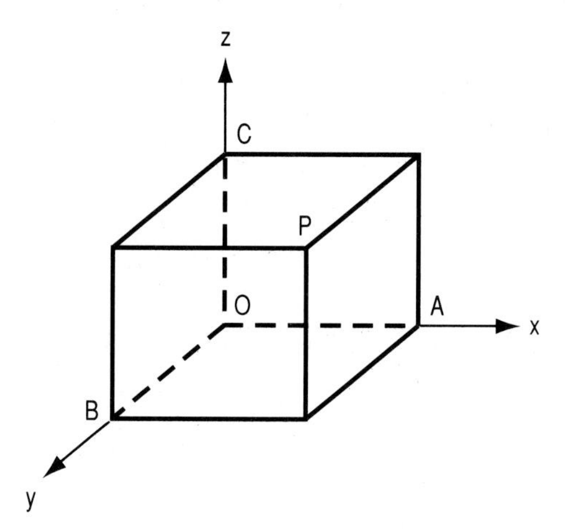

FIGURE 3.24 Rectangular coordinates of a point in three-dimensional space.

a coordinate of P is simply the length of the directed line OP as projected on the corresponding coordinate axis. The designation $P(x_1, y_1, z_1)$ identifies a point having the specified coordinates.

3.3.4 Distance between Two Points

Let d denote the distance between points $P_1(x_1, y_1, z_1)$ and $P_2(x_2, y_2, z_2)$. Then

$$d = \sqrt{(x_2 - x_1)^2 + (y_2 - y_1)^2 + (z_2 - z_1)^2} \tag{3.38}$$

3.3.5 Direction Cosines and Direction Components

Let L denote a directed line in space that passes through the origin. The angle between the positive side of a coordinate axis and the positive direction of L is a *direction angle* of L, and the cosine of this angle is a *direction cosine* of L. These definitions are illustrated in Fig. 3.25, where P is an arbitrary point on L and α, β, and γ are the direction angles of L with reference to the x, y, and z axes, respectively. The direction cosines of L and their designations are as follows:

$$\cos\alpha = l = \frac{OA}{OP} \qquad \cos\beta = m = \frac{OB}{OP} \qquad \cos\gamma = n = \frac{OC}{OP} \tag{3.39}$$

Now let L denote a directed line that does not pass through the origin. We draw a directed line L' through the origin parallel to L and having the same positive direction as L. The direction angles of L' are also those of L.

Let $P_1(x_1, y_1, z_1)$ and $P_2(x_2, y_2, z_2)$ denote arbitrary points on the directed line L, where P_1P_2 has the positive direction. Then

$$l = \frac{x_2 - x_1}{P_1P_2} \qquad m = \frac{y_2 - y_1}{P_1P_2} \qquad n = \frac{z_2 - z_1}{P_1P_2} \tag{3.40}$$

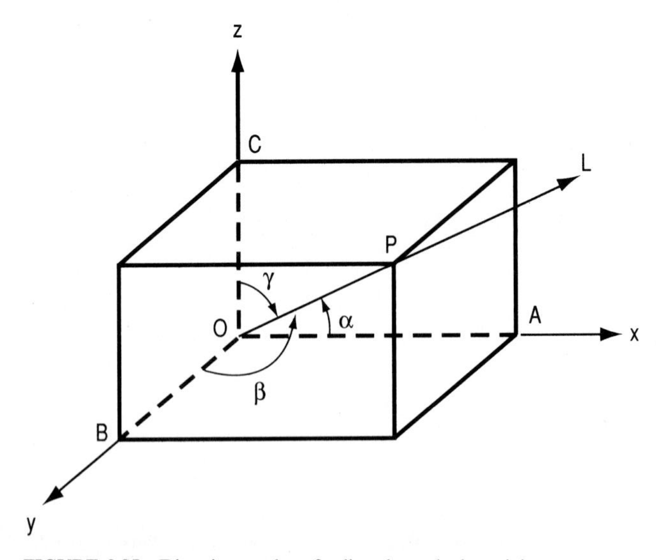

FIGURE 3.25 Direction angles of a line through the origin.

From Eq. (3.38), we have

$$l^2 + m^2 + n^2 = 1 \tag{3.41}$$

Let a, b, and c denote a set of numbers that have the same ratios as l, m, and n; that is, $a : b : c = l : m : n$. The numbers a, b, and c are termed *direction components* (or *direction numbers*) of L. The direction components provide a simple means of expressing the inclination of a line. For example, assume that a line has direction components 14, -6, 9. These numbers inform us that if we proceed from one point on the line to another so that x increases by 14 units, y decreases by 6 units and z increases by 9 units. A line can have an infinite number of sets of direction components, but the numbers within each set have constant ratios to one another. Assume that the direction components have the same algebraic signs as their corresponding direction cosines. Let

$$k = \sqrt{a^2 + b^2 + c^2} \tag{3.42}$$

It follows that

$$l = \frac{a}{k} \qquad m = \frac{b}{k} \qquad n = \frac{c}{k} \tag{3.43}$$

EXAMPLE 3.8 A line has direction components 12, 4, -3. Compute the direction angles of this line.

SOLUTION By Eq. (3.42), $k = 13$. Then

$$\cos\alpha = \frac{12}{13} \qquad \alpha = 22.62^\circ$$

$$\cos\beta = \frac{4}{13} \qquad \beta = 72.08^\circ$$

$$\cos\gamma = \frac{-3}{13} \qquad \gamma = 103.34^\circ$$

3.3.6 Angle between Two Lines

Let θ denote the angle between the directed lines L_1 and L_2. If these lines intersect, θ is taken as the angle having the positive sides of these lines as its sides, as shown in Fig. 3.26. If these lines do not intersect, θ is defined in this manner: Through the origin, draw lines L_1' and L_2' parallel to L_1 and L_2, respectively, with the same positive directions as L and L'. Angle θ equals the angle between L_1' and L_2'. Thus,

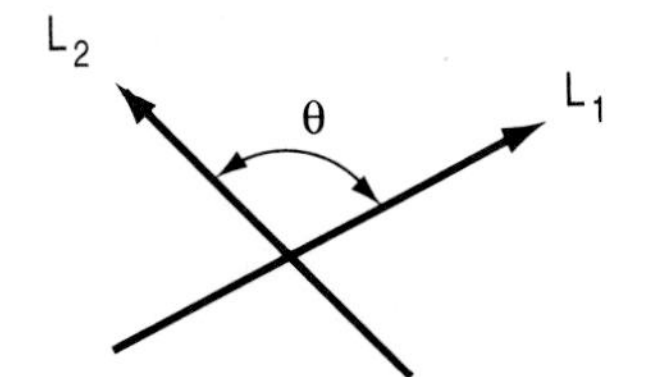

FIGURE 3.26 Normal view of lines L_1 and L_2.

θ measures the difference between the positive directions of L_1 and L_2, regardless of whether or not they intersect.

Applying the notation for direction cosines and direction components in Art. 3.3.5, we have the following:

$$\cos\theta = l_1 l_2 + m_1 m_2 + n_1 n_2 \tag{3.44a}$$

$$\cos\theta = \frac{a_1 a_2 + b_1 b_2 + c_1 c_2}{k_1 k_2} \tag{3.44b}$$

Since $\cos 90° = 0$, it follows that two lines are perpendicular to each other if and only if the expressions at the right in Eq. (3.44) are 0.

3.3.7 Equation of a Plane

In Fig. 3.27, L is a directed line through the origin perpendicular to a given plane Q, and L intersects Q at N. The directed line ON is the *normal* to Q. Let l, m, and n denote the direction cosines of the normal, and let $p = ON$. The equation of Q is

$$lx + my + nz - p = 0 \tag{3.45}$$

If we now multiply both sides of Eq. (3.45) by a constant K, we obtain an equation of the form

$$Ax + By + Cz + D = 0 \tag{3.46}$$

This is the *general form* of the equation of a plane. In this form, A, B, and C are direction components of the normal, and

$$D = -pK \tag{3.47}$$

From Eq. (3.41),

$$K = \pm\sqrt{A^2 + B^2 + C^2} \tag{3.48}$$

and the algebraic sign of K is opposite to that of D.

Let x_i, y_i, and z_i denote the directed distances from the origin to the points at which plane Q intersects the x, y, and z axes, respectively, as shown in Fig. 3.28. These distances are known as the *intercepts* of Q. For any point on Q,

$$\frac{x}{x_i} + \frac{y}{y_i} + \frac{z}{z_i} = 1 \tag{3.49}$$

This is the *intercept form* of the equation of a plane.

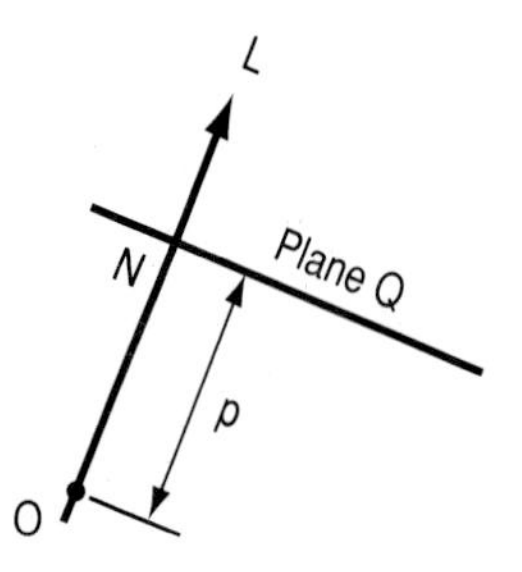

FIGURE 3.27 Edge view of plane Q.

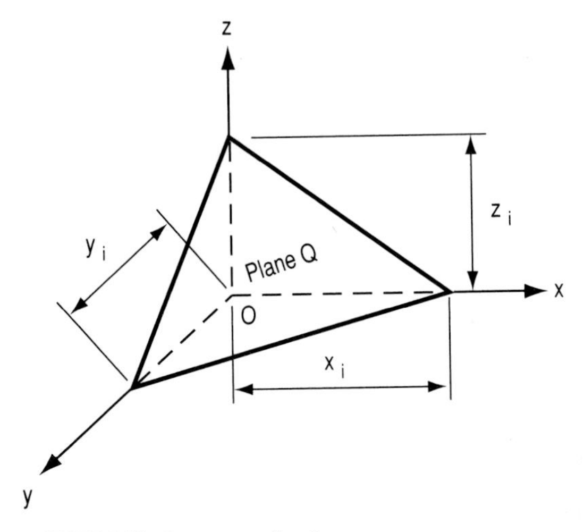

FIGURE 3.28 Intercepts of a plane.

If a plane is parallel to a coordinate plane, the corresponding coefficient in Eq. (3.46) is 0. For example, if a plane is parallel to the x axis, its equation is simply $By + Cz + D = 0$. Similarly, if a plane is perpendicular to the xy plane, it is parallel to the z axis, and its equation is $Ax + By + D = 0$.

3.3.8 Distance from a Point to a Plane

Let Q denote a plane having the equation $Ax + By + Cz + D = 0$, let $P(x_1, y_1, z_1)$ denote a point outside Q, and let d denote the perpendicular distance from P to Q. Now let

$$e = Ax_1 + By_1 + Cz_1 + D \tag{3.50}$$

If C and e have the same algebraic sign, P lies above Q; if they have opposite signs, P lies below Q. Then

$$d = \left| \frac{e}{K} \right| \tag{3.51}$$

where K is defined by Eq. (3.48).

EXAMPLE 3.9 A plane has the equation $16x - 21y + 12z + 67 = 0$. Find the perpendicular distance from $P(-9, 3, 2)$ to this plane.

SOLUTION By Eq. (3.48), $K = -29$.

$$e = 16(-9) - 21 \times 3 + 12 \times 2 + 67 = -116$$

$$d = \frac{116}{29} = 4$$

Since C and e have opposite signs, P lies below the plane.

3.3.9 Angle between Two Planes

In Fig. 3.29, planes Q_1 and Q_2 intersect along line L. We select an arbitrary point P on L and draw lines PA and PB in Q_1 and Q_2, respectively, both perpendicular to L. The angle θ between the planes is considered to be the angle between PA and PB.

Let $A_1x + B_1y + C_1z + D_1 = 0$ and $A_2x + B_2y + C_2z + D_2 = 0$ be the equations of Q_1 and Q_2, respectively. Angle θ equals the angle between the normals of the planes, and it follows that

$$\cos\theta = \pm\frac{A_1A_2 + B_1B_2 + C_1C_2}{K_1K_2} \tag{3.52}$$

where K is defined by Eq. (3.48). It is usually a simple matter to determine on which side of each plane the acute angle is located.

Planes Q_1 and Q_2 are perpendicular to each other if and only if

$$A_1A_2 + B_1B_2 + C_1C_2 = 0 \tag{3.53}$$

and they are parallel to each other if and only if

$$\frac{A_1}{A_2} = \frac{B_1}{B_2} = \frac{C_1}{C_2} \tag{3.54}$$

EXAMPLE 3.10 Compute the angle between the planes having the equations $9x + 8y - 12z - 85 = 0$ and $24x - 32y + 9z + 51 = 0$.

SOLUTION

$$K_1 = \sqrt{9^2 + 8^2 + 12^2} = 17$$

$$K_2 = -\sqrt{24^2 + 32^2 + 9^2} = -41$$

$$\cos\theta = \frac{9 \times 24 + 8(-32) + (-12)9}{17(-41)} = \frac{-148}{-697}$$

The acute angle between the planes is 77.74°.

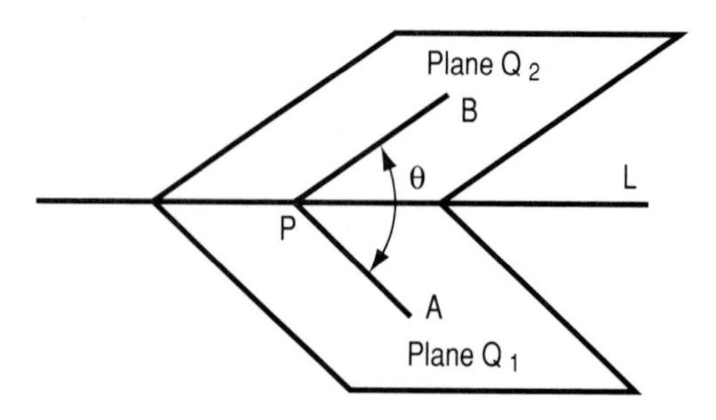

FIGURE 3.29 Angle between two planes.

3.3.10 Intersection of Given Plane with a Coordinate Plane

It is often advantageous to define a given plane by expressing its relationship to a coordinate plane, and we select the xy plane for this purpose. The line of intersection of a given plane and a coordinate plane is called the *trace* of the first plane on the second.

Again let Q denote a plane having the equation $Ax + By + Cz + D = 0$, let T denote its trace on the xy plane, and let ϕ denote the angle between Q and the xy plane. Since $z = 0$ for every point on T, the equation of T is $Ax + By + D = 0$. From Eq. (3.52), we obtain

$$\cos\phi = \pm\frac{C}{K} \tag{3.55}$$

where K is defined by Eq. (3.48). The solution for C is

$$C = \pm\frac{\sqrt{A^2 + B^2}}{\tan\phi} \tag{3.56}$$

Where the trace of Q and the value of ϕ are given, the equation of Q can be formulated readily.

EXAMPLE 3.11 Plane Q in Fig. 3.30 intersects the coordinate axes at R, S, and T, where $OR = 12$ and $OS = 5$. The angle between plane Q and the xy plane is 62.85°. Formulate the equation of Q.

SOLUTION Draw OU perpendicular to RS, and then draw TU. Since TU is the shortest line that can be drawn from T to RS, TU is also perpendicular to RS, and $\angle OUT = \phi = 62.85°$. The length of RS is 13, and

$$OU = 12\sin\alpha = 12\left(\frac{5}{13}\right) \qquad OT = (OU)\tan 62.85° = 9$$

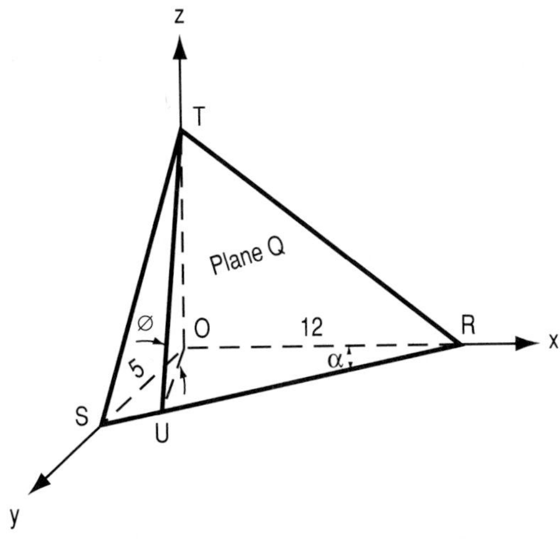

FIGURE 3.30

Applying Eq. (3.49) and then multiplying through by 180, we obtain the following:

$$\frac{x}{12} + \frac{y}{5} + \frac{z}{9} - 1 = 0$$

$$15x + 36y + 20z - 180 = 0$$

The last result is the equation of Q in general form, and we shall test it for compliance with Eq. (3.55).

$$K = \sqrt{15^2 + 36^2 + 20^2} = 43.8292$$

$$\cos\phi = \frac{20}{43.8292} \qquad \phi = 62.85^\circ$$

Thus, the equation of Q is valid.

3.3.11 Equation of a Straight Line

Assume that a given plane is not parallel to any coordinate axis. In selecting a point in this plane, we can assign arbitrary values to two of the three coordinates. Therefore, the plane is identified completely by means of one equation containing three variables. On the other hand, in selecting a point on a given straight line, we can assign an arbitrary value to only one of the three coordinates. Therefore, the line can be identified fully only by means of a *composite* equation that consists of two equations containing three variables. Each equation in the set may be interpreted as the equation of a plane that contains the given line.

Let the composite equation of line L be

$$\begin{aligned} A_1x + B_1y + C_1z + D_1 &= 0 \\ A_2x + B_2y + C_2z + D_2 &= 0 \end{aligned} \tag{3.57}$$

This is the *general form* of the composite equation of the line.

Now let a, b, and c denote, respectively, the x, y, and z direction components of line L. One set of values is as follows:

$$\begin{aligned} a &= B_1C_2 - B_2C_1 \\ b &= C_1A_2 - C_2A_1 \\ c &= A_1B_2 - A_2B_1 \end{aligned} \tag{3.58}$$

Consider that the line L having these direction components contains point $P_1(x_1, y_1, z\ )$. Then

$$\frac{x - x_1}{a} = \frac{y - y_1}{b} = \frac{z - z_1}{c} \tag{3.59}$$

This is the *symmetric form* of the composite equation of the line.

Now consider that a line contains the points $P_1(x_1, y_1, z_1)$ and $P_2(x_2, y_2, z_2)$. Then

$$\frac{x - x_1}{x_2 - x_1} = \frac{y - y_1}{y_2 - y_1} = \frac{z - z_1}{z_2 - z_1} \tag{3.60}$$

This is the *two-point form* of the equation of the line.

3.3.12 Equation of Family of Planes through a Line

If Eq. (3.57) is the composite equation of a line L, the entire family of planes that contain L has the general equation

$$A_1x + B_1y + C_1z + D_1 + k(A_2x + B_2y + C_2z + D_2) = 0 \tag{3.61}$$

where k has a unique value for each plane.

If a plane contains L and has some other known characteristic, the problem of finding its equation becomes one of finding the value of k for this particular plane.

EXAMPLE 3.12 Planes Q_1 and Q_2 have the equations $5x + 3y - 7z - 19 = 0$ and $3x - 8y + 4z + 115 = 0$, respectively. Plane R contains the line of intersection of Q_1 and Q_2 and point $P(-3, 4, -6)$. Establish the equation of R.

SOLUTION By Eq. (3.61), plane R has an equation of the form $5x + 3y - 7z - 19 + k(3x - 8y + 4z + 115) = 0$. The coordinates of P satisfy this equation. Therefore, we replace x, y, and z with -3, 4, and -6, respectively, to obtain $20 + 50k = 0$, or $k = -2/5$. Replacing k in the general equation with this value and then multiplying through by 5, we obtain $19x + 31y - 43z - 325 = 0$ as the equation of R.

The result can be verified by selecting an arbitrary point on the line of intersection of Q_1 and Q_2 and demonstrating that its coordinates satisfy the equation of R.

3.3.13 Projecting Planes

A plane that contains line L and is perpendicular to a coordinate plane is termed a *projecting plane* of L. Finding the equation of a projecting plane is a simple matter.

Let Q_{xy} denote the plane through L that is perpendicular to the xy plane. In accordance with the discussion in Art. 3.3.7, $C = 0$ is the general equation for this plane. Therefore, in Eq. (3.61), we set $C_1 + kC_2 = 0$, or $k = -C_1/C_2$. Similarly, for Q_{xz} we set $k = -B_1/B_2$, and for Q_{yz} we set $k = -A_1/A_2$.

3.3.14 Parallelism and Perpendicularity of a Line and Plane

Let Q denote a plane having the equation $Ax + By + Cz + D = 0$. Let L denote a line having direction components a, b, and c, and let ON denote the normal to Q. If L is parallel to Q, it is perpendicular to ON; if L is perpendicular to Q, it is parallel to ON.

From Eq. (3.44b), we deduce that L is parallel to Q if and only if

$$aA + bB + cC = 0 \tag{3.62}$$

Similarly, L is perpendicular to Q if and only if

$$\frac{a}{A} = \frac{b}{B} = \frac{c}{C} \tag{3.63}$$

3.3.15 Cylindrical and Spherical Coordinates

It is often advantageous to locate a point in space by using a coordinate system different from the rectangular system described in Art. 3.3.3. Let P denote a point having the rectangular coordinates x, y, and z; let P' denote the projection of P on the xy plane (i.e., the foot of the perpendicular from P to the xy plane).

In Fig. 3.31*a*, let r and θ be the polar coordinates of P' as defined in Art. 3.1.9. The quantities r, θ, and z constitute the *cylindrical coordinates* of P, and the point is identified by the notation $P(r, \theta, z)$. The relationship between the two sets of coordinates is given by Eqs. (3.23) and (3.24).

In Fig. 3.31*b*, let $OP = \rho$, let θ denote the angle from the x axis to OP', and let ϕ denote the angle from the z axis to OP. The quantities ρ, θ, and ϕ constitute the *spherical coordinates* of P, and the point is identified by the notation $P(\rho, \theta, \phi)$. The relationships are as follows:

$$\begin{aligned} x &= \rho \sin \phi \cos \theta \\ y &= \rho \sin \phi \sin \theta \\ z &= \rho \cos \phi \\ \rho &= \sqrt{x^2 + y^2 + z^2} \end{aligned} \tag{3.64}$$

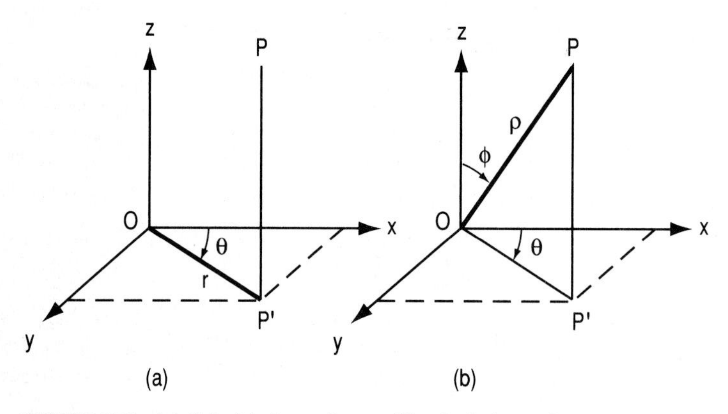

FIGURE 3.31 (*a*) Cylindrical coordinates; (*b*) spherical coordinates.

3.4 QUADRIC SURFACES

3.4.1 Definitions

Two straight lines that neither intersect nor are parallel to each other are called *skew lines*. It is impossible to pass a plane through both lines. Two nonskew lines, by contrast, can be joined by a plane.

A three-dimensional surface that has an equation of the second degree is termed a *quadric surface*. Such surfaces are of two types: a *ruled surface*, on which it is possible to draw straight lines, and a *double-curved surface*, on which it is impossible to do so. For example, a cone is a ruled surface and a sphere is a double-curved surface. A *surface of revolution* is generated by a line that revolves about a fixed axis while retaining its original shape. The surface is ruled or double-curved according to whether the revolving line is straight or curved, respectively. Ruled surfaces are widely used for manufactured objects and structures because they can be formed with relative ease.

It is advantageous to conceive that a ruled surface is generated by a straight line that moves in some prescribed manner. This line is the *generating line* or *generatrix*. Each position occupied by the generating line is an *element* of the surface. If the generating line is constrained to intersect a fixed line that is straight or a plane curve, the latter is a *directing line* or *directrix*; if the generating line is constrained to remain parallel to a fixed plane, the latter is a *directing plane*. A ruled surface is *single-curved* if two successive elements are nonskew lines, and it is *warped* if two successive elements are skew lines. If the surface can be generated by using either of two distinct lines as the directing line, the surface is *double-ruled*.

Since it is possible to pass a plane through two successive elements of a single-curved surface, this surface can be transformed to a plane surface without distortion by cutting it along an element and unfolding it. The plane surface that results is the *development* of the curved surface. In sheet-metal construction, a ruled surface is built by first forming its development and then folding the latter into position.

In Art. 3.2.1, we defined the standard position of a conic section, and we shall extend the definition to encompass quadric surfaces.

3.4.2 Visualizing the Shape of a Surface

When the equation of a quadric surface is given, it becomes necessary to visualize its shape. Two devices are available to aid visualization: finding the sections of the surface, and identifying planes of symmetry.

Let S denote a quadric surface and Q denote a plane that is coincident with or parallel to a coordinate plane. The line of intersection of S and Q is the *section* of S on Q. For brevity, we shall call it the section on or parallel to the particular coordinate plane. To illustrate the use of sections, consider the surface having the equation $ax - by^2 - cz = 0$, where a, b, and c are positive constants. A part of this surface is shown in Fig. 3.32. The section of the surface on or parallel to the xy plane is found by setting $z = k$, where k is a constant, and rearranging terms in the given equation to obtain $x = (by^2 + ck)/a$. The section is a parabola having its principal axis in the xz plane. Similarly, the section of this surface on the xz plane is found by setting $y = 0$ to obtain $ax - cz = 0$. The section is the straight line L in Fig. 3.32. As we shall find in Art. 3.4.3, this surface is an oblique parabolic cylindrical surface.

Planes of symmetry are identified readily. As an illustration, consider again the surface having the equation $ax - by^2 - cz = 0$. Since each positive y value can be replaced with its corresponding negative value without changing the x and z values, the surface is symmetric about the xz plane.

3.4.3 The Cylinder

If a generating line is constrained to remain parallel to a fixed line and to intersect a directing curve, a *cylindrical surface* is formed. If the directing curve is closed, the cylindrical surface is termed a *cylinder*. A cylindrical surface is named according to

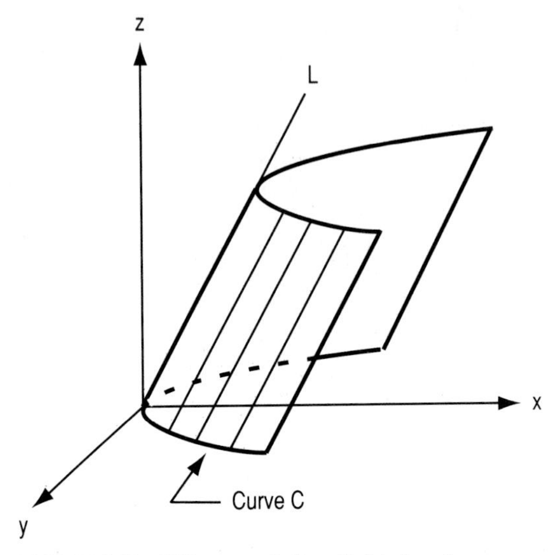

FIGURE 3.32 Oblique parabolic cylindrical surface.

the form of its directing curve; for example, an elliptic cylinder is one that has an ellipse as its directing curve. If the generating line is perpendicular to the plane of the directing curve, the cylindrical surface is *right*; otherwise, it is *oblique*. A right circular cylinder can also be generated by a straight line that revolves about a fixed line parallel to itself, and consequently this cylinder is a surface of revolution.

With reference to Fig. 3.32, we may consider that the surface is generated by a straight line that remains parallel to line L in the xz plane and intersects the parabolic curve C in the xy plane. Therefore, it is an oblique parabolic cylindrical surface.

Assume that a right cylindrical surface has its directing curve in the xy plane. Since the values of x and y do not establish the value of z, the equation of the surface is simply that of its directing curve. For example, if the directing curve is a circle of radius r with its center at the origin, the equation of the cylinder is $x^2 + y^2 = r^2$.

3.4.4 The Cone

If a generating line is constrained to pass through a fixed point and to intersect a directing curve, a *conical surface* is formed. The fixed point is the *vertex* of the surface. If the surface is considered to extend beyond the vertex, two nappes are formed. If the directing curve is closed, the conical surface is termed a *cone*. A conical surface is also named according to the form of its directing curve. In general, if the directing curve is a conic, the surface is described as *quadric conical*.

Assume that a quadric cone is circular or elliptic. The straight line connecting the vertex and the center of the directing curve is the *axis* of the cone. As shown in Fig. 3.33, the cone is *right* if the axis is perpendicular to the plane of the directing curve, and *oblique* if such is not the case.

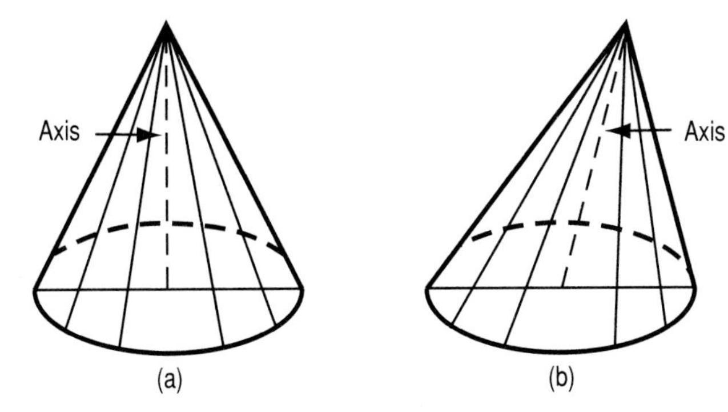

FIGURE 3.33 (*a*) Right cone; (*b*) oblique cone.

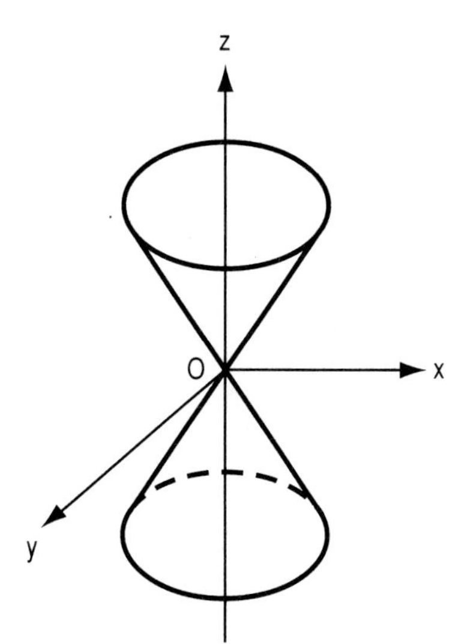

FIGURE 3.34 Quadric cone.

The right quadric cone in Fig. 3.34 has its vertex at the origin and its axis on the z axis. The equation of this cone is

$$\frac{x^2}{a^2} + \frac{y^2}{b^2} - \frac{z^2}{c^2} = 0 \tag{3.65}$$

If $a \neq b$, the section on a plane parallel to the xy plane is an ellipse; if $a = b$, the section is a circle.

3.4.5 The Sphere

A sphere is the locus of all points that lie at a fixed distance from a fixed point. The fixed point is the *center*; the fixed distance is the *radius*. Let $C(h, k, l)$ be the center, and let r denote the radius. The equation of the sphere is

$$(x - h)^2 + (y - k)^2 + (z - l)^2 = r^2 \tag{3.66}$$

We may conceive that the sphere is generated by a circle that revolves about a diameter through 180°.

Let Q denote a plane that is tangent to a sphere at point P. The radius to point P is perpendicular to Q.

EXAMPLE 3.13 Find the radius of a sphere that has its center at $C(10, -2, 3)$ and is tangent to the plane having the equation $11x - 36y + 48z - 204 = 0$.

SOLUTION The radius equals the perpendicular distance from C to the plane. Applying Eqs. (3.48), (3.50), and (3.51) in turn, we obtain the following:

$$K = \sqrt{11^2 + (-36)^2 + 48^2} = 61$$

$$e = 11 \times 10 + (-36)(-2) + 48 \times 3 - 204 = 122$$

$$\text{Radius} = \frac{122}{61} = 2$$

3.4.6 The Ellipsoid

Figure 3.35 shows an ellipsoid in standard position, and its equation is

$$\frac{x^2}{a^2} + \frac{y^2}{b^2} + \frac{z^2}{c^2} = 1 \qquad (3.67)$$

The ellipsoid is symmetric about all three coordinate planes. The section of the surface on each coordinate plane is an ellipse having its major and minor axes on coordinate axes.

We may conceive that the ellipsoid is generated in this manner: We start with the ellipse $ABCD$ in the xz plane and revolve it about the z axis through 180°. Simultaneously, we transform its shape in such manner that each point on the

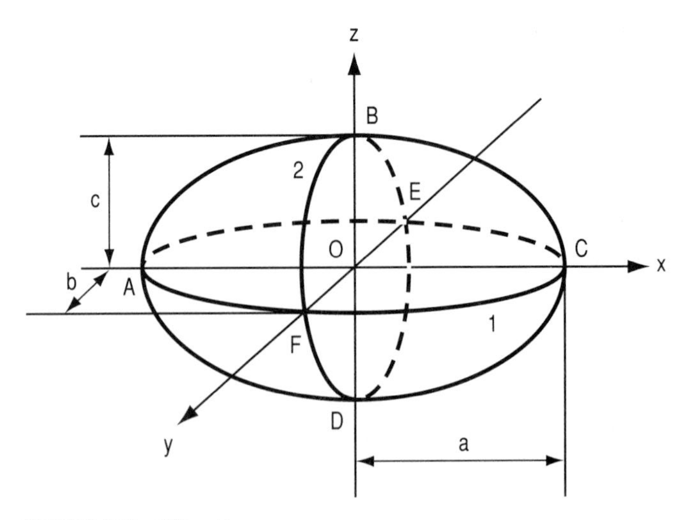

FIGURE 3.35 Ellipsoid.

ellipse describes an elliptic path about the z axis. Thus, when $ABCD$ has been revolved through 90°, it has the shape $EBFD$.

If any two of the three parameters a, b, and c are equal, the ellipsoid is a surface of revolution. For example, if $b = c$, the ellipsoid in Fig. 3.35 can be generated by revolving ellipse 1 about the x axis. Curve 2 is then a circle. An ellipsoid of revolution is termed a *prolate spheroid* or *oblate spheroid*, respectively, according to whether the generating ellipse revolves about its major or minor axis. The sphere may be regarded as a degenerate form of ellipsoid in which $a = b = c$.

3.4.7 The Hyperboloid of One Sheet

Figure 3.36 shows a hyperboloid of one sheet (or nappe) in standard position, and its equation is

$$\frac{x^2}{a^2} + \frac{y^2}{b^2} - \frac{z^2}{c^2} = 1 \tag{3.68}$$

The hyperboloid is symmetric about all three coordinate planes, and its x and y intercepts are $\pm a$ and $\pm b$, respectively. The surface has the following sections: parallel to the xz plane, a hyperbola; parallel to the yz plane, a hyperbola; parallel to the xy plane, an ellipse.

We may conceive that the hyperboloid is generated in this manner: We start with the semihyperbola ABC in the xz plane and revolve it about the z axis. Simultaneously, we transform its shape in such manner that each point on this curve describes an elliptic path about the z axis. Thus, after a 90° rotation, hyperbola 1 has been transformed to hyperbola 2.

If $a = b$, hyperbolas 1 and 2 have identical shape, and the section parallel to the xy plane is a circle. In this special case, the hyperboloid can be generated by revolving hyperbola 1 about the z axis, and the hyperboloid is a surface of revolution.

Let Q denote an arbitrary plane that is perpendicular to the xy plane and at a

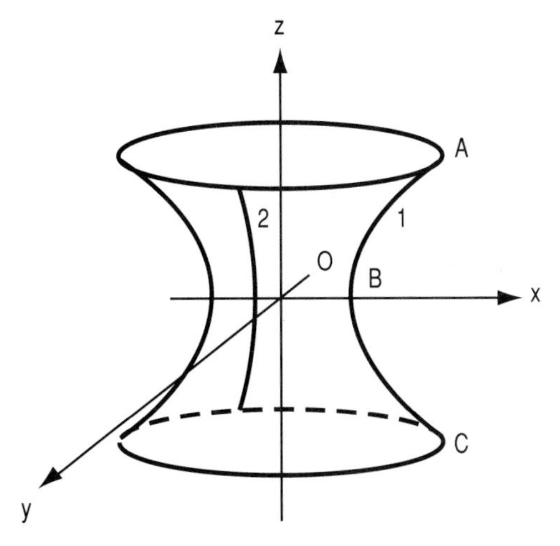

FIGURE 3.36 Hyperboloid of one sheet.

distance a from the origin. The section of the hyperboloid of revolution on this plane consists of two straight lines. Therefore, the hyperboloid of revolution is a double-ruled surface. It is used extensively for observation towers and cooling towers.

The hyperboloid of revolution can be generated by a straight line in this manner: Figure 3.37 contains three circles that are centered on the z axis. Circles 1 and 3 have a radius R, they are parallel to the xy plane, and they are at the indicated locations. Circle 2 has a radius r, where $r < R$, and it lies in the xy plane. The inclined line L is tangent to these circles at A, B, and C, as shown. Line L revolves about the z axis, and the surface it generates is a hyperboloid of revolution in which $a^2 = r^2$ and $c^2 = r^2h^2/(R^2 - r^2)$. If L is replaced with a line L' that is symmetric with L about the yz plane, line L' generates the same surface as L. Two successive positions of L are skew lines. Thus, the hyperboloid of revolution is both double-ruled and warped.

With reference to the circles in Fig. 3.37, if r approaches R, the surface approaches a right circular cylinder. If r approaches 0, the surface approaches a right circular cone. Therefore, from our present perspective, we may view the hyperboloid of revolution as a surface that is intermediate between these two extreme types of surfaces.

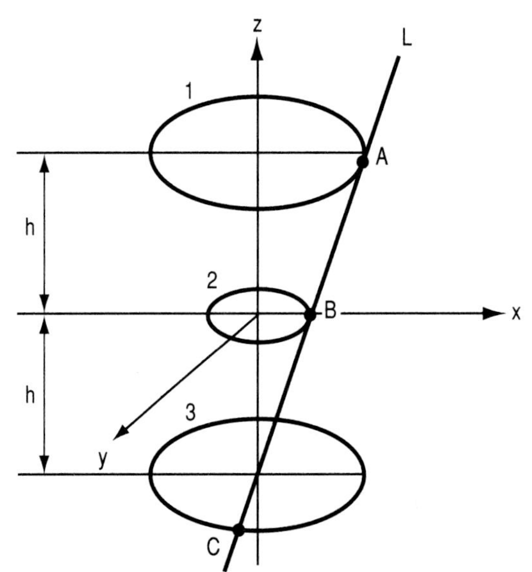

FIGURE 3.37 Method of generating a hyperboloid of revolution with a straight line.

3.4.8 The Hyperboloid of Two Sheets

Figure 3.38 shows a hyperboloid of two sheets in standard position, and its equation is

$$\frac{x^2}{a^2} - \frac{y^2}{b^2} - \frac{z^2}{c^2} = 1 \qquad (3.69)$$

This hyperboloid is also symmetric about all three coordinate planes, and its x intercepts are $\pm a$. The surface has the following sections: parallel to the xy plane, a hyperbola; parallel to the xz plane, a hyperbola; parallel to the yz plane, an ellipse.

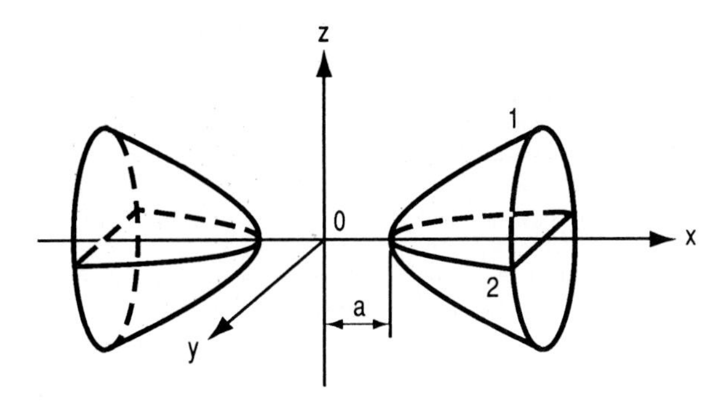

FIGURE 3.38 Hyperboloid of two sheets.

The hyperboloid of two sheets differs from that of one sheet in this respect: In Fig. 3.36, hyperbolas 1 and 2 have a common *conjugate* axis; in Fig. 3.38, the hyperbolas have a common *transverse* axis. (The terms are defined in Art. 3.2.6.)

If $b = c$, hyperbolas 1 and 2 have identical shape, the section parallel to the yz plane is a circle, and the hyperboloid is a surface of revolution.

3.4.9 The Elliptic Paraboloid

Figure 3.39 shows an elliptic paraboloid in standard position, and its equation is

$$\frac{x^2}{a^2} + \frac{y^2}{b^2} - z = 0 \tag{3.70}$$

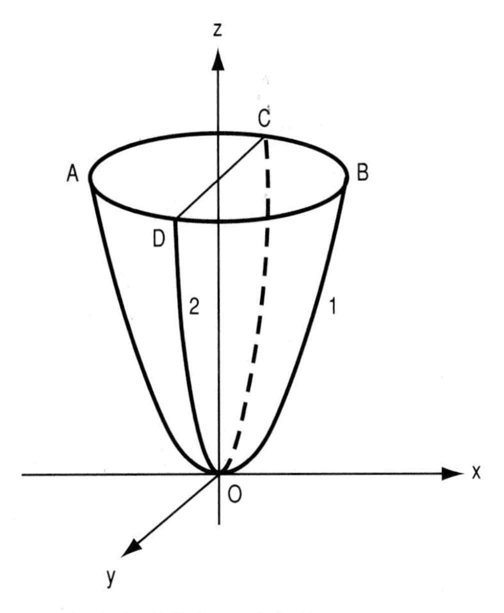

FIGURE 3.39 Elliptic paraboloid.

The surface is symmetric about the xz and yz planes, and it is tangent to the xy plane at the origin. The surface has the following sections: parallel to the xz plane, a parabola; parallel to the yz plane, a parabola; parallel to the xy plane, an ellipse.

We may conceive that the elliptic paraboloid is generated in this manner: We start with the parabola AOB in the xz plane and revolve it about the z axis through 180°. Simultaneously, we transform its shape in such manner that each point on this curve describes an elliptic path about the z axis. Thus, when AOB has been revolved through 90°, it has the shape COD.

If $a = b$, parabolas 1 and 2 have identical shape, the section parallel to the xy plane is a circle, and the paraboloid is a surface of revolution.

3.4.10 The Hyperbolic Paraboloid

The shape of the hyperbolic paraboloid can be visualized most effectively by conceiving a method of generating the surface. In Fig. 3.40, the parabola AOB lies in the xz plane and has the equation $a^2z = x^2$. Parabola CDE moves along the x axis in such manner that its plane remains parallel to the yz plane and its vertex remains on AOB. When the moving parabola lies in the yz plane, its equation is $b^2z = -y^2$. The surface generated by the moving parabola CDE is a hyperbolic paraboloid, and its equation is

$$\frac{x^2}{a^2} - \frac{y^2}{b^2} - z = 0 \tag{3.71}$$

The surface is symmetric about the xz and yz planes. Its sections are as follows: parallel to the xy plane, a hyperbola; parallel to the xz plane, a parabola; parallel to the yz plane, a parabola; on the xy plane, two straight lines through the origin. As Fig. 3.40 reveals, the surface opens upward on the xz plane and downward on each plane perpendicular thereto. Thus, the surface is saddle-shaped.

If $a = b$, the hyperbolic paraboloid is *rectangular*, and its equation may be written as $z = (c/2)(x^2 - y^2)$, where $c = 2/a^2$. We now rotate the x and y axes about

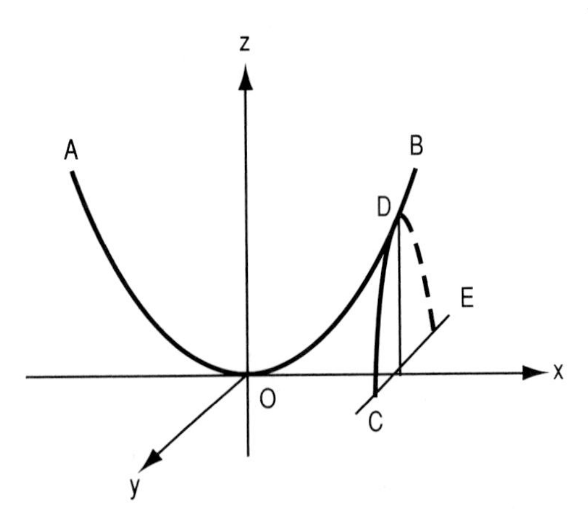

FIGURE 3.40 Method of generating a hyperbolic paraboloid.

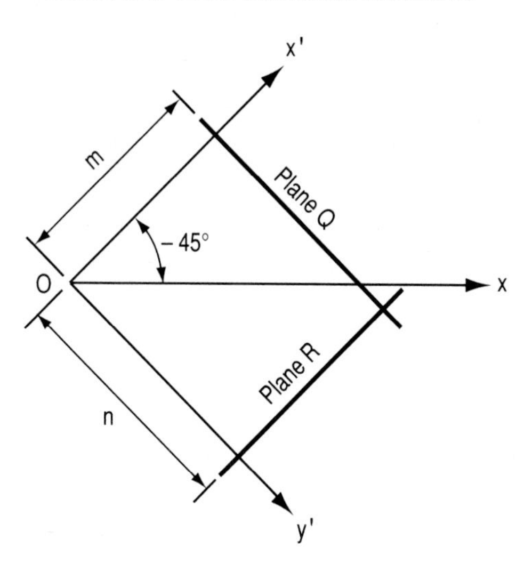

FIGURE 3.41 Normal view of xy plane.

the z axis through an angle of $-45°$, as shown in Fig. 3.41. Applying Eq. (3.15), we obtain the transformed equation

$$z = cx'y' \tag{3.71a}$$

In Fig. 3.41, we pass planes Q and R perpendicular to the x' and y' axes, respectively, and at the indicated locations, where m and n are arbitrary constants. The equation of the section of the surface is as follows: on Q, $z = cmy'$; on R, $z = cnx'$. Thus, the section is a straight line on both planes, and the surface is double-ruled. Since the slope of the section varies with its distance from the x' or y' axis, successive elements are nonparallel, and the surface is warped. We may conceive that the surface is generated by a straight line that moves in such manner that it remains parallel to the $y'z$ plane and it intersects the x' axis and the straight-line section on plane R.

In general, the rectangular hyperbolic paraboloid may be generated in this manner: In Fig. 3.42, we start with the rectangular block and draw the skew lines

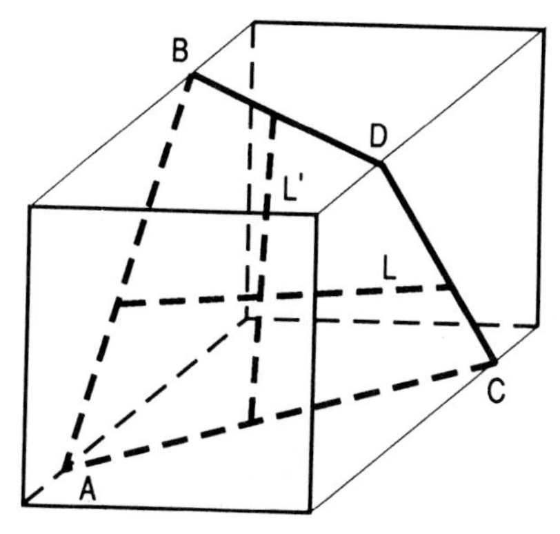

FIGURE 3.42 Method of generating a rectangular hyperbolic paraboloid with a straight line.

AB and CD in the left and right faces, respectively. Consider that a straight line L moves in such manner that it remains parallel to the top and bottom faces and intersects lines AB and CD. Line L generates a rectangular hyperbolic paraboloid, and AC and BD are elements of the surface. Now consider that a straight line L' moves in such manner that it remains parallel to the left and right faces and intersects lines AC and BD. Line L' generates the same surface as L. Because the rectangular hyperbolic paraboloid is relatively simple to construct and has considerable stiffness, it is used extensively for roofs, bridge portals, the wing walls of dams, and the bows of ships.

SECTION 4
PERMUTATIONS AND COMBINATIONS

4.1 PROPERTIES OF PERMUTATIONS AND COMBINATIONS

4.1.1 Law of Multiplication

The study of permutations and combinations rests on the following principle, which is known as the *law of multiplication*:

Theorem 4.1. Assume that n acts are to be performed in sequence. If the first act can be performed in m_1 alternative ways, the second act in m_2 alternative ways, . . . , the nth act in m_n alternative ways, the entire set of n acts can be performed in $m_1 m_2 \cdots m_n$ alternative ways.

In the subsequent material, we shall apply factorial numbers, which are defined in Art. 1.16.

4.1.2 Ordinary Permutations

An arrangement of a group of items or individuals in which the order or rank is of significance is called a *permutation*. The arrangement may contain the entire group of items or only part of the group. For example, the following are permutations of the first 3 letters of the alphabet taken 2 at a time: *ab*, *ba*, *ac*, *ca*, *bc*, *cb*.

EXAMPLE 4.1 How many permutations can be formed of the first 7 letters of the alphabet taken 3 at a time?

SOLUTION The first position in the permutation can be assigned to any one of 7 letters; the second position can then be assigned to any one of 6 letters; the third position can then be assigned to any one of 5 letters. Therefore, by Theorem 4.1, the number of possible permutations is $7 \cdot 6 \cdot 5 = 210$.

Let ${}_nP_r$ denote the number of possible permutations of n items taken r at a time. By generalizing from Example 4.1, we arrive at this result:

$$ {}_nP_r = n(n-1)(n-2) \cdots (n-r+1) = \frac{n!}{(n-r)!} \tag{4.1} $$

In the special case where $r = n$, we have

$$_nP_n = n! \tag{4.1a}$$

4.1.3 Permutations with Indistinguishable Items

If a permutation contains several items that are indistinguishable from one another, a rearrangement of these items alone fails to produce a new permutation. Assume that we have formed all permutations of the 8 letters of the word *parabola*, taken all at a time, and let X denote the number of permutations in the set. If we make the 3 *a*'s distinguishable in some manner and then rearrange them, we transform each permutation in the original set to 3! permutations in a new set. Since the new set contains 8! permutations, it follows that $3!X = 8!$, and $X = 8!/3! = 6720$. Similarly, for the word *parallel*, $X = 8!/(2!3!) = 3360$; for the word *architecture*, $X = 12!/(2!2!2!2!)$.

In general, consider that a group contains n items, that j items are of class A and k items of class B, and that all items within a class are indistinguishable from one another. Let $_nP_{n(j,k)}$ denote the number of possible permutations of all n items. Then

$$_nP_{n(j,k)} = \frac{n!}{j!k!} \tag{4.2}$$

Now assume that the group consists exclusively of two classes of items. Then $k = n - j$, and Eq. (4.2) becomes

$$_nP_{n(j,n-j)} = \frac{n!}{j!(n-j)!} \tag{4.2a}$$

In the subsequent material, it is understood that all items in a permutation are distinguishable if nothing is stated to the contrary.

4.1.4 Circular Permutations

If items are arranged in a closed loop and we are concerned solely with their *relative* order, the items form a *circular permutation*. Let $_nCP_r$ denote the number of possible circular permutations of n items taken r at a time. Then

$$_nCP_r = \frac{n!}{r[(n-r)!]} \tag{4.3}$$

In the special case where $r = n$, we have

$$_nCP_n = (n-1)! \tag{4.3a}$$

EXAMPLE 4.2 A group consists of 6 individuals. Four individuals will be selected at random and seated at a round table. If it is only the relative order of these individuals that is significant, how many seating arrangements can be devised?

SOLUTION

$$\text{No. of arrangements} = \frac{6!}{4(2!)} = 90$$

4.1.5 Transpose Permutations

Assume that a permutation of the first n integers is transformed to a new permutation by interchanging the integers and their respective position numbers. The resulting permutation is the *transpose* of the first.

As an illustration, consider the permutation 43152. We record it in the folllowing manner:

Position number	1	2	3	4	5
Given permutation	4	3	1	5	2
Transpose permutation	3	5	2	1	4

The transpose permutation is constructed by placing integer 1 in the fourth position, integer 2 in the third position, integer 3 in the first position, integer 4 in the fifth position, and integer 5 in the second position. Thus, the transpose permutation is 35214.

If two objects bear an identical relationship to each other, this relationship is described as *reflexive*. For example, if line A is perpendicular to line B, the reverse is also true, and therefore perpendicularity is a reflexive relationship.

Consider that we start with a given permutation $P1$ and form its transpose $P1^{\mathrm{T}}$. If we now form the transpose of $P1^{\mathrm{T}}$, we revert to $P1$. Thus, the transpose relationship is reflexive.

4.1.6 Standard and Aberrant Permutations

Consider that the entire set of permutations of n items taken all at a time has been formed. One permutation in this set may be regarded as the *standard permutation*. The position that an item occupies in the standard permutation is its *standard position*. An *aberrant permutation* is one in which all items are in nonstandard position. For example, if *abcde* is the standard permutation of the first 5 letters of the alphabet, then *caebd* is an aberrant permutation but *caedb* is not.

Let A_n denote the number of possible aberrant permutations of n items. The value of A_n can be obtained recursively by applying either of the following equations:

$$A_n = nA_{n-1} + (-1)^n \tag{4.4a}$$

$$A_n = (n-1)(A_{n-2} + A_{n-1}) \tag{4.4b}$$

Thus, starting with the known value $A_1 = 0$ and applying Eq. (4.4*a*), we obtain $A_2 = 1$, $A_3 = 2$, $A_4 = 9$, $A_5 = 44$, $A_6 = 265$, $A_7 = 1854$, $A_8 = 14{,}833$, etc.

The following is an independent equation for A_n:

$$A_n = \frac{n!}{2!} - \frac{n!}{3!} + \frac{n!}{4!} - \frac{n!}{5!} + \cdots + (-1)^n \frac{n!}{n!} \tag{4.4c}$$

EXAMPLE 4.3 A building contains 5 rooms, and a unique set of environmental conditions is maintained in each one. In an experiment in horticulture, a specimen of a particular type of plant is placed in each room. At the end of 1 week, each specimen will be transferred to a different room in the building. In how many ways can the specimens be rearranged?

SOLUTION The present arrangement of the specimens by room may be regarded as their standard permutation, and each new arrangement is an aberrant permutation. The number of possible new arrangements is $A_5 = 44$.

4.1.7 Combinations

A grouping of items in which the order or rank is of no significance, or in which the order or rank is predetermined, is termed a *combination*. Thus, in forming a combination, we are concerned solely with the *identity* of the items selected.

Let ${}_nC_r$ denote the number of possible combinations of n items taken r at a time. By reasoning similar to that in Art. 4.1.3, we obtain $(r!)_nC_r = {}_nP_r$, and therefore

$$ {}_nC_r = \frac{n!}{r!(n-r)!} = \binom{n}{r} \tag{4.5} $$

(Refer to Art. 1.22 for the notation.) Thus, Eqs. (1.26) to (1.32), inclusive, may be restated in terms of ${}_nC_r$. For mathematical consistency, we set ${}_nC_0 = 1$. From Eq. (4.2*a*), we have

$$ {}_nC_r = {}_nP_{n(r,\, n-r)} \tag{4.6} $$

EXAMPLE 4.4 Five-place numbers are to be formed by using the digits 1 to 8, inclusive, with the digits arranged in descending order of magnitude. For example, 76431 is acceptable but 76413 is unacceptable. How many such numbers can be formed?

SOLUTION Since the order of the digits is predetermined, each number represents a *combination* of 8 digits taken 5 at a time. The quantity of such numbers is ${}_8C_5 = 8!/(5!3!) = 56$.

4.1.8 Composite Permutations and Combinations

Where a permutation or combination is built up of smaller permutations and combinations, the number of possibilities is found by applying Theorem 4.1.

EXAMPLE 4.5 Box A contains 5 objects and box B contains 6 objects. All 11 objects are unique. A group of 4 objects will be formed by drawing any 2 objects from box A and any 2 objects from box B, and these 4 objects will be placed in a row. How many arrangements are possible if (*a*) the objects can be mingled; (*b*) the objects must be segregated according to the box from which they were drawn?

SOLUTION *Part a:* The number of groups that can be formed is ${}_5C_2 \times {}_6C_2 = 10 \times 15 = 150$. The number of permutations corresponding to each group is ${}_4P_4 = 24$. Therefore, the number of possible arrangements is $150 \times 24 = 3600$.

Part b: The number of possible arrangements of the objects from box A and box B, respectively, is ${}_5P_2 = 20$ and ${}_6P_2 = 30$. Since the box A objects can either precede or follow the box B objects, the number of possible arrangements is $2 \times 20 \times 30 = 1200$.

4.1.9 Permutations and Combinations Subject to Restrictions

Where restrictions are imposed on the manner in which a permutation or combination may be formed, the number of acceptable permutations or combinations can be found by simple analysis.

EXAMPLE 4.6 Six individuals are to be seated at a round table, but Adams and Barnes cannot be seated alongside each other. If we are concerned solely with the relative order of these individuals, how many seating arrangements can be devised?

SOLUTION *Method 1:* Consider that we first assign a seat to Adams, then to Barnes, and then to the remaining 4 individuals. After a seat is assigned to Adams, we can assign any one of 3 seats to Barnes. The remaining seats can then be assigned in ${}_4P_4 = 24$ different ways. Therefore, by Theorem 4.1, the number of acceptable seating arrangements is $3 \times 24 = 72$.

Method 2: In the absence of any restrictions, the number of possible seating arrangements would be ${}_6CP_6 = 5! = 120$. We can form an unacceptable arrangement by seating Barnes to the right or left of Adams. The 4 remaining individuals can then be seated in 24 different ways. Therefore, the number of unacceptable arrangements is $2 \times 24 = 48$, and the number of acceptable arrangements is $120 - 48 = 72$.

4.1.10 Inversions and Character of a Permutation

In our present discussion, we shall deal with a permutation of n items taken all at a time. In Art. 4.1.6, we defined a standard permutation, and for illustrative purposes we shall take *abcdefg* as the standard permutation of the first 7 letters of the alphabet.

If the relative order of two items in a given permutation is contrary to that in the standard permutation, an *inversion* exists. For example, the permutation *cadbfge* contains 5 inversions, for these reasons: c precedes a and b; d precedes b; f and g both precede e. A permutation is designated as *even* or *odd*, respectively, according to whether it contains an even or odd number of inversions. The state of being even or odd is termed the *character* of the permutation. The following principles pertain to the number of inversions and character of a permutation:

Theorem 4.2. The maximum number of inversions that a permutation can contain is $n(n-1)/2$.

This quantity is the number of pairs of items that can be formed. The number of inversions is maximum if each pair of items is in reverse order. Thus, the permutation *gfedcba* contains 21 inversions, and that number is maximum when $n = 7$.

Theorem 4.3. If two items in a permutation are interchanged, the resulting permutation has a character contrary to that of the original one.

For example, the permutation *cadbfge* has 5 inversions, and it is odd. If we interchange d and g, we obtain *cagbfde*. The second permutation has 8 inversions, and it is even.

Theorem 4.4. If all permutations of a group of items are formed, the number of even permutations equals the number of odd permutations.

This principle stems from Theorem 4.3 because every even permutation can be transformed to an odd permutation, and vice versa.

We shall now apply the following notation with reference to a given item:

Let $i =$ its position number in the given permutation
$h =$ its position number in the standard permutation
$S = i + h$

For example, consider the permutation *fcdbgae*. With reference to the letter c, $i = 2$, $h = 3$, and $S = 2 + 3 = 5$.

Theorem 4.5. If a given item in a permutation is deleted, the character of the resulting permutation is the same as or contrary to that of the original permutation according to whether the S value of the deleted item is even or odd, respectively.

For example, the permutation *cadbfge* is odd. With reference to b, $i = 4$, $h = 2$, and $S = 6$. If we delete b, the reduced permutation is *cadfge*, and it is also odd.

Theorem 4.6. Transpose permutations contain the same number of inversions. Consequently, they have the same character.

As an illustration, let 12345 be the standard permutation of the first 5 digits. Consider the transpose permutations 43152 and 35214 referred to in Art. 4.1.5. They both contain 6 inversions.

If the order of the items in a permutation is inverted, the resulting permutation is the *converse* of the given one. For example, the converse of *bfcegad* is *dagecfb*.

Theorem 4.7. The total number of inversions in a pair of converse permutations is $n(n-1)/2$.

This principle stems from the fact that each pair of items forms an inversion in one and only one of the two permutations. The number of pairs of items is the specified quantity.

Theorem 4.8. If all permutations of a group of items are formed, the number of permutations that contain r inversions equals the number of permutations that contain $n(n-1)/2 - r$ inversions.

This principle stems from Theorem 4.7.

Consider that the entire set of permutations of n items taken all at a time has been formed, and let ${}_nI_r$ denote the number of permutations that contain r inversions. Starting with ${}_1I_0 = 1$, we can obtain all values of ${}_nI_r$ recursively by applying the following equations: If $r < n$,

$$ {}_nI_r = {}_nI_{r-1} + {}_{n-1}I_r \qquad (4.7a)$$

If $r \geq n$,

$$ {}_nI_r = {}_nI_{r-1} + {}_{n-1}I_r - {}_{n-1}I_{r-n} \qquad (4.7b)$$

Table 4.1 presents values of ${}_nI_r$ where n ranges from 1 to 6. When r becomes

TABLE 4.1 Values of $_nI_r$

Number of inversions, r	Number of items, n					
	1	2	3	4	5	6
0	1	1	1	1	1	1
1		1	2	3	4	5
2			2	5	9	14
3			1	6	15	29
4				5	20	49
5				3	22	71
6				1	20	90
7					15	101
8					9	101
9					4	90
10					1	71
11						49
12						29
13						14
14						5
15						1
Total	1	2	6	24	120	720

greater than $n(n-1)/4$, we may apply Theorem 4.8, thereby obviating the need for additional calculations. The total in each column is $_nP_n = n!$.

Again consider that the entire set of permutations of n items taken all at a time has been formed, and let V_n denote the total number of inversions in the set. Then

$$V_n = \frac{n(n-1)}{4} n! \tag{4.8}$$

This equation yields the following recursive equation:

$$V_n = \frac{n^2}{n-2} V_{n-1} \tag{4.9}$$

where $n > 2$. Thus, starting with $V_2 = 1$, we obtain the following:

$$V_3 = \left(\frac{9}{1}\right)1 = 9 \qquad V_4 = \left(\frac{16}{2}\right)9 = 72$$

$$V_5 = \left(\frac{25}{3}\right)72 = 600 \qquad V_6 = \left(\frac{36}{4}\right)600 = 5400$$

4.1.11 Repeated Permutations and Combinations

Assume the following: We have a set of items; these items can be classified into several categories on the basis of size, color, model, etc.; all items of the same category are considered to be indistinguishable. We now form a permutation or combination of items in this set, and there is no restriction on the frequency with which items of a given category can appear. A permutation or combination of this type is described as *repeated.*

Let n = number of categories in the set
r = number of items in the permutation or combination
${}_nRP_r$ = number of possible distinguishable repeated permutations
${}_nRC_r$ = number of possible distinguishable repeated combinations

We have the following:

$$ {}_nRP_r = n^r \tag{4.10} $$

$$ {}_nRC_r = {}_{n+r-1}C_r = \frac{(n+r-1)!}{(n-1)!r!} \tag{4.11} $$

Equation (4.10) stems from the fact that each of the r positions in the sequence can be filled in n different ways. Equation (4.11) can be established by mathematical induction.

EXAMPLE 4.7 A set consists of 7 red, 6 green, and 9 yellow spheres. Four spheres will be selected and placed in a row. How many color arrangements are possible?

SOLUTION Since there is no restriction on the frequency with which a given color can appear, each arrangement is a repeated permutation with $n = 3$ and $r = 4$. The number of possible color arrangements is ${}_3RP_4 = 3^4 = 81$.

EXAMPLE 4.8 Six-place numbers are to be formed by using the digits 2, 4, and 7, under this condition: A digit can be equal to but not greater than the digit to its right. How many such numbers can be formed?

SOLUTION Since each digit can appear repeatedly and the order of the digits is predetermined, each number is a repeated combination with $n = 3$ and $r = 6$. The quantity of such numbers is ${}_3RC_6 = 8!/(2!6!) = 28$. The numbers are recorded in Table 4.2 in ascending order of magnitude.

TABLE 4.2

222 222	222 447	227 777	444 444
222 224	222 477	244 444	444 447
222 227	222 777	244 447	444 477
222 244	224 444	244 477	444 777
222 247	224 447	244 777	447 777
222 277	224 477	247 777	477 777
222 444	224 777	277 777	777 777

4.2 METHOD OF GENERATING PERMUTATIONS

4.2.1 Introduction

In Arts. 4.1.6 and 4.1.10, we defined the terms *standard permutation* and *inversion*, respectively. We shall apply these terms in the subsequent material. It is understood that the items to be permuted are all distinguishable.

In many instances, it is necessary to generate all possible permutations of a group of items taken all at a time, to assign an identifying number to each permutation, and to determine the number of inversions in each permutation. We shall present a methodical procedure for performing these tasks, one that can be implemented readily by computer.

4.2.2 Expressing Integers in Factorial Form

Every positive integer can be expressed as the sum of factorial numbers. For example, the number 93 can be expressed in this manner:

$$93 = 3(4!) + 3(3!) + 1(2!) + 1(1!) + 0(0!)$$

The practice of expressing numbers in factorial form is subject to certain rules. Let n and k denote positive integers such that $k < n!$ The expression for k in terms of n has the following form:

$$k = c_1(n-1)! + c_2(n-2)! + \cdots + c_j(n-j)! + \cdots + c_{n-1}(1!) + c_n(0!) \tag{4.12}$$

The rules are as follows: The coefficients $c_1, c_2, \ldots, c_n$ are positive integers or 0; $0 \le c_j \le n-j$; all n terms appearing in Eq. (4.12) must be included in the expression, even if some coefficients are 0.

Taken in the order indicated, the coefficients in Eq. (4.12) constitute the *factorial progression* of k. If the right side of Eq. (4.12) is divided by $(n-1)!$, the quotient is c_1; if the remainder is then divided by $(n-2)!$, the quotient is c_2; etc. Consequently, the factorial progression of a given number can be constructed by successive division.

EXAMPLE 4.9 Construct the factorial progression of 549, using $n = 7$.

SOLUTION The results obtained in the successive divisions are recorded in Table 4.3. The factorial progression appears in the third column, and it is 0, 4, 2, 3, 1, 1, 0. Therefore, $549 = 0(6!) + 4(5!) + 2(4!) + 3(3!) + 1(2!) + 1(1!) + 0(0!)$.

TABLE 4.3 Constructing Factorial Progression of 549

Step	Division	Quotient	Remainder
1	$\frac{549}{6!} = \frac{549}{720}$	0	549
2	$\frac{549}{5!} = \frac{549}{120}$	4	69
3	$\frac{69}{4!} = \frac{69}{24}$	2	21
4	$\frac{21}{3!} = \frac{21}{6}$	3	3
5	$\frac{3}{2!} = \frac{3}{2}$	1	1
6	$\frac{1}{1!} = \frac{1}{1}$	1	0
7	$\frac{0}{0!} = \frac{0}{1}$	0	0

4.2.3 Formation of a Permutation by Use of a Sequencing Progression

Consider that a permutation of n items taken all at a time is formed by transferring items from the standard permutation to the new permutation. At each stage, the items that are awaiting transfer constitute a *remnant permutation*. We shall formulate a sequence in which items are to be transferred.

Let $s_1, s_2, \ldots, s_n$ denote n positive integers having the property $1 \leq s_j \leq n - j + 1$. Taken in the order indicated, these integers constitute a *sequencing progression*. The number of possible sequencing progressions is $n!$, and the number of possible permutations of n items is also $n!$. Therefore, each progression can be associated with a permutation, and the progression can be applied as the sequence in which items are transferred to the new permutation. The rule is as follows: The jth item to be transferred in forming the new permutation is the one that occupies the s_jth position in the remnant permutation (or in the standard permutation if $j = 1$).

EXAMPLE 4.10 Form a permutation of the first 6 letters of the alphabet by taking *abcdef* as the standard permutation and applying the sequencing progression 2, 5, 3, 3, 2, 1.

SOLUTION Setting $n = 6$, we observe that each number in this progression falls within its allowable range. The work is performed in Table 4.4. We record the standard permutation in the first column, and then we record the sequencing progression in vertical formation in the second column. In step 1, we transfer b from the standard permutation to the new permutation because it occupies the second position in the former. A remnant permutation *acdef* now exists. In step 2, we transfer f to the new permutation because it occupies the fifth position in the remnant permutation. A remnant permutation *acde* now exists. Continuing in this manner, we obtain the results shown in Table 4.4, and the new permutation is *bfdeca*.

A comparison of c_j in the factorial progression and s_j in the sequencing progression reveals that the boundary values of s_j exceed those of c_j by 1.

TABLE 4.4 Formation of Permutation by Sequencing Progression 2, 5, 3, 3, 2, 1

Standard or remnant permutation	Number in sequencing progression	Item to be transferred
abcdef	2	*b*
a cdef	5	*f*
a cde	3	*d*
a c e	3	*e*
a c	2	*c*
a	1	*a*

4.2.4 Identifying Number of a Permutation

It now becomes a simple matter to assign an identifying number to each permutation in a set. However, we shall reverse the order by starting with a given identifying number and then forming the corresponding permutation. Let m denote the number of a permutation, where $1 \le m \le n!$. The steps are as follows:

1. Set $k = m - 1$.
2. Construct the factorial progression of k in terms of n.
3. Increase each number in the factorial progression by 1 to transform this to a sequencing progression.
4. Form the permutation by applying this sequencing progression.

EXAMPLE 4.11 Form the 214th permutation of the first 6 letters of the alphabet, taking *abcdef* as the standard permutation.

SOLUTION

$$n = 6 \qquad m = 214 \qquad k = m - 1 = 213$$

Proceeding as in Table 4.3, we find that the factorial progression of 213 with $n = 6$ is 1, 3, 3, 1, 1, 0; that is, $213 = 1(5!) + 3(4!) + 3(3!) + 1(2!) + 1(1!) + 0(0!)$. The corresponding sequencing progression is 2, 4, 4, 2, 2, 1. Proceeding as in Table 4.4, we find that the 214th permutation is *befcda*.

To generate the complete set of permutations of n items, we simply allow m to range from 1 to $n!$. Since there is a unique sequencing progression associated with each permutation, there is also a unique factorial progression associated with each permutation.

4.2.5 Number of Inversions in a Permutation

Where permutations have been generated by the formula we have presented, the number of inversions in each permutation can be found instantly by applying the following principle:

Theorem 4.9. The number of inversions in a permutation equals the sum of the terms in its associated factorial progression.

For example, the permutation *befcda* in Example 4.11 has the factorial progression 1, 3, 3, 1, 1, 0. The sum of these terms is 9, and that is the number of inversions in the permutation.

Theorem 4.9 stems from the fact that, when we transfer the jth item in the remnant permutation to the new permutation, we create $j - 1$ inversions in the latter. This theorem provides an alternative method of proving Theorems 4.2 and 4.8.

SECTION 5

MATRIX ALGEBRA

5.1 BASIC CONCEPTS

5.1.1 Definitions and Notation

A *matrix* is a rectangular array of numbers (or of letters that represent numbers). The array is enclosed in brackets to indicate its extent, and each matrix is assigned an uppercase boldface letter for identification.

The numbers that compose the matrix are termed its *elements* or *members*. Elements that lie on a horizontal line constitute a *row*, and those that lie on a vertical line constitute a *column*. The rows and columns are assigned identifying numbers by the following convention: Number the rows from top to bottom; number the columns from left to right.

The *size* of a matrix is referred to as its *order* or *dimension*, and it is expressed by specifying the number of rows and the number of columns composing the matrix, in that sequence. Thus, a matrix having m rows and n columns is said to be of order $m \times n$, or it is described as an $m \times n$ matrix.

To illustrate the foregoing definitions, consider the following matrix:

$$\mathbf{A} = \begin{bmatrix} 7 & 5 & 6 & 3 \\ -11 & 8 & -3 & 2 \\ 4 & 9 & 0 & -15 \end{bmatrix}$$

This matrix contains 3 rows and 4 columns; therefore, it is a 3×4 matrix. The third row of the matrix is $4, 9, 0, -15$; the second column is $5, 8, 9$.

Where the elements of a matrix are represented by algebraic symbols, the practice is as follows: Each element is assigned the lowercase italicized letter corresponding to the matrix label, and two subscripts are appended to specify the location of the element. The first subscript is the row number; the second is the column number. For example, the symbol b_{25} (read "*b* sub two five") identifies the element in matrix **B** that lies in the second row and fifth column. Thus, with reference to the foregoing matrix **A**, we have $a_{13} = 6$ and $a_{32} = 9$.

A matrix that consists of a single row is called a *row vector*, and one that consists of a single column is called a *column vector*. A matrix in which all elements are zero is termed a *null matrix*, and it is designated as **0**.

Consider that we have two sets of items, A and B. The item in A and the item in B that have the same identifying number are said to *correspond* to each other.

Two matrices are *equal* to each other if they are identical in all respects, that is, if they are of the same order and if corresponding elements in the two matrices are equal to each other. If matrices **A** and **B** are equal, we write **A** = **B**.

Consider that one or more rows or columns of a matrix **A** are deleted, and let **B** denote the matrix that remains. Then **B** is a *submatrix* of **A**. For example, let

$$\mathbf{A} = \begin{bmatrix} -3 & 2 & 9 & -12 \\ 11 & 7 & -4 & 1 \\ 5 & 8 & 6 & 3 \end{bmatrix}$$

If the third row and second and fourth columns of **A** are deleted, the submatrix is

$$\mathbf{B} = \begin{bmatrix} -3 & 9 \\ 11 & -4 \end{bmatrix}$$

A row or column of a matrix is described as *nonzero* if it contains at least one nonzero element.

5.1.2 Transpose and Negative Matrices

Consider that a given matrix **A** is transformed by converting its rows to columns and its columns to rows, without disturbing the relative order of the rows and columns. Thus, the ith row becomes the ith column, and the jth column becomes the jth row. The matrix that results is known as the *transpose* of **A**, and it is denoted by $\mathbf{A}^T$ or $\mathbf{A}'$. For example, if

$$\mathbf{A} = \begin{bmatrix} 3 & 9 \\ 2 & 6 \\ 4 & 1 \\ 8 & 5 \end{bmatrix} \quad \text{then} \quad \mathbf{A}^T = \begin{bmatrix} 3 & 2 & 4 & 8 \\ 9 & 6 & 1 & 5 \end{bmatrix}$$

Now consider that a given matrix **A** is transformed by changing the algebraic sign of each element. The matrix that results is the *negative* of **A**, and it is denoted by $-\mathbf{A}$.

In matrix algebra (as in real-number algebra), parentheses are applied to specify the sequence for performing operations. For example, the notation $(-\mathbf{A})^T$ constitutes the following instruction: Form the negative of **A**, and then form the transpose of the negative.

If the transpose of $\mathbf{A}^T$ is formed, the result is the original matrix **A**. Expressed symbolically,

$$(\mathbf{A}^T)^T = \mathbf{A} \tag{5.1}$$

Therefore, in accordance with the definition in Art. 4.1.5, the transpose relationship is reflexive.

Similarly, we have

$$-(-\mathbf{A}) = \mathbf{A} \tag{5.2}$$

and therefore the negative relationship is reflexive.

Consider that a given matrix **A** is transformed to its transpose and the latter is then transformed to its negative. The matrix that evolves is called the *negative* *transpose* of **A**, and it is denoted by $-\mathbf{A}^T$. For example, if

$$\mathbf{A} = \begin{bmatrix} 3 & 7 & -4 \\ 9 & -11 & 6 \end{bmatrix} \qquad \text{then} \qquad -\mathbf{A}^T = \begin{bmatrix} -3 & -9 \\ -7 & 11 \\ 4 & -6 \end{bmatrix}$$

Manifestly, the sequence of operations can be reversed. Thus,

$$-(\mathbf{A}^T) = (-\mathbf{A})^T = -\mathbf{A}^T \tag{5.3}$$

We also have

$$-(-\mathbf{A}^T)^T = \mathbf{A} \tag{5.4}$$

and therefore the negative-transpose relationship is reflexive.

5.1.3 Square Matrices

A matrix in which the number of rows equals the number of columns is termed a *square matrix*. Let n denote this number. The matrix is said to be of the nth order, or of order n.

The diagonal line extending from the upper-left corner to the lower-right corner of a square matrix is termed the *principal* or *main diagonal*. Let **A** denote a square matrix. The elements $a_{11}, a_{22}, \ldots, a_{nn}$ that lie on this diagonal are termed the *diagonal elements*. The sum of the diagonal elements is the *trace* of the matrix.

Several types of square matrices are highly significant, and therefore they have been assigned suitable names. The following matrices will serve as illustrations:

$$\mathbf{A} = \begin{bmatrix} 9 & 5 & 4 & 2 \\ 0 & 1 & 8 & 0 \\ 0 & 0 & -2 & 3 \\ 0 & 0 & 0 & 7 \end{bmatrix} \qquad \mathbf{B} = \begin{bmatrix} 9 & 0 & 0 & 0 \\ 0 & 4 & 0 & 0 \\ 0 & 0 & 8 & 0 \\ 0 & 0 & 0 & 7 \end{bmatrix}$$

Triangular Diagonal

$$\mathbf{C} = \begin{bmatrix} 6 & 0 & 0 \\ 0 & 6 & 0 \\ 0 & 0 & 6 \end{bmatrix} \qquad \mathbf{D} = \mathbf{I} = \begin{bmatrix} 1 & 0 & 0 \\ 0 & 1 & 0 \\ 0 & 0 & 1 \end{bmatrix}$$

Scalar Unit (identity)

$$\mathbf{E} = \begin{bmatrix} 2 & 6 & 3 \\ 6 & 9 & 8 \\ 3 & 8 & 4 \end{bmatrix} \qquad \mathbf{F} = \begin{bmatrix} 0 & -6 & 3 \\ 6 & 0 & -8 \\ -3 & 8 & 0 \end{bmatrix}$$

Symmetric Skew-symmetric

A matrix such as **A** in which all elements above or below the principal diagonal have zero value is a *triangular matrix*. Moreover, a triangular matrix is classified as *lower-triangular* or *upper-triangular*, respectively, according to whether the zero elements lie above or below the principal diagonal. Thus, **A** is an upper-triangular matrix. However, in a triangular matrix, some zero elements may be interspersed with the nonzero elements, as is true of **A**.

A matrix such as **B** in which all nondiagonal elements have zero value is a *diagonal matrix*. Thus, $b_{ij} = 0$ if $i \neq j$. A diagonal matrix such as **C** in which all diagonal elements are equal is a *scalar matrix*.

A scalar matrix such as **D** in which all diagonal elements are 1 is a *unit* or *identity matrix*, and it is denoted by **I**. The order of the matrix can be specified by appending a subscript. For example, $\mathbf{I}_8$ denotes the unit matrix of order 8. The unit matrix can be defined by means of the *Kronecker-delta symbol* δ_{ij}, which has these values:

$$\delta_{ij} \begin{cases} = 0 & \text{if } i \neq j \\ = 1 & \text{if } i = j \end{cases} \tag{5.5}$$

Thus, for the unit matrix, $a_{ij} = \delta_{ij}$.

A matrix such as **E** in which $e_{ij} = e_{ji}$ is termed a *symmetric matrix*. Corresponding rows and columns are identical, and therefore $\mathbf{E}^T = \mathbf{E}$.

A matrix such as **F** in which $f_{ij} = -f_{ji}$ is termed a *skew-symmetric* or *antisymmetric matrix*. The diagonal elements must perforce have zero value, and $-\mathbf{F}^T = \mathbf{F}$.

5.2 ALGEBRAIC OPERATIONS WITH MATRICES

5.2.1 Introduction

We shall define the arithmetic operations that may be applied to two matrices, or to a matrix and a real number. However, it is imperative to recognize at the outset that the arithmetic operations of matrix algebra and those of real-number algebra are wholly distinct, despite a similarity of nomenclature, because they apply to things that differ in kind. Consequently, a theorem or operation of real-number algebra does not necessarily have a counterpart in matrix algebra. For example, one theorem of real-number algebra states that the order of multiplication is immaterial; that is, $ab = ba$. There is no theorem corresponding to this in matrix algebra. Similarly, the operation of dividing one real number by another has no counterpart in matrix algebra because the division of one matrix by another is simply not defined.

5.2.2 Addition and Subtraction of Matrices

Let **A** and **B** denote two matrices of identical order. Consider that the corresponding elements of these matrices are added together and that the sums are placed in the corresponding positions in a third matrix **C**. This operation is called the *addition* of **A** and **B**, and **C** is termed their matrix *sum*. The operation is expressed symbolically as $\mathbf{C} = \mathbf{A} + \mathbf{B}$. Thus, the defining equation for addition is

$$c_{ij} = a_{ij} + b_{ij} \tag{5.6}$$

where i and j assume all possible values. This definition can be extended to include the addition of three or more matrices.

Analogously, matrix *subtraction* entails the subtraction of corresponding elements. Let $\mathbf{D} = \mathbf{A} - \mathbf{B}$. Then

$$d_{ij} = a_{ij} - b_{ij} \tag{5.7}$$

and **D** is called the *difference* between **A** and **B**.

To illustrate the foregoing definitions, again let $\mathbf{C}=\mathbf{A}+\mathbf{B}$ and $\mathbf{D}=\mathbf{A}-\mathbf{B}$, where

$$\mathbf{A}=\begin{bmatrix} 3 & 6 \\ 5 & 8 \\ -2 & 9 \end{bmatrix} \qquad \mathbf{B}=\begin{bmatrix} -6 & 1 \\ 0 & 9 \\ 8 & 3 \end{bmatrix}$$

We have the following:

$$\mathbf{C}=\begin{bmatrix} -3 & 7 \\ 5 & 17 \\ 6 & 12 \end{bmatrix} \qquad \mathbf{D}=\begin{bmatrix} 9 & 5 \\ 5 & -1 \\ -10 & 6 \end{bmatrix}$$

The following laws pertaining to matrix addition and subtraction follow directly from the definitions:

$$\mathbf{A}+\mathbf{0}=\mathbf{A} \tag{5.8}$$

$$\mathbf{A}+\mathbf{B}=\mathbf{B}+\mathbf{A} \tag{5.9}$$

$$\mathbf{A}+(\mathbf{B}+\mathbf{C})=(\mathbf{A}+\mathbf{B})+\mathbf{C} \tag{5.10}$$

$$(\mathbf{A}+\mathbf{B})^{\mathrm{T}}=\mathbf{A}^{\mathrm{T}}+\mathbf{B}^{\mathrm{T}} \tag{5.11}$$

If $\mathbf{A}$ is a square matrix,

$$(\mathbf{A}+\mathbf{A}^{\mathrm{T}})^{\mathrm{T}}=\mathbf{A}+\mathbf{A}^{\mathrm{T}} \tag{5.12}$$

$$(\mathbf{A}-\mathbf{A}^{\mathrm{T}})^{\mathrm{T}}=-(\mathbf{A}-\mathbf{A}^{\mathrm{T}}) \tag{5.13}$$

Equations (5.9) and (5.10) state that matrix addition is *commutative* and *associative*, respectively. Equation (5.12) states in effect that the sum of a square matrix and its transpose is a symmetric matrix, since only a symmetric matrix is identical with its transpose. Similarly, Eq. (5.13) states in effect that the difference between a square matrix and its transpose is a skew-symmetric matrix.

5.2.3 Multiplication and Division of a Matrix by a Scalar

A real number is often referred to as a *scalar quantity*, or simply a *scalar*, to distinguish it clearly from a matrix.

Consider that all elements of matrix $\mathbf{A}$ are multiplied by a scalar α and that the products are placed in the corresponding positions in a second matrix $\mathbf{B}$. This operation is called the *multiplication* of $\mathbf{A}$ and α, and it is expressed symbolically as $\mathbf{B}=\alpha\mathbf{A}$. Thus, the defining equation for $\mathbf{B}$ is

$$b_{ij}=\alpha a_{ij} \tag{5.14}$$

The *division* of a matrix by a scalar is defined in an analogous manner.

To illustrate the foregoing definitions, let

$$\mathbf{A}=\begin{bmatrix} 8 & -2 & 14 \\ 6 & 16 & 12 \end{bmatrix}$$

Consider that **A** is multiplied by 3 to form **B** and that **A** is divided by 2 to form **C**. Then

$$\mathbf{B} = 3\mathbf{A} = \begin{bmatrix} 24 & -6 & 42 \\ 18 & 48 & 36 \end{bmatrix} \qquad \mathbf{C} = \frac{\mathbf{A}}{2} = \begin{bmatrix} 4 & -1 & 7 \\ 3 & 8 & 6 \end{bmatrix}$$

The following laws stem directly from Eq. (5.14):

$$(\alpha + \beta)\mathbf{A} = \alpha\mathbf{A} + \beta\mathbf{A} \tag{5.15}$$

$$\alpha(\mathbf{A} + \mathbf{B}) = \alpha\mathbf{A} + \alpha\mathbf{B} \tag{5.16}$$

where α and β denote scalars. These equations state that the multiplication of matrices by scalars is *distributive* with respect to addition.

If a unit matrix is multiplied by a scalar α, the result is a scalar matrix in which the diagonal elements have the value α. Therefore, a scalar matrix whose diagonal elements are α can be denoted by $\alpha\mathbf{I}$.

5.2.4 Multiplication of Matrices

Let **A** and **B** denote two matrices that have this characteristic: The number of columns in **A** equals the number of rows in **B**. These matrices are *conformable* with respect to one another, and they can be combined by an operation known as the *multiplication* of matrices. Let **C** denote the matrix formed by multiplying **A** by **B**. Matrix **C** is the *product* of **A** and **B**, and the operation is expressed symbolically as $\mathbf{C} = \mathbf{AB}$.

The formula for constructing **C** is given by this equation:

$$c_{ij} = a_{i1}b_{1j} + a_{i2}b_{2j} + a_{i3}b_{3j} + \cdots + a_{in}b_{nj} \tag{5.17}$$

where n denotes the number of columns in **A** and the number of rows in **B**, and i and j assume all possible values. Expressed verbally, c_{ij} is formed by taking the ith row of **A** and the jth column of **B**, multiplying corresponding elements, and summing the products.

As an illustration of matrix multiplication, consider the following matrices:

$$\mathbf{A} = \begin{bmatrix} 4 & 1 & 9 \\ 6 & 2 & 8 \\ 7 & 3 & 5 \\ 11 & 10 & 12 \end{bmatrix} \qquad \mathbf{B} = \begin{bmatrix} 2 & 9 \\ 5 & 12 \\ 8 & 10 \end{bmatrix}$$

These matrices are conformable, and $n = 3$. By allowing i and j in Eq. (5.17) to vary from 1 to 4 and from 1 to 2, respectively, we obtain the following results:

$$c_{11} = 4 \times 2 + 1 \times 5 + 9 \times 8 = 85$$

$$c_{12} = 4 \times 9 + 1 \times 12 + 9 \times 10 = 138$$

$$c_{21} = 6 \times 2 + 2 \times 5 + 8 \times 8 = 86$$

$$c_{22} = 6 \times 9 + 2 \times 12 + 8 \times 10 = 158$$

$$c_{31} = 7 \times 2 + 3 \times 5 + 5 \times 8 = 69$$

$$c_{32} = 7 \times 9 + 3 \times 12 + 5 \times 10 = 149$$

$$c_{41} = 11 \times 2 + 10 \times 5 + 12 \times 8 = 168$$

$$c_{42} = 11 \times 9 + 10 \times 12 + 12 \times 10 = 339$$

The matrix product is

$$\mathbf{C} = \mathbf{AB} = \begin{bmatrix} 85 & 138 \\ 86 & 158 \\ 69 & 149 \\ 168 & 339 \end{bmatrix}$$

Several computer languages contain a function that yields the product of two matrices directly.

In general, if $\mathbf{C} = \mathbf{AB}$, matrix $\mathbf{C}$ has the same number of rows as $\mathbf{A}$ and the same number of columns as $\mathbf{B}$. Thus, if $\mathbf{A}$ is an $m \times n$ matrix and $\mathbf{B}$ is an $n \times p$ matrix, $\mathbf{C}$ is an $m \times p$ matrix. We shall indicate the order of each factor and of the product by this notation:

$$[m \times n][n \times p] = [m \times p] \tag{5.18}$$

In matrix multiplication, it is mandatory that the matrices to be multiplied be listed in the proper order, since the symbols **AB** and **BA** have entirely disparate meanings. To eliminate all possibility of confusion, it is said that, in forming the product **AB**, matrix **A** is *postmultiplied* by **B** and **B** is *premultiplied* by **A**. Moreover, **A** is called the *prefactor* and **B** the *postfactor* of the product **AB**. The definition of matrix multiplication may be extended to include the multiplication of three or more matrices.

EXAMPLE 5.1 If **A** is a 10×9 matrix, **C** has twice as many columns as rows, and **ABC** is a square matrix, what is the order of **B**?

SOLUTION Equation (5.18) discloses that the number of rows in **B** is 9. Let p denote the number of columns in **B**. For the product **AB**, we have $[10 \times 9][9 \times p] = [10 \times p]$. Now let r denote the number of rows in **C**. For the product **ABC**, we have $[10 \times p][r \times 2r] = [10 \times 2r]$. From Eq. (5.18), $r = p$. Since **ABC** is square, $2r = 10$ or $r = 5$ and $p = 5$. Thus, **B** is a 9×5 matrix.

EXAMPLE 5.2 If **A** is a 5×8 matrix, **B** has twice as many rows as columns, and **ABC** has three times as many columns as rows, what is the order of **C**?

SOLUTION Matrix **B** has 8 rows and 4 columns, and **AB** is a 5×4 matrix. Therefore, **C** has 4 rows. Let p denote the number of columns in **C**. Then **ABC** is a $5 \times p$ matrix, and $p = 3 \times 5 = 15$. Thus, **C** is a 4×15 matrix.

The laws pertaining to the multiplication of matrices are as follows:

$$\alpha(\mathbf{AB}) = (\alpha\mathbf{A})\mathbf{B} = \mathbf{A}(\alpha\mathbf{B}) \tag{5.19}$$

where α denotes a scalar

$$\mathbf{A}(\mathbf{BC}) = (\mathbf{AB})\mathbf{C} \tag{5.20}$$

$$(\mathbf{A}+\mathbf{B})\mathbf{C} = \mathbf{AC} + \mathbf{BC} \tag{5.21a}$$

$$\mathbf{C}(\mathbf{A}+\mathbf{B}) = \mathbf{CA} + \mathbf{CB} \tag{5.21b}$$

$$(\mathbf{ABC})^{\mathrm{T}} = \mathbf{C}^{\mathrm{T}}\mathbf{B}^{\mathrm{T}}\mathbf{A}^{\mathrm{T}} \tag{5.22}$$

$$(\mathbf{AA}^{\mathrm{T}})^{\mathrm{T}} = \mathbf{AA}^{\mathrm{T}} \tag{5.23}$$

Equation (5.20) states that matrix multiplication is *associative*, and Eq. (5.21) states that it is *distributive*. Equation (5.22) is a statement of the *reversal rule* for the transpose of a matrix product, and it can be extended to include the product of any number of matrices. Equation (5.23) states in effect that the product of a matrix and its transpose is a symmetric matrix, since only a symmetric matrix is identical with its transpose.

The multiplications **AB** and **BA** are both possible only if **A** and **B** are matrices of reverse order or if they are both square matrices of identical order. First assume that **A** is an $m \times n$ matrix and **B** is an $n \times m$ matrix. Then **AB** is a square matrix of order m, and **BA** is a square matrix of order n. Now assume that **A** and **B** are square matrices of identical order. In the general case, **AB** and **BA** differ. It follows that matrix multiplication is generally not commutative.

EXAMPLE 5.3 Form the products **AB** and **BA**, where

$$\mathbf{A} = \begin{bmatrix} 6 & -5 \\ 4 & 10 \end{bmatrix} \qquad \mathbf{B} = \begin{bmatrix} 3 & 7 \\ 8 & 5 \end{bmatrix}$$

SOLUTION The products are as follows:

$$\mathbf{AB} = \begin{bmatrix} -22 & 17 \\ 92 & 78 \end{bmatrix} \qquad \mathbf{BA} = \begin{bmatrix} 46 & 55 \\ 68 & 10 \end{bmatrix}$$

The two products are unequal.

As we shall soon discover, there are several special cases where matrix multiplication is commutative.

5.2.5 Multiplication of Special Matrices

It is instructive to establish the form of a matrix product where a factor is of a special type. We start with the following composite statement:

Theorem 5.1. When a matrix **A** is (premultiplied, postmultiplied) by a diagonal matrix **B**, the effect is to multiply the mth (row, column) of **A** by b_{mm}.

As an illustration, let

$$\mathbf{A}_1 = \begin{bmatrix} 4 & 9 & 1 \\ 8 & 2 & 6 \end{bmatrix} \qquad \mathbf{A}_2 = \begin{bmatrix} 5 & 2 \\ 7 & 1 \\ 3 & 4 \end{bmatrix}$$

$$\mathbf{B} = \begin{bmatrix} 8 & 0 & 0 \\ 0 & 5 & 0 \\ 0 & 0 & 3 \end{bmatrix}$$

The matrix products are as follows:

$$\mathbf{A}_1\mathbf{B} = \begin{bmatrix} 32 & 45 & 3 \\ 64 & 10 & 18 \end{bmatrix} \qquad \mathbf{B}\mathbf{A}_2 = \begin{bmatrix} 40 & 16 \\ 35 & 5 \\ 9 & 12 \end{bmatrix}$$

Now consider that **B** is a scalar matrix, and let β denote its diagonal elements. By Theorem 5.1, if a matrix **A** is multiplied by **B**, the entire matrix **A** is multiplied by the scalar β. Moreover, if **A** is a square matrix, we have

$$\mathbf{AB} = \mathbf{BA} = \beta\mathbf{A} \qquad (a)$$

Thus, matrix multiplication is commutative in this special case.

Now consider that **A** is a square matrix and **B** is the unit matrix. Equation (a) becomes

$$\mathbf{AI} = \mathbf{IA} = \mathbf{A} \qquad (5.24)$$

This equation reveals that the role played by the unit matrix in the multiplication of matrices is analogous to the role played by the number 1 in the multiplication of real numbers.

Finally, consider that both **A** and **B** are diagonal matrices. From Theorem 5.1, we deduce the following: Matrix multiplication is commutative in the present case, the matrix product is itself a diagonal matrix, and each diagonal element of the product equals the product of the corresponding elements of **A** and **B**.

The following matrices serve as an illustration:

$$\mathbf{A} = \begin{bmatrix} 9 & 0 & 0 \\ 0 & 5 & 0 \\ 0 & 0 & -2 \end{bmatrix} \qquad \mathbf{B} = \begin{bmatrix} 12 & 0 & 0 \\ 0 & 6 & 0 \\ 0 & 0 & 7 \end{bmatrix}$$

$$\mathbf{AB} = \mathbf{BA} = \begin{bmatrix} 108 & 0 & 0 \\ 0 & 30 & 0 \\ 0 & 0 & -14 \end{bmatrix}$$

5.2.6 Pauli Matrices

The following matrices, which are called *Pauli matrices*, are applied in the study of atomic physics:

$$\mathbf{A} = \begin{bmatrix} 1 & 0 \\ 0 & 1 \end{bmatrix} \qquad \mathbf{B} = \begin{bmatrix} 0 & 1 \\ -1 & 0 \end{bmatrix} \qquad \mathbf{C} = \begin{bmatrix} 0 & -1 \\ 1 & 0 \end{bmatrix}$$

$$\mathbf{D} = \begin{bmatrix} -1 & 0 \\ 0 & -1 \end{bmatrix} \qquad \mathbf{E} = \begin{bmatrix} i & 0 \\ 0 & -i \end{bmatrix} \qquad \mathbf{F} = \begin{bmatrix} -i & 0 \\ 0 & i \end{bmatrix}$$

$$\mathbf{G} = \begin{bmatrix} 0 & -i \\ -i & 0 \end{bmatrix} \qquad \mathbf{H} = \begin{bmatrix} 0 & i \\ i & 0 \end{bmatrix}$$

We shall consider that matrices **A** through **D** constitute *set 1*, and matrices **E** through **H** constitute *set 2*. Pauli matrices have the following properties:

1. The product of two Pauli matrices is itself a Pauli matrix. Therefore, Pauli matrices constitute a *group* with respect to multiplication.
2. Let **M** denote a set 1 matrix and let **N** and **P** denote set 2 matrices such that $\mathbf{MN} = \mathbf{P}$. Then $\mathbf{PM} = \mathbf{N}$. For example, $\mathbf{BE} = \mathbf{G}$, and $\mathbf{GB} = \mathbf{E}$. Similarly, $\mathbf{CH} = \mathbf{F}$, and $\mathbf{FC} = \mathbf{H}$.

5.2.7 Positive Powers of a Matrix

If a square matrix **A** is multiplied by itself, the product is a square matrix of the same order as **A**; it is denoted by $\mathbf{A}^2$. If $\mathbf{A}^2$ is postmultiplied by **A**, the product is denoted by $\mathbf{A}^3$; etc. Let n denote a positive integer. Taking $\mathbf{A}^1 = \mathbf{A}$ as a starting point, we may tentatively define $\mathbf{A}^n$ by the recursive equation

$$\mathbf{A}^n = \mathbf{A}^{n-1}\mathbf{A}$$

Matrix $\mathbf{A}^n$ is called a *power matrix* of **A**. Equation (5.20) discloses that $\mathbf{A}^n$ can also be formed by successive premultiplication by **A**. Therefore, the power matrix is more properly defined by the following composite equation:

$$\mathbf{A}^n = \mathbf{A}\mathbf{A}^{n-1} = \mathbf{A}^{n-1}\mathbf{A} \tag{5.25}$$

From the definition of a power matrix, it follows that

$$\mathbf{A}^m\mathbf{A}^n = \mathbf{A}^n\mathbf{A}^m = \mathbf{A}^{m+n} \tag{5.26}$$

where m and n are positive integers.

As an illustration, let

$$\mathbf{A} = \begin{bmatrix} 2 & -4 \\ -3 & 5 \end{bmatrix}$$

We obtain the following:

$$\mathbf{A}^2 = \begin{bmatrix} 16 & -28 \\ -21 & 37 \end{bmatrix} \qquad \mathbf{A}^3 = \begin{bmatrix} 116 & -204 \\ -153 & 269 \end{bmatrix}$$

$$\mathbf{A}^5 = \mathbf{A}^2\mathbf{A}^3 = \mathbf{A}^3\mathbf{A}^2 = \begin{bmatrix} 6{,}140 & -10{,}796 \\ -8{,}097 & 14{,}237 \end{bmatrix}$$

If Eq. (5.26) is extended to include the case where $m = 0$, it becomes $\mathbf{A}^0\mathbf{A}^n = \mathbf{A}^n$. However, $\mathbf{IA}^n = \mathbf{A}^n$, and it follows that

$$\mathbf{A}^0 = \mathbf{I} \tag{5.27}$$

The analogy between **I** and the number 1 is again apparent.

Equation (5.22) yields the following:

$$(\mathbf{A}^n)^{\mathrm{T}} = (\mathbf{A}^{\mathrm{T}})^n \tag{5.28}$$

For example, with reference to the foregoing matrix **A**,

$$\mathbf{A}^{\mathrm{T}} = \begin{bmatrix} 2 & -3 \\ -4 & 5 \end{bmatrix} \qquad (\mathbf{A}^{\mathrm{T}})^5 = \begin{bmatrix} 6{,}140 & -8{,}097 \\ -10{,}796 & 14{,}237 \end{bmatrix}$$

and the last matrix is the transpose of $\mathbf{A}^5$.

5.2.8 Expressing a Set of Equations in Matrix Form

Where it is necessary to present a set of equations rather than an individual equation, the presentation can often be made most effectively by using matrix multiplication. To illustrate this mode of presentation, consider that x, y, and z are functions of the independent variables a, b, and c, the relationships being as follows:

$$x = 8a + 3b - 2c \qquad y = 5a - 7b \qquad z = -a - 4b + 9c$$

These equations can be expressed in either of the following forms:

$$[x \quad y \quad z] = [a \quad b \quad c] \begin{bmatrix} 8 & 5 & -1 \\ 3 & -7 & -4 \\ -2 & 0 & 9 \end{bmatrix}$$

$$\begin{bmatrix} x \\ y \\ z \end{bmatrix} = \begin{bmatrix} 8 & 3 & -2 \\ 5 & -7 & 0 \\ -1 & -4 & 9 \end{bmatrix} \begin{bmatrix} a \\ b \\ c \end{bmatrix}$$

As another illustration, consider Eq. (3.14) for the transformation of coordinates, which we repeat for convenience:

$$x' = x \cos\theta + y \sin\theta$$

$$y' = -x \sin\theta + y \cos\theta$$

These equations can be expressed in this form:

$$\begin{bmatrix} x' \\ y' \end{bmatrix} = \begin{bmatrix} \cos\theta & \sin\theta \\ -\sin\theta & \cos\theta \end{bmatrix} \begin{bmatrix} x \\ y \end{bmatrix}$$

5.3 TRANSFORMATION OF MATRICES

5.3.1 Elementary Operations and Equivalent Matrices

Consider that a given matrix is transformed to a new matrix by operating upon it in some prescribed manner. There are three operations that are of particular significance, and they are termed *elementary operations*. Let h and k denote nonzero constants. The elementary operations and their symbolic designations are as follows:

	Operation	Designation
1*a*	Interchange the *m*th and *n*th rows.	R(m, n)
1*b*	Interchange the *m*th and *n*th columns.	C(m, n)
2*a*	Multiply each element in the *m*th row by h.	hRm
2*b*	Multiply each element in the *m*th column by h.	hCm
3*a*	Increase each element in the *m*th row by k times the corresponding element in the *n*th row.	R$m + k$Rn
3*b*	Increase each element in the *m*th column by k times the corresponding element in the *n*th column.	C$m + k$Cn

Where a series of elementary operations is to be performed upon a matrix, the operations must be performed *in the sequence specified.* Associated with every elementary operation is an *inverse operation* such that, if the two operations are performed in sequence, the matrix reverts to its original form. For example, the inverse of hRm is $(1/h)$Rm, and the inverse of R$m + k$Rn is R$m - k$Rn. We see that the inverse of an elementary operation is itself an elementary operation.

If a matrix **A** is transformed to a matrix **B** by a series of elementary operations, **B** is said to be *equivalent* to **A**. Moreover, if the operations affected only the rows or columns of **A**, matrix **B** is *row-equivalent* or *column-equivalent* to **A**, respectively.

If matrix **B** is derived from **A** by a series of elementary operations, **A** can be derived from **B** by replacing each operation with its inverse and then reversing the sequence of operations. Therefore, if **B** is equivalent to **A**, then **A** is equivalent to **B**, and matrix equivalence is a reflexive relationship. Figure 5.1 shows this relationship.

To illustrate the elementary operations and their symbolism, we shall apply the following matrix:

$$\mathbf{A} = \begin{bmatrix} 3 & 7 & 9 & 4 \\ 2 & -5 & 8 & 1 \\ 6 & 12 & 16 & 10 \end{bmatrix}$$

This matrix is to be transformed according to this formula: Triple each element in the first row; interchange the second and third rows; increase each element in the first column by twice the corresponding element in the second column. The first and second operations may be performed concurrently because they are independent of each other.

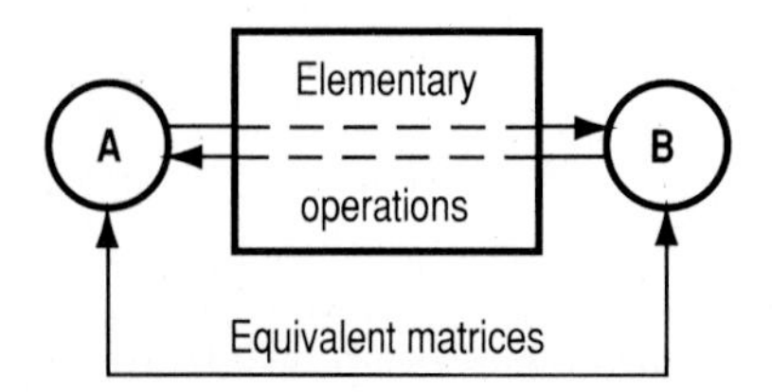

FIGURE 5.1

We shall record the matrices that arise from these operations in this manner: Each matrix will be placed after its predecessor, proceeding from left to right; the matrices will be numbered in the order in which they arise, the number being placed directly below the matrix; the operation by which a given matrix is derived from its predecessor will be recorded symbolically above the matrix.

Starting with the original matrix, the transformation is as follows:

$$\underset{①}{\begin{bmatrix} 3 & 7 & 9 & 4 \\ 2 & -5 & 8 & 1 \\ 6 & 12 & 16 & 10 \end{bmatrix}} \qquad \overset{\substack{3R1 \\ R(2,3)}}{\underset{②}{\begin{bmatrix} 9 & 21 & 27 & 12 \\ 6 & 12 & 16 & 10 \\ 2 & -5 & 8 & 1 \end{bmatrix}}}$$

$$\overset{C1+2C2}{\underset{③}{\begin{bmatrix} 51 & 21 & 27 & 12 \\ 30 & 12 & 16 & 10 \\ -8 & -5 & 8 & 1 \end{bmatrix}}}$$

These three matrices are equivalent to one another. Moreover, matrices 1 and 2 are row-equivalent, and matrices 2 and 3 are column-equivalent.

We have the following principle:

Theorem 5.2. If **A** and **B** are equivalent to each other, $\mathbf{A}^T$ and $\mathbf{B}^T$ are also equivalent to each other.

5.3.2 Reduction of Square Matrix to Simpler Form

It is frequently necessary to transform a given square matrix to an equivalent triangular matrix or diagonal matrix if possible. The given matrix is then said to be reduced to its triangular or diagonal form; alternatively, the given matrix is said to be *triangularized* or *diagonalized.*

It should be emphasized, however, that usually the purpose of making this transformation is not simply to identify the equivalent matrix but also to identify the operations that are needed to achieve this transformation. Consequently, the total solution to a problem in matrix transformation encompasses both the final matrix and a compilation of the steps leading to that matrix.

EXAMPLE 5.4 Transform the following matrix to an equivalent diagonal matrix, using row operations exclusively.

$$\begin{bmatrix} 18 & 12 & -16 \\ 9 & 0 & 4 \\ -54 & -18 & 16 \end{bmatrix}$$

SOLUTION The transformation consists of a series of cycles, in each of which the nondiagonal elements in a particular *column* are reduced to zero. We shall start with the first column and then proceed one column to the right in each subsequent cycle, in this manner:

Cycle 1
R2 − (1/2)R1
R3 + 3R1

$$\begin{bmatrix} 18 & 12 & -16 \\ 9 & 0 & 4 \\ -54 & -18 & 16 \end{bmatrix} \qquad \begin{bmatrix} 18 & 12 & -16 \\ 0 & -6 & 12 \\ 0 & 18 & -32 \end{bmatrix}$$

① ②

Cycle 2
R1 + 2R2
R3 + 3R2

Cycle 3
R1 − 2R3
R2 − 3R3

$$\begin{bmatrix} 18 & 0 & 8 \\ 0 & -6 & 12 \\ 0 & 0 & 4 \end{bmatrix} \qquad \begin{bmatrix} 18 & 0 & 0 \\ 0 & -6 & 0 \\ 0 & 0 & 4 \end{bmatrix}$$

③ ④

The diagonal matrix 4 is equivalent to the given matrix 1.

Alternatively, we may start with the last column and then proceed one column to the left in each subsequent cycle. The process culminates in the following matrix:

$$\begin{bmatrix} -6 & 0 & 0 \\ 0 & 4.5 & 0 \\ 0 & 0 & 16 \end{bmatrix}$$

In summary, assume that a square matrix is reduced to diagonal form through elementary row operations and that the reduction process starts with the first column and proceeds to the right. The ith cycle consists of reducing all nondiagonal elements in the ith column to zero, the formula being

$$\mathrm{R}j - \left(\frac{a_{ji}}{a_{ii}}\right)\mathrm{R}i$$

where j assumes all possible values except i. With respect to the ith cycle, the ith row is termed the *pivot row*, and element a_{ii} is termed the *pivot element*. The process of reducing the nondiagonal elements in a given column to zero is called *elimination*.

As Example 5.4 illustrates, the reduction of a square matrix to diagonal form does not yield a unique matrix; in general, there are multiple diagonal matrices that are equivalent to the given one. Since the *total* solution to a problem in matrix transformation consists of both the final matrix and the operations that yield that matrix, a change in the operations produces a change in the final matrix.

It is interesting to observe that the two alternative diagonal matrices in Example 5.4 have this common characteristic: The product of their diagonal elements is −432. The reason for this equality will become evident in Art. 5.5.4.

If a square matrix **A** can be transformed to an equivalent diagonal matrix **B**, it can also be reduced to the unit matrix by dividing each row of **B** by its diagonal element to reduce that element to unity. However, for solution by computer, it is preferable to modify the sequence of operations by reducing the pivot element to unity before elimination is undertaken. The operation of reducing the pivot element to unity is called *normalization*. Thus, each cycle in the transformation of the given matrix to the unit matrix consists of two steps: normalization and elimination, in that order. The formula is

$$\left(\frac{1}{a_{ii}}\right)\mathrm{R}i$$

$$\mathrm{R}j - a_{ji}\,\mathrm{R}i$$

EXAMPLE 5.5 Transform the following matrix to the unit matrix, using row operations exclusively.

$$\mathbf{A} = \begin{bmatrix} 6 & 30 & -66 \\ 4 & 23 & -53 \\ -1 & -3 & 7 \end{bmatrix}$$

SOLUTION

Cycle 1

Normalization (1/6)R1

$$\begin{bmatrix} 1 & 5 & -11 \\ 4 & 23 & -53 \\ -1 & -3 & 7 \end{bmatrix} \quad (2)$$

Elimination R2 − 4R1, R3 + R1

$$\begin{bmatrix} 1 & 5 & -11 \\ 0 & 3 & -9 \\ 0 & 2 & -4 \end{bmatrix} \quad (3)$$

Cycle 2

Normalization (1/3)R2

$$\begin{bmatrix} 1 & 5 & -11 \\ 0 & 1 & -3 \\ 0 & 2 & -4 \end{bmatrix} \quad (4)$$

Elimination R1 − 5R2, R3 − 2R2

$$\begin{bmatrix} 1 & 0 & 4 \\ 0 & 1 & -3 \\ 0 & 0 & 2 \end{bmatrix} \quad (5)$$

Cycle 3

Normalization (1/2)R3

$$\begin{bmatrix} 1 & 0 & 4 \\ 0 & 1 & -3 \\ 0 & 0 & 1 \end{bmatrix} \quad (6)$$

Elimination R1 − 4R3, R2 + 3R3

$$\begin{bmatrix} 1 & 0 & 0 \\ 0 & 1 & 0 \\ 0 & 0 & 1 \end{bmatrix} \quad (7)$$

5.3.3 Multiplication by Elementary Matrices

If a unit matrix is transformed to an equivalent matrix, the latter is called an *elementary matrix*, as shown in Fig. 5.2. Moreover, the equivalent matrix is termed *row-elementary* if the transformation involved solely row operations, and *column-elementary* if it involved solely column operations.

As an illustration, consider that $\mathbf{I}_4$ is transformed by the following formula: R(1, 2); −3R2; R3 + 2R4. The matrix that results is

$$\begin{bmatrix} 0 & 1 & 0 & 0 \\ -3 & 0 & 0 & 0 \\ 0 & 0 & 1 & 2 \\ 0 & 0 & 0 & 1 \end{bmatrix}$$

This matrix is row-elementary.

If a square matrix **A** can be transformed to the unit matrix by elementary operations, the unit matrix in turn can be transformed to **A** by elementary operations, and the following principle applies:

Theorem 5.3. A square matrix that can be transformed to the unit matrix by elementary operations is an elementary matrix.

EXAMPLE 5.6 Determine whether the following matrix is elementary:

$$\begin{bmatrix} 2 & -2 & -6 \\ -3 & 10 & 23 \\ -5 & 6 & 17 \end{bmatrix}$$

SOLUTION Proceeding as in Example 5.5, we attempt to reduce this matrix to $\mathbf{I}_3$ by row operations, and we arrive at the following matrix:

$$\begin{bmatrix} 1 & 0 & -1 \\ 0 & 1 & 2 \\ 0 & 0 & 0 \end{bmatrix}$$

Since the third row has vanished, no further reduction is possible, and therefore the given matrix fails the test. The given matrix is not elementary, and it cannot be derived by performing elementary operations upon $\mathbf{I}_3$.

The following principle pertains to multiplication by elementary matrices:

Theorem 5.4. Let $\mathbf{J}_r$ and $\mathbf{J}_c$ denote row-elementary and column-elementary matrices, respectively, and let $\mathbf{B}_r = \mathbf{J}_r\mathbf{A}_r$ and $\mathbf{B}_c = \mathbf{A}_c\mathbf{J}_c$, where $\mathbf{A}_r$ and $\mathbf{A}_c$ are

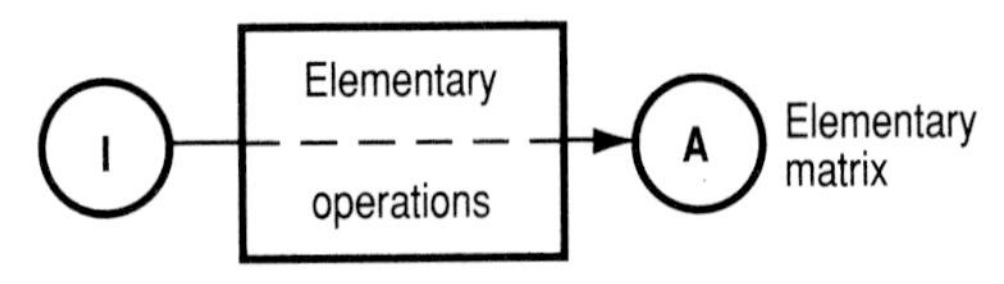

FIGURE 5.2

conformable with respect to the indicated multiplication. The operations that transformed **I** to $\mathbf{J}_r$ can also transform $\mathbf{A}_r$ to $\mathbf{B}_r$, and the operations that transformed **I** to $\mathbf{J}_c$ can also transform $\mathbf{A}_c$ to $\mathbf{B}_c$. Therefore, $\mathbf{B}_r$ and $\mathbf{B}_c$ are equivalent to $\mathbf{A}_r$ and $\mathbf{A}_c$, respectively.

Figure 5.3 exhibits these relationships in pictorial form.

EXAMPLE 5.7 Matrix **A** shown below is to be transformed to an equivalent matrix **B** by the following formula: R1 − 4R3, R(1, 2). Indicate how this transformation can be performed by matrix multiplication.

$$\mathbf{A} = \begin{bmatrix} 8 & 6 & 1 & 4 \\ 0 & 9 & 2 & 3 \\ 7 & 5 & 10 & -11 \end{bmatrix}$$

SOLUTION When $\mathbf{I}_3$ is transformed in the prescribed manner, the result is

$$\mathbf{J}_r = \begin{bmatrix} 0 & 1 & 0 \\ 1 & 0 & -4 \\ 0 & 0 & 1 \end{bmatrix}$$

Performing the multiplication, we obtain the following:

$$\mathbf{B} = \mathbf{J}_r\mathbf{A} = \begin{bmatrix} 0 & 9 & 2 & 3 \\ -20 & -14 & -39 & 48 \\ 7 & 5 & 10 & -11 \end{bmatrix}$$

Upon investigation, we find that the given formula will lead from **A** to **B**.

EXAMPLE 5.8 Matrix **A** shown below is to be transformed to an equivalent matrix **B** by the following formula: 2C3, C3 − C1. Indicate how this transformation can be performed by matrix multiplication.

$$\mathbf{A} = \begin{bmatrix} 6 & 0 & 9 & -5 \\ 1 & -4 & 3 & 8 \end{bmatrix}$$

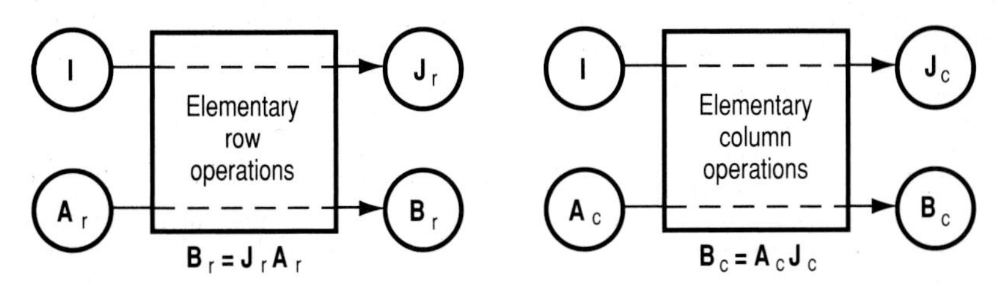

FIGURE 5.3

SOLUTION When $\mathbf{I}_4$ is transformed in the prescribed manner, the result is

$$\mathbf{J}_c = \begin{bmatrix} 1 & 0 & -1 & 0 \\ 0 & 1 & 0 & 0 \\ 0 & 0 & 2 & 0 \\ 0 & 0 & 0 & 1 \end{bmatrix}$$

$$\mathbf{B} = \mathbf{A}\mathbf{J}_c = \begin{bmatrix} 6 & 0 & 12 & -5 \\ 1 & -4 & 5 & 8 \end{bmatrix}$$

We again find that the given formula will lead from **A** to **B**.

5.3.4 Effect of Transforming a Matrix Factor

Where two matrices are to be multiplied, it is instructive to consider how a transformation of one matrix affects the matrix product. The following principle applies:

Theorem 5.5. When two matrices are multiplied, a row transformation of the prefactor matrix or a column transformation of the postfactor matrix produces an identical transformation of the matrix product.

As an illustration, consider the following multiplication:

$$\begin{bmatrix} 5 & 8 \\ -1 & 3 \\ 7 & 4 \end{bmatrix} \begin{bmatrix} 6 & 3 & 2 & 0 \\ 8 & 9 & -5 & 7 \end{bmatrix} = \begin{bmatrix} 94 & 87 & -30 & 56 \\ 18 & 24 & -17 & 21 \\ 74 & 57 & -6 & 28 \end{bmatrix}$$

Now consider that the prefactor is transformed by the formula R3 + 4R2 and the postfactor is transformed by the formula C2 − 2C1, C3 + C4. The multiplication is transformed to the following:

$$\begin{bmatrix} 5 & 8 \\ -1 & 3 \\ 3 & 16 \end{bmatrix} \begin{bmatrix} 6 & -9 & 2 & 0 \\ 8 & -7 & 2 & 7 \end{bmatrix} = \begin{bmatrix} 94 & -101 & 26 & 56 \\ 18 & -12 & 4 & 21 \\ 146 & -139 & 38 & 112 \end{bmatrix}$$

We find that the second matrix product can be derived from the first by the formula R3 + 4R2, C2 − 2C1, C3 + C4, and this relationship is in accord with Theorem 5.5.

5.4 COMPLEX MATRICES

5.4.1 Definitions and Notation

In Art. 1.8, we defined a complex number and its conjugate. If A denotes a complex number, A^* denotes its conjugate. The negative of the conjugate of a complex number is called the *negative conjugate* of that number. The given number and its negative conjugate differ solely with respect to the sign of the real number. For example, if $A = 8 + 3i$, then $-A^* = -(8 - 3i) = -8 + 3i$.

A matrix that has complex numbers as its elements is termed a *complex matrix*.

A complex matrix **A** may be viewed as the matrix sum $\mathbf{A}_1 + \mathbf{B}_1 i$, where $\mathbf{A}_1$ and $\mathbf{B}_1$ are real matrices of identical order as **A**. For example, let

$$\mathbf{A} = \begin{bmatrix} 6-2i & 7+5i \\ -2+i & 19 \\ -4i & 8-3i \end{bmatrix}$$

Decomposing this matrix, we have

$$\mathbf{A} = \begin{bmatrix} 6 & 7 \\ -2 & 19 \\ 0 & 8 \end{bmatrix} + i \begin{bmatrix} -2 & 5 \\ 1 & 0 \\ -4 & -3 \end{bmatrix}$$

5.4.2 Properties of Conjugate Matrices

If the elements of a complex matrix **A** are replaced with their conjugates, the resulting matrix is termed the *conjugate* of **A**, and it is denoted by $\mathbf{A}^*$. The following matrices are illustrative:

$$\mathbf{A} = \begin{bmatrix} 3+2i & 4-9i \\ 5i & 8 \end{bmatrix} \qquad \mathbf{A}^* = \begin{bmatrix} 3-2i & 4+9i \\ -5i & 8 \end{bmatrix}$$

Since $(\mathbf{A}^*)^* = \mathbf{A}$, the conjugate relationship is reflexive.

The following laws pertain to conjugate matrices:

$$(\mathbf{A}+\mathbf{B})^* = \mathbf{A}^* + \mathbf{B}^* \tag{5.29}$$

$$(\mathbf{AB})^* = \mathbf{A}^*\mathbf{B}^* \tag{5.30}$$

If the elements of a complex matrix **A** are replaced with their negative conjugates, the resulting matrix is termed the *negative conjugate* of **A**. Since the resulting matrix is also the negative of $\mathbf{A}^*$, it is denoted by $-\mathbf{A}^*$. The following matrices are illustrative:

$$\mathbf{A} = \begin{bmatrix} 4+7i & -6+5i \\ 9-2i & 8i \end{bmatrix} \qquad -\mathbf{A}^* = \begin{bmatrix} -4+7i & 6+5i \\ -9-2i & 8i \end{bmatrix}$$

The negative-conjugate relationship is reflexive because $-(-\mathbf{A}^*)^* = \mathbf{A}$.

5.4.3 Hermitian and Negative Hermitian Adjoints

Let **A** denote a complex matrix. If $\mathbf{A}^*$ is transposed, the resulting matrix is called the *Hermitian adjoint* of **A**, and it is denoted by $\mathbf{A}^{\mathrm{H}}$. Thus, by definition, $(\mathbf{A}^*)^{\mathrm{T}} = \mathbf{A}^{\mathrm{H}}$. A matrix and its Hermitian adjoint are of reverse order; that is, if **A** is an $m \times n$ matrix, $\mathbf{A}^{\mathrm{H}}$ is an $n \times m$ matrix. The following matrices are illustrative:

$$\mathbf{A} = \begin{bmatrix} 5+2i & -7i \\ -3+4i & 3-6i \\ 2-9i & 8 \end{bmatrix}$$

$$\mathbf{A}^* = \begin{bmatrix} 5-2i & 7i \\ -3-4i & 3+6i \\ 2+9i & 8 \end{bmatrix}$$

$$\mathbf{A}^{\mathrm{H}} = \begin{bmatrix} 5-2i & -3-4i & 2+9i \\ 7i & 3+6i & 8 \end{bmatrix}$$

The conjugate of the transpose of a matrix is equal to the Hermitian adjoint of the matrix. Expressed symbolically,

$$(\mathbf{A}^{\mathrm{T}})^* = (\mathbf{A}^*)^{\mathrm{T}} = \mathbf{A}^{\mathrm{H}} \tag{5.31}$$

Thus, if a matrix is to be subjected to conjugation and transposition in succession, the sequence in which these operations are performed is immaterial.

Figure 5.4 shows diagrammatically the manner in which the four matrices are related to one another. Since conjugation and transposition are reversible operations, we may start at any node in the loop and proceed in either direction. As this diagram discloses, $\mathbf{A}$ and $\mathbf{A}^{\mathrm{H}}$ are Hermitian adjoints of each other, and therefore the Hermitian-adjoint relationship is reflexive. The diagram also discloses that $\mathbf{A}^{\mathrm{T}}$ and $\mathbf{A}^*$ constitute a pair of Hermitian adjoints.

We have the following laws:

$$(\mathbf{A} + \mathbf{B})^{\mathrm{H}} = \mathbf{A}^{\mathrm{H}} + \mathbf{B}^{\mathrm{H}} \tag{5.32}$$

$$(\mathbf{AB})^{\mathrm{H}} = \mathbf{B}^{\mathrm{H}}\mathbf{A}^{\mathrm{H}} \tag{5.33}$$

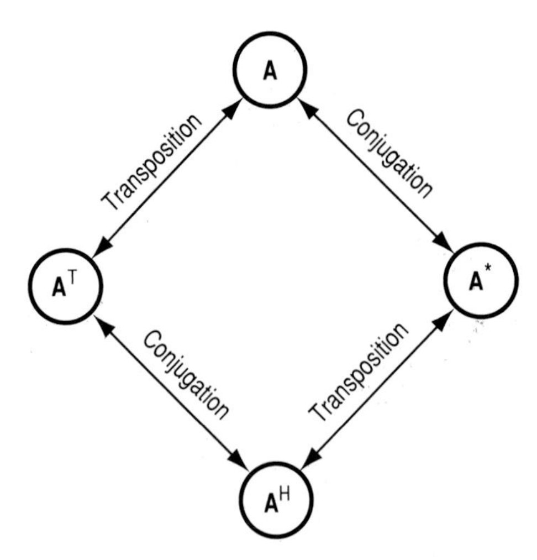

FIGURE 5.4

Again let **A** denote a complex matrix. If $-\mathbf{A}^*$ is transposed, the resulting matrix is called the *negative Hermitian adjoint* of **A**. Since the resulting matrix is also the negative of $\mathbf{A}^{\mathrm{H}}$, it is denoted by $-\mathbf{A}^{\mathrm{H}}$.

The conjugate of the negative transpose of a matrix is equal to the negative Hermitian adjoint of the matrix. Expressed symbolically,

$$(-\mathbf{A}^{\mathrm{T}})^* = -(\mathbf{A}^*)^{\mathrm{T}} = -\mathbf{A}^{\mathrm{H}} \tag{5.34}$$

This equation is the analog of Eq. (5.31), and Fig. 5.5 shows diagrammatically the manner in which the four matrices are related to one another.

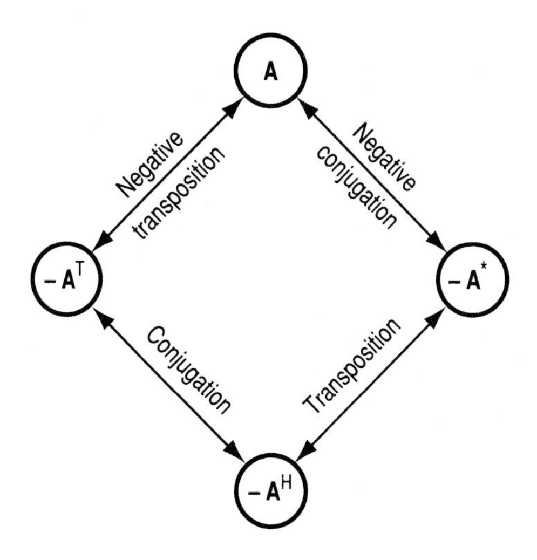

FIGURE 5.5

5.4.4 Hermitian and Skew-Hermitian Matrices

A square matrix **A** that is identical with its Hermitian adjoint is referred to as a *Hermitian matrix*. From the definition, it follows that

$$a_{pq} = a_{qp}^* \tag{5.35}$$

Thus, elements that are symmetrically located with respect to the principal diagonal are conjugates of each other, and the diagonal elements are perforce restricted to real numbers.

The following is a Hermitian matrix:

$$\begin{bmatrix} 6 & 4+2i & 7i \\ 4-2i & 3 & 8+3i \\ -7i & 8-3i & 15 \end{bmatrix}$$

The diagram in Fig. 5.4 assumes the form in Fig. 5.6*a* in the case of a Hermitian

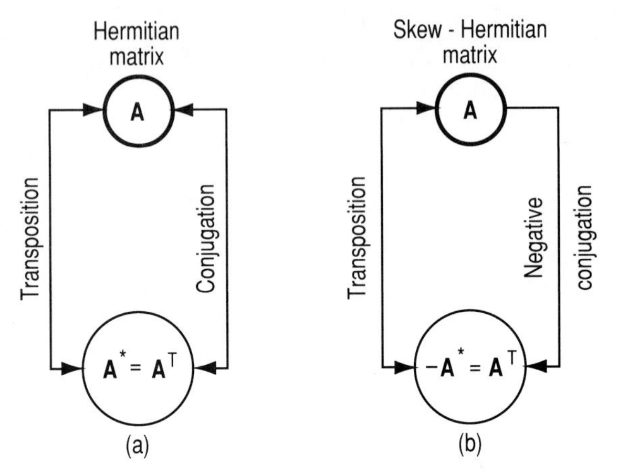

FIGURE 5.6 Relationships for (*a*) Hermitian matrix; (*b*) skew-Hermitian matrix.

matrix. Matrix **A** is the transpose of **A***, but it is also the transpose of $\mathbf{A}^T$. The following principle results:

Theorem 5.6. The transpose and the conjugate of a Hermitian matrix are identical.

A square matrix **A** that is identical with its negative Hermitian adjoint is referred to as a *skew-Hermitian matrix*. From the definition, it follows that

$$a_{pq} = -a^*_{qp} \tag{5.36}$$

Thus, elements that are symmetrically located with respect to the principal diagonal are negative conjugates of each other, and the diagonal elements are perforce restricted to imaginary numbers.

The following is a skew-Hermitian matrix:

$$\begin{bmatrix} 5i & 4-3i & 2+5i \\ -4-3i & 18i & -6+3i \\ -2+5i & 6+3i & 2i \end{bmatrix}$$

The diagram in Fig. 5.5 assumes the form in Fig. 5.6*b* in the case of a skew-Hermitian matrix. Matrix **A** is the transpose of $-\mathbf{A}^*$, but it is also the transpose of $\mathbf{A}^T$. The following principle results:

Theorem 5.7. The transpose and negative conjugate of a skew-Hermitian matrix are identical.

We also have the following principles:

Theorem 5.8. The sum of a square matrix and its Hermitian adjoint is a Hermitian matrix.

Theorem 5.9. The difference between a square matrix and its Hermitian adjoint (or the sum of a square matrix and its negative Hermitian adjoint) is a skew-Hermitian matrix.

Theorem 5.10. Every square matrix may be considered to be the sum of a Hermitian matrix and a skew-Hermitian matrix.

Theorem 5.11. The product of a matrix and its Hermitian adjoint is a Hermitian matrix, regardless of the order in which the matrices are multiplied.

Theorem 5.12. The product of a matrix and its negative Hermitian adjoint is a Hermitian matrix, regardless of the order in which the matrices are multiplied.

To illustrate Theorems 5.8 to 5.10, let

$$\mathbf{A} = \begin{bmatrix} 4+2i & 7-3i \\ 5+8i & -9i \end{bmatrix}$$

We now have the following:

$$\mathbf{A}^{\mathrm{H}} = \begin{bmatrix} 4-2i & 5-8i \\ 7+3i & 9i \end{bmatrix} \qquad \mathbf{A}+\mathbf{A}^{\mathrm{H}} = \begin{bmatrix} 8 & 12-11i \\ 12+11i & 0 \end{bmatrix}$$

$$\mathbf{A}-\mathbf{A}^{\mathrm{H}} = \begin{bmatrix} 4i & 2+5i \\ -2+5i & -18i \end{bmatrix}$$

Matrix $\mathbf{A}+\mathbf{A}^{\mathrm{H}}$ is Hermitian, and matrix $\mathbf{A}-\mathbf{A}^{\mathrm{H}}$ is skew-Hermitian. Dividing the last two matrices by 2 and summing the results, we obtain the following:

$$\begin{bmatrix} 4 & 6-5.5i \\ 6+5.5i & 0 \end{bmatrix} + \begin{bmatrix} 2i & 1+2.5i \\ -1+2.5i & -9i \end{bmatrix} = \begin{bmatrix} 4+2i & 7-3i \\ 5+8i & -9i \end{bmatrix} = \mathbf{A}$$

5.5 DETERMINANTS

Associated with every square matrix is a numerical value called its *determinant*. We shall now define the determinant and investigate its properties.

5.5.1 Definition and Notation

In Art. 4.1.10, we defined even and odd permutations, and we shall now have occasion to apply the definition. We shall consider 1, 2, 3, . . . , m to be the standard permutation of the first m integers. In the subsequent material, n denotes the order of a square matrix.

Consider that we form all possible combinations of the elements of a square matrix by applying this rule: Select one and only one element from each row and each column. Thus, each combination contains n elements. For example, if $\mathbf{A}$ is a square matrix of fourth order, the following is a satisfactory combination:

$a_{12}, a_{23}, a_{31}, a_{44}$. We now form the product of the elements in each combination, and we prefix an algebraic sign to each product in accordance with a rule that we shall state momentarily. We now form the algebraic sum of these products. This sum is the determinant of the matrix.

The rule for establishing the algebraic sign that is prefixed to the product is as follows: Arrange the elements in each combination in such manner that the first subscripts (row numbers) appear in ascending order of magnitude. The second subscripts (column numbers) form a permutation of the first n integers taken all at a time. By Eq. (4.1a), the number of combinations thus formed is $n!$ The algebraic sign that is prefixed to each product is plus or minus according to whether the permutation of the second subscripts is even or odd, respectively. For example, the expression $a_{12}a_{23}a_{31}a_{44}$ contains two inversions of the second subscripts (because both 2 and 3 precede 1). Therefore, the permutation of the second subscripts is even, and the algebraic sign prefixed to the product is plus.

The determinant of a matrix **A** is denoted by $|\mathbf{A}|$. Where reference is made to the determinant of a matrix rather than the matrix itself, the array of numbers is enclosed within vertical lines.

Symbolically, the determinant may be defined in this manner:

$$|\mathbf{A}| = \Sigma\,(-1)^v\, a_{1i}a_{2j}a_{3k} \cdots a_{nt} \tag{5.37}$$

where the second subscripts are all distinct, the number of terms is $n!$, and v denotes the number of inversions of the second subscripts.

For a second-order matrix **A**, the determinant is

$$|\mathbf{A}| = a_{11}a_{22} - a_{12}a_{21} \tag{5.38}$$

The determinant of a third-order matrix can readily be obtained by this device: Expand the matrix by repeating the second and third columns, placing them at the right. Form the products of all elements that lie on a diagonal that contains 3 elements. The algebraic sign prefixed to a product is plus or minus according to whether the diagonal line inclines downward to the right or downward to the left, respectively.

EXAMPLE 5.9 Compute the determinant of the following matrix:

$$\mathbf{A} = \begin{bmatrix} 12 & 6 & -9 \\ 3 & 8 & 15 \\ 4 & 11 & 5 \end{bmatrix}$$

SOLUTION The expanded matrix is

$$\begin{bmatrix} 12 & 6 & -9 & 12 & 6 \\ 3 & 8 & 15 & 3 & 8 \\ 4 & 11 & 5 & 4 & 11 \end{bmatrix}$$

Proceeding in the prescribed manner, we obtain

$$|\mathbf{A}| = 12 \times 8 \times 5 + 6 \times 15 \times 4 + (-9)3 \times 11 - (-9)8 \times 4$$
$$- 12 \times 15 \times 11 - 6 \times 3 \times 5 = -1239$$

When the expression for a determinant is written in accordance with Eq. (5.37), as it was in Example 5.9, the determinant is said to be *expanded*. When $n > 3$, the expansion of a determinant becomes cumbersome, and we shall subsequently present alternative methods of calculating the determinant.

If a determinant has zero value, its matrix is described as *singular*; if a determinant has a nonzero value, its matrix is described as *regular* or *nonsingular*.

5.5.2 Properties of Determinants

In the following material, n again denotes the order of the given matrix or determinant.

Theorem 5.13. The determinant of a matrix equals that of its transpose; that is, $|\mathbf{A}| = |\mathbf{A}^T|$.

Theorem 5.14. If all elements in a given row or column are zero, the matrix is singular.

Theorem 5.15. If a matrix is multiplied by a scalar α, its determinant is multiplied by α^n.

Theorem 5.16. If all elements in a given row or column are multiplied by a constant c, the determinant is multiplied by c.

Theorem 5.17. If two rows or two columns are interchanged, the determinant changes sign.

Theorem 5.18. If two rows or two columns are identical, the matrix is singular.

Theorem 5.19. If two rows or two columns are proportional, the matrix is singular.

As an illustration, consider the matrix

$$\begin{bmatrix} 8 & -7 & 6 \\ 24 & 11 & 18 \\ 20 & 16 & 15 \end{bmatrix}$$

Each element in the third column is three-fourths of the corresponding element in the first column, and the determinant of this matrix is zero.

Theorem 5.20. If two matrices have $n - 1$ identical rows or columns, the sum of their determinants equals the determinant of the matrix formed by retaining the identical rows or columns and adding the rows or columns that differ.

The following matrices illustrate this principle:

$$\mathbf{A} = \begin{bmatrix} -19 & 17 & 4 \\ 13 & 21 & 32 \\ -10 & 9 & 6 \end{bmatrix} \qquad \mathbf{B} = \begin{bmatrix} 12 & -6 & 15 \\ 13 & 21 & 32 \\ -10 & 9 & 6 \end{bmatrix}$$

The second and third rows are identical. Forming the third matrix in the prescribed manner, we obtain

$$\mathbf{C} = \begin{bmatrix} -7 & 11 & 19 \\ 13 & 21 & 32 \\ -10 & 9 & 6 \end{bmatrix}$$

Then $|\mathbf{A}| = -2380$, $|\mathbf{B}| = 5349$, and $|\mathbf{C}| = |\mathbf{A}| + |\mathbf{B}| = 2969$.

Theorem 5.21. If each element in one row or column is increased by the corresponding element in another row or column multiplied by a constant, the determinant remains constant.

For example, consider the matrix

$$\begin{bmatrix} -18 & 23 \\ 30 & 14 \end{bmatrix}$$

If each element in the second column is increased by the corresponding element in the first column multiplied by $-1/3$, the resulting matrix is

$$\begin{bmatrix} -18 & 29 \\ 30 & 4 \end{bmatrix}$$

The determinant of each matrix is -942.

Theorem 5.22. The determinant of the product of two matrices equals the product of the determinants of the factors. Expressed symbolically, if $\mathbf{C} = \mathbf{AB}$, then $|\mathbf{C}| = |\mathbf{A}| \times |\mathbf{B}|$.

As a corollary of Theorem 5.22, we have

$$|\mathbf{A}^n| = |\mathbf{A}|^n \tag{5.39}$$

Theorem 5.23. If two matrices are equivalent to each other, they are either both singular or both regular.

If a given row of a matrix consists of multiples of preceding rows, it is *secondary* with respect to those rows; otherwise, it is *primary*. A secondary and primary column are defined in an analogous manner. As an illustration, consider the matrix

$$\begin{bmatrix} 9 & 3 & -7 \\ 2 & -8 & 4 \\ 13 & -13 & 1 \end{bmatrix}$$

Each element in the third row equals the corresponding element in the first row plus twice the corresponding element in the second row. Therefore, the third row is secondary, but the first and second rows are primary. If all the rows or columns of a matrix are primary, these rows or columns are said to be *linearly independent* of one another.

Theorem 5.24. A square matrix is regular if and only if all its rows and all its columns are linearly independent.

5.5.3 Minors and Cofactors

Let **A** denote a square matrix of order n. Now consider that the ith row and jth column of **A** are deleted, and let $\mathbf{S}_{ij}$ denote the submatrix that remains. This is of $(n-1)$th order. The determinant of $\mathbf{S}_{ij}$ is termed the *minor* of element a_{ij}, and it is denoted by $|\mathbf{M}_{ij}|$. The *cofactor* of element a_{ij}, which is denoted by $|\mathbf{A}_{ij}|$, is defined in this manner:

$$|\mathbf{A}_{ij}| = (-1)^{i+j}|\mathbf{M}_{ij}| \tag{5.40}$$

Thus, the cofactor and minor are coincident if $i+j$ is even, and they are of opposite algebraic sign if $i+j$ is odd. Figure 5.7 is presented as an aid in memorizing these definitions.

The following principle affords a practical means of evaluating a determinant when n is relatively large:

Theorem 5.25. If all elements in a given row or column of a square matrix are multiplied by their cofactors, the sum of these products is equal to the determinant of the matrix.

This principle can be expressed symbolically in either of these alternative forms:

$$|\mathbf{A}| = a_{i1}|\mathbf{A}_{i1}| + a_{i2}|\mathbf{A}_{i2}| + \cdots + a_{in}|\mathbf{A}_{in}| \tag{5.41a}$$

$$|\mathbf{A}| = a_{1j}|\mathbf{A}_{1j}| + a_{2j}|\mathbf{A}_{2j}| + \cdots + a_{nj}|\mathbf{A}_{nj}| \tag{5.41b}$$

EXAMPLE 5.10 Solve Example 5.9 by applying Theorem 5.25.

SOLUTION We shall compute the cofactors of the elements in the third row. We apply Eq. (5.38), and the cofactors are as follows:

$$|\mathbf{A}_{31}| = \begin{vmatrix} 6 & -9 \\ 8 & 15 \end{vmatrix} = 6 \times 15 - (-9)8 = 162$$

$$|\mathbf{A}_{32}| = -\begin{vmatrix} 12 & -9 \\ 3 & 15 \end{vmatrix} = -[12 \times 15 - (-9)3] = -207$$

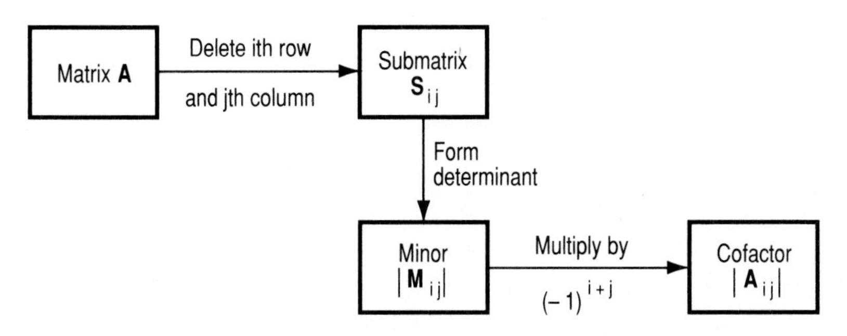

FIGURE 5.7 Method of forming cofactors.

$$|\mathbf{A}_{33}| = \begin{vmatrix} 12 & 6 \\ 3 & 8 \end{vmatrix} = 12 \times 8 - 6 \times 3 = 78$$

$$|\mathbf{A}| = 4 \times 162 + 11(-207) + 5 \times 78 = -1239$$

This result coincides with that in Example 5.9.

The cofactor method of evaluating a determinant may be applied to determinants of higher order. However, it requires a cyclic procedure in which cofactors are successively reduced to lower order.

Theorem 5.26. If all elements in a given row or column of a square matrix are multiplied by the cofactors of the corresponding elements in any other row or column, the sum of these products is equal to zero.

5.5.4 Evaluating Determinants by Triangularization

When the determinant of a triangular matrix is expanded, the only term that has a nonzero value is that which contains the diagonal elements. In this term, the number of inversions of the second subscripts is zero. We thus arrive at the following conclusion:

Theorem 5.27. The determinant of a triangular matrix is equal to the product of the diagonal elements.

As Example 5.4 indicates, a square matrix **A** can be transformed to an equivalent triangular matrix **B** by successively adding multiples of one row to another row. As a consequence of Theorem 5.21, $|\mathbf{A}| = |\mathbf{B}|$. This equality suggests a third method of calculating a determinant: reducing the given matrix to triangular form and then applying Theorem 5.27.

EXAMPLE 5.11 Solve Example 5.9 by applying the triangularization method.

SOLUTION The original matrix and the ensuing matrices are as follows:

$$\begin{bmatrix} 12 & 6 & -9 \\ 3 & 8 & 15 \\ 4 & 11 & 5 \end{bmatrix}$$
$$\text{①}$$

R2 − (1/4)R1

R3 − (1/3)R1 $\qquad\qquad$ R3 − (18/13)R2

$$\begin{bmatrix} 12 & 6 & -9 \\ 0 & \dfrac{13}{2} & \dfrac{69}{4} \\ 0 & 9 & 8 \end{bmatrix} \qquad \begin{bmatrix} 12 & 6 & -9 \\ 0 & \dfrac{13}{2} & \dfrac{69}{4} \\ 0 & 0 & -\dfrac{413}{26} \end{bmatrix}$$
$$\text{②} \qquad\qquad\qquad \text{③}$$

$$|\mathbf{A}| = 12(13/2)(-413/26) = -1239$$

5.5.5 Applications of Theorem 5.27

The fact that the determinant of a triangular matrix is equal to the product of the diagonal elements has many consequences, and we shall now present them. In the subsequent material, it is understood that the transformation of a matrix is accomplished by means of elementary operations.

Theorem 5.28. The determinant of a diagonal matrix is equal to the product of the diagonal elements.

Theorem 5.29. If the diagonal elements of a scalar matrix have the value α, the determinant is equal to α^n.

Theorem 5.30. The determinant of a unit matrix is equal to 1.

Theorem 5.31. A triangular matrix is regular if and only if all its diagonal elements have nonzero values.

Theorem 5.32. If a triangular matrix is regular, it can be transformed to the unit matrix.

Refer to Fig. 5.8*a*. A regular matrix **A** can be transformed to a regular triangular matrix **T** in the manner illustrated in Example 5.11. The triangular matrix in turn can be transformed to the unit matrix **I**. Therefore, **A** can be transformed to **I**. Conversely, **I** can be transformed to **A**. In Art 5.3.3, we defined an elementary matrix as one that results from the transformation of the unit matrix. We thus arrive at the following principle:

Theorem 5.33. Every regular matrix is elementary.

As shown in Fig. 5.8*b*, a regular matrix **A** can be transformed to **I**, and **I** can be transformed to an arbitrary regular matrix **B** of the same order. Therefore, **A** can be transformed to **B**, and the following principle results:

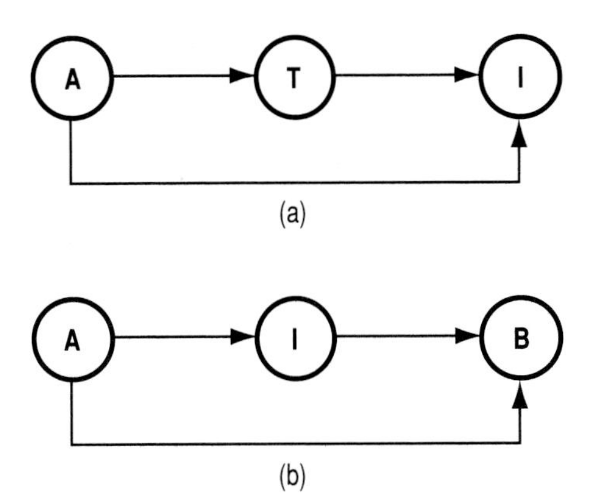

FIGURE 5.8 (*a*) Reduction of regular matrix to unit matrix; (*b*) relationship among regular matrices of identical order.

Theorem 5.34. All regular matrices of identical order are equivalent to one another.

5.5.6 Rank of a Matrix

Consider that we have an $m \times n$ matrix, where $m > n$. By eliminating rows and columns, we can form square submatrices of the nth, $(n-1)$th, $(n-2)$th, . . . , order. Consider that all possible square submatrices are formed. The *rank* of the given matrix is the highest order of a regular submatrix.

As an illustration, consider a 5×4 matrix. The number of inherent square submatrices is as follows: fourth-order, 5; third-order, 40; etc. Assume that all fourth-order submatrices have zero determinants but at least one third-order submatrix has a nonzero determinant. The rank of the given matrix is 3.

Theorem 5.35. Equivalent matrices are of equal rank.

Theorem 5.36. The rank of a matrix is equal to the number of linearly independent rows or linearly independent columns it contains, whichever of these two numbers is smaller.

The rank of a square matrix can readily be determined by transforming the matrix to an equivalent triangular matrix and applying the following principle:

Theorem 5.37. The rank of a square matrix is equal to the number of nonzero rows in its equivalent upper-triangular matrix or the number of nonzero columns in its equivalent lower-triangular matrix.

EXAMPLE 5.12 Determine the rank of the following matrix:

$$\mathbf{A} = \begin{bmatrix} 6 & -18 & 54 & 42 & 30 \\ 3 & -5 & 35 & 5 & 27 \\ -2 & 8 & -11 & -4 & -7 \\ 5 & -11 & 55 & 31 & 35 \\ -1 & 1 & -9 & 25 & -15 \end{bmatrix}$$

SOLUTION By subjecting this matrix to elementary column operations, we arrive at the following equivalent lower-triangular matrix:

$$\mathbf{B} = \begin{bmatrix} 6 & 0 & 0 & 0 & 0 \\ 3 & 4 & 0 & 0 & 0 \\ -2 & 2 & 3 & 0 & 0 \\ 5 & 4 & 2 & 0 & 0 \\ -1 & -2 & 4 & 0 & 0 \end{bmatrix}$$

Since **B** contains 3 nonzero columns, **A** is of rank 3. It follows that **A** is singular and every fourth-order submatrix that is inherent in **A** is also singular.

5.5.7 Relationship between Equivalence Transformations and Matrix Multiplication

Theorem 5.33 enables us to restate Theorem 5.4 in the following manner:

Theorem 5.38. If a given matrix is multiplied by a regular matrix, the matrix product is equivalent to the given matrix.

It follows that a given matrix **A** can be transformed to an equivalent matrix **B** indirectly by multiplying **A** by a specific regular matrix, and this was the procedure followed in Examples 5.7 and 5.8. Conversely, the product of a given matrix **A** and a regular matrix can be obtained indirectly by subjecting **A** to a specific set of elementary operations.

EXAMPLE 5.13 The following matrices are given:

$$\mathbf{A} = \begin{bmatrix} 2 & -9 \\ 7 & 6 \\ 4 & 1 \end{bmatrix} \qquad \mathbf{C} = \begin{bmatrix} 4 & 20 \\ 2 & 13 \end{bmatrix}$$

If $\mathbf{B} = \mathbf{AC}$, determine how **B** can be derived from **A** by elementary operations.

SOLUTION If we perform the multiplication directly, we obtain

$$\mathbf{B} = \begin{bmatrix} -10 & -77 \\ 40 & 218 \\ 18 & 93 \end{bmatrix}$$

We shall now obtain **B** by elementary operations. Figure 5.3 (or Theorem 5.4) reveals that **C** must be reduced to $\mathbf{I}_2$ by column operations, and these operations are as follows: (1/4)C1, C2 − 20C1, (1/3)C2, C1 − (1/2)C2. Therefore, **B** can be derived from **A** by the following formula: C1 + (1/2)C2, 3C2, C2 + 20C1, 4C1. Upon investigation, we find that this formula does lead from **A** to **B**.

Since the process of raising a square matrix to a positive integral power involves successive multiplication, it is evident that the task can also be performed through the successive application of elementary operations if the given matrix is regular. In Fig. 5.3*a*, consider that we replace $\mathbf{J}_r$ with **A**, where **A** is a regular matrix, and replace **A** with $\mathbf{A}^n$, where n is a positive integer. Then $\mathbf{B} = \mathbf{A}^{n+1}$, and the following principle results:

Theorem 5.39. A set of elementary row or column operations that transforms the unit matrix to a regular matrix **A** also transforms $\mathbf{A}^n$ to $\mathbf{A}^{n+1}$.

Figure 5.9 depicts this relationship. It follows that a regular matrix can be raised to the nth power by applying the appropriate set of operations $n-1$ times.

EXAMPLE 5.14 The following matrix is given:

$$\mathbf{A} = \begin{bmatrix} 6 & -12 \\ 4 & -5 \end{bmatrix}$$

Construct $\mathbf{A}^3$ by applying elementary column operations.

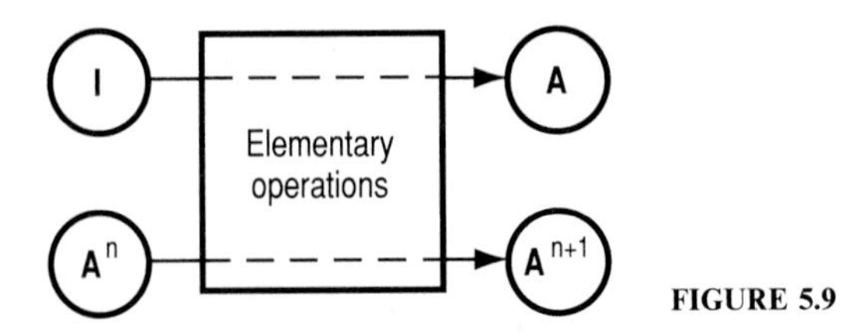

FIGURE 5.9

SOLUTION By multiplication, we obtain the following:

$$\mathbf{A}^2 = \begin{bmatrix} -12 & -12 \\ 4 & -23 \end{bmatrix} \qquad \mathbf{A}^3 = \begin{bmatrix} -120 & 204 \\ -68 & 67 \end{bmatrix}$$

The column operations that reduce **A** to **I** are as follows: (1/6)C1, C2 + 12C1, (1/3)C2, C1 − (2/3)C2. The inverse of this formula transforms **I** to **A**. Therefore, the transformation of **A** to $\mathbf{A}^3$ is as follows:

$$\underset{\text{②}}{\overset{\text{C1} + (2/3)\text{C2}}{\begin{bmatrix} -2 & -12 \\ \frac{2}{3} & -5 \end{bmatrix}}} \quad \underset{\text{③}}{\overset{3\text{C2}}{\begin{bmatrix} -2 & -36 \\ \frac{2}{3} & -15 \end{bmatrix}}} \quad \underset{\text{④}}{\overset{\text{C2} - 12\text{C1}}{\begin{bmatrix} -2 & -12 \\ \frac{2}{3} & -23 \end{bmatrix}}} \quad \underset{\text{⑤}}{\overset{6\text{C1}}{\begin{bmatrix} -12 & -12 \\ 4 & -23 \end{bmatrix}}}$$

$$\underset{\text{⑥}}{\overset{\text{C1} + (2/3)\text{C2}}{\begin{bmatrix} -20 & -12 \\ -\frac{34}{3} & -23 \end{bmatrix}}} \quad \underset{\text{⑦}}{\overset{3\text{C2}}{\begin{bmatrix} -20 & -36 \\ -\frac{34}{3} & -69 \end{bmatrix}}} \quad \underset{\text{⑧}}{\overset{\text{C2} - 12\text{C1}}{\begin{bmatrix} -20 & 204 \\ -\frac{34}{3} & 67 \end{bmatrix}}} \quad \underset{\text{⑨}}{\overset{6\text{C1}}{\begin{bmatrix} -120 & 204 \\ -68 & 67 \end{bmatrix}}}$$

The fifth and ninth matrices are $\mathbf{A}^2$ and $\mathbf{A}^3$, respectively.

Article 5.2.5 revealed that matrix multiplication is commutative in certain special cases. The relationship between an equivalence transformation and matrix multiplication enables us to view the subject of commutativity with a clearer perspective. For example, with reference to Fig. 5.3, let $\mathbf{J}_r = \mathbf{J}_c = \mathbf{C}$, where **C** is a scalar matrix having the diagonal elements α. The unit matrix is transformed to **C** by multiplying each row by α or by multiplying each column by α. Since these two sets of operations produce identical effects upon every square matrix, it follows that **CA** = **AC**, and therefore the multiplication of a square matrix **A** by a scalar matrix **C** is commutative.

5.6 INVERSE MATRIX AND ITS APPLICATIONS

5.6.1 Definition of Inverse Matrix

If m is a real number, we have $mm^{-1} = 1$. Since the matrix **I** is analogous to the number 1, we write the analogous equation $\mathbf{AA}^{-1} = \mathbf{I}$, where **A** is a regular matrix. The matrix $\mathbf{A}^{-1}$ defined by this equation is termed the *inverse* or *reciprocal* of **A**.

As we shall find, the inverse of a regular matrix is of fundamental significance in matrix algebra, for it is a tool in solving matrix equations. Since matrix division is not defined, it is impossible to divide a matrix expression by a given matrix. However, the desired effect can be achieved by multiplying the expression by the inverse of the given matrix.

5.6.2 Matrix Inversion

The process of forming the inverse of a given matrix is known as *matrix inversion*. Several computer languages have functions that yield the inverse of a matrix directly; nevertheless, it is of prime importance that the individual who applies matrix algebra be familiar with the technique.

Several methods of matrix inversion are available, but we shall confine ourselves to the one that is simplest to apply. It rests upon the following principle, which is a reversal of Theorem 5.39:

Theorem 5.40. A set of elementary row or column operations that reduces a regular matrix $\mathbf{A}$ to the unit matrix also transforms the unit matrix to $\mathbf{A}^{-1}$.

Figure 5.10 depicts this relationship.

EXAMPLE 5.15 Construct the inverse of the following matrix; verify the result.

$$\mathbf{A} = \begin{bmatrix} 2 & 6 & -14 \\ -3 & -5 & 9 \\ 2 & 8 & -16 \end{bmatrix}$$

SOLUTION The following formula, which involves row operations, transforms the given matrix to the unit matrix: (1/2)R1, R2 + 3R1, R3 − 2R1, (1/4)R2, R1 − 3R2, R3 − 2R2, (1/4)R3, R1 − 2R3, R2 + 3R3. When this formula is applied to $\mathbf{I}_3$, the following matrix results:

$$\mathbf{A}^{-1} = \begin{bmatrix} \frac{1}{4} & -\frac{1}{2} & -\frac{1}{2} \\ -\frac{15}{16} & -\frac{1}{8} & \frac{3}{4} \\ -\frac{7}{16} & -\frac{1}{8} & \frac{1}{4} \end{bmatrix} = \frac{1}{16}\begin{bmatrix} 4 & -8 & -8 \\ -15 & -2 & 12 \\ -7 & -2 & 4 \end{bmatrix}$$

By multiplication, we find that $\mathbf{AA}^{-1} = \mathbf{I}_3$, and the result is thus confirmed.

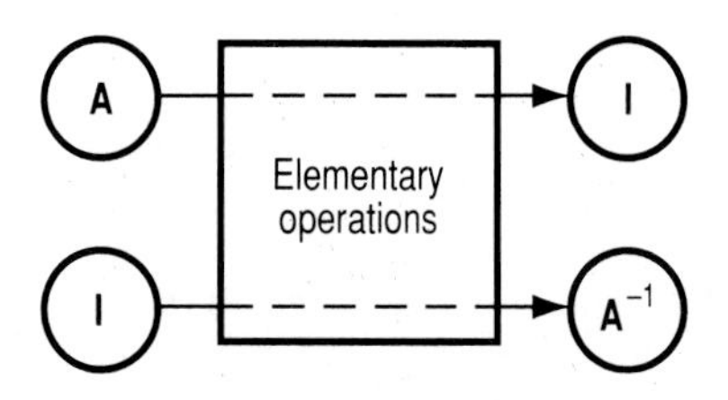

FIGURE 5.10 Method of constructing the inverse matrix.

For a regular second-order matrix **A**, we have

$$\mathbf{A}^{-1} = \frac{1}{|\mathbf{A}|}\begin{bmatrix} a_{22} & -a_{12} \\ -a_{21} & a_{11} \end{bmatrix} \tag{5.42}$$

EXAMPLE 5.16 Construct the inverse of the following matrix:

$$\mathbf{A} = \begin{bmatrix} 7 & -13 \\ 9 & 2 \end{bmatrix}$$

SOLUTION

$$|\mathbf{A}| = 7 \times 2 - (-13)9 = 131$$

$$\mathbf{A}^{-1} = \frac{1}{131}\begin{bmatrix} 2 & 13 \\ -9 & 7 \end{bmatrix}$$

5.6.3 Properties of the Inverse Matrix

The following principles pertain to the inverse of a matrix:

Theorem 5.41. The multiplication of a matrix and its inverse is commutative; that is, $\mathbf{AA}^{-1} = \mathbf{A}^{-1}\mathbf{A} = \mathbf{I}$.

Theorem 5.42. The inverse of a triangular matrix is itself a triangular matrix of the same form (lower- or upper-triangular).

The following equation is analogous to Eq. (5.22):

$$(\mathbf{ABC})^{-1} = \mathbf{C}^{-1}\mathbf{B}^{-1}\mathbf{A}^{-1} \tag{5.43}$$

This equation can be extended to include the product of any number of regular matrices. Theorem 5.22 yields

$$|\mathbf{A}^{-1}| = \frac{1}{|\mathbf{A}|} \tag{5.44}$$

5.6.4 Solution of Matrix Equations

The following example illustrates the use of inverse matrices in solving matrix equations.

EXAMPLE 5.17 In the following equations, **B** and **C** are regular matrices. Solve each equation for **A**.

$\mathbf{ABC} = \mathbf{D}$ (*b*) $\mathbf{BAC} = \mathbf{D}$ (*c*) $\mathbf{BAC}^{-1} = \mathbf{D}$ (*d*)

SOLUTION Equation (*b*): Postmultiply both sides of the equation by $\mathbf{C}^{-1}$, giving $\mathbf{ABCC}^{-1} = \mathbf{DC}^{-1}$. But $\mathbf{ABCC}^{-1} = \mathbf{AB}(\mathbf{CC}^{-1}) = \mathbf{ABI} = \mathbf{AB}$. Thus, $\mathbf{AB} = \mathbf{DC}^{-1}$. Now postmultiply both sides of the last equation by $\mathbf{B}^{-1}$, giving $\mathbf{ABB}^{-1} = \mathbf{DC}^{-1}\mathbf{B}^{-1}$. But $\mathbf{ABB}^{-1} = \mathbf{A}(\mathbf{BB}^{-1}) = \mathbf{AI} = \mathbf{A}$. Thus, $\mathbf{A} = \mathbf{DC}^{-1}\mathbf{B}^{-1}$.

Equation (*c*): Premultiply both sides of the equation by $\mathbf{B}^{-1}$, and then postmultiply both sides of the resulting equation by $\mathbf{C}^{-1}$. The result is $\mathbf{A} = \mathbf{B}^{-1}\mathbf{DC}^{-1}$.

Equation (d): Premultiply both sides of the equation by $\mathbf{B}^{-1}$, and then post-multiply both sides of the resulting equation by $\mathbf{C}$. The result is $\mathbf{A} = \mathbf{B}^{-1}\mathbf{DC}$.

In Art. 5.1.1, we defined a null matrix $\mathbf{0}$ as a matrix in which all elements are zero. Let $\mathbf{A}$ and $\mathbf{B}$ denote a square matrix and rectangular matrix, respectively, such that $\mathbf{AB} = \mathbf{0}$ or $\mathbf{BA} = \mathbf{0}$. If we assume that $\mathbf{A}$ is regular, we arrive at the solution $\mathbf{B} = \mathbf{0}$. If this solution is incorrect, $\mathbf{A}$ must be singular.

Theorem 5.43. If the product of a square matrix $\mathbf{A}$ and a rectangular matrix $\mathbf{B}$ is a null matrix, then either $\mathbf{A}$ is singular or $\mathbf{B}$ is a null matrix.

As an illustration, consider the multiplication

$$\begin{bmatrix} 10 & -6 \\ 20 & -12 \end{bmatrix} \begin{bmatrix} 3 \\ 5 \end{bmatrix} = \begin{bmatrix} 0 \\ 0 \end{bmatrix}$$

Theorem 5.43 states that the square matrix is singular, and we find that this is true. It is to be emphasized that Theorem 5.43 does not have a converse.

5.6.5 Negative Powers of a Matrix

Let $\mathbf{A}$ denote a regular matrix and n denote a positive integer. By definition, $\mathbf{A}^{-n}$ is the nth power of $\mathbf{A}^{-1}$; that is,

$$\mathbf{A}^{-n} = (\mathbf{A}^{-1})^n \tag{5.45}$$

The following equations apply:

$$\mathbf{A}^{-m}\mathbf{A}^{-n} = \mathbf{A}^{-(m+n)} \tag{5.46a}$$

$$\mathbf{A}^{m}\mathbf{A}^{-n} = \mathbf{A}^{m-n} \tag{5.46b}$$

where m is also a positive integer.

$$\mathbf{A}^{-n} = (\mathbf{A}^{n})^{-1} \tag{5.47}$$

Expressed verbally, $\mathbf{A}^{-n}$ is the inverse of $\mathbf{A}^{n}$.

5.6.6 Continuity among Matrix Powers

Again let $\mathbf{A}$ denote a regular matrix. By Eq. (5.27), $\mathbf{A}^0 = \mathbf{I}$. With reference to Figs. 5.9 and 5.10, if we replace $\mathbf{A}$ with $\mathbf{A}^1$ and $\mathbf{I}$ with $\mathbf{A}^0$, the continuity among the powers of a regular matrix is highlighted. Let O denote a set of elementary operations that transforms $\mathbf{I}$ to $\mathbf{A}$, and let O′ denote the corresponding set of elementary operations that transforms $\mathbf{A}$ to $\mathbf{I}$. The set O′ is the inverse of O; every operation in O has an inverse in O′, and the sequence of operations in O′ is the reverse of that in O.

Now let m denote any integer, positive, negative, or 0. If we apply 0 to $\mathbf{A}^m$, we form $\mathbf{A}^{m+1}$; if we apply 0′ to $\mathbf{A}^m$, we form $\mathbf{A}^{m-1}$. Thus, 0 produces a *progression* among matrix powers, and 0′ products a *regression*.

5.6.7 Orthogonal Matrices

If a regular matrix $\mathbf{A}$ has the property that $\mathbf{AA}^T = \mathbf{I}$, it is described as *orthogonal*. The following is an orthogonal matrix:

$$\mathbf{A} = \begin{bmatrix} \frac{9}{15} & \frac{12}{15} & 0 \\ -\frac{12}{15} & \frac{9}{15} & 0 \\ 0 & 0 & 1 \end{bmatrix}$$

Theorem 5.44. The transpose and inverse of an orthogonal matrix are identical. Conversely, if the transpose and inverse of a matrix are identical, the matrix is orthogonal.

Theorem 5.45. The multiplication of an orthogonal matrix and its transpose is commutative.

Theorem 5.46. The transpose of an orthogonal matrix is itself an orthogonal matrix.

Theorem 5.47. The product of two orthogonal matrices is an orthogonal matrix.

Theorem 5.48. The determinant of an orthogonal matrix is ± 1.

An orthogonal matrix is described as *proper* or *improper* according to whether its determinant is 1 or -1, respectively. Theorem 5.48 does not have a converse, for a matrix may have a determinant of 1 or -1 without being orthogonal.

Theorem 5.49. The sum of the squares of the elements in a given row or column of an orthogonal matrix is equal to 1. If corresponding elements in two rows or columns of an orthogonal matrix are multiplied, the sum of the products is zero.

Theorem 5.50. Conversely, a square matrix is orthogonal if the following conditions *both* exist: The sum of the squares of the elements in every row or column is equal to 1; the sum of the products of corresponding elements in every pair of rows or columns is equal to zero.

5.6.8 Similar Matrices

In accordance with Theorem 5.38, assume that a square matrix $\mathbf{A}$ is transformed to an equivalent matrix $\mathbf{B}$ by the following multiplication:

$$\mathbf{CAC}^{-1} = \mathbf{B} \tag{5.48}$$

where $\mathbf{C}$ is a regular matrix of the same order as $\mathbf{A}$. Matrix $\mathbf{B}$ is said to be *similar* to $\mathbf{A}$ through $\mathbf{C}$, the transformation of $\mathbf{A}$ to $\mathbf{B}$ is called a *similarity transformation*, and $\mathbf{C}$ is called the *similarity matrix*.

EXAMPLE 5.18 With reference to Eq. (5.48), find $\mathbf{B}$ if

$$\mathbf{A} = \begin{bmatrix} 8 & 6 \\ 5 & -3 \end{bmatrix} \qquad \mathbf{C} = \begin{bmatrix} 5 & 8 \\ 7 & 11 \end{bmatrix}$$

SOLUTION We find that $|\mathbf{C}| = -1$, and Eq. (5.42) yields

$$\mathbf{C}^{-1} = \begin{bmatrix} -11 & 8 \\ 7 & -5 \end{bmatrix}$$

$$\mathbf{B} = \mathbf{CAC}^{-1} = \begin{bmatrix} 5 & 8 \\ 7 & 11 \end{bmatrix} \begin{bmatrix} 8 & 6 \\ 5 & -3 \end{bmatrix} \begin{bmatrix} -11 & 8 \\ 7 & -5 \end{bmatrix}$$

$$\mathbf{B} = \begin{bmatrix} -838 & 610 \\ -1158 & 843 \end{bmatrix}$$

Equation (5.48) may be rewritten in any of the following forms:

$$\mathbf{CA} = \mathbf{BC} \qquad \mathbf{AC}^{-1} = \mathbf{C}^{-1}\mathbf{B} \qquad \mathbf{A} = \mathbf{C}^{-1}\mathbf{BC} \tag{5.49}$$

Theorem 5.51. If $\mathbf{B}$ is similar to $\mathbf{A}$ through $\mathbf{C}$, then $\mathbf{A}$ is similar to $\mathbf{B}$ through $\mathbf{C}^{-1}$, $\mathbf{A}^T$ is similar to $\mathbf{B}^T$ through $\mathbf{C}^T$, $\mathbf{B}^{-1}$ is similar to $\mathbf{A}^{-1}$ through $\mathbf{C}$, $\mathbf{B}^n$ is similar to $\mathbf{A}^n$ through $\mathbf{C}$, where n is an integer, and $\mathbf{B} + \alpha\mathbf{I}$ is similar to $\mathbf{A} + \alpha\mathbf{I}$ through $\mathbf{C}$, where α is a scalar.

Theorem 5.52. If $\mathbf{B}_1$ and $\mathbf{B}_2$ are similar to $\mathbf{A}_1$ and $\mathbf{A}_2$, respectively, through $\mathbf{C}$, then $\mathbf{B}_1 + \mathbf{B}_2$ is similar to $\mathbf{A}_1 + \mathbf{A}_2$ through $\mathbf{C}$.

Theorem 5.53. A scalar matrix is similar to itself and to no other matrix.

Theorem 5.54. The determinants of similar matrices are equal.

5.7 SOLUTION OF SIMULTANEOUS LINEAR EQUATIONS

Many problems in applied mathematics require the solution of a system of simultaneous linear equations. A system containing two unknowns is solved readily by the method presented in Art. 1.12, and a system containing three unknowns can be solved directly by a special-purpose calculator. A system containing more than three unknowns can be solved very effectively by matrix algebra, which provides a compact notation for recording the equations, a direct method of testing the equations for consistency, and a rapid method of solution.

5.7.1 Definitions and Notation

The basic notation for simultaneous linear equations is illustrated by the following system of three equations:

$$\begin{aligned} a_{11}x_1 + a_{12}x_2 + a_{13}x_3 &= c_1 \\ a_{21}x_1 + a_{22}x_2 + a_{23}x_3 &= c_2 \\ a_{31}x_1 + a_{32}x_2 + a_{33}x_3 &= c_3 \end{aligned} \tag{5.50}$$

The symbols x_1, x_2, and x_3 denote unknown quantities, and the remaining symbols denote constants. Each coefficient of an unknown is assigned a double subscript; the first corresponds to the equation number, and the second coincides with the subscript of the unknown.

The *solution* of a system of simultaneous equations has two meanings: a set of values of the unknowns that satisfies the given equations, and the process of finding this set of values. The system may have one solution, no solution, or an infinite number of solutions. A system that has one solution is *determinate.*

The *degree of freedom* of a system equals the number of unknowns in the system that may be assigned arbitrary values. As an illustration, consider the system

$$5x_1 + 2x_2 + 8x_3 = 15$$

$$-x_1 + 6x_2 + 4x_3 = 17$$

We may assign an arbitrary value to one of the unknowns. Let us set $x_1 = 5$. The two remaining unknowns then acquire the specific values $x_2 = 5.4$ and $x_3 = -2.6$. Therefore, this system has 1 degree of freedom. A system of equations for which the degree of freedom exceeds zero has an infinite number of solutions.

If the constants $c_1, c_2, \ldots$, on the right side of Eq. (5.50) are all zero, the system of equations is *homogeneous*; otherwise, it is *nonhomogeneous.* A system of equations containing as many equations as unknowns is said to be *square.* We shall not consider systems in which the number of equations exceeds the number of unknowns.

5.7.2 Criterion for Determinateness

A given equation in a system is called *secondary* if it can be formed by taking multiples of preceding equations, either singly or in combination; otherwise, it is called *primary.* The primary equations of a system are said to be *linearly independent* of one another.

The equations composing a system are *inconsistent* or *incompatible* with one another if it is possible to form a secondary equation that is identical with a given equation except with respect to the constant at the right. For example, consider the following system:

$$3x_1 + 7x_2 - 12x_3 = 85 \qquad (e)$$

$$4x_1 - 6x_2 + 14x_3 = 32 \qquad (f)$$

$$5x_1 + 4x_2 - 5x_3 = 96 \qquad (g)$$

If we multiply Eq. (f) by 1/2 and add the result to Eq. (e), we obtain the following secondary equation:

$$5x_1 + 4x_2 - 5x_3 = 101 \qquad (h)$$

Since Eq. (h) conflicts with Eq. (g), it follows that Eqs. (e), (f), and (g) are inconsistent.

A system of equations in n unknowns is determinate if the system contains n equations that are linearly independent and consistent.

5.7.3 Matrix Notation for Simultaneous Equations

Consider again the system of simultaneous equations in Eq. (5.50). This system can

$$\begin{bmatrix} a_{11} & a_{12} & a_{13} \\ a_{21} & a_{22} & a_{23} \\ a_{31} & a_{32} & a_{33} \end{bmatrix} \begin{bmatrix} x_1 \\ x_2 \\ x_3 \end{bmatrix} = \begin{bmatrix} c_1 \\ c_2 \\ c_3 \end{bmatrix}$$

With appropriate designations for the matrices, this equation my be written in this form:

$$\mathbf{AX} = \mathbf{C} \tag{5.51}$$

In this notation, **A** is the *matrix of coefficients*, **X** is the *column vector of unknowns*, and **C** is the *column vector of constants*.

It will be convenient to juxtapose **C** with **A** to form an augmented matrix, in this manner:

$$\begin{bmatrix} a_{11} & a_{12} & a_{13} & c_1 \\ a_{21} & a_{22} & a_{23} & c_2 \\ a_{31} & a_{32} & a_{33} & c_3 \end{bmatrix} = [\mathbf{A} \quad \mathbf{C}]$$

A and **C** are called, respectively, the *coefficient submatrix* and the *constant submatrix* of the augmented matrix. Manifestly, it is possible to perform elementary operations upon the augmented matrix without violating the given equations.

In Art. 5.5.6, we defined the rank of a square matrix and demonstrated that the rank can be established by transforming the matrix to an equivalent triangular matrix. We have the following principle:

Theorem 5.55. In a system of simultaneous linear equations, the number of equations that are linearly independent and consistent is equal to the rank of its matrix of coefficients.

EXAMPLE 5.19 Test the following system of equations for linear independence and consistency.

$$3x_1 + 9x_2 - 6x_3 = 66$$

$$2x_1 + 10x_2 + 4x_3 = 76$$

$$4x_1 + 8x_2 - 16x_3 = 118$$

SOLUTION The augmented matrix is as follows:

$$\begin{bmatrix} 3 & 9 & -6 & 66 \\ 2 & 10 & 4 & 76 \\ 4 & 8 & -16 & 118 \end{bmatrix}$$

We shall reduce the coefficient submatrix to upper-triangular form, with the diagonal elements equal to 1. For the first two cycles, the formula is as follows: (1/3)R1, R2 − 2R1, R3 − 4R1, (1/4)R2, R3 + 4R2. At this point, the augmented matrix is

$$\begin{bmatrix} 1 & 3 & -2 & 22 \\ 0 & 1 & 2 & 8 \\ 0 & 0 & 0 & 62 \end{bmatrix}$$

The third row now contains zero coefficients but a nonzero constant. This condition signals an inconsistency. Upon investigation, we find that the left side of the third equation, but not the right side, is obtained by doubling the first equation and subtracting the second. The coefficient submatrix is of rank 2 and the augmented matrix is of rank 3.

EXAMPLE 5.20 Test the following system of equations for linear independence and consistency.

$$2x_1 + 4x_2 + 6x_3 + 4x_4 = 56$$

$$x_1 + 5x_2 + 9x_3 + 14x_4 = 127$$

$$-x_1 + x_2 + 3x_3 + 10x_4 = 71$$

$$3x_1 + 9x_2 + 18x_3 + 24x_4 = 228$$

SOLUTION The augmented matrix is as follows:

$$\begin{bmatrix} 2 & 4 & 6 & 4 & 56 \\ 1 & 5 & 9 & 14 & 127 \\ -1 & 1 & 3 & 10 & 71 \\ 3 & 9 & 18 & 24 & 228 \end{bmatrix}$$

Proceeding as in Example 5.19, we arrive at the following matrix at the end of the second cycle:

$$\begin{bmatrix} 1 & 2 & 3 & 2 & 28 \\ 0 & 1 & 2 & 4 & 33 \\ 0 & 0 & 0 & 0 & 0 \\ 0 & 0 & 3 & 6 & 45 \end{bmatrix}$$

The disappearance of the third row signals that the third equation is secondary, and investigation discloses that this equation is obtained by subtracting the first equation from the second. By interchanging the third and fourth rows and completing the transformation, we obtain the following matrix:

$$\begin{bmatrix} 1 & 2 & 3 & 2 & 28 \\ 0 & 1 & 2 & 4 & 33 \\ 0 & 0 & 1 & 2 & 15 \\ 0 & 0 & 0 & 0 & 0 \end{bmatrix}$$

No inconsistencies have appeared. Both the coefficient submatrix and the augmented matrix are of rank 3, and the system therefore consists of 3 linearly independent, consistent equations. It has 1 degree of freedom.

On the basis of the results obtained in Examples 5.19 and 5.20, it is apparent that Theorem 5.55 can now be amplified to the following:

Theorem 5.56. The difference between the number of equations in a system and the rank of its augmented matrix is equal to the number of secondary equations that are present. The difference between the rank of the augmented matrix and the rank of the coefficient submatrix is equal to the number of inconsistencies that are present.

5.7.4 Solution of System by Reduction to Simpler Form

A highly efficient method of solving a determinate system of simultaneous linear equations consists of reducing the augmented matrix of the system to a simpler form. Specifically, if the coefficient submatrix is reduced to diagonal form, the technique is known as the *Gauss-Jordan method.* We shall extend this method by reducing the coefficient submatrix to unit form.

EXAMPLE 5.21 Solve the following system of equations:

$$6x_1 + 30x_2 + 78x_3 = 456$$
$$3x_1 + 18x_2 + 48x_3 = 273$$
$$4x_1 + 22x_2 + 62x_3 = 350$$

SOLUTION The augmented matrix is as follows:

$$\begin{bmatrix} 6 & 30 & 78 & 456 \\ 3 & 18 & 48 & 273 \\ 4 & 22 & 62 & 350 \end{bmatrix}$$

The coefficient submatrix is reduced to unit form by the following formula: (1/6)R1, R2 − 3R1, R3 − 4R1, (1/3)R2, R1 − 5R2, R3 − 2R2, (1/4)R3, R1 + 2R3, R2 − 3R3. The matrix that culminates is

$$\begin{bmatrix} 1 & 0 & 0 & 9 \\ 0 & 1 & 0 & 3 \\ 0 & 0 & 1 & 4 \end{bmatrix}$$

Translating this matrix to a system of equations, we have

$$x_1 = 9 \qquad x_2 = 3 \qquad x_3 = 4$$

5.7.5 Solution of System by Determinants

A determinate system of simultaneous linear equations can also be solved by this procedure: Convert the coefficient matrix $\mathbf{A}$ to matrix $\mathbf{B}_i$ by replacing the coefficients of x_i with the constants, where i assumes all possible values. Then

$$x_i = \frac{|\mathbf{B}_i|}{|\mathbf{A}|} \tag{5.52}$$

The method of solution embodied in Eq. (5.52) is known as *Cramer's rule.*

EXAMPLE 5.22 Solve the following system of equations:

$$3x_1 - 7x_2 + 2x_3 = 47$$

$$9x_1 + 4x_2 - 5x_3 = 19$$

$$10x_1 - 6x_2 + 8x_3 = 90$$

SOLUTION

$$\mathbf{A} = \begin{bmatrix} 3 & -7 & 2 \\ 9 & 4 & -5 \\ 10 & -6 & 8 \end{bmatrix} \qquad \mathbf{B}_1 = \begin{bmatrix} 47 & -7 & 2 \\ 19 & 4 & -5 \\ 90 & -6 & 8 \end{bmatrix}$$

Evaluating the determinants, we obtain $|\mathbf{A}| = 672$ and $|\mathbf{B}_1| = 3360$. Then $x_1 = 3360/672 = 5$. In like manner, we obtain $x_2 = -4$ and $x_3 = 2$.

5.7.6 Incomplete Systems

A system of simultaneous equations is *incomplete* or *deficient* if the number of unknowns exceeds the number of linearly independent equations. Let n denote the number of unknowns and r the number of linearly independent equations, where $n > r$. The degree of freedom of the system is $n - r$.

We may assign a set of arbitrary values to $n - r$ unknowns and then calculate the corresponding values of the remaining r unknowns. This procedure is tantamount to considering $n - r$ unknowns as *independent variables* and the remaining r unknowns as *dependent variables*, the latter being functions of the former. The arbitrary values thus assigned are called *arbitrary parameters*. Every set of arbitrary parameters yields a unique solution of the system. In many instances, our interest centers about the *relative* values of the unknowns, and the following principle applies:

Theorem 5.57. Every solution of an incomplete nonhomogeneous system of equations contains a unique set of relative values of the unknowns.

For example, consider the system

$$9x_1 - 4x_2 + 3x_3 = 64$$

$$7x_1 + 3x_2 + 4x_3 = 97$$

If we set $x_3 = 10$, we obtain $x_1 = 6$, $x_2 = 5$, and $x_1/x_2 = 1.2$. If we now set $x_3 = -1$, we obtain $x_1 = 11$, $x_2 = 8$, and $x_1/x_2 = 1.375$.

5.7.7 Homogeneous Systems

As stated in Art. 5.7.1, a system of simultaneous equations is homogeneous if all constants at the right have zero value. For this type of system, Eq. (5.51) reduces to $\mathbf{AX} = \mathbf{0}$. One solution that immediately suggests itself is $\mathbf{X} = \mathbf{0}$; that is, all the unknowns have zero value. This is known as the *trivial solution* of the system. The following principles pertain to a homogeneous system:

Theorem 5.58

a. If the coefficient submatrix of a homogeneous system is regular, only the trivial solution exists. If the coefficient submatrix is singular, an infinite number of nontrivial solutions exists.

b. If the difference between the number of equations and the rank of the coefficient submatrix is 1, all nontrivial solutions contain the same relative values of the unknowns.

c. If the difference between the number of equations and the rank of the coefficient submatrix exceeds 1, the nontrivial solutions can be divided into sets in such manner that all solutions within a set contain the same relative values of the unknowns.

5.8 EIGENVALUES AND EIGENVECTORS

5.8.1 Definition of Eigenvalues

Let **A** denote a regular matrix, and let c denote a scalar having this characteristic: When c is deducted from each diagonal element of **A**, the resulting matrix is singular.

In general, if **A** is of order n, we have

$$\begin{vmatrix} a_{11}-c & a_{12} & \cdots & a_{1n} \\ a_{21} & a_{22}-c & \cdots & a_{2n} \\ \cdot & \cdot & \cdots & \cdot \\ \cdot & \cdot & \cdots & \cdot \\ \cdot & \cdot & \cdots & \cdot \\ a_{n1} & a_{n2} & \cdots & a_{nn}-c \end{vmatrix} = 0 \tag{5.53a}$$

Since the process of deducting c from each diagonal element of **A** is tantamount to deducting the scalar matrix $c\mathbf{I}$ from **A**, the foregoing equation may be written in this form:

$$|\mathbf{A} - c\mathbf{I}| = 0 \tag{5.53b}$$

When this determinant is expanded, the resulting expression is a polynomial function in the nth degree with respect to c, and therefore Eq. (5.53) has n roots. These roots will be designated as $c_1, c_2, \ldots, c_n$.

As an illustration, consider the following second-order regular matrix:

$$\mathbf{A} = \begin{bmatrix} 13 & 5 \\ 2 & 4 \end{bmatrix}$$

Deducting c from each diagonal element and applying Eq. (5.38), we obtain the following:

$$\begin{vmatrix} 13-c & 5 \\ 2 & 4-c \end{vmatrix} = 0$$

$$(13-c)(4-c) - 5 \times 2 = 0$$

This equation becomes

$$42 - 17c + c^2 = 0$$

By factoring or applying Eq. (1.39), we obtain $c_1 = 3$ and $c_2 = 14$.

The problem of transforming a regular matrix **A** to a singular matrix by deducting c from the diagonal elements is referred to as the *eigenvalue problem*. Equation (5.53) is called the *eigenvalue equation* or *characteristic equation* of the problem, and the roots of this equation are termed the *eigenvalues, characteristic roots*, or *latent roots* of the given matrix.

The definition of an eigenvalue can now be extended to encompass singular as well as regular matrices. A singular matrix of order n also has n eigenvalues, but one has a zero value. For example, the singular matrix

$$\begin{bmatrix} 4 & 6 \\ 6 & 9 \end{bmatrix}$$

has the eigenvalues $c_1 = 0$ and $c_2 = 13$.

5.8.2 Properties of Eigenvalues

For a matrix of order n, the characteristic equation may be written in the following polynomial form:

$$f(c) = a_0c^n + a_1c^{n-1} + a_2c^{n-2} + \cdots + a_{n-1}c + a_n = 0 \qquad (5.54)$$

In Art. 5.1.3, we defined the trace of a square matrix as the sum of the diagonal elements. Let t_1 denote the trace of **A**, t_2 the trace of $\mathbf{A}^2, \ldots, t_n$ the trace of $\mathbf{A}^n$. It can be demonstrated that the coefficients in Eq. (5.54) have the following values:

$$\begin{aligned} a_0 &= 1 \\ a_1 &= -t_1 \\ a_2 &= -\frac{1}{2}(a_1t_1 + t_2) \\ &\vdots \\ a_n &= -\frac{1}{n}(a_{n-1}t_1 + a_{n-2}t_2 + \cdots + a_1t_{n-1} + t_n) \end{aligned} \qquad (5.55)$$

EXAMPLE 5.23 Test the validity of Eq. (5.55) by applying it to the following matrix:

$$\mathbf{A} = \begin{bmatrix} 17 & 12 \\ 9 & 20 \end{bmatrix}$$

SOLUTION

$$\mathbf{A}^2 = \begin{bmatrix} 397 & 444 \\ 333 & 508 \end{bmatrix}$$

$$t_1 = 17 + 20 = 37 \qquad t_2 = 397 + 508 = 905$$

The characteristic equation of **A** is

$$(17 - c)(20 - c) - 12 \times 9 = 0$$

or

$$c^2 - 37c + 232 = 0$$

In this equation, $a_0 = 1$, $a_1 = -37$, $a_2 = 232$. By Eq. (5.55),

$$a_0 = 1 \qquad a_1 = -37 \qquad a_2 = -\left(\frac{1}{2}\right)[(-37)(37) + 905] = 232$$

The equation has yielded correct results.

Theorem 5.59. The sum of the eigenvalues of a matrix is equal to the trace of the matrix.

Theorem 5.60. The diagonal elements of a diagonal matrix are the eigenvalues of that matrix.

The foregoing principle stems from the fact that the determinant of a diagonal matrix equals the product of the diagonal elements. Therefore, if each diagonal element is reduced by a_{ii}, where i can assume any value, the determinant vanishes. In Art. 5.6.8, we defined similar matrices, and we now have the following principles:

Theorem 5.61. Similar matrices have equal eigenvalues.

Theorem 5.62. If a diagonal matrix **D** is similar to a regular matrix **A**, the diagonal elements of **D** are the eigenvalues of **A**. Conversely, if the diagonal elements of **D** are the eigenvalues of **A**, then **D** is similar to **A**.

Theorem 5.62 follows as a consequence of Theorems 5.60 and 5.61.

5.8.3 Canonical Form of a Matrix

Assume that **A** is of order n and has the eigenvalues $c_1, c_2, \ldots, c_n$. The diagonal matrix **D** that is similar to **A** is

$$\mathbf{D} = \begin{bmatrix} c_1 & 0 & \cdots & 0 \\ 0 & c_2 & \cdots & 0 \\ \cdot & \cdot & \cdots & \cdot \\ 0 & 0 & \cdots & c_n \end{bmatrix}$$

Matrix **D** is called the *canonical form* of **A**.

Assume that the n eigenvalues of **A** are all distinct. Since the subscripting of the eigenvalues is arbitrary, a particular canonical form represents a permutation of n items taken all at a time, and therefore **A** has $n!$ alternative canonical forms.

When the eigenvalues of a given matrix **A** are known, a canonical form **D** can be constructed merely by placing these eigenvalues on the principal diagonal. However, in many instances it is necessary to identify the similarity matrix **C** that will transform **A** to **D** by the multiplication $\mathbf{CAC}^{-1} = \mathbf{D}$. Article 5.8.5 provides a method of doing this.

5.8.4 Definition and Properties of Eigenvectors

Let **A** denote a matrix of order n, let c_i denote an eigenvalue of **A**, and let $\mathbf{X}_i$ denote a column vector of n rows. Since $\mathbf{A} - c_i\mathbf{I}$ is a singular matrix, it follows from Theorem 5.58*a* that there exist vectors $\mathbf{X}_i$ that satisfy the following homogeneous equation:

$$(\mathbf{A} - c_i\mathbf{I})\mathbf{X}_i = \mathbf{0} \tag{5.56}$$

A vector $\mathbf{X}_i$ that meets this requirement is called an *eigenvector*, *characteristic vector*, or *latent vector* of **A**. For every eigenvalue c_i, Eq. (5.56) constitutes a system of n simultaneous homogeneous equations, and every system of equations has an infinite number of solutions. Thus, corresponding to every eigenvalue c_i is a set of eigenvectors $\mathbf{X}_i$, the number of eigenvectors in the set being infinite. Equation (5.56) can be recast in this form:

$$\mathbf{AX}_i = c_i\mathbf{X}_i \tag{5.57}$$

EXAMPLE 5.24 The matrix

$$\mathbf{A} = \begin{bmatrix} 5 & 2 & 4 \\ -3 & 6 & 2 \\ 3 & -3 & 1 \end{bmatrix}$$

has the eigenvalues $c_1 = 2$, $c_2 = 3$, and $c_3 = 7$, as can readily be verified. Find an eigenvector corresponding to each eigenvalue.

SOLUTION Starting with c_1, we have

$$(\mathbf{A} - c_1\mathbf{I})\mathbf{X}_1 = \begin{bmatrix} 3 & 2 & 4 \\ -3 & 4 & 2 \\ 3 & -3 & -1 \end{bmatrix} \begin{bmatrix} x_1 \\ x_2 \\ x_3 \end{bmatrix} = \begin{bmatrix} 0 \\ 0 \\ 0 \end{bmatrix}$$

Performing the multiplication and taking only the first two equations that result, we obtain

$$3x_1 + 2x_2 + 4x_3 = 0$$

$$-3x_1 + 4x_2 + 2x_3 = 0$$

Arbitrarily setting $x_1 = 2$, we obtain $x_2 = 3$ and $x_3 = -3$. Proceeding to c_2, we have

$$(\mathbf{A} - c_2\mathbf{I})\mathbf{X}_2 = \begin{bmatrix} 2 & 2 & 4 \\ -3 & 3 & 2 \\ 3 & -3 & -2 \end{bmatrix} \begin{bmatrix} x_1 \\ x_2 \\ x_3 \end{bmatrix} = \begin{bmatrix} 0 \\ 0 \\ 0 \end{bmatrix}$$

$$2x_1 + 2x_2 + 4x_3 = 0$$

$$-3x_1 + 3x_2 + 2x_3 = 0$$

Arbitrarily setting $x_1 = 2$, we obtain $x_2 = 4$ and $x_3 = -3$. Proceeding to c_3, we have

$$(\mathbf{A} - c_3\mathbf{I})\mathbf{X}_3 = \begin{bmatrix} -2 & 2 & 4 \\ -3 & -1 & 2 \\ 3 & -3 & -6 \end{bmatrix} \begin{bmatrix} x_1 \\ x_2 \\ x_3 \end{bmatrix} = \begin{bmatrix} 0 \\ 0 \\ 0 \end{bmatrix}$$

$$-2x_1 + 2x_2 + 4x_3 = 0$$

$$-3x_1 - x_2 + 2x_3 = 0$$

Arbitrarily setting $x_1 = 1$, we obtain $x_2 = -1$ and $x_3 = 1$.

We have thus formed the following eigenvectors of $\mathbf{A}$:

$$\mathbf{X}_1 = \begin{bmatrix} 2 \\ 3 \\ -3 \end{bmatrix} \qquad \mathbf{X}_2 = \begin{bmatrix} 2 \\ 4 \\ -3 \end{bmatrix} \qquad \mathbf{X}_3 = \begin{bmatrix} 1 \\ -1 \\ 1 \end{bmatrix}$$

The given matrix $\mathbf{A}$ is regular, and in all instances $\mathbf{A} - c_i\mathbf{I}$ is of rank 2. It follows from Theorem 5.57*b* that all elements within a given set of eigenvectors have constant relative values. Therefore, we may generalize our solution to this problem by introducing a constant k, where k can assume any value whatever. Our eigenvectors then assume the following form:

$$\mathbf{X}_1 = \begin{bmatrix} 2k \\ 3k \\ -3k \end{bmatrix} \qquad \mathbf{X}_2 = \begin{bmatrix} 2k \\ 4k \\ -3k \end{bmatrix} \qquad \mathbf{X}_3 = \begin{bmatrix} k \\ -k \\ k \end{bmatrix}$$

The following principles pertain to eigenvectors:

Theorem 5.63. Eigenvectors that correspond to distinct eigenvalues are linearly independent of one another.

Theorem 5.64. If a square matrix $\mathbf{A}$, a column vector $\mathbf{X}$, and a scalar c satisfy the equation $\mathbf{AX} = c\mathbf{X}$, then c is an eigenvalue of $\mathbf{A}$ and $\mathbf{X}$ is an eigenvector of $\mathbf{A}$ corresponding to c.

Theorem 5.63 is illustrated by Example 5.24. The three eigenvalues of $\mathbf{A}$ are all distinct, and the three eigenvectors are linearly independent.

Equation (5.57) can be extended to the following general form:

$$\mathbf{A}^n\mathbf{X}_i = c_i^n\mathbf{X}_i \tag{5.58}$$

where n is any positive integer.

5.8.5 Reduction of Matrix to Canonical Form

Let $\mathbf{A}$ denote a regular matrix of order n that has n distinct eigenvalues. Let $\mathbf{X}_1, \mathbf{X}_2, \mathbf{X}_3, \ldots, \mathbf{X}_n$ denote particular eigenvectors corresponding, respectively, to the eigenvalues $c_1, c_2, c_3, \ldots, c_n$ of **A**. Consider that a matrix **Q** is formed having these eigenvectors as its columns. Thus,

$$\mathbf{Q} = [\mathbf{X}_1 \quad \mathbf{X}_2 \quad \mathbf{X}_3 \cdots \mathbf{X}_n]$$

Matrix **Q** is termed a *modal matrix* of **A**. As an illustration, the following is a modal matrix of matrix **A** in Example 5.24:

$$\mathbf{Q} = \begin{bmatrix} 2 & 2 & 1 \\ 3 & 4 & -1 \\ -3 & -3 & 1 \end{bmatrix}$$

Theorem 5.65. If a regular matrix **A** of order n has n distinct eigenvalues, it can be reduced to a canonical form **D** by applying the inverse of its modal matrix **Q** as a similarity matrix. Expressed symbolically,

$$\mathbf{Q}^{-1}\mathbf{A}\mathbf{Q} = \mathbf{D} \tag{5.59}$$

As an illustration, consider again the matrix **A** in Example 5.24, which has the eigenvalues $c_1 = 2$, $c_2 = 3$, and $c_3 = 7$. Taking its modal matrix **Q** displayed above, we obtain the following:

$$\mathbf{Q}^{-1}\mathbf{A}\mathbf{Q} = \begin{bmatrix} 0.2 & -1.0 & -1.2 \\ 0 & 1.0 & 1.0 \\ 0.6 & 0 & 0.4 \end{bmatrix} \begin{bmatrix} 5 & 2 & 4 \\ -3 & 6 & 2 \\ 3 & -3 & 1 \end{bmatrix} \begin{bmatrix} 2 & 2 & 1 \\ 3 & 4 & -1 \\ -3 & -3 & 1 \end{bmatrix}$$

$$= \begin{bmatrix} 2 & 0 & 0 \\ 0 & 3 & 0 \\ 0 & 0 & 7 \end{bmatrix} = \begin{bmatrix} c_1 & 0 & 0 \\ 0 & c_2 & 0 \\ 0 & 0 & c_3 \end{bmatrix}$$

5.8.6 Hamilton-Cayley Theorem

Let **A** denote a regular matrix of order n, and consider that its characteristic equation in the form given by Eq. (5.54) has been established. Now consider that the polynomial function $f(c)$ in this equation is transformed to a matrix function $f(\mathbf{A})$ by replacing the scalar c with **A** while retaining the coefficients. Since $a_n = a_n c^0$ and $\mathbf{A}^0 = \mathbf{I}$, the last term in the transformed equation is $a_n\mathbf{I}$. The *Hamilton-Cayley theorem*, which is of far-reaching importance, states that $f(\mathbf{A})$ is a null matrix; that is,

$$f(\mathbf{A}) = a_0\mathbf{A}^n + a_1\mathbf{A}^{n-1} + a_2\mathbf{A}^{n-2} + \cdots + a_{n-1}\mathbf{A} + a_n\mathbf{I} = \mathbf{0} \tag{5.60}$$

The Hamilton-Cayley theorem may be tersely expressed in this manner:

Theorem 5.66. A square matrix satisfies its own characteristic equation.

EXAMPLE 5.25 Test the validity of the Hamilton-Cayley theorem by applying it to the following matrix:

$$\mathbf{A} = \begin{bmatrix} 7 & 4 \\ 8 & -3 \end{bmatrix}$$

SOLUTION

$$\mathbf{A}^2 = \begin{bmatrix} 81 & 16 \\ 32 & 41 \end{bmatrix}$$

Forming the characteristic equation and then replacing c with $\mathbf{A}$, we obtain the following:

$$(7 - c)(-3 - c) - 4 \times 8 = 0 \qquad \text{or} \qquad c^2 - 4c - 53 = 0$$

$$f(\mathbf{A}) = \mathbf{A}^2 - 4\mathbf{A} - 53\mathbf{I}$$

$$f(\mathbf{A}) = \begin{bmatrix} 81 & 16 \\ 32 & 41 \end{bmatrix} - 4\begin{bmatrix} 7 & 4 \\ 8 & -3 \end{bmatrix} - 53\begin{bmatrix} 1 & 0 \\ 0 & 1 \end{bmatrix}$$

$$= \begin{bmatrix} 81 & 16 \\ 32 & 41 \end{bmatrix} - \begin{bmatrix} 28 & 16 \\ 32 & -12 \end{bmatrix} - \begin{bmatrix} 53 & 0 \\ 0 & 53 \end{bmatrix} = \begin{bmatrix} 0 & 0 \\ 0 & 0 \end{bmatrix}$$

The Hamilton-Cayley theorem provides another method of obtaining the inverse of a square matrix. If we multiply both sides of Eq. (5.60) by $\mathbf{A}^{-1}$ and then solve the resulting equation for $\mathbf{A}^{-1}$, we obtain

$$\mathbf{A}^{-1} = -\frac{1}{a_n}(a_0\mathbf{A}^{n-1} + a_1\mathbf{A}^{n-2} + a_2\mathbf{A}^{n-3} + \cdots + a_{n-1}\mathbf{I}) \tag{5.61}$$

Thus, by forming the power matrices through $\mathbf{A}^{n-1}$, forming the characteristic equation of $\mathbf{A}$, and then applying Eq. (5.61), we obtain the inverse of $\mathbf{A}$.

SECTION 6
BOOLEAN ALGEBRA

In 1847, the English mathematician George Boole formulated a new system of algebra to serve as a tool in the study of logic. In 1938, C. E. Shannon demonstrated that boolean algebra is highly useful in the design of switching circuits, and since then the subject has become an integral part of modern mathematics.

6.1 CONCEPTS AND NOTATION

The concepts that underlie boolean algebra can be developed most effectively by relating them directly to a set of tangible objects. In this system of algebra, our interest centers about the *state* of an object, and each object has two possible states. Thus, the object may be open or closed, luminous or dark, horizontal or vertical, red or green, etc. These objects, which we shall term *components*, are arranged to form a *system*, and the system also has two possible states.

The state of a component or of the system is expressed by use of a code consisting of the numerals 0 and 1. For example, if the possible states are horizontal and vertical, 0 can denote that the component is horizontal and 1 that it is vertical. The expression $A = 0$ states that component A is at the state represented by 0. For brevity, we say that A has the *value* 0, although we are representing a state of being rather than a numerical quantity. When referring to the system, the numeral 1 represents the required state of the system; when referring to a component, 1 represents the state of the component that allows the system to attain its required state.

Assume that a system consists of two components: A and B. The state of the system is determined by the states of A and B and of the manner in which these components are arranged. Two arrangements are possible. Under the first arrangement, the system is 1 if *either* A *or* B is 1 (or if both are 1); under the second arrangement, the system is 1 if and only if *both* A *and* B are 1. We shall refer to these relationships of the components as the OR and AND relationships, respectively.

As an illustration, refer to Fig. 6.1*a*, where components A and B are arranged in parallel. Assume that we are currently at point m and wish to reach point n. To do this, we must pass through either A or B. Now assume that each component is either passable or impassable. The system itself is passable if it allows movement from m to n, and it is impassable if such is not the case. Manifestly, the system is passable if either A or B is passable (or if both are passable).

Now refer to Fig. 6.1*b*, where components A and B are arranged in series. To

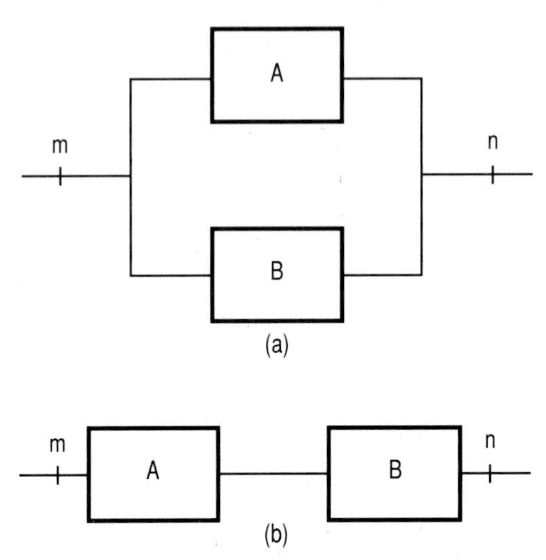

FIGURE 6.1 Passage through system. (*a*) Components in parallel; (*b*) components in series.

move from m to n, we must pass through both A and B. Under this arrangement, the system is passable only if both A and B are passable.

In summary, the OR relationship corresponds to the parallel arrangement of the components, and the AND relationship corresponds to the series arrangement. We shall apply this concept of movement across the system repeatedly in the following material.

The OR and AND relationships are denoted by the symbols for addition and multiplication, respectively, that are used in ordinary algebra. Thus, the expression $A + B$ (read "A or B") signifies that the system is 1 if either A or B is 1 (or if both are 1); the expression AB (read "A and B") signifies that the system is 1 if both A and B are 1.

Although we are applying the symbols for addition and multiplication, it is to be emphasized that the relationships of boolean algebra are completely divorced from the operations of ordinary algebra. The laws of ordinary algebra have no place in boolean algebra because the two subjects are unrelated.

A system consisting of three or more components may be considered to contain *subsystems*, and the composition of a subsystem can be described by the use of parentheses. As an illustration, refer to Fig. 6.2. This system contains two subsystems. One consists of A and B in parallel, and the other consists of C and D in parallel. The two subsystems are arranged in series. Thus, to move from m to n, we must first pass through either A or B, and then through either C or D. The expression for this system is $(A + B)(C + D)$.

A component is described as a *variable* or a *constant* according to whether its state is alterable or unalterable, respectively. The expression $A0$ describes a system in which A is in series with a component that is always at state 0. Similarly, the expression $A + 1$ describes a system in which A is in parallel with a component that is always at state 1. Two or more components are said to be *identical* if they are always at the same state, and they are represented by an identical symbol. For example, the expression AA describes a system in which two identical components are arranged in series.

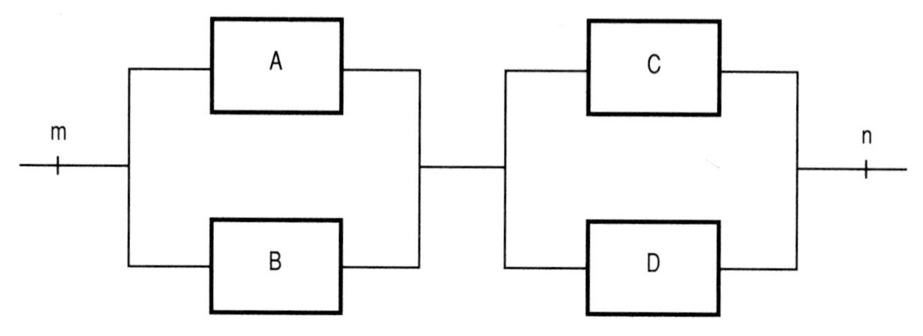

FIGURE 6.2 System consisting of subsystems arranged in series.

6.2 EQUIVALENCE

Assume that two systems contain some identical components. The two systems are *equivalent* to one another if they are always at the same state, and equivalence is expressed by use of an equals sign. For example, the statement $AA = A$ means that a system consisting of two components A in series is equivalent to a system having a single component A. A statement of equivalence is called an *equation*.

The following equations are self-evident:

$$A + 0 = A \tag{6.1}$$

$$A + 1 = 1 \tag{6.2}$$

$$A + A = A \tag{6.3}$$

$$A0 = 0 \tag{6.4}$$

$$A1 = A \tag{6.5}$$

$$AA = A \tag{6.6}$$

$$A + B = B + A \tag{6.7}$$

$$AB = BA \tag{6.8}$$

Equations (6.7) and (6.8) state that components A and B are *commutative*.

If three components are arranged in parallel or in series, any two may be considered to form a subsystem. Therefore, the components are *associative*. Expressed symbolically,

$$(A + B) + C = A + (B + C) \tag{6.9}$$

$$(AB)C = A(BC) \tag{6.10}$$

Consider the expression $A(B + C)$, which describes the system in Fig. 6.3*a*. In moving from m to n, we have two alternative paths: A and B, and A and C. Therefore, the system is passable if both A and B are passable, or if both A and C are passable. It follows that the system in Fig. 6.3*a* is equivalent to that in Fig. 6.3*b*, and we have

$$A(B + C) = AB + AC \tag{6.11}$$

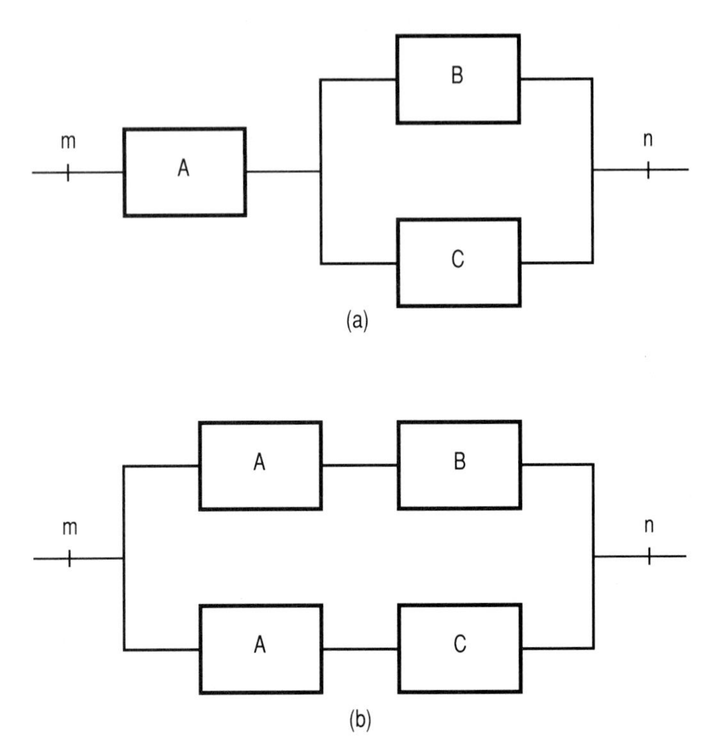

FIGURE 6.3 Equivalent systems. (*a*) System for $A(B + C)$; (*b*) system for $AB + AC$.

Therefore, in the expression at the left, A is *distributive*. Reversing the procedure, we see that it is possible to *factor* the repeating variable A in the expression at the right, thereby obtaining the expression at the left.

Equation (6.11) is extendible. Consider the expression $(A + B)(C + D)$, which describes the system in Fig. 6.2. In moving from m to n, we have four alternative paths: A and C, A and D, B and C, and B and D. It follows that

$$(A + B)(C + D) = AC + AD + BC + BD$$

Alternatively, this equation can be derived from Eq. (6.11), in this manner:

$$\begin{aligned}(A + B)(C + D) &= (A + B)C + (A + B)D \\ &= AC + BC + AD + BD \\ &= AC + AD + BC + BD\end{aligned}$$

Similarly, we have

$$(A + B + C)(D + E) = AD + AE + BD + BE + CD + CE$$

6.3 GENERALIZED TERMINOLOGY

The equations of Art. 6.2 were devised by visualizing a specific application of boolean algebra and invoking an illustrative system consisting of passable and impassable components. We have now reached a point where reference to a specific application is no longer essential, and we may therefore modify our terminology to make it more general. We may now speak of variables and constants, rather than components, of functions, rather than systems, and of terms in an expression, rather than subsystems. In our present terminology, Eq. (6.11) acquires the following meaning: The expression $A(B + C)$ is equivalent to the expression $AB + AC$.

A function is denoted by the letter f. The statement

$$f = AB + CD$$

means that f assumes the value 1 if either of these conditions exists: A and B are both 1; C and D are both 1. Thus, the expression at the right is a set of specifications; it lists the requirements that must be satisfied for f to become 1.

In the following material, the word *expression* refers to the expression for a function, and the value of the expression is the value of the function itself.

6.4 COMPLEMENTARY VARIABLES AND EXPRESSIONS

The symbol A' (read "not A") denotes a variable that is always at the state different from that of A. Thus, if $A = 0$, $A' = 1$, and vice versa. It follows at once that

$$A + A' = 1 \qquad (6.12)$$

For this reason, A and A' are said to be *complementary* to one another. Similarly, we have

$$AA' = 0 \qquad (6.13)$$

The symbol A'' denotes the complement of A', or $(A')'$. Then

$$A'' = A \qquad (6.14)$$

We shall consider A to be the independent variable and A' the dependent variable.

The NOT notation also applies to expressions as well as single variables. For example, the expression $(AB)'$ is always at the state different from that of AB. Thus, if $AB = 0$, $(AB)' = 1$, and vice versa. The expressions AB and $(AB)'$ are also said to be complementary to one another.

The conditions that make a given expression 1 are those that make its complementary expression 0. First consider the expression $(AB)'$. This expression is 1 when the expression AB is 0. For the latter condition, the requirement is simply that either A or B (or both) be 0. Expressed alternatively, the requirement is that either A' or B' (or both) be 1. Therefore,

$$(AB)' = A' + B' \qquad (6.15)$$

This equation is extendible. For example,

$$(ABC)' = A' + B' + C'$$

Now consider the expression $(A + B)'$. This expression is 1 when the expression

$A + B$ is 0. For the latter condition, the requirement is that both A and B be 0. Expressed alternatively, the requirement is that both A' and B' be 1. Therefore,

$$(A + B)' = A'B' \tag{6.16}$$

This equation is also extendible. For example,

$$(A + B + C)' = A'B'C'$$

Equations (6.15) and (6.16) are referred to as *De Morgan's laws.*

Consider the expression $(AB + AC + BC)'$. In the complementary expression $AB + AC + BC$, each term contains two of the three variables, and the variables are connected by the AND relationship. Therefore, the second expression is 0 if any two variables are 0. For example, if $A = 0$ and $B = 0$, all three terms are 0, and it follows that the second expression is 0. Then

$$(AB + AC + BC)' = A'B' + A'C' + B'C' \tag{6.17}$$

and

$$(AB + AC + BC)' = (A' + B')(A' + C')(B' + C') \tag{6.18}$$

An expression may contain both a given variable and its complementary variable. The expression $A + A'B$ is an illustration. This expression is 1 if either of the following conditions exists: $A = 1$; $A = 0$ and $B = 1$. Thus, it is simply necessary that either A or B be 1, and

$$A + A'B = A + B \tag{6.19}$$

6.5 ALGEBRAIC DERIVATION OF EQUATIONS

The equations of Arts. 6.2 and 6.4 were obtained through the use of pure logic, and they constitute the foundation of boolean algebra. Other equations can now be derived by purely algebraic methods. At each point in the process, we can manipulate the previously compiled equations, both basic and nonbasic. Nevertheless, in a relatively simple case, it may be preferable to derive a nonbasic equation by pure logic rather than by algebra. Moreover, where the logical method is followed, the illustrative system that we introduced in Art. 6.1 helps us considerably in visualizing the logical relationships. We shall now derive several equations both by pure logic and by algebra.

First consider the expression $A + AB$, which describes the system in Fig. 6.4. If

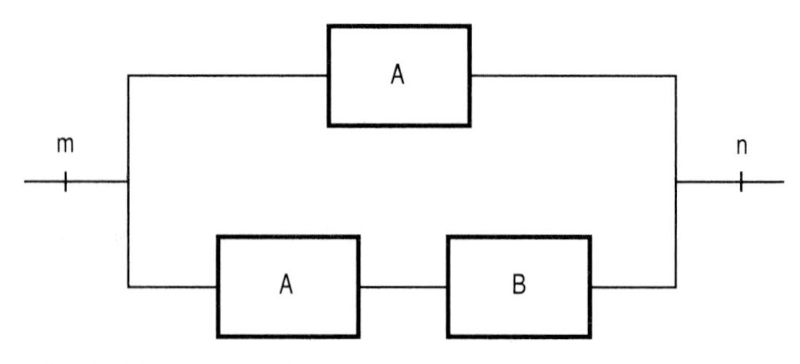

FIGURE 6.4 System for $A + AB$.

we are to traverse the system, it is mandatory that A be passable; whether B is passable or impassable is inconsequential. Therefore,

$$A + AB = A \tag{6.20}$$

The algebraic derivation of Eq. (6.20) is as follows: Let

$$f = A + AB$$

By Eq. (6.5),

$$f = A1 + AB$$

By factoring,

$$f = A(1 + B)$$

By Eq. (6.2),

$$f = A1$$

By Eq. (6.5),

$$f = A$$

Equation (6.20) is extendible. For example,

$$A + AB + AC = A$$

Now consider the expression $A(A + B)$, which describes the system in Fig. 6.5. If we are to traverse the system, it is mandatory that A be passable; whether B is passable or impassable is inconsequential. Therefore,

$$A(A + B) = A \tag{6.21}$$

The algebraic derivation of Eq. (6.21) is as follows: Let

$$f = A(A + B)$$

By distributive law,

$$f = AA + AB$$

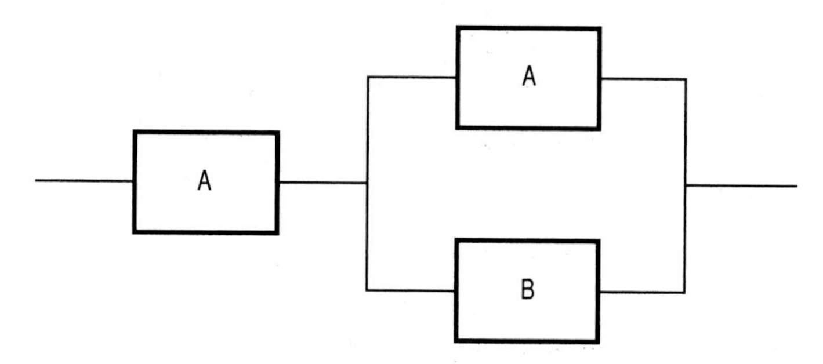

FIGURE 6.5 System for $A(A + B)$.

By Eq. (6.6),

$$f = A + AB$$

By Eq. (6.20),

$$f = A$$

Now consider the expression $(A + B)(A + C)$, which describes the system in Fig. 6.6*a*. We can traverse the system if either of the following conditions exists (or if both exist): A is passable; both B and C are passable. Therefore, the system is equivalent to that in Fig. 6.6*b*, and we have

$$(A + B)(A + C) = A + BC \tag{6.22}$$

The algebraic derivation of Eq. (6.22) is as follows: Let

$$f = (A + B)(A + C)$$

By distributive law,

$$f = AA + AC + AB + BC$$

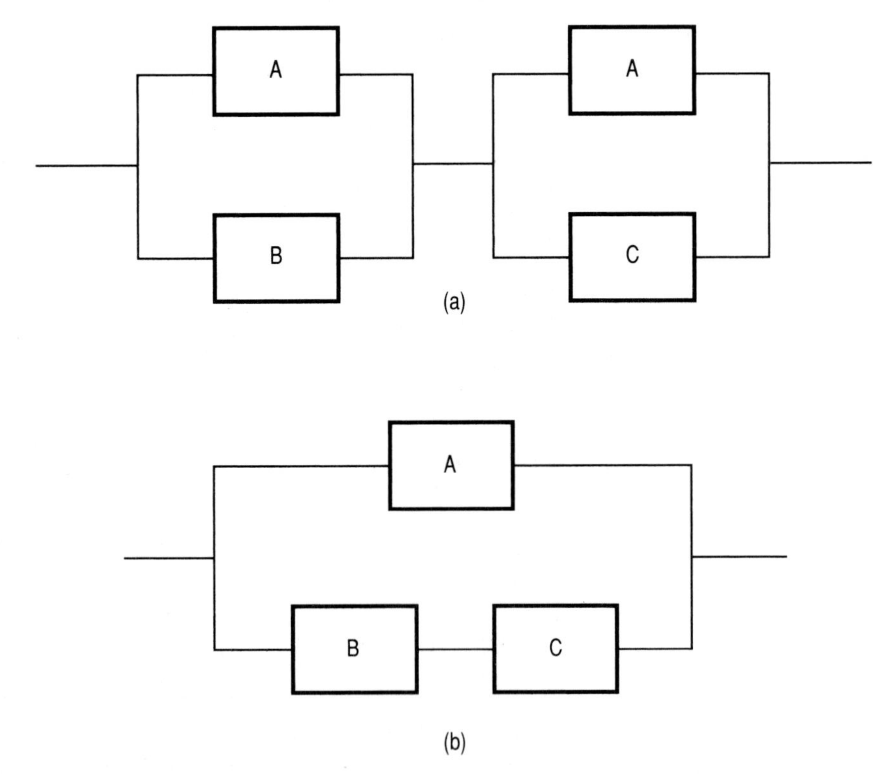

FIGURE 6.6 Equivalent systems. (*a*) System for $(A + B)(A + C)$; (*b*) system for $A + BC$.

By Eq. (6.6),

$$f = A + AC + AB + BC$$

By Eq. (6.20),

$$f = A + BC$$

The expression $A(A' + B)$ requires that $A = 1$ and, correspondingly, $A' = 0$. Therefore, it is mandatory that $B = 1$. Thus,

$$A(A' + B) = AB \tag{6.23}$$

The algebraic derivation of Eq. (6.23) is as follows:

$$A(A' + B) = AA' + AB = 0 + AB = AB$$

The expression $AB + AB'$ requires that $A = 1$. The value of B is inconsequential because either B or B' is 1. Thus,

$$AB + AB' = A \tag{6.24}$$

The algebraic derivation of Eq. (6.24) is as follows:

$$AB + AB' = A(B + B') = A1 = A$$

6.6 LAWS OF DUALITY

Boolean algebra is characterized by a duality that imparts an image to each expression and each equation. This duality can be exploited to considerable advantage. There are two laws of duality, and we shall discuss each one in turn.

Theorem 6.1. A given expression is transformed to its complementary expression if the OR and AND relationships are interchanged and each variable is replaced with its complementary variable.

This principle stems from Eqs. (6.15) and (6.16). Through repeated application of these equations, it can be shown to apply to any expression, however complex. As an illustration, let

$$f = AB'(C + D'E)$$

By Eq. (6.15),

$$f' = A' + B + (C + D'E)'$$

By Eq. (6.16),

$$f' = A' + B + C'(D'E)'$$

By Eq. (6.15)

$$f' = A' + B + C'(D + E')$$

Thus, $$[AB'(C + D'E)]' = A' + B + C'(D + E')$$

Theorem 6.2. A given equation is transformed to a corresponding equation if the OR and AND relationships are interchanged and the constants 0 and 1 are interchanged.

The corresponding equation is called the *dual* of the first. By selecting equations and making the prescribed interchanges, we find, for example, that Eq. (6.5) is the dual of Eq. (6.1), and Eq. (6.23) is the dual of Eq. (6.19). The proof of this principle lies in the fact that each basic equation has a dual. Since a nonbasic equation results from a combination of basic equations, it follows that every nonbasic equation also has a dual.

6.7 USE OF TRUTH TABLES

It is often advantageous to construct a table in which we record every possible combination of values of the independent variables and the corresponding value of an expression. This table is referred to as a *truth table*. If the expression contains n independent variables, the number of combinations of values is 2^n. The rows of the table are numbered consecutively, starting with 0. By convention, the combinations are recorded in such sequence that the digits in the combination form the row number in the binary system. As an illustration, refer to Table 6.1, which is the truth table for the expressions $A + B$ and AB.

Truth tables have many uses, one of which is to confirm a given equation. As an illustration, we shall confirm the following equation, which results from the first law of duality:

$$(A + B'C)' = A'(B + C') \qquad (a)$$

Refer to Table 6.2. We find that $A + B'C$ and $A'(B + C')$ always differ in value, and Eq. (a) is thus confirmed.

TABLE 6.1

Row	A	B	$A + B$	AB
0	0	0	0	0
1	0	1	1	0
2	1	0	1	0
3	1	1	1	1

TABLE 6.2

Row	A	B	C	$B'C$	$B + C'$	$A + B'C$	$A'(B + C')$
0	0	0	0	0	1	0	1
1	0	0	1	1	0	1	0
2	0	1	0	0	1	0	1
3	0	1	1	0	1	0	1
4	1	0	0	0	1	1	0
5	1	0	1	1	0	1	0
6	1	1	0	0	1	1	0
7	1	1	1	0	1	1	0

6.8 TYPES OF EXPRESSIONS

It is possible to assign names to relatively simple boolean expressions and terms on the basis of the manner in which they are constructed. Expressions such as $A + B' + C$ and ABC' are called OR and AND expressions (or terms), respectively. Where a single variable stands alone, it may be viewed as either an OR or an AND term, since $A = A + 0$ and $A = A1$.

An AND-to-OR expression is one in which AND terms are connected to one another by the OR relationship. Thus, $AB' + BCD + CE'$ and $A' + BCD' + B'E$ are AND-to-OR expressions. An OR-to-AND expression is one in which OR terms are connected to one another by the AND relationship. Thus, $(A + B)(B + C' + D)$ and $A(A' + B)$ are OR-to-AND expressions.

We have found that it is possible to convert a given AND-to-OR expression to an equivalent OR-to-AND expression, and vice versa, by applying the first law of duality. However, in many instances a conversion can be performed through repeated application of Eq. (6.22), which we repeat in this form:

$$A + BC = (A + B)(A + C)$$

As an illustration, let $f = AB + CD$. If we apply Eq. (6.22) three times with the appropriate substitutions, we obtain the following:

$$\begin{aligned} f &= (AB + C)(AB + D) \\ &= [(A + C)(B + C)][(A + D)(B + D)] \end{aligned}$$

Thus,

$$AB + CD = (A + C)(A + D)(B + C)(B + D) \tag{6.25}$$

Unfortunately, the terminology concerning boolean expressions has not been standardized. In some texts, what we have called AND-to-OR and OR-to-AND expressions are called *disjunctive* and *conjunctive* expressions, respectively.

6.9 STANDARD EXPRESSIONS

Assume that f is a function of n independent variables. In the expression for f, a term is said to be in *standard form* if it is either an OR or AND term and each variable (or its complement) appears in the term once and only once. The following are standard terms for a four-variable function:

$$AB'C'D \qquad A'BCD \qquad A + B + C + D'$$

$$A' + B' + C + D$$

The expression for f is said to be in *standard, elemental,* or *canonical form* if it is composed of standard terms. The following are standard four-variable expressions:

$$AB'CD' + A'BC'D' + AB'C'D \qquad ABCD + AB'C'D$$

$$(A + B' + C + D)(A' + B + C + D')$$

Every nonstandard AND-to-OR expression can be expanded to standard form. As an illustration, let

$$f = A'BC + AB' + B$$

The first term is standard, the second term lacks C (or C'), and the third term lacks both A and C (or A' and C'). We therefore apply Eq. (6.12) and write the following:

$$f = A'BC + AB'(C + C') + (A + A')B(C + C')$$
$$= A'BC + AB'C + AB'C' + ABC + ABC' + A'BC + A'BC'$$

Since the first and sixth terms are identical and $A + A = A$, we delete the sixth term, and the final expression is

$$f = A'BC + AB'C + AB'C' + ABC + ABC' + A'BC'$$

From the principle of duality, it follows that every nonstandard OR-to-AND expression can also be expanded to standard form.

Since each standard term contains either a given independent variable or its complement and the number of independent variables is n, the number of standard terms that can be formed is 2^n. If we expand our definition of a standard expression to include the degenerate cases where f is a constant, we may say that a standard expression is a combination of m standard terms, where $0 < m < 2^n$. Therefore, in accordance with Eqs. (4.5) and (1.29), the number of possible standard expressions is 2 raised to the 2^nth power. Each standard expression corresponds to a unique function of the 6 independent variables.

As an illustration, let $n = 2$. The number of possible standard expressions is $2^4 = 16$, and the AND-to-OR forms are recorded in Table 6.3. Since the last expression encompasses all possible terms, it corresponds to the case $f = 1$.

In some texts, standard AND and OR terms are called *minterms* and *maxterms*, respectively. Standard AND-to-OR and OR-to-AND expressions are called *minterm forms* and *maxterm forms*, respectively.

TABLE 6.3 Standard AND-to-OR Expressions for Two-Variable Functions

0	$A'B + AB'$
$A'B'$	$A'B + AB$
$A'B$	$AB' + AB$
AB'	$A'B' + A'B + AB'$
AB	$A'B' + A'B + AB$
$A'B' + A'B$	$A'B' + AB' + AB$
$A'B' + AB'$	$A'B + AB' + AB$
$A'B' + AB$	$A'B' + A'B + AB' + AB$

6.10 REDUNDANCIES

To justify its presence, each variable or term that appears in a boolean expression must either introduce a new opportunity or impose a new requirement for the expression to assume the value 1. If we find upon examination that a particular variable or term fails to perform either of these functions, that variable or term is redundant, and it may be eliminated.

From our present point of view, many equations that we have presented may be interpreted as statements of redundancy. For example, Eq. (6.21) states that the term $A + B$ is redundant in the expression $A(A + B)$, and Eq. (6.23) states that the variable A' is redundant in the expression $A(A' + B)$.

We shall now consider another case of redundancy. Let

$$f = AB + A'C + BC$$

If $A = 1$ and $B = 1$, the first term is 1. If $A = 0$ and $C = 1$, the second term is 1. Therefore, $f = 1$ if *either* B *or* C is 1, and allowing both B and C to be 1 does not introduce a new opportunity for f to be 1. Thus, the term BC is redundant, and we have

$$AB + A'C + BC = AB + A'C \tag{6.26}$$

The algebraic derivation of Eq. (6.26) is as follows:

$$\begin{aligned} AB + A'C + BC &= AB1 + A'C1 + BC1 \\ &= AB1 + A'C1 + BC(A + A') \\ &= AB1 + A'C1 + ABC + A'BC \\ &= AB(1 + C) + A'C(1 + B) \\ &= AB1 + A'C1 \\ &= AB + A'C \end{aligned}$$

6.11 REDUCTION OF EXPRESSIONS

A letter that represents a variable is termed a *literal.* In counting literals, we include duplicates. For example, the expression $AB' + AB'CD + CDE'$ contains nine literals.

A given expression is said to be *reduced to simpler form* when it is transformed to an equivalent expression with fewer literals. Before presenting the methods of reduction, we must define two terms.

Two AND terms are said to be *compatible* with one another if one term can be obtained from the other by replacing one and only one variable with its complement. For example, the terms $AB'D'G$ and $ABD'G$ are compatible because the second term can be obtained from the first by replacing B' with B. A given AND term is said to *include* a longer AND term if the variables that appear in the former also appear in the latter. For example, the term $AC'D$ includes the term $AB'C'DEF'$ because the latter contains A, C', and D. It will be convenient to extend the definition of inclusion by saying that a given AND term includes itself.

A given expression can be reduced to simpler form by applying any of the following methods, or a combination of these methods:

1. *Consolidation of terms:* If two compatible terms are connected by the OR relationship, they can be combined in accordance with Eq. (6.24), or an extension of this equation. For example,

$$\begin{aligned} AB'D'G + ABD'G &= AD'G(B' + B) \\ &= AD'G1 = AD'G \end{aligned}$$

Thus, the combined term contains solely the variables that are common to the compatible terms.

2. *Elimination of redundancies:* For example, Eq. (6.19) reduces the expression $A + AB'$, which contains three literals, to $A + B$, which contains only two.
3. *Factoring:* For example, the equation $AB + AC = A(B + C)$ reduces an expression with four literals to an expression with three.

In analyzing an AND-to-OR expression for compatibility, we may find that a given term is compatible with several terms in the expression. In this case, it is permissible to combine each compatible pair of terms. The justification is that $A + A = A$, and therefore this duplication of terms does not inject any error. For example, let

$$f = AB'CD + AB'C'D + A'B'CD$$

Terms 1 and 2 combine to form $AB'D$; terms 1 and 3 combine to form $B'CD$. Thus, the expression for f has been reduced to

$$f = AB'D + B'CD$$

On the other hand, the consolidation of terms can inadvertently induce redundancies. As an illustration, let

$$f = ABC' + ABC + A'BC + A'B'C$$

Terms 1 and 2 combine to form AB; terms 3 and 4 combine to form $A'C$; terms 2 and 3 combine to form BC. Thus, the expression for f has been reduced to

$$f = AB + A'C + BC$$

However, Eq. (6.26) states that BC is redundant, and consequently it was unnecessary to combine terms 2 and 3.

On the basis of the foregoing discussion, we arrive at this conclusion: In reducing an AND-to-OR expression through consolidation of terms, it is essential that each term in the original expression be included in at least one term in the reduced expression. However, if a term in the original expression is included in more than one term in the reduced expression, a redundancy *may* exist.

The reduction of a given expression to simpler form is a prime objective in many applications of boolean algebra, and we shall now consider how the task can be performed methodically. For generality, we start with a standard AND-to-OR expression. The simplest expression to which this can be reduced through consolidation of terms and elimination of redundancies is called its *optimal expression.* If the optimal expression can be reduced through factoring, the expression that emerges is called the *minimal expression* corresponding to the original expression. As we shall find, a given expression may have multiple minimal expressions, since there may be a choice in the variables to be factored.

6.12 ALGEBRAIC METHOD OF REDUCTION

In 1952, W. V. Quine presented a formalized algebraic procedure for reducing a standard AND-to-OR expression to its optimal form. The terminology of this method and the principles on which it rests have already been discussed in Art. 6.11. We shall add one definition. If a given term in an AND-to-OR expression is incompatible with any other term in the expression, it is called a *prime implicant.*

To illustrate the Quine procedure, let

$$f = AB'C'D + A'BC'D + A'B'C'D + ABCD' + A'B'C'D' + AB'C'D' + A'BCD \qquad (b)$$

We shall initially reduce this expression by consolidating terms, proceeding in cycles until all possibilities have been exhausted. In cycle 1, we number the terms in (b) and record them in columnar form in Table 6.4. Starting with term 1, we compare it with each subsequent term for compatibility. We find that term 1 is compatible with both term 3 and term 6. Each compatible pair of terms is identified in Table 6.4 by placing the mark X in a unique column and on the same row as the term. We next take term 2, and we find that it is compatible with both term 3 and term 7. Similarly, term 3 is compatible with term 5, and term 5 is compatible with term 6. Term 4 is a prime implicant; since it remains uncombined, it must be carried into the subsequent expression for f.

In Table 6.5, we record the numbers of the terms that are compatible and the combined term they form. For example, terms 1 and 3 form $B'C'D$, and terms 1 and 6 form $AB'C'$. At the end of cycle 1, the expression for f has been reduced to the following:

$$f = ABCD' + B'C'D + AB'C' + A'C'D + A'BD + A'B'C' + B'C'D' \qquad (c)$$

However, it is seen at once that the expression in (c) also contains compatible terms, and therefore a second cycle of consolidation is needed.

In cycle 2, we number all terms in (c) beyond the first and record them in Table 6.4. Proceeding as before, we find that term 1 is compatible with term 6, term

TABLE 6.4 Identity of Compatible Pairs of Terms

Cycle 1								Cycle 2			
1	$AB'C'D$	X	X					1	$B'C'D$	X	
2	$A'BC'D$			X	X			2	$AB'C'$		X
3	$A'B'C'D$	X		X		X		3	$A'C'D$		
4	$ABCD'$							4	$A'BD$		
5	$A'B'C'D'$					X	X	5	$A'B'C'$		X
6	$AB'C'D'$		X				X	6	$B'C'D'$	X	
7	$A'BCD$				X						

TABLE 6.5

Cycle	Compatible terms	Combined term
1	1 and 3	$B'C'D$
	1 and 6	$AB'C'$
	2 and 3	$A'C'D$
	2 and 7	$A'BD$
	3 and 5	$A'B'C'$
	5 and 6	$B'C'D'$
2	1 and 6	$B'C'$
	2 and 5	$B'C'$

2 is compatible with term 5, and terms 3 and 4 are prime implicants. Returning to Table 6.5, we find that both compatible pairs form $B'C'$. Since $A + A = A$, we take $B'C'$ only once. At the end of cycle 2, the expression for f has been reduced to the following:

$$f = ABCD' + A'C'D + A'BD + B'C' \qquad (d)$$

The terms in (d) are incompatible with one another, and the consolidation process is now complete.

The second and final step in the Quine procedure is to investigate (d) for redundancies and to eliminate any that may exist. This is accomplished by ascertaining how the terms in (b) are included in those of (d). In Table 6.6, we again record the terms in (b) in columnar form at the left, and we record the terms in (d) in a horizontal row across the top. We now take each term in (d) and compare it with each term in (b) with respect to inclusion. We find that $ABCD'$ includes only term 4, and we indicate this condition by placing the mark X at the indicated location. We next find that $A'C'D$ includes terms 2 and 3, $A'BD$ includes terms 2 and 7, etc. In effect, the information offered by Table 6.6 is a duplication of that offered by Table 6.5, but in a more suitable form.

We now proceed to construct the optimal expression for f. As stated in Art. 6.11, each term in the original expression must be included at least once in a reduced expression. Therefore, the optimal expression must contain $ABCD'$ because this is the only term in (d) that includes term 4 in (b). It will be helpful to circle term 4 to signify that it is now embodied in the optimal expression. Similarly, the optimal expression must contain $B'C'$ because this is the only term in (d) that includes terms 1, 5, and 6 in (b). Finally, the optimal expression must contain $A'BD$ because this is the only term in (d) that contains term 7 in (b). We now conclude that $A'C'D$ is redundant because terms 2 and 3 are already embodied in the optimal expression. Therefore, the optimal expression is

$$f = ABCD' + B'C' + A'BD$$

By factoring, we now obtain the minimal expression

$$f = B(ACD' + A'D) + B'C'$$

In the second step, we discovered that

$$B'C' + A'BD + A'C'D = B'C' + A'BD$$

TABLE 6.6

	Terms in (d)			
Terms in (b)	$ABCD'$	$A'C'D$	$A'BD$	$B'C'$
1 $AB'C'D$				X
2 $A'BC'D$		X	X	
3 $A'B'C'D$		X		X
4 $ABCD'$	X			
5 $A'B'C'D'$				X
6 $AB'C'D'$				X
7 $A'BCD$			X	

This equation can readily be proved by a method similar to that in the proof of Eq. (6.26).

6.13 CONSTRUCTION AND PROPERTIES OF KARNAUGH MAPS

The algebraic method of reducing an expression that was developed in Art. 6.12 leads to a graphical method that is highly efficient. In the graphical method, each possible term in a standard AND-to-OR expression is represented by an exclusive region of space, and regions that represent compatible terms fall into adjacent positions. As a result of this adjacency, compatible terms are instantly recognizable, and the consolidation of terms proceeds very rapidly.

The graphical method involves construction of a *Karnaugh map* (or *truth map*), which is a rectangular array of squares or *cells*. Each cell corresponds to a specific standard AND term. If n denotes the number of independent variables, the number of cells in the map is 2^n, in accordance with the statement in Art. 6.9. In this array, the horizontal rows are numbered from the top down, and the vertical columns are numbered from left to right, starting with the number 1 in both cases. A cell is identified by specifying its row and column, in that order. For example, cell 25 (read "two five") lies in row 2 and column 5.

To describe the construction of a Karnaugh map, we start with an expression containing three independent variables: A, B, and C. The map for this expression will contain $2^3 = 8$ cells. Refer to Fig. 6.7, which is an array consisting of 2 rows and 4 columns. Each variable, both independent and dependent, must be represented by at least 1 row or column. We have made the following assignments: row 1 to A; row 2 to A'; columns 1 and 2 to B; columns 3 and 4 to B'; columns 2 and 3 to C; columns 1 and 4 to C'. These assignments are recorded by placing the appropriate labels along the periphery of the array, as shown. It is advantageous to visualize that the map is inscribed on a torus rather than a plane, the result being that it is circular in both directions. Thus, columns 1 and 4 are adjacent to one another.

As a result of the row and column assignments, there is a change of only one variable as we move from one row or column to an adjacent row or column, and each cell corresponds to a unique term. For example, cell 11 corresponds to ABC', cell 21 corresponds to $A'BC'$, and cell 23 corresponds to $A'B'C$. Moreover, adjacent cells correspond to compatible terms. Thus, cells 12 and 22 correspond, respectively, to ABC and $A'BC$, and these terms are compatible. Similarly, cells 21 and 24 (which are adjacent) correspond, respectively, to $A'BC'$ and $A'B'C'$, and these terms are compatible.

	B	B	B'	B'
A		1		1
A'		1		
	C'	C	C	C'

FIGURE 6.7 Karnaugh map for three-variable expression.

Now let $f = ABC + AB'C' + A'BC$. These terms are represented by cells 12, 14, and 22, and the expression is represented by placing the numeral 1 in each of these cells. Thus, each arrangement of 1's that can be devised corresponds to a unique three-variable expression. The cells marked 1 are referred to as *p cells*.

We now turn to an expression containing four independent variables. The map for this expression will contain $2^4 = 16$ cells. Refer to Fig. 6.8, where the rows and columns are assigned in the manner indicated. Again, we visualize the map as being

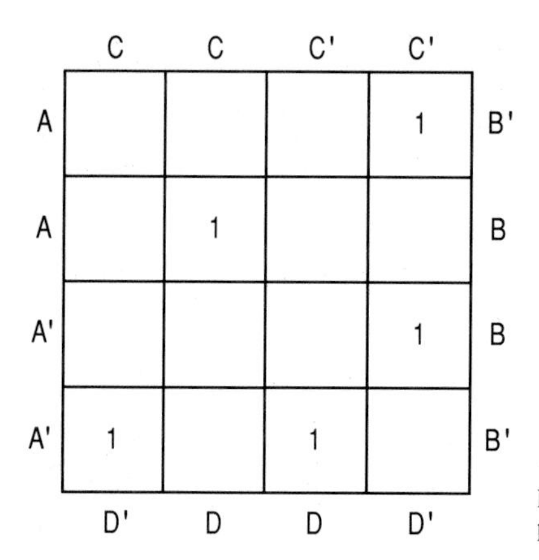

FIGURE 6.8 Karnaugh map for four-variable expression.

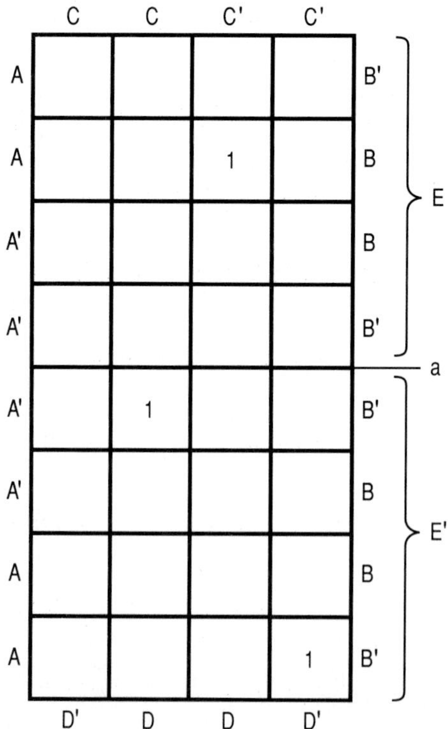

FIGURE 6.9 Karnaugh map for five-variable expression.

circular in both directions; as a result, rows 1 and 4 are adjacent, and columns 1 and 4 are adjacent. Now let

$$f = A'B'CD' + A'BC'D' + ABCD + AB'C'D' + A'B'C'D$$

These terms are represented by cells 41, 34, 22, 14, and 43, respectively, and we place the numeral 1 in each of these cells.

An expression containing five independent variables requires $2^5 = 32$ cells. Refer to Fig. 6.9, where we have an array consisting of 8 rows and 4 columns. We may visualize that the subarray for E' is obtained by revolving the subarray for E about line a. As we move from row 4 to row 5, E changes to E'; as we move from row 8 to the adjacent row 1, E' changes to E. Figure 6.9 represents the expression

$$f = A'B'CDE' + ABC'DE + AB'C'D'E'$$

It is apparent that if we continually double the number of rows or columns, we can theoretically construct Karnaugh maps to accommodate expressions with any number of independent variables. In all instances, caution must be exercised to ensure that only one variable changes to its complementary variable as we move from one row or column to an adjacent row or column.

In the following material, we shall append to the symbol f a subscript that corresponds to the number of the drawing in which its expression is mapped.

6.14 REDUCTION BY USE OF KARNAUGH MAPS

We shall now demonstrate the remarkable utility of a Karnaugh map in reducing a standard AND-to-OR expression to optimal form. Let m denote a positive integer, and assume that a map contains 2^m adjacent p cells. These cells are said to form an *mth-order block*. The terms that correspond to the cells in this block can be combined into a single term that contains solely the variables within which the block is confined.

As an illustration, let

$$f_{10} = ABCD' + ABCD + A'BCD' + A'BCD$$

This expression is mapped in Fig. 6.10. The p cells form a second-order block, and

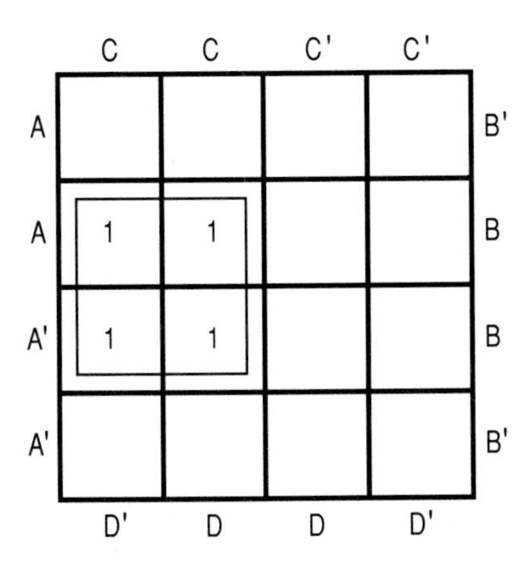

FIGURE 6.10 Karnaugh map with second-order block.

this block is confined within the rows for B and the columns for C. Therefore, the expression for f_{10} reduces to $f_{10} = BC$. The proof is as follows:

$$f_{10} = BC(AD' + AD + A'D' + A'D)$$
$$= BC[A(D' + D) + A'(D' + D)]$$
$$= BC(A + A') = BC$$

Now refer to Fig. 6.11. Since columns 1 and 4 are adjacent, the p cells in this map form a third-order block. This block is confined solely within the columns for D', and therefore the standard expression that is mapped in this drawing has D' as its optimal expression.

In general, if the standard expression contains n independent variables and the block is of the mth order, the combined term corresponding to this block contains $n - m$ variables. It is convenient to view an isolated p cell as a zero-order block, the justification being that $2^0 = 1$. The extent of a block is shown by enclosing it with light lines.

We now turn to the expression that was reduced algebraically in Art. 6.12. Let

$$f_{12} = AB'C'D + A'BC'D + A'B'C'D + ABCD' + A'B'C'D' + AB'C'D' + A'BCD$$

This expression is mapped in Fig. 6.12. Since rows 1 and 4 are adjacent, cells 13, 14, 43, and 44 form a second-order block, and their terms combine to form $B'C'$. Cells 32 and 33 form a first-order block, and their terms combine to form $A'BD$. Cell 21 is a zero-order block, and its term is $ABCD'$. At this point, all p cells have been taken into account, and the optimal expression is complete. It is true that cells 33 and 43 constitute a block, but both these cells have already been taken into account. Therefore, carrying the term $A'C'D$ into the reduced expression would be redundant. Thus, the optimal expression for f_{12} is

$$f_{12} = B'C' + A'BD + ABCD'$$

This result coincides with that obtained algebraically in Art. 6.12.

Refer to Fig. 6.13. Blocks are formed in the manner indicated, and the expression that is mapped in this drawing has the optimal form

$$f_{13} = BC + ACD + ABD$$

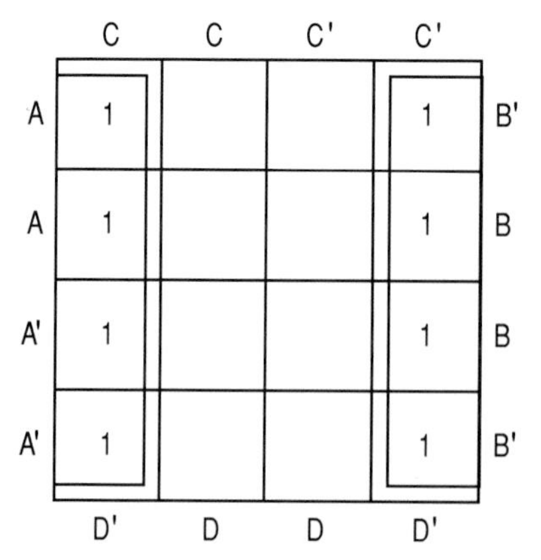

FIGURE 6.11 Karnaugh map with third-order block.

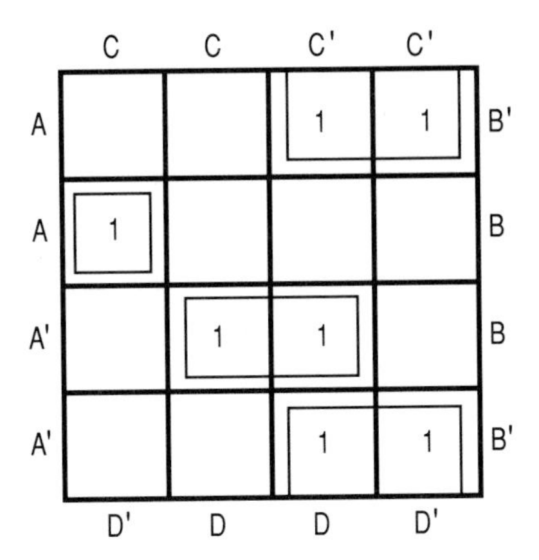

FIGURE 6.12 Map with distinctive blocks.

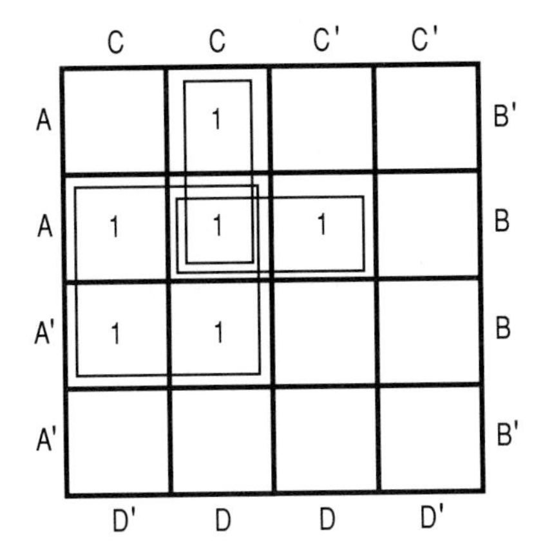

FIGURE 6.13 Map with overlapping blocks.

Cell 22 is located in all three blocks, but this duplication is necessary. Now refer to Fig. 6.14. It is possible to combine the p cells to form one 4-cell block and two 2-cell blocks. The resulting expression would have $2+3+3=8$ literals. On the other hand, it is possible to form 4-cell blocks exclusively, in the manner shown, and the resulting expression is

$$f_{14} = AD + BC + BD$$

This expression contains only six literals, and it is the optimal expression.

We can now formulate the procedure to be followed in reducing a given expression to its optimal form through use of a Karnaugh map. It consists of the following steps: Combine the p cells to form the largest possible blocks, proceeding from blocks of maximum order to blocks of progressively lower order. Then record the terms corresponding to these blocks, and connect them with the OR relationship.

After the optimal expression has been constructed, the minimal expression can readily be obtained by factoring. For example, consider the optimal expression for

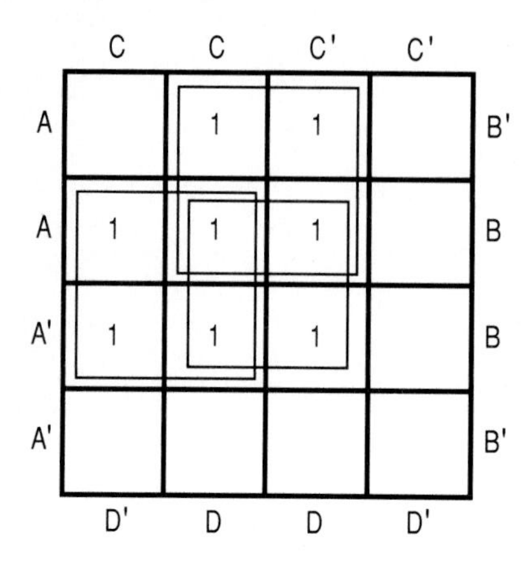

FIGURE 6.14 Map with overlapping second-order blocks.

f_{14} that was previously constructed. We obtain the following alternative minimal expressions:

$$f_{14} = D(A + B) + BC$$

$$f_{14} = AD + B(C + D)$$

Each expression contains five literals.

6.15 USE OF INCOMPLETE BLOCKS

Figure 6.15 is the map of the expression

$$f_{15} = AB'C + A'BC + A'B'C$$

Combining the cells in the manner shown, we obtain the optimal expression

$$f_{15} = A'C + B'C$$

By factoring, we then obtain the minimal expression

$$f_{15} = C(A' + B')$$

We shall now apply this elementary function to illustrate an important principle.

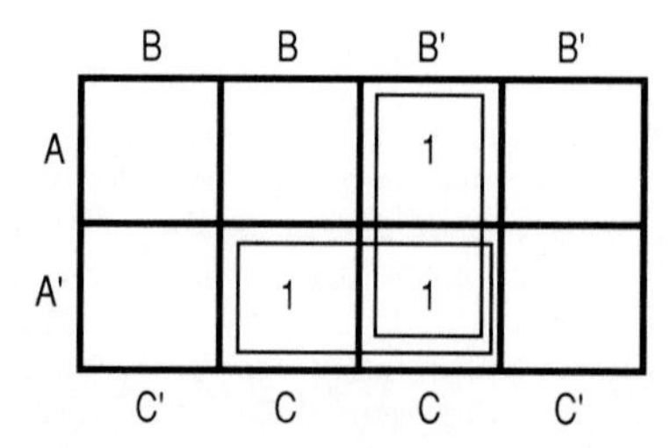

FIGURE 6.15 Map with an incomplete second-order block.

Assume that the p cells in a Karnaugh map would form a block of the mth order if a relatively small number of adjacent cells were transformed to p cells. The true p cells are said to constitute an *incomplete block* of the mth order. Since our policy is to form the largest blocks possible, incomplete blocks can be used to considerable advantage. For example, in Fig. 6.15, the p cells would form a second-order block if cell 12 were transformed to a p cell. Therefore, we tentatively make this transformation, and the expression corresponding to this 4-cell block is C. We now apply a correction by imposing the additional requirement that ABC be 0. At this point, our expression for f_{15} has been transformed to

$$f_{15} = C(ABC)'$$

By Eq. (6.15),

$$f_{15} = C(A' + B' + C')$$

Since $CC' = 0$, the last expression reduces to

$$f_{15} = C(A' + B')$$

Thus, through use of an incomplete block, we can arrive directly at the minimal expression corresponding to a given expression. This possibility is of considerable value where the given expression is relatively long.

Now refer to Fig. 6.16. Forming the second-order and first-order blocks indicated, we obtain the optimal expression

$$f_{16} = CD + B'C'D$$

By factoring,

$$f_{16} = D(C + B'C')$$

By Eq. (6.19),

$$f_{16} = D(B' + C)$$

This is the minimal expression, and we shall now obtain it through use of an incomplete block. We tentatively transform cells 23 and 33 to p cells, thereby

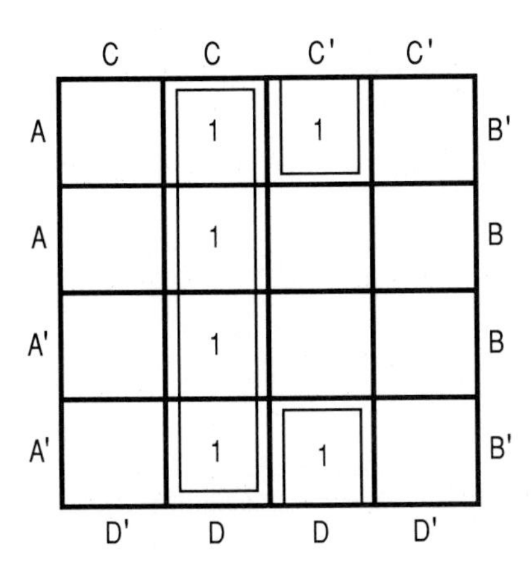

FIGURE 6.16 Map with an incomplete third-order block.

forming a third-order block having D as its expression. Cells 23 and 33 form a first-order block having $BC'D$ as its term. Imposing the additional requirement that $BC'D$ be 0, we obtain

$$f_{16} = D(BC'D)' = D(B' + C + D') = D(B' + C)$$

In the present case, use of an incomplete block provided a far more direct route to the minimal expression.

6.16 REDUCTION OF COMPLEMENTARY EXPRESSION

In constructing the Karnaugh map for a given AND-to-OR expression, we placed the numeral 1 in each cell corresponding to a term in the expression. Now consider that we place the numeral 0 in all remaining cells. The map of a given expression can be instantly transformed to that of its complementary expression merely by changing the numeral in each cell. The rationale is as follows:

Assume that a given expression contains n independent variables. As previously stated, there are 2^n possible standard AND terms. When a set of values is assigned to the n variables, one and only one term acquires the value 1. For example, if we set $A = 0$, $B = 1$, $C = 1$, $D = 0$, the term $A'BCD'$ becomes 1, and all other terms become 0. If the given expression contains $A'BCD'$, it becomes 1; otherwise, it becomes 0. Thus, we may say that each cell in the map corresponds to a specific set of values of the variables. The numeral appearing in the cell gives the value that the expression will acquire if the set of values corresponding to that cell materializes.

As an illustration, consider again the expression that is mapped in Fig. 6.16. The complementary expression is mapped in Fig. 6.17 by applying the transformation rule. Forming the blocks shown, we obtain

$$f_{17} = f'_{16} = D' + BC'$$

No factoring is possible, and this is the minimal expression for f'_{16}. Comparing it with the minimal expression for f_{16}, we find that the two expressions are consonant with the first law of duality.

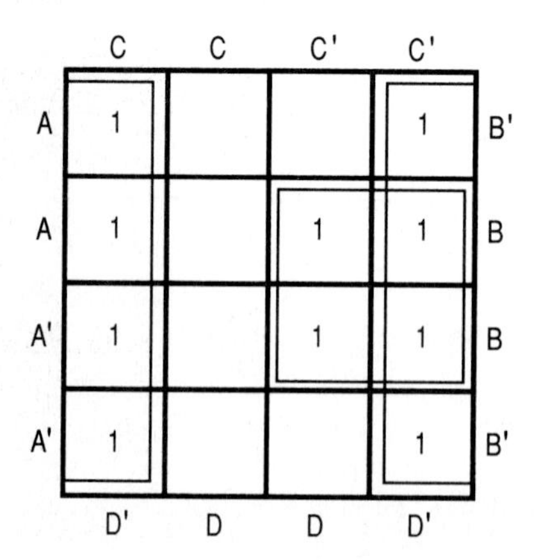

FIGURE 6.17 Map of expression that is complementary to expression mapped in Fig. 6.16.

6.17 CHANGE OF FORM OF EXPRESSION

Where the standard AND-to-OR expression for a function is given, it may be desirable to transform this to a standard OR-to-AND expression. This transformation can readily be achieved.

To illustrate the procedure, refer to Fig. 6.18, which is the map of the function

$$f_{18} = ABC' + AB'C' + A'BC + A'B'C$$

The expression for the complementary function is

$$f'_{18} = ABC + AB'C + A'BC' + A'B'C'$$

Applying the first law of duality to the last result, we obtain

$$f_{18} = (A' + B' + C')(A' + B + C')(A + B' + C)(A + B + C)$$

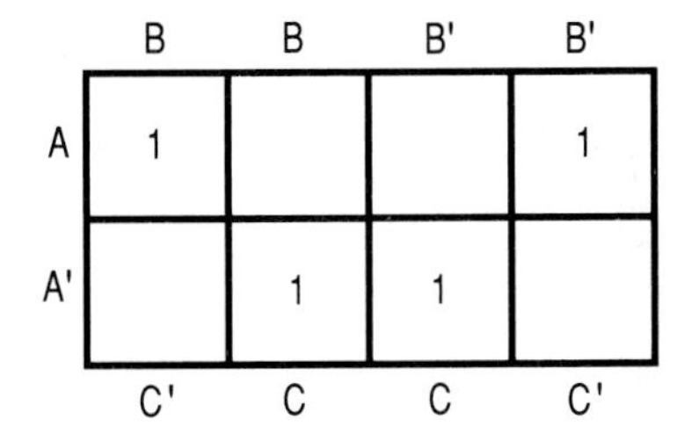

FIGURE 6.18 Use of Karnaugh map to change form of a standard expression.

6.18 IRRELEVANT TERMS

As previously stated, if there are n independent variables, there are normally 2^n possible sets of values of these variables, and each set of values corresponds to a specific standard AND term. However, in practice certain sets of values may never occur, and the AND terms corresponding to these impossible sets of values are called *irrelevant* or *don't-care terms*. It is permissible to enlarge the true expression for the function to include irrelevant terms if doing so assists us in reducing the expression. Since irrelevant terms can never become 1, their presence in the expression does not introduce any error.

As a very simple illustration, let f_{19} denote a function of the variables A, B, and C. In Table 6.7, we have recorded the sets of values that make $f_{19} = 1$ and those that

TABLE 6.7

Condition	A	B	C
	1	1	0
$f = 1$	1	0	1
	0	1	0
	1	1	1
$f = 0$	1	0	0
	0	1	1

make $f_{19} = 0$. The two remaining sets of values are precluded by physical conditions. Thus, the true expression for f_{19} is

$$f_{19} = ABC' + AB'C + A'BC'$$

Now refer to Fig. 6.19. The cells corresponding to $f_{19} = 1$ and $f_{19} = 0$ are marked with the corresponding numeral, and the two remaining cells are marked with an X. If we exclude cell 23 from consideration, we combine the three *p* cells in the manner shown to obtain

$$f_{19} = AB'C + BC' \qquad (e)$$

On the other hand, if we take cell 23 into account, we combine this with cell 13 to obtain

$$f_{19} = B'C + BC' \qquad (f)$$

The expression in (f) is preferable to that in (e) because it contains four literals instead of five. We have excluded cell 24 because it cannot contribute toward reducing the expression for f_{19}.

Now refer to Fig. 6.20. The cells marked 1 or X may be combined to form the blocks recorded in Table 6.8, and the optimal expression for this function becomes

$$f_{20} = AC + AD + BD + B'CD'$$

Where irrelevant terms are present, the problem of reducing a standard AND-to-OR expression to optimal form may be ambiguous, for there may be alternative ways of forming blocks in the Karnaugh map. For example, with reference to

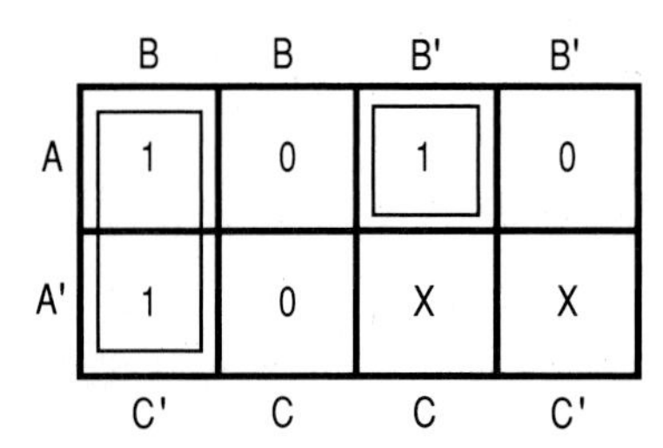

FIGURE 6.19 Method of identifying irrelevant terms on a map.

	C	C	C'	C'	
A	1	X	1	0	B'
A	1	X	X	0	B
A'	0	1	1	X	B
A'	1	0	X	X	B'
	D'	D	D	D'	

FIGURE 6.20 Formation of blocks containing irrelevant terms.

TABLE 6.8

Cells in block	Term
11, 12, 21, 22	AC
12, 13, 22, 23	AD
22, 23, 32, 33	BD
11, 41	$B'CD'$

Fig. 6.20, it is possible to combine cell 41 with cell 44 rather than cell 11. The optimal expression for this function then becomes

$$f_{20} = AC + AD + BD + A'B'D'$$

6.19 THE EXCLUSIVE OR RELATIONSHIP

The expression $A + B$ is 1 if either A or B is 1, or if both are 1. However, in some applications of boolean algebra it is necessary to impose a more stringent requirement: The expression is to be 1 if *one and only one* variable is 1. This requirement is referred to as the EXCLUSIVE OR relationship, and it is denoted by the notation $A \oplus B$. With reference to Table 6.1, the condition $A \oplus B = 1$ corresponds to rows 1 and 2 but not to row 3.

Since the expression $A \oplus B$ is 1 if one variable is 1 and the other is 0, it follows that

$$A \oplus B = (A + B)(A' + B') \tag{6.27}$$

and

$$A \oplus B = AB' + A'B \tag{6.28}$$

Equation (6.28) can be derived from Eq. (6.27), but it is self-evident. We also have the following:

$$A \oplus 0 = A \tag{6.29}$$

$$A \oplus 1 = A' \tag{6.30}$$

$$A \oplus A = 0 \tag{6.31}$$

Consider the expression $(A \oplus B)'$. The complementary expression $A \oplus B$ is 0 if both variables are 0, or if both variables are 1. Therefore,

$$(A \oplus B)' = AB + A'B' \tag{6.32}$$

and

$$(A \oplus B)' = (A + B')(A' + B) \tag{6.33}$$

Equations (6.32) and (6.33) can also be obtained from Eqs. (6.27) and (6.28), respectively, by applying the first law of duality.

6.20 EXPRESSION FOR COMPLEX SYSTEM

Where a boolean system is relatively complex, the expression for the system can be formed in stages. The procedure consists of starting with simple subsystems and

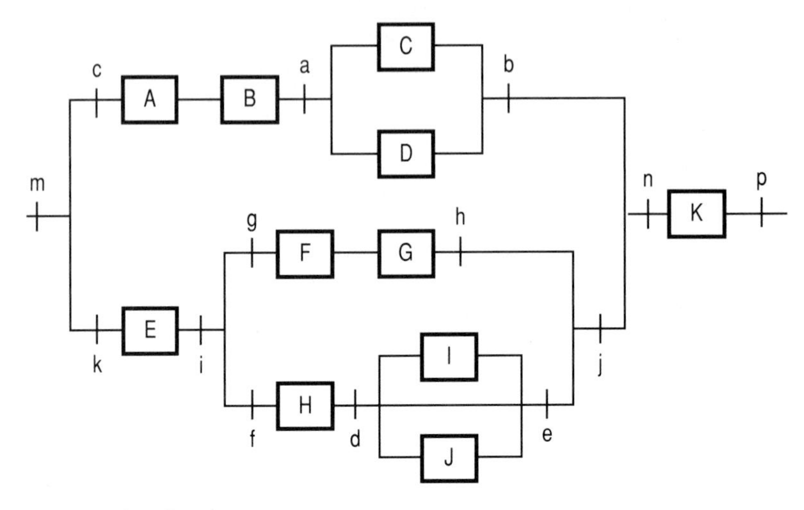

FIGURE 6.21 Complex system.

then gradually enlarging each subsystem by introducing more components, continuing the process until the entire system is encompassed.

As an illustration, refer to Fig. 6.21. We wish to traverse the system, and therefore it is necessary to specify the conditions that make the system passable. Let E(xy) denote the expression for the subsystem that lies between points x and y. We have the following:

$$\mathrm{E}(ab) = C + D$$

$$\mathrm{E}(cb) = AB(C + D) = ABC + ABD$$

$$\mathrm{E}(de) = I + J$$

$$\mathrm{E}(fe) = H(I + J) = HI + HJ$$

$$\mathrm{E}(gh) = FG$$

$$\mathrm{E}(ij) = \mathrm{E}(gh) + \mathrm{E}(fe) = FG + HI + HJ$$

$$\mathrm{E}(kj) = E[\mathrm{E}(ij)] = EFG + EHI + EHJ$$

$$\begin{aligned} \mathrm{E}(mn) &= \mathrm{E}(cb) + \mathrm{E}(kj) \\ &= ABC + ABD + EFG + EHI + EHJ \end{aligned}$$

$$\mathrm{E}(mp) = [\mathrm{E}(mn)]K$$

$$\mathrm{E}(mp) = ABCK + ABDK + EFGK + EHIK + EHJK$$

Since each term in the expression for the system corresponds to a path leading from m to p, it would have been possible to write this expression directly by listing all the paths. However, there is a possibility of overlooking a particular path, and the methodical procedure that we have followed is preferable.

SECTION 7

TRANSFORMATION OF SPACE IN COMPUTER GRAPHICS

Assume that we are to construct a picture of an object. For the picture to be meaningful, two requirements must generally be met. First, the object must be oriented in such manner than the picture discloses all salient features of the object. Second, the shape of the object must be modified somewhat so that the picture will conform reasonably well with the image that the object presents to the eye of the observer. These effects can be achieved most effectively by considering that the space in which the object is located undergoes a transformation, such as a rotation or a compression. Therefore, the practice of computer graphics requires a thorough understanding of the mathematics of the transformation of space. This branch of mathematics applies matrix algebra extensively, and the latter subject is presented in Sec. 5.

7.1 INTRODUCTION

7.1.1 Definitions and Notation

Let P denote an arbitrary point in space, and consider that a transformation of space causes this point to be displaced to P'. Point P is said to be *transformed to* or *mapped onto* P', and P' is called the *image* of P. Similarly, if the transformation causes a line L to assume the position L', then L is transformed to L', and L' is the image of L. A point that remains stationary during a transformation is called a *fixed point*. If all points on a given line or plane remain fixed, that line or plane is described as fixed.

A geometric property that is unchanged by a spatial transformation is said to be *invariant* under that transformation. For example, if parallel lines remain parallel, then parallelism is invariant. An *invariant line* or *plane* is one that remains stationary, although points on that line or plane may undergo displacement along the line or plane.

In our present study, the coordinate x, y, and z axes have the positions shown in Fig. 7.1. Point P is an arbitrary point in space, and lines PA, PB, and PC are perpendicular to the x, y, and z axes, respectively. The coordinates of P are as follows: $x = OA$, $y = OB$, $z = OC$. The coordinate axes are considered to remain

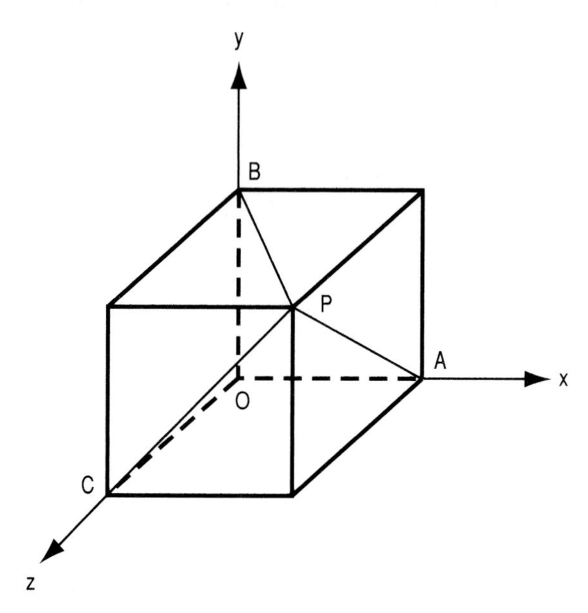

FIGURE 7.1 Location of coordinate axes.

fixed under all transformations, since they serve merely as reference axes. Thus, our present method differs from that in Art. 3.1.6, where we simply displaced the coordinate axes without disturbing the surrounding space.

Consider that point $P(x, y, z)$ is transformed to $P'(x', y', z')$ under a given transformation. The transformation is described as *linear* if the coordinates of P' are linear functions of the coordinates of P. For example, assume the following:

$$\begin{aligned} x' &= 4x + 7y - 3z + 2 \\ y' &= 8x - 2y + 5z \\ z' &= 3x - 5y - 9z - 4 \end{aligned} \qquad (a)$$

This transformation is linear.

If space undergoes a bodily displacement, all points remain fixed distances apart. A transformation of this type is termed a *rigid-motion transformation* or *isometry*. Thus, the rotation of space about any axis is a rigid-motion transformation.

If P' is the image of a single point P, the transformation is called a *one-to-one* (or *singular*) *mapping*. On the other hand, if P' is the image of an infinite number of points, the transformation is a *many-to-one mapping*. The first type of transformation is reversible; the second type is irreversible. We shall first study one-to-one mappings.

Let T denote a one-to-one mapping. Associated with T is an *inverse transformation* that can nullify the effects of T. Thus, if P is mapped onto P' under T, then P' is mapped onto P under the inverse of T. For example, if T consists of rotating space about the z axis through a *clockwise* angle of 38°, the inverse of T consists of rotating space about the z axis through a *counterclockwise* angle of 38°. A transformation of space that coincides with its inverse is termed an *involution*.

In computer graphics, the coordinates of a point are displayed in the form of a row vector, which is termed the *position vector* of the point. For example, the following are the position vectors of $P_1(7, -3, 2)$ and $P_2(-9, -5, 6)$, respectively:

$$[7 \quad -3 \quad 2] \qquad [-9 \quad -5 \quad 6]$$

For purely mathematical reasons, it is necessary in some instances to assign a fourth coordinate to every point in space. This fourth coordinate is labeled w, and the coordinates are recorded in this order: x, y, z, w. These four coordinates constitute the *homogeneous coordinates* of the point.

7.1.2 Transformation by Matrix Multiplication

Let T denote the transformation of space that maps $P(x, y, z)$ onto $P'(x', y', z')$. Under a linear transformation, it is possible to transform the coordinates of P to those of P' by postmultiplying the position vector of P by a square matrix $\mathbf{T}$, which is termed the *transformation matrix* corresponding to T. Thus, if three coordinates suffice, we have

$$[x' \quad y' \quad z'] = [x \quad y \quad z]\mathbf{T} \tag{7.1}$$

As an illustration, consider Eq. (a) in Art 7.1.1. Applying homogeneous coordinates with $w = 1$, we can express the transformation in the following manner:

$$[x' \quad y' \quad z' \quad 1] = [x \quad y \quad z \quad 1]\begin{bmatrix} 4 & 8 & 3 & 0 \\ 7 & -2 & -5 & 0 \\ -3 & 5 & -9 & 0 \\ 2 & 0 & -4 & 1 \end{bmatrix}$$

The matrix at the right is the transformation matrix. It follows that there is a transformation matrix associated with every linear transformation.

Since a one-to-one mapping has an inverse, its transformation matrix also has an inverse. On the other hand, since a many-to-one mapping has no inverse, its transformation matrix also has no inverse. The following principles emerge:

Theorem 7.1. The transformation matrix of a one-to-one mapping is regular, and that of a many-to-one mapping is singular.

Theorem 7.2. If transformations T and U are inverses of each other, the corresponding transformation matrices $\mathbf{T}$ and $\mathbf{U}$ are also inverses of each other.

Theorem 7.3. The transformation matrix of an involution is coincident with its inverse matrix.

If space is subjected to several transformations in series, the transformation matrix corresponding to the composite transformation is obtained by concatenating the transformation matrices of the individual transformations. For example, assume that space is subjected to a transformation T_1 and then a transformation T_2. Let $\mathbf{T}_1$ and $\mathbf{T}_2$ denote the corresponding transformation matrices. Then

$$[x' \quad y' \quad z'] = [x \quad y \quad z]\mathbf{T}_1\mathbf{T}_2$$

Since matrix multiplication is generally not commutative, it is imperative that the individual transformation matrices be multiplied in the proper sequence.

7.2 REVERSIBLE TRANSFORMATIONS

7.2.1 Types of Transformations of the Plane

We shall first investigate the transformation of a plane and then extend our investigation to the transformation of three-dimensional space. The plane that is to be transformed is considered to be the xy plane. Since $z = 0$ for each point in the plane, this coordinate is omitted in the position vector of a point. The following are the basic types of transformations of the plane:

1. Displacement of the plane in a specified direction. This displacement is termed a *translation.*
2. *Rotation* of the plane about an axis perpendicular to the plane.
3. Inversion of the plane about an axis within or perpendicular to the plane. This transformation is called a *reflection*. The inversion may be considered to result from rotating the plane about this axis through an angle of 180°.
4. Distortion of the plane by distending or compressing it about an axis in the plane. This transformation is called a *scaling.*
5. Distortion of the plane by distending or compressing it about an axis perpendicular to the plane. The first type of transformation is called a *dilation*, and the second type is called a *compaction*.
6. Distortion of the plane in a manner that causes every line perpendicular to an axis in the plane to rotate through a given angle. This transformation is called a *shear*.

We shall study each of these transformations in turn. The first three transformations are of the rigid-motion type, and we shall find that all six transformations are linear.

Each of the foregoing transformations is a one-to-one mapping. Consequently, each transformation has a regular transformation matrix. The inverse of a given transformation matrix can be found by applying the method of matrix inversion presented in Art. 5.6.2, or simply by inverting the transformation.

A transformation of the plane is *elemental* if it can be described exclusively in terms of the coordinate axes. For example, if the plane is reflected about the x, y, or z axis, the transformation is elemental; if the plane is reflected about some other axis, the transformation is nonelemental. We shall first study elemental transformations of the plane, and then we shall demonstrate that every nonelemental transformation may be viewed as a composite of elemental transformations.

7.2.2 Translation of the Plane

With reference to Fig. 7.2*a*, consider that the plane is translated a distance d_x in the positive x direction and a distance d_y in the positive y direction. Point $P(x, y)$ is mapped onto $P'(x', y')$. Then

$$x' = x + d_x \qquad y' = y + d_y \tag{7.2}$$

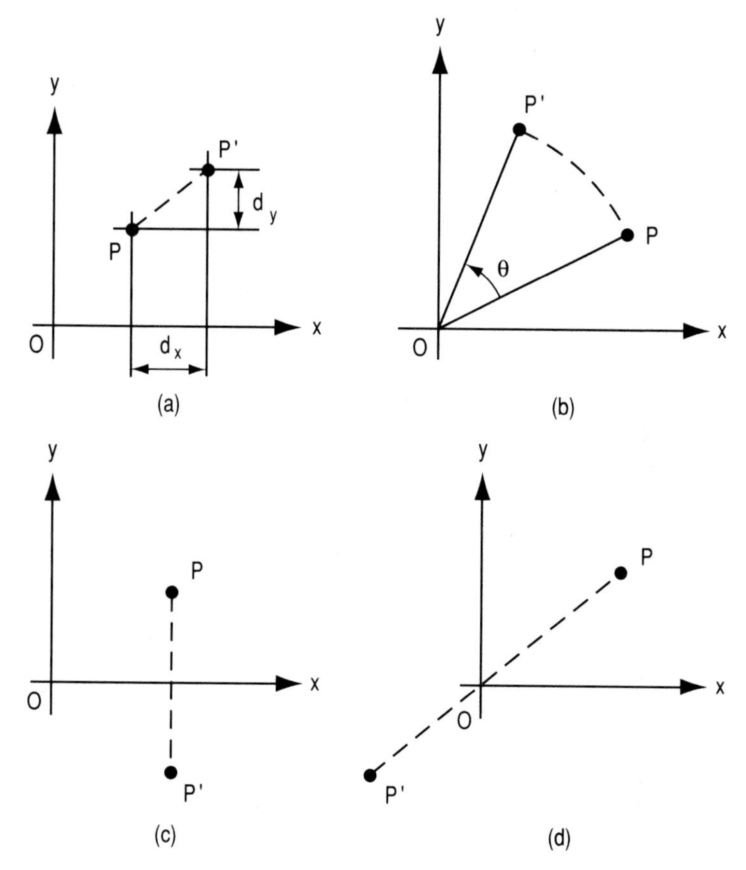

FIGURE 7.2 Rigid-motion transformations of the plane. (*a*) Translation; (*b*) rotation about *z* axis; (*c*) reflection about *x* axis; (*d*) reflection about *z* axis.

Applying homogeneous coordinates with $w = 1$, we have the following:

$$[x' \quad y' \quad 1] = [x \quad y \quad 1]\begin{bmatrix} 1 & 0 & 0 \\ 0 & 1 & 0 \\ d_x & d_y & 1 \end{bmatrix} \tag{7.3}$$

The transformation matrix at the right is denoted by $\mathbf{T}_{\text{trans}}$.

7.2.3 Rotation of the Plane

When the plane is rotated about an axis perpendicular to the plane, the angle of rotation is considered positive if the rotation is counterclockwise.

With reference to Fig. 7.2*b*, consider that the plane is rotated about the *z* axis through an angle θ. Then

$$x' = x \cos \theta - y \sin \theta$$
$$y' = x \sin \theta + y \cos \theta \tag{7.4}$$

Applying matrix multiplication, we have

$$[x' \quad y'] = [x \quad y]\begin{bmatrix} \cos \theta & \sin \theta \\ -\sin \theta & \cos \theta \end{bmatrix} \tag{7.5}$$

The transformation matrix at the right is denoted by $\mathbf{T}_{\text{rot}}$.

7.2.4 Reflection of the Plane

With reference to Fig. 7.2*c*, consider the plane to be reflected about the x axis. This reflection causes point P to assume the corresponding position P' on the opposite side of the x axis, and the reflection can be effected by revolving the plane about the x axis through an angle of 180°. Then

$$x' = x \qquad y' = -y \tag{7.6}$$

$$[x' \quad y'] = [x \quad y]\begin{bmatrix} 1 & 0 \\ 0 & -1 \end{bmatrix} \tag{7.7}$$

The transformation matrix at the right is denoted by $\mathbf{T}_{\text{refl},x}$. If the plane is again reflected about the x axis, P' is mapped onto P. It follows that a reflection is an involution, and $\mathbf{T}_{\text{refl},x}$ coincides with its inverse.

For a reflection of the plane about the y axis, we have

$$x' = -x \qquad y' = y \tag{7.8}$$

$$[x' \quad y'] = [x \quad y]\begin{bmatrix} -1 & 0 \\ 0 & 1 \end{bmatrix} \tag{7.9}$$

The transformation matrix at the right is denoted by $\mathbf{T}_{\text{refl},y}$.

Now consider that the plane is reflected about both the x axis and the y axis, in either order, thereby transforming point P in Fig. 7.2*d* to P'. This composite reflection produces reflection of the plane about the z axis, with

$$x' = -x \qquad y' = -y \tag{7.10}$$

$$[x' \quad y'] = [x \quad y]\begin{bmatrix} -1 & 0 \\ 0 & -1 \end{bmatrix} \tag{7.11}$$

The transformation matrix at the right is denoted by $\mathbf{T}_{\text{refl},z}$, and $\mathbf{T}_{\text{refl},z} = \mathbf{T}_{\text{refl},x}\mathbf{T}_{\text{refl},y}$. However, reflection of the plane about the z axis is equivalent to rotation of the plane about the z axis through an angle of 180°. Thus, $\mathbf{T}_{\text{refl},z}$ is actually a special form of $\mathbf{T}_{\text{rot}}$, with $\theta = 180°$.

7.2.5 Scaling of the Plane

With reference to Fig. 7.3*a*, consider that the plane is stretched outward from the x axis, thereby transforming P and Q to P' and Q', respectively. Then

$$x' = x \qquad y' = s_x y \tag{7.12}$$

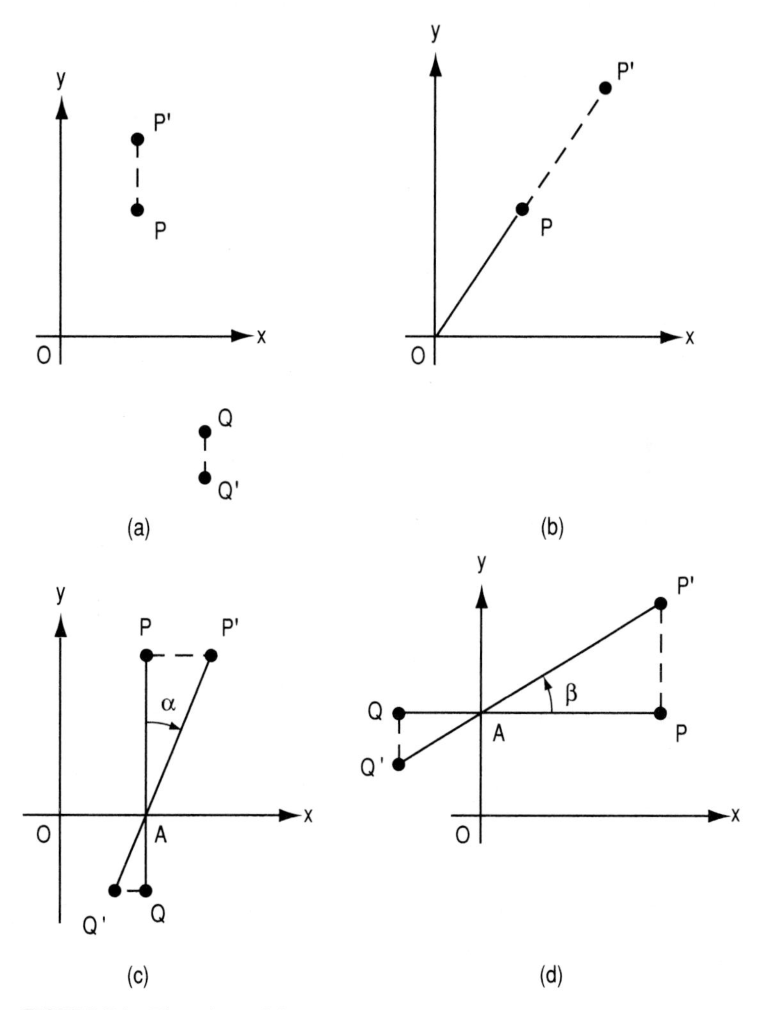

FIGURE 7.3 Distortions of the plane. (*a*) Scaling about x axis; (*b*) dilation about z axis; (*c*) shear about x axis; (*d*) shear about y axis.

where s_x is a constant greater than 1. This type of transformation is a *scaling*, and s_x is the *scale factor*. Alternatively, the plane may be *compressed* about the x axis, in which case $0 < s_x < 1$. When expressed by matrix multiplication, the transformation is as follows:

$$[x' \quad y'] = [x \quad y]\begin{bmatrix} 1 & 0 \\ 0 & s_x \end{bmatrix} \tag{7.13}$$

The transformation matrix at the right is denoted by $\mathbf{T}_{\text{scal},x}$.

Similarly, for scaling about the y axis with a factor s_y, we have the following:

$$x' = s_y x \qquad y' = y \tag{7.14}$$

$$[x' \quad y'] = [x \quad y]\begin{bmatrix} s_y & 0 \\ 0 & 1 \end{bmatrix} \tag{7.15}$$

If the plane is subjected to scaling about both the x axis and the y axis, in either order, the results are as follows:

$$x' = s_y x \qquad y' = s_x y \tag{7.16}$$

$$[x' \quad y'] = [x \quad y]\begin{bmatrix} s_y & 0 \\ 0 & s_x \end{bmatrix} \tag{7.17}$$

The transformation matrix at the right is denoted by $\mathbf{T}_{\text{scal},xy}$.

7.2.6 Dilation and Compaction of the Plane

With reference to Fig. 7.3*b*, consider that the plane is stretched outward from the z axis, causing the radial distance OP to be multiplied by a constant k, where $k > 1$. The plane is said to be dilated about the z axis. Then

$$x' = kx \qquad y' = ky \tag{7.18}$$

$$[x' \quad y'] = [x \quad y]\begin{bmatrix} k & 0 \\ 0 & k \end{bmatrix} \tag{7.19}$$

The transformation matrix at the right is denoted by $\mathbf{T}_{\text{dil},z}$. Alternatively, the plane may be *compacted* about the z axis, in which case $0 < k < 1$. The constant k is called the *dilation* or *compaction factor*, whichever applies.

The dilation or compaction of the plane may be considered to result from a scaling of the plane in both the x and y directions with $s_x = s_y = k$. Thus, the matrix $\mathbf{T}_{\text{dil},z}$ is a special form of $\mathbf{T}_{\text{scal},xy}$.

7.2.7 Shear of the Plane

If the plane is subjected to shear about the x axis, a point is displaced parallel to the x axis, the magnitude of the displacement being directly proportional to its distance from that axis. Let h_x denote the constant of proportionality. Then

$$x' = x + h_x y \qquad y' = y \tag{7.20}$$

$$[x' \quad y'] = [x \quad y]\begin{bmatrix} 1 & 0 \\ h_x & 1 \end{bmatrix} \tag{7.21}$$

The constant h_x is called the *shear factor*.

In Fig. 7.3*c*, line QAP is parallel to the y axis. If h_x is positive, points P and Q are displaced to the right and to the left, respectively. Thus, line QAP increases in length and rotates through a *clockwise* angle α, as shown, where $\tan \alpha = h_x$. On the other hand, if h_x is negative, line QAP rotates through a *counterclockwise* angle α.

Similarly, if the plane is subjected to shear about the y axis with a shear factor h_y, we have the following:

$$x' = x \qquad y' = y + h_y x \tag{7.22}$$

$$[x' \quad y'] = [x \quad y]\begin{bmatrix} 1 & h_y \\ 0 & 1 \end{bmatrix} \tag{7.23}$$

In Fig. 7.3*d*, line QAP is parallel to the x axis. If h_y is positive, this line rotates through a *counterclockwise* angle β, as shown, where $\tan \beta = h_y$. On the other hand, if h_y is negative, line QAP rotates through a *clockwise* angle β.

If the plane is subjected to *simultaneous* shear about the x axis with a factor h_x and shear about the y axis with a factor h_y, we have the following:

$$x' = x + h_x y \qquad y' = y + h_y x \tag{7.24}$$

$$[x' \quad y'] = [x \quad y]\begin{bmatrix} 1 & h_y \\ h_x & 1 \end{bmatrix} \tag{7.25}$$

The transformation matrices in Eqs. (7.21), (7.23), and (7.25) are denoted by $\mathbf{T}_{\text{sh},x}$, $\mathbf{T}_{\text{sh},y}$, and $\mathbf{T}_{\text{sh},xy}$, respectively.

7.2.8 Nonelemental Transformations of the Plane about Axes through the Origin

As previously stated, a nonelemental transformation of the plane may be viewed as a composite of elemental transformations. We shall illustrate this principle with a specific case. At present, we shall confine our study to transformations about axes that pass through the origin.

EXAMPLE 7.1 Line L in Fig. 7.4*a* has the indicated position. The plane is to be reflected about L. Establish the corresponding transformation of coordinates.

SOLUTION The transformation will be accomplished in three steps, each step involving an elemental transformation. Where necessary, we shall assign a distinctive designation to the transformation matrix in a particular step. As before, let $P(x, y)$ denote an arbitrary point in the plane. The steps are as follows:

1. Rotate the plane about the z axis in the clockwise direction through an angle α, thereby placing line L on the x axis and mapping P onto $P_1(x_1, y_1)$, as shown in Fig. 7.4*b*. Applying Eq. (7.5) with $\theta = -\alpha$, we have

$$[x_1 \quad y_1] = [x \quad y]\mathbf{T}_a = [x \quad y]\begin{bmatrix} \cos\alpha & -\sin\alpha \\ \sin\alpha & \cos\alpha \end{bmatrix}$$

2. Reflect the plane about the x axis, thereby mapping P_1 onto $P_2(x_2, y_2)$, as shown in Fig. 7.4*c*. Applying Eq. (7.7), we have

$$[x_2 \quad y_2] = [x_1 \quad y_1]\mathbf{T}_{\text{refl},x} = [x_1 \quad y_1]\begin{bmatrix} 1 & 0 \\ 0 & -1 \end{bmatrix}$$

3. Rotate the plane about the z axis in the counterclockwise direction through an angle α, thereby restoring L to its original position and mapping P_2 onto P', as shown in Fig. 7.4*d*. Applying Eq. (7.5) with $\theta = \alpha$, we have

$$[x' \quad y'] = [x_2 \quad y_2]\mathbf{T}_b = [x_2 \quad y_2]\begin{bmatrix} \cos\alpha & \sin\alpha \\ -\sin\alpha & \cos\alpha \end{bmatrix}$$

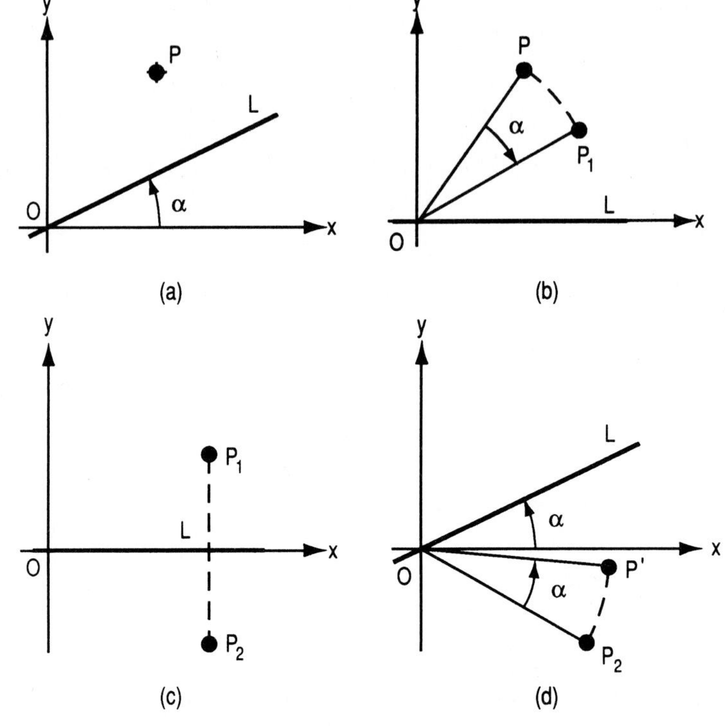

FIGURE 7.4 Reflection of plane about line through origin. (*a*) Initial position of point; (*b*) transformation in step 1; (*c*) transformation in step 2; (*d*) transformation in step 3.

The reflection about line L is now completed, and the original and final coordinates are related in this manner:

$$[x' \quad y'] = [x \quad y]\mathbf{T}_a\mathbf{T}_{\mathrm{refl},x}\mathbf{T}_b$$

Performing the indicated matrix multiplication and applying Eqs. (2.11) and (2.12), we obtain the following:

$$[x' \quad y'] = [x \quad y]\begin{bmatrix} \cos 2\alpha & \sin 2\alpha \\ \sin 2\alpha & -\cos 2\alpha \end{bmatrix} \qquad (b)$$

$$x' = x \cos 2\alpha + y \sin 2\alpha$$

$$y' = x \sin 2\alpha - y \cos 2\alpha$$

The procedure for effecting reflection about line L can be modified by first rotating the plane in the counterclockwise direction through an angle of $90° - \alpha$, thereby placing L on the y axis, and then reflecting the plane about the y axis. However, this modified procedure will also yield Eq. (*b*), as it must.

The transformation matrix in Eq. (*b*) is the reflection matrix for reflection about an *arbitrary* axis through the origin, and we shall denote it by $\mathbf{T}_{\mathrm{refl}}$. The matrices

$\mathbf{T}_{\text{refl},x}$ and $\mathbf{T}_{\text{refl},y}$ in Eqs. (7.7) and (7.9), respectively, may be regarded as special forms of $\mathbf{T}_{\text{refl}}$ in which $\alpha = 0$ and $\alpha = 90°$, respectively. Since reflection is an involution, $\mathbf{T}_{\text{refl}}$ coincides with its inverse, and $\mathbf{T}^2_{\text{refl}} = \mathbf{I}$.

There are two particular axes of reflection that are of major importance. The first is line L_1 in Fig. 7.5*a*. This makes an angle of 45° with the positive x axis, and its equation is $y = x$. With $\alpha = 45°$,

$$[x' \quad y'] = [x \quad y]\begin{bmatrix} 0 & 1 \\ 1 & 0 \end{bmatrix} \tag{7.26}$$

$$x' = y \qquad y' = x$$

Thus, the coordinates of a point exchange their values. The second significant axis of reflection is line L_2 in Fig. 7.5*b*. This makes an angle of 135° with the positive x axis, and its equation is $y = -x$. With $\alpha = 135°$,

$$[x' \quad y'] = [x \quad y]\begin{bmatrix} 0 & -1 \\ -1 & 0 \end{bmatrix} \tag{7.27}$$

$$x' = -y \qquad y' = -x$$

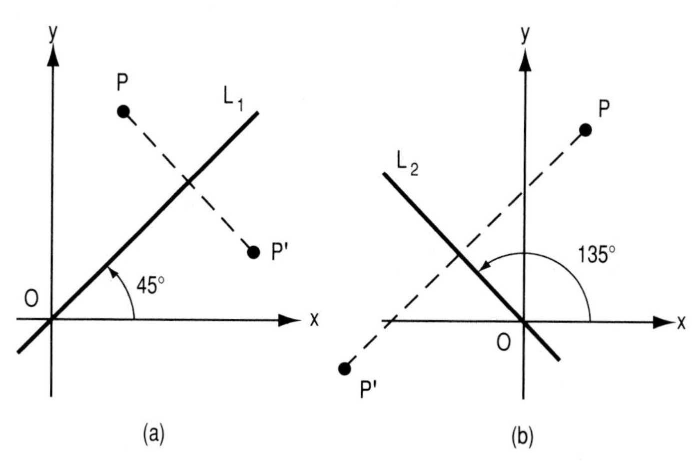

FIGURE 7.5 Reflection of plane about line having the equation (*a*) $y = x$ and (*b*) $y = -x$.

7.2.9 General Second-Order Transformation Matrix

Consider that the plane is subjected to a scaling about the x and y axes and then to simultaneous shears about these axes. Applying Eqs. (7.17) and (7.25), we obtain the following:

$$[x' \quad y'] = [x \quad y]\begin{bmatrix} s_y & 0 \\ 0 & s_x \end{bmatrix}\begin{bmatrix} 1 & h_y \\ h_x & 1 \end{bmatrix}$$

Performing the multiplication, we obtain

$$[x' \quad y'] = [x \quad y]\begin{bmatrix} s_y & s_y h_y \\ s_x h_x & s_x \end{bmatrix} \tag{7.28}$$

If the order of operations is inverted by subjecting the plane to the shears and then the scaling, the transformation of coordinates is as follows:

$$[x' \quad y'] = [x \quad y]\begin{bmatrix} s_y & s_x h_y \\ s_y h_x & s_x \end{bmatrix} \tag{7.29}$$

Now consider that the composite transformations represented by Eqs. (7.28) and (7.29) are followed by reflection of the plane about the x or y axis, or both. The effect is to change the algebraic signs of some elements in the transformation matrix. Thus, any composite transformation of the plane that consists of scalings, shears, and reflections will yield a regular second-order transformation matrix in which the elements are functions of the scale and shear factors. Since we can assign arbitrary values to these factors (with due regard for algebraic signs), we can manipulate these factors to produce an arbitrary regular second-order transformation matrix. Reversing our point of view, we arrive at this principle:

Theorem 7.4. Every regular second-order matrix may be regarded as the transformation matrix corresponding to a transformation of the plane that is composed of scalings and shears about the x and y axes, or scalings, shears, and reflections about those axes.

EXAMPLE 7.2 Devise a composite transformation of the plane for which the following matrix serves as the transformation matrix:

$$\begin{bmatrix} 1.20 & 0.84 \\ -0.21 & 0.70 \end{bmatrix}$$

SOLUTION By comparing each element in the given matrix with the corresponding element in the matrix in Eq. (7.28), we obtain the following values of the factors:

$$s_x = 0.70 \qquad s_y = 1.20$$

$$h_x = \frac{-0.21}{0.70} = -0.30 \qquad h_y = \frac{0.84}{1.20} = 0.70$$

Thus, the composite transformation corresponding to the given matrix is as follows: Scaling about the x and y axes with the factors $s_x = 0.70$ and $s_y = 1.20$, followed by simultaneous shear about the x and y axes with the factors $h_x = -0.30$ and $h_y = 0.70$.

Alternatively, by comparing each element in the given matrix with the corresponding element in the matrix in Eq. (7.29), we obtain the following values:

$$s_x = 0.70 \qquad s_y = 1.20 \qquad h_x = -0.175 \qquad h_y = 1.20$$

Thus, an alternative composite transformation is as follows: Simultaneous shear about the x and y axes with the factors $h_x = -0.175$ and $h_y = 1.20$, followed by scaling about the x and y axes with the factors $s_x = 0.70$ and $s_y = 1.20$.

The two alternative composite transformations that we have devised are of course equivalent to each other because they produce identical effects.

EXAMPLE 7.3 Devise a composite transformation of the plane for which the following matrix serves as the transformation matrix:

$$\begin{bmatrix} 1.60 & -0.64 \\ -0.72 & -0.90 \end{bmatrix}$$

Verify the result.

SOLUTION If we apply Eq. (7.28) directly, we obtain $s_x = -0.90$. This result is invalid, since a scale factor is restricted to positive values. However, this obstacle can be surmounted by considering that the plane is subjected to a reflection about the x axis, thereby causing the ordinate of a point to change sign, and then a scaling about the x axis. The order in which the reflection and scaling occur is immaterial. We now have the following:

$$s_x = 0.90 \qquad s_y = 1.60$$

$$h_x = \frac{-0.72}{-0.90} = 0.80 \qquad h_y = \frac{-0.64}{1.60} = -0.40$$

Thus, a composite transformation of the plane corresponding to the given matrix is as follows: Reflection about the x axis and scaling about the x and y axes with the factors $s_x = 0.90$ and $s_y = 1.60$, followed by shear about the x and y axes with the factors $h_x = 0.80$ and $h_y = -0.40$.

To verify the result, assume that the reflection is performed first. By applying Eqs. (7.7), (7.17), and (7.25) with the appropriate numerical values, we obtain the following:

$$\begin{aligned}[x' \quad y'] &= [x \quad y]\begin{bmatrix} 1 & 0 \\ 0 & -1 \end{bmatrix}\begin{bmatrix} 1.60 & 0 \\ 0 & 0.90 \end{bmatrix}\begin{bmatrix} 1 & -0.40 \\ 0.80 & 1 \end{bmatrix} \\ &= [x \quad y]\begin{bmatrix} 1.60 & -0.64 \\ -0.72 & -0.90 \end{bmatrix}\end{aligned}$$

It is possible to devise an alternative composite transformation, as was done in Example 7.2.

7.2.10 Transformation of Straight Lines

In a rigid-motion transformation of the plane, all relationships among points and lines remain invariant. We shall now investigate the manner in which straight lines change when the plane is distorted. For this purpose, consider that the plane is subjected to the following general linear transformation, which encompasses all possible elemental transformations except displacement:

$$[x' \quad y'] = [x \quad y]\begin{bmatrix} a & b \\ c & d \end{bmatrix} \tag{7.30}$$

This transformation matrix will be denoted by $\mathbf{T}_{\text{gen}}$. Performing the multiplication, we obtain

$$x' = ax + cy \qquad y' = bx + dy \tag{7.31}$$

Let L denote a straight line in the original plane having the equation $y = mx + B$. In accordance with the statement in Art. 3.1.2, m is the slope of the line and B is its y intercept. Now let $P(x, y)$ denote a point on this line. By solving Eqs. (7.31) for x and y and then substituting these expressions in the equation of L, we find that the transformed coordinates of P have this relationship:

$$y' = \left(\frac{b + md}{a + mc}\right)x' + \left(\frac{ad - bc}{a + mc}\right)B \tag{7.32}$$

This equation has the form $y' = m'x' + B'$; therefore, it is the equation of a straight line L' that has a slope m' and a y intercept of B'. Thus, all points that were on L now lie on L', and the following principle emerges:

Theorem 7.5. When the plane is subjected to any linear transformation, the image of a straight line is also a straight line.

To express this in another manner, points that are collinear remain collinear under a linear transformation.

EXAMPLE 7.4 A straight line L has the equation $y = 4x - 7$. The plane is subjected to shear about the y axis with a factor of 1.4, scaling about the x axis with a factor of 0.6, and reflection about the y axis, in the specified sequence. Find the equation of L', the image of L.

SOLUTION With respect to L, $m = 4$ and $B = -7$. Multiplying the transformation matrices in Eqs. (7.23), (7.13), and (7.9), in that order, we obtain the following transformation matrix:

$$\begin{bmatrix} 1 & 1.4 \\ 0 & 1 \end{bmatrix} \begin{bmatrix} 1 & 0 \\ 0 & 0.6 \end{bmatrix} \begin{bmatrix} -1 & 0 \\ 0 & 1 \end{bmatrix} = \begin{bmatrix} -1 & 0.84 \\ 0 & 0.60 \end{bmatrix}$$

In this matrix, $a = -1$, $b = 0.84$, $c = 0$, and $d = 0.60$. Substituting in Eq. (7.32), we obtain $y' = -3.24x' - 4.2$ as the equation of L'.

EXAMPLE 7.5 Line L in Fig. 7.6 has the end points $P_1(3, 8)$ and $P_2(10, 13)$. This line is to be transformed to line L', which has the end points $P_1'(4, -2)$ and $P_2'(15, 6)$. Construct the transformation matrix.

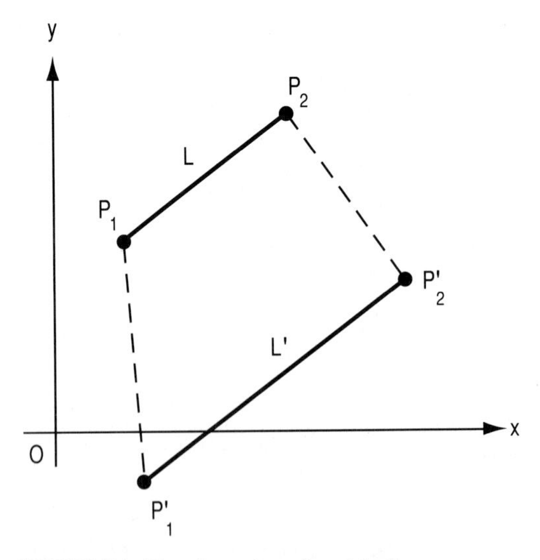

FIGURE 7.6 Transformation of straight line.

SOLUTION Applying Eq. (7.31) to the transformation of P_1 to P'_1, we obtain

$$4 = 3a + 8c \tag{c}$$

$$-2 = 3b + 8d \tag{d}$$

Applying Eq. (7.31) to the transformation of P_2 to P'_2, we obtain

$$15 = 10a + 13c \tag{e}$$

$$6 = 10b + 13d \tag{f}$$

Solving the simultaneous equations (*c*) and (*e*), we obtain $a = 68/41$, $c = -5/41$. Solving the simultaneous equations (*d*) and (*f*), we obtain $b = 74/41$, $d = -38/41$. Then

$$[x' \quad y'] = [x \quad y]\left(\frac{1}{41}\right)\begin{bmatrix} 68 & 74 \\ -5 & -38 \end{bmatrix}$$

As Example 7.5 demonstrates, it is possible to transform a straight line to an arbitrary straight line in the same plane by means of a linear transformation.

In Fig. 7.7*a*, points P_1, P_2, and P_3 lie on line L. Under a linear transformation of the plane, these points are mapped onto the indicated points on line L'. By applying Eq. (7.31), we find that

$$\frac{P'_1P'_2}{P'_2P'_3} = \frac{P_1P_2}{P_2P_3} \tag{7.33}$$

In Fig. 7.7*b*, L_1 and L_2 are straight lines intersecting at Q. When a linear transformation occurs, Q is mapped onto a point Q' that lies on each of the two image lines. Thus, Q' is the point of intersection of the images of L_1 and L_2.

Theorem 7.6. When the plane is subjected to any linear transformation, the point of intersection of two straight lines is mapped onto the point of intersection of the images of those lines.

Again let $y = mx + B$ be the equation of a straight line, where m is the slope of the line. Consider that two lines in the original plane are parallel to each other. Since these lines have an identical slope, their equations have an identical m value. Equation (7.32) discloses that the images of these lines have an identical m' value under a linear transformation of the plane; therefore, the images have an identical slope.

Now consider that two lines in the original plane are perpendicular to each other. As stated in Art. 3.1.3, their slopes have a product of -1. Equation (7.32) discloses that the slopes of their images have a product of -1 only if the following relationships both exist:

$$b^2 - d^2 = c^2 - a^2 \qquad bd = -ac \tag{7.34}$$

The transformation matrices for the rotation and reflection satisfy both these equations (as they must), but the transformation matrices for scaling and shear satisfy them only in the special cases where $s_x = s_y$ and $h_x = -h_y$. We thus arrive at the following principles:

Theorem 7.7. When the plane is subjected to any linear transformation, parallel lines remain parallel.

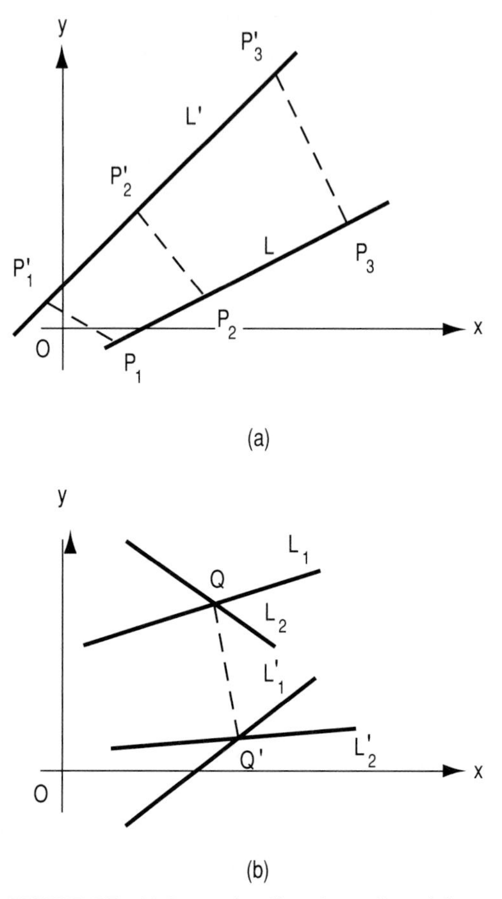

FIGURE 7.7 (*a*) Proportionality of transformed line segments; (*b*) transformation of point of intersection of lines.

Theorem 7.8. When the plane is subjected to scalings or shears, or both, perpendicular lines remain perpendicular only if the two scale factors are equal and the two shear factors are equal in absolute value but differ in their algebraic signs.

Thus, under scalings and shears, parallelism is invariant, but perpendicularity generally is not.

7.2.11 Transformation of Areas

When the plane is distorted, an area in the plane changes its magnitude, and we shall now investigate this change.

In Fig. 7.8, $OABC$ is a rectangle located on the x and y axes and having a width m and length n, as shown. The coordinates of its vertices are as follows: $A(m, 0)$;

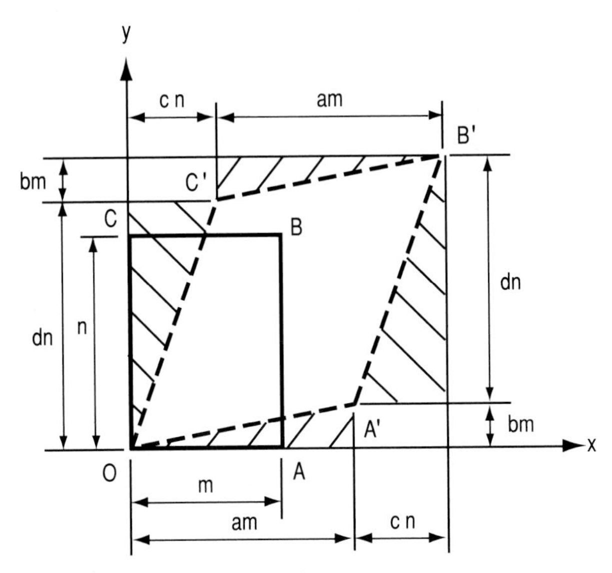

FIGURE 7.8 Transformation of area.

$B(m, n)$; $C(0, n)$. Now consider that the plane is subjected to the transformation described by Eq. (7.30). Under this transformation, the vertices of the rectangle are mapped onto the following points: $A'(am, bm)$; $B'(am + cn, bm + dn)$; $C'(cn, dn)$. Thus, the rectangle $OABC$ is transformed to the parallelogram $OA'B'C'$ shown in the drawing.

Let Q denote the area of $OABC$; then $Q = mn$. Now let Q' denote the area of the image of the rectangle. Then

$$\begin{aligned} Q' &= (am + cn)(bm + dn) - 2(\tfrac{1}{2})(am)(bm) - 2(\tfrac{1}{2})(cn)(dn) - 2(bm)(cn) \\ &= (ad - bc)mn \end{aligned}$$

Thus,

$$Q' = (ad - bc)Q \tag{7.35}$$

The expression in parentheses is the determinant of $\mathbf{T}_{gen}$, and we can write

$$Q' = Q\begin{vmatrix} a & b \\ c & d \end{vmatrix} \tag{7.36}$$

Since an area is always considered positive, we apply the absolute value of the determinant.

Now let Q_1 denote the area in Fig. 7.9 that is bounded by lines L_1, L_2, V_1, and V_2, where V_1 and V_2 are parallel to the y axis. We divide this space into four strips of uniform width and form the rectangles shown. The total area of the rectangles is unequal to Q_1. However, if we successively halve the width of the strips (and thereby successively double the number of strips), the total area of the rectangles approaches Q_1 as a limit. Under the transformation given by Eq. (7.30), the area of each rectangle, and therefore the sum of these areas, changes in accordance with Eq. (7.35). It follows that Q_1 is transformed in accordance with Eq. (7.35), and this equation is universal with respect to linear transformations.

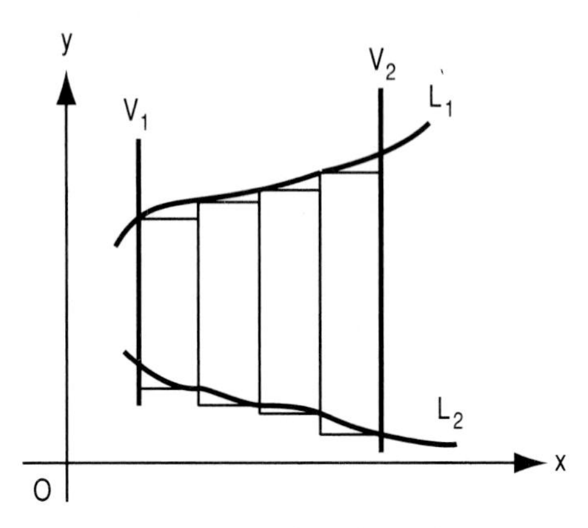

FIGURE 7.9 Division of area into strips.

EXAMPLE 7.6 The area bounded by a closed curve has a magnitude of 36 units. Find the magnitude of the area after the plane is subjected to the following composite transformation, performed in the indicated sequence: a shear about the y axis with a factor of 0.8; a scaling about the x axis with a factor of 2.0; a reflection about the y axis; a shear about the x axis with a factor of -0.6.

SOLUTION Applying the transformation matrix corresponding to each elemental transformation, we obtain the following matrix for the composite transformation:

$$\begin{bmatrix} 1 & 0.8 \\ 0 & 1 \end{bmatrix}\begin{bmatrix} 1 & 0 \\ 0 & 2 \end{bmatrix}\begin{bmatrix} -1 & 0 \\ 0 & 1 \end{bmatrix}\begin{bmatrix} 1 & 0 \\ -0.6 & 1 \end{bmatrix}=\begin{bmatrix} -1.96 & 1.60 \\ -1.20 & 2.00 \end{bmatrix}$$

$$ad - bc = (-1.96)(2.00) - (1.60)(-1.20) = -2.00$$

Disregarding the minus sign, we obtain $36(2.00) = 72$ units as the magnitude of the area following the composite transformation.

The reflection about the y axis did not affect the magnitude of the transformed area; it merely changed the algebraic sign of the determinant.

7.2.12 General Nonelemental Transformation of the Plane

We shall now extend our study of nonelemental transformations of the plane to encompass transformations about arbitrary axes. For this purpose, it is necessary to enlarge the transformation matrix $\mathbf{T}_{gen}$ in Eq. (7.30) to a third-order matrix, which we denote by $\mathbf{T}_{gen,3}$. This enlargement can be accomplished by using homogeneous coordinates with $w = 1$. Equation (7.30) then becomes

$$[x' \quad y' \quad 1] = [x \quad y \quad 1]\begin{bmatrix} a & b & 0 \\ c & d & 0 \\ 0 & 0 & 1 \end{bmatrix}$$

EXAMPLE 7.7 Line L in Fig. 7.10 has the indicated position. The plane is to be reflected about L. Establish the corresponding transformation of coordinates.

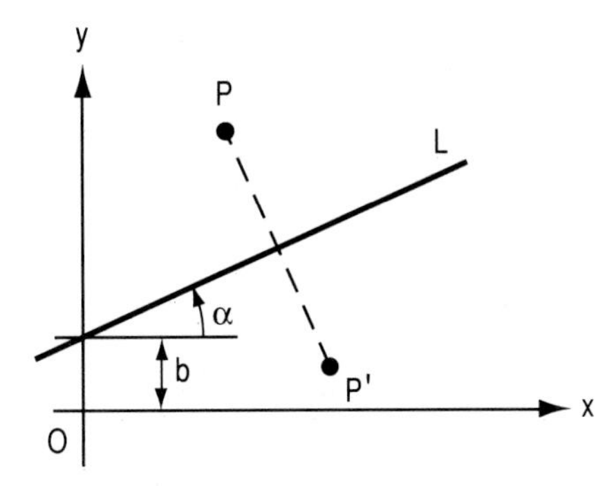

FIGURE 7.10 Reflection of plane about arbitrary axis.

SOLUTION This problem is an extension of Example 7.1. The transformation is accomplished in the following steps:

1. Translate the plane a distance b in the negative y direction. This downward displacement of the plane causes the axis of reflection to pass through the origin.
2. Transform the plane in the manner described in Example 7.1.
3. Translate the plane a distance b in the positive y direction. The axis of reflection is now restored to its original position.

Applying the transformation matrix $\mathbf{T}_{\text{trans}}$ in Eq. (7.3) and $\mathbf{T}_{\text{refl}}$ in Example 7.1, we obtain the following transformation matrix for the present case:

$$\begin{bmatrix} 1 & 0 & 0 \\ 0 & 1 & 0 \\ 0 & -b & 1 \end{bmatrix} \begin{bmatrix} \cos 2\alpha & \sin 2\alpha & 0 \\ \sin 2\alpha & -\cos 2\alpha & 0 \\ 0 & 0 & 1 \end{bmatrix} \begin{bmatrix} 1 & 0 & 0 \\ 0 & 1 & 0 \\ 0 & b & 1 \end{bmatrix}$$

$$= \begin{bmatrix} \cos 2\alpha & \sin 2\alpha & 0 \\ \sin 2\alpha & -\cos 2\alpha & 0 \\ -b \sin 2\alpha & b \cos 2\alpha + b & 1 \end{bmatrix} \qquad (g)$$

This transformation matrix is the *general reflection matrix*, and it is denoted by $\mathbf{T}_{\text{refl,gen}}$. Matrix $\mathbf{T}_{\text{refl}}$ is a special form of $\mathbf{T}_{\text{refl,gen}}$ in which $b = 0$. We now set

$$[x' \quad y' \quad 1] = [x \quad y \quad 1]\mathbf{T}_{\text{refl,gen}}$$

$$x' = x \cos 2\alpha + (y - b) \sin 2\alpha$$

$$y' = x \sin 2\alpha - (y - b) \cos 2\alpha + b$$

To test these equations, consider a point that lies on line L. Its coordinates have the relationship $y = x \tan \alpha + b$. Upon substitution, we find that $x' = x$ and $y' = y$. Thus, according to these equations, the point remains fixed, and it does. The equations thus appear to be valid.

EXAMPLE 7.8 The plane is to be rotated through an angle θ in the counterclockwise direction about an axis passing through $Q(a, b)$. Establish the corresponding transformation of coordinates.

SOLUTION Refer to Fig. 7.11. Proceeding as in Example 7.7, we transform the plane in these steps: Translate the plane to make the axis of rotation coincident with the z axis; rotate the plane about the z axis; then translate the plane to restore the axis of rotation to its original position. The transformation matrix is as follows:

$$\begin{bmatrix} 1 & 0 & 0 \\ 0 & 1 & 0 \\ -a & -b & 1 \end{bmatrix}\begin{bmatrix} \cos\theta & \sin\theta & 0 \\ -\sin\theta & \cos\theta & 0 \\ 0 & 0 & 1 \end{bmatrix}\begin{bmatrix} 1 & 0 & 0 \\ 0 & 1 & 0 \\ a & b & 1 \end{bmatrix}$$

$$= \begin{bmatrix} \cos\theta & \sin\theta & 0 \\ -\sin\theta & \cos\theta & 0 \\ -a\cos\theta + b\sin\theta + a & -a\sin\theta - b\cos\theta + b & 1 \end{bmatrix}$$

This transformation matrix is the *general rotation matrix*, and we shall denote it by $\mathbf{T}_{\text{rot,gen}}$. Then

$$[x' \quad y' \quad 1] = [x \quad y \quad 1]\mathbf{T}_{\text{rot,gen}}$$

$$x' = (x - a)\cos\theta - (y - b)\sin\theta + a$$

$$y' = (x - a)\sin\theta + (y - b)\cos\theta + b$$

According to these equations, point Q remains fixed, and it does.

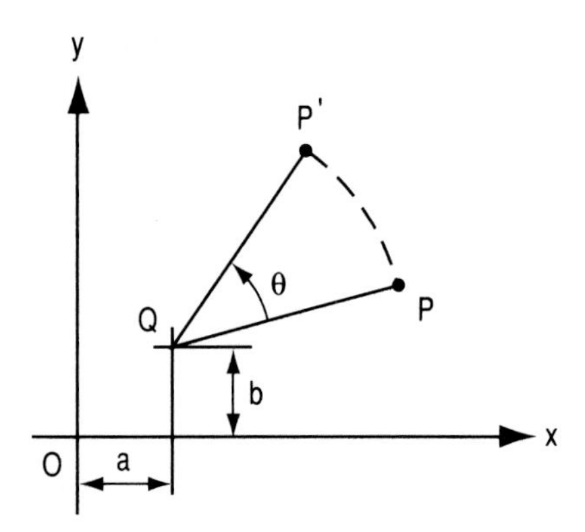

FIGURE 7.11 Rotation of plane about arbitrary axis.

7.2.13 Elemental Transformations of Three-Dimensional Space

The six linear transformations of the plane that we have investigated have their analogs in the transformation of three-dimensional space. In describing these transformations, we shall first record the coordinates of the image point P' and then display the corresponding matrix multiplication. The elemental transformations are as follows:

Translation. Consider that space is translated the distances d_x, d_y, and d_z in the positive x, y, and z directions, respectively. Then

$$x' = x + d_x \qquad y' = y + d_y \qquad z' = z + d_z$$

Applying homogeneous coordinates with $w = 1$, we have

$$[x' \quad y' \quad z' \quad 1] = [x \quad y \quad z \quad 1]\begin{bmatrix} 1 & 0 & 0 & 0 \\ 0 & 1 & 0 & 0 \\ 0 & 0 & 1 & 0 \\ d_x & d_y & d_z & 1 \end{bmatrix} \tag{7.37}$$

Rotation. Consider that space is rotated about a coordinate axis. We consider that the observer is located at the positive side of the axis and that the line of sight is toward the origin, as shown in Fig. 7.12. The angle of rotation θ is positive if space is rotated about this axis in a counterclockwise direction as viewed by the observer. For rotation about the x axis,

$$x' = x \qquad y' = y \cos\theta - z \sin\theta \qquad z' = y \sin\theta + z \cos\theta$$

$$[x' \quad y' \quad z'] = [x \quad y \quad z]\begin{bmatrix} 1 & 0 & 0 \\ 0 & \cos\theta & \sin\theta \\ 0 & -\sin\theta & \cos\theta \end{bmatrix} \tag{7.38}$$

For rotation about the y axis,

$$x' = x \cos\theta + z \sin\theta \qquad y' = y \qquad z' = -x \sin\theta + z \cos\theta$$

$$[x' \quad y' \quad z'] = [x \quad y \quad z]\begin{bmatrix} \cos\theta & 0 & -\sin\theta \\ 0 & 1 & 0 \\ \sin\theta & 0 & \cos\theta \end{bmatrix} \tag{7.39}$$

Rotation of space about the z axis is merely an extension of the rotation of the xy plane about that axis.
Then

$$x' = x \cos\theta - y \sin\theta \qquad y' = x \sin\theta + y \cos\theta \qquad z' = z$$

$$[x' \quad y' \quad z'] = [x \quad y \quad z]\begin{bmatrix} \cos\theta & \sin\theta & 0 \\ -\sin\theta & \cos\theta & 0 \\ 0 & 0 & 1 \end{bmatrix} \tag{7.40}$$

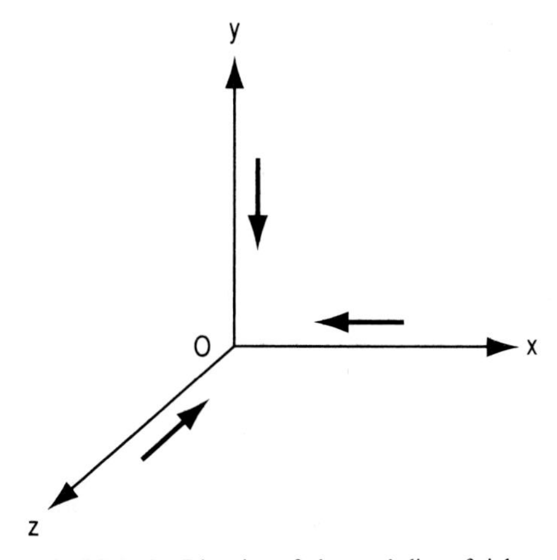

FIGURE 7.12 Direction of observer's line of sight.

Reflection. The reflection of the plane about an axis was assumed to result from revolving the plane about that axis through an angle of 180°. On the other hand, the reflection of space about a given plane must simply be accepted as a possibility, without attempting to visualize the mode of its accomplishment. For a reflection about the xy plane,

$$x' = x \qquad y' = y \qquad z' = -z$$

$$[x' \quad y' \quad z'] = [x \quad y \quad z]\begin{bmatrix} 1 & 0 & 0 \\ 0 & 1 & 0 \\ 0 & 0 & -1 \end{bmatrix} \tag{7.41}$$

Reflections about the xz and yz planes are similar.

Consider that space is reflected about a plane that contains the x axis and lies midway between the y and z axes. Then

$$x' = x \qquad y' = z \qquad z' = y$$

$$[x' \quad y' \quad z'] = [x \quad y \quad z]\begin{bmatrix} 1 & 0 & 0 \\ 0 & 0 & 1 \\ 0 & 1 & 0 \end{bmatrix} \tag{7.42}$$

Scaling. When space is scaled about a given plane, the distance of a point from that plane is multiplied by the scale factor. Consider that space is scaled about the xy, xz, and yz planes with the factors s_1, s_2, and s_3, respectively. Then

$$x' = s_3 x \qquad y' = s_2 y \qquad z' = s_1 z$$

$$[x' \quad y' \quad z'] = [x \quad y \quad z]\begin{bmatrix} s_3 & 0 & 0 \\ 0 & s_2 & 0 \\ 0 & 0 & s_1 \end{bmatrix} \tag{7.43}$$

Dilation or Compaction. Consider that space is dilated or compacted about the x axis by a factor k. Then

$$x' = x \qquad y' = ky \qquad z' = kz$$

$$[x' \quad y' \quad z'] = [x \quad y \quad z]\begin{bmatrix} 1 & 0 & 0 \\ 0 & k & 0 \\ 0 & 0 & k \end{bmatrix} \tag{7.44}$$

This transformation is equivalent to dilation or compaction of space about the xy and xz planes with a factor k.

Similarly, the dilation or compaction of space about the origin is equivalent to dilation or compaction about all three coordinate planes with a factor k.

Shear. When space undergoes a shear about a given plane, the displacement of a point has these characteristics: It is parallel to that plane, it is in a specified direction, and its magnitude is directly proportional to the distance of the point from that plane. Consider that space undergoes a shear about the xy plane in the

x direction with a factor h_1. Then

$$x' = x + h_1 z \qquad y' = y \qquad z' = z$$

$$[x' \quad y' \quad z'] = [x \quad y \quad z]\begin{bmatrix} 1 & 0 & 0 \\ 0 & 1 & 0 \\ h_1 & 0 & 1 \end{bmatrix} \tag{7.45}$$

A line perpendicular to the xy plane rotates in a plane parallel to the xz plane.

Now consider that space undergoes a shear about the xy plane in the y direction with a factor h_2. Then

$$x' = x \qquad y' = y + h_2 z \qquad z' = z$$

$$[x' \quad y' \quad z'] = [x \quad y \quad z]\begin{bmatrix} 1 & 0 & 0 \\ 0 & 1 & 0 \\ 0 & h_2 & 1 \end{bmatrix} \tag{7.46}$$

Shears about the remaining coordinate planes are similar.

The following principles pertaining to the transformation of space are analogous to those for the transformation of the plane:

Theorem 7.9. When space is subjected to any linear transformation, the image of a straight line is also a straight line, and the image of a plane is also a plane.

Theorem 7.10. When space is subjected to any linear transformation, the point of intersection of two straight lines is mapped onto the point of intersection of the images of those lines, and the line of intersection of two planes is mapped onto the line of intersection of the images of those planes.

Theorem 7.11. When space is subjected to any linear transformation, parallel lines remain parallel, and parallel planes remain parallel.

7.2.14 Nonelemental Transformations of Space

A nonelemental transformation of three-dimensional space may also be viewed as a composite of elemental transformations. We shall first consider cases where the axis of transformation passes through the origin.

In Art. 3.3.5, we defined the direction angles and direction cosines of a directed line. We shall now assign new designations to the direction cosines to conform to those used in computer graphics. In Fig. 7.13, L is a line through the origin having the indicated positive direction. The angles between the positive side of L and the positive sides of the x, y, and z axes are denoted by α, β, and γ, respectively, as shown. These are the direction angles of L, and the cosines of these angles are the direction cosines of the line. The new designations are these:

$$n_1 = \cos\alpha \qquad n_2 = \cos\beta \qquad n_3 = \cos\gamma$$

Equation (3.41) becomes

$$n_1^2 + n_2^2 + n_3^2 = 1$$

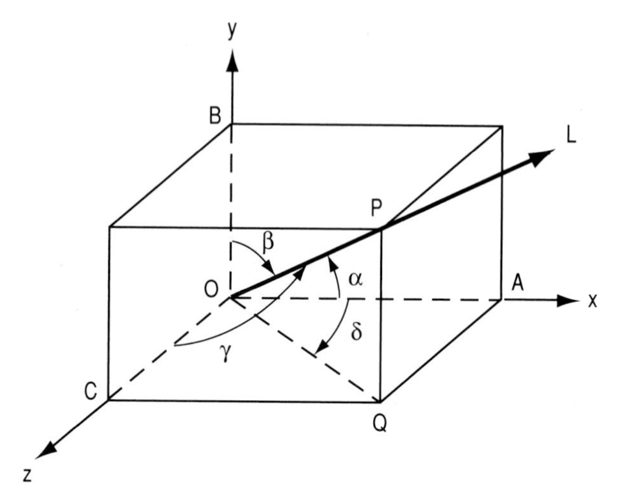

FIGURE 7.13 Direction angles of a line through the origin.

In Fig. 7.13, $P(x, y, z)$ is an arbitrary point on L. Then

$$n_1^2 = \frac{x^2}{(OP)^2} \qquad n_2^2 = \frac{y^2}{(OP)^2} \qquad n_3^2 = \frac{z^2}{(OP)^2}$$

We now draw PQ perpendicular to the xz plane, where Q lies on that plane, and we then draw QA perpendicular to the x axis. Let δ denote angle QOA. Then

$$\sin^2 \delta = \frac{z^2}{x^2 + z^2} = \frac{n_3^2}{n_1^2 + n_3^2}$$

$$\cos^2 \delta = \frac{x^2}{x^2 + z^2} = \frac{n_1^2}{n_1^2 + n_3^2}$$

Now consider that L is an axis of transformation; i.e., space is to be subjected to a rotation, reflection, scaling, or shear about L. This nonelemental transformation of space can be achieved in the following steps:

1. Rotate space about the y axis through an angle δ. The axis of transformation now lies in the xy plane.
2. Rotate space about the z axis through an angle β. The axis of transformation now lies on the y axis.
3. Perform the transformation specified (rotation, reflection, etc.) with respect to the y axis.
4. Rotate space about the z axis through an angle $-\beta$.
5. Rotate space about the y axis through an angle $-\delta$.

Steps 4 and 5 restore line L to its original position. The transformation matrix corresponding to steps 1 and 2 is as follows:

$$\begin{bmatrix} \cos\delta & 0 & -\sin\delta \\ 0 & 1 & 0 \\ \sin\delta & 0 & \cos\delta \end{bmatrix} \begin{bmatrix} \cos\beta & \sin\beta & 0 \\ -\sin\beta & \cos\beta & 0 \\ 0 & 0 & 1 \end{bmatrix}$$

$$= \begin{bmatrix} \cos\delta\cos\beta & \cos\delta\sin\beta & -\sin\delta \\ -\sin\beta & \cos\beta & 0 \\ \sin\delta\cos\beta & \sin\delta\sin\beta & \cos\delta \end{bmatrix} = \mathbf{T}_1$$

The transformation matrix corresponding to steps 4 and 5 is the inverse (as well as the transpose) of $\mathbf{T}_1$. It is as follows:

$$\begin{bmatrix} \cos\delta\cos\beta & -\sin\beta & \sin\delta\cos\beta \\ \cos\delta\sin\beta & \cos\beta & \sin\delta\sin\beta \\ -\sin\delta & 0 & \cos\delta \end{bmatrix} = \mathbf{T}_1^{-1}$$

Now let $\mathbf{M}$ denote the specific transformation matrix corresponding to step 3. The transformation matrix $\mathbf{A}$ corresponding to the nonelemental transformation is

$$\mathbf{A} = \mathbf{T}_1 \mathbf{M} \mathbf{T}_1^{-1} \tag{7.47}$$

To illustrate the procedure, assume that space is to be rotated about line L in Fig. 7.13 through a counterclockwise angle θ, the line of sight of the observer being from P to O. (This is a highly important problem in computer graphics.) Step 3 now consists of rotating space about the y axis through a positive angle θ. We set $\mathbf{M}$ equal to the transformation matrix in Eq. (7.39) and perform the multiplication. We shall express the product $\mathbf{A}$ in terms of the direction cosines of L. Moreover, since $\mathbf{A}$ contains lengthy expressions, we shall record the elements of this matrix in running form.

$$\begin{aligned}
a_{11} &= n_1^2 + (1 - n_1^2)\cos\theta \\
a_{12} &= n_1 n_2 (1 - \cos\theta) + n_3 \sin\theta \\
a_{13} &= n_1 n_3 (1 - \cos\theta) - n_2 \sin\theta \\
a_{21} &= n_1 n_2 (1 - \cos\theta) - n_3 \sin\theta \\
a_{22} &= n_2^2 + (1 - n_2^2)\cos\theta \\
a_{23} &= n_2 n_3 (1 - \cos\theta) + n_1 \sin\theta \\
a_{31} &= n_1 n_3 (1 - \cos\theta) + n_2 \sin\theta \\
a_{32} &= n_2 n_3 (1 - \cos\theta) - n_1 \sin\theta \\
a_{33} &= n_3^2 + (1 - n_3^2)\cos\theta
\end{aligned} \tag{7.48}$$

The matrix in Eq. (7.48) has a highly interesting structure. This can be discerned by dividing the elements into the following groups: a_{11}, a_{22}, and a_{33}; a_{12}, a_{23}, and a_{31}; a_{13}, a_{21}, and a_{32}. The elements in a group have the same form but differ in their subscripts. As we proceed from one element to the next, the subscripts change in this manner: 1 becomes 2, 2 becomes 3, and 3 becomes 1.

If the axis of transformation (line L in Fig. 7.13) does not pass through the origin, we extend the procedure by translating space to cause this axis to pass through the origin, performing the specified transformation, and then displacing space again to restore the axis of transformation to its original position.

7.2.15 General Third-Order Transformation Matrix

We shall apply the following notation for the general third-order transformation matrix:

$$\begin{bmatrix} a & b & c \\ d & e & f \\ g & h & i \end{bmatrix}$$

This transformation matrix may result from a transformation of space that consists of scalings and shears, or scalings, shears, and reflections. If the diagonal elements are all positive, the transformation contains scalings and shears only, and the diagonal elements are the scale factors.

EXAMPLE 7.9 A plane contains the following points: $P_1(2, 3, -7)$, $P_2(-12, 18, 1)$, and $P_3(4, 10, 5)$. The transformed plane is to contain the following corresponding points: $P'_1(-16, -43, 32)$, $P'_2(4, -19, -33)$, and $P'_3(52, 1, 11)$. Construct the transformation matrix.

SOLUTION For convenience, we shall combine the position vectors of the three points into a single matrix. Then

$$\begin{bmatrix} -16 & -43 & 32 \\ 4 & -19 & -33 \\ 52 & 1 & 33 \end{bmatrix} = \begin{bmatrix} 2 & 3 & -7 \\ -12 & 18 & 1 \\ 4 & 10 & 5 \end{bmatrix} \begin{bmatrix} a & b & c \\ d & e & f \\ g & h & i \end{bmatrix}$$

Taking the elements in the first column at the left, we obtain

$$-16 = 2a + 3d - 7g$$

$$4 = -12a + 18d + g$$

$$52 = 4a + 10d + 5g$$

The solution of this system of equations is $a = 3$, $d = 2$, $g = 4$. Continuing in this manner, we obtain $b = -1$, $e = -2$, $h = 5$, $c = 4$, $f = 1$, $i = -3$. Therefore, the transformation matrix is

$$\begin{bmatrix} 3 & -1 & 4 \\ 2 & -2 & 1 \\ 4 & 5 & -3 \end{bmatrix}$$

7.3 PROJECTIONS

7.3.1 Definitions

We shall now investigate transformations of space that are many-to-one mappings. In Fig. 7.14*a*, A is a reference point and J is a reference line. Every straight line that

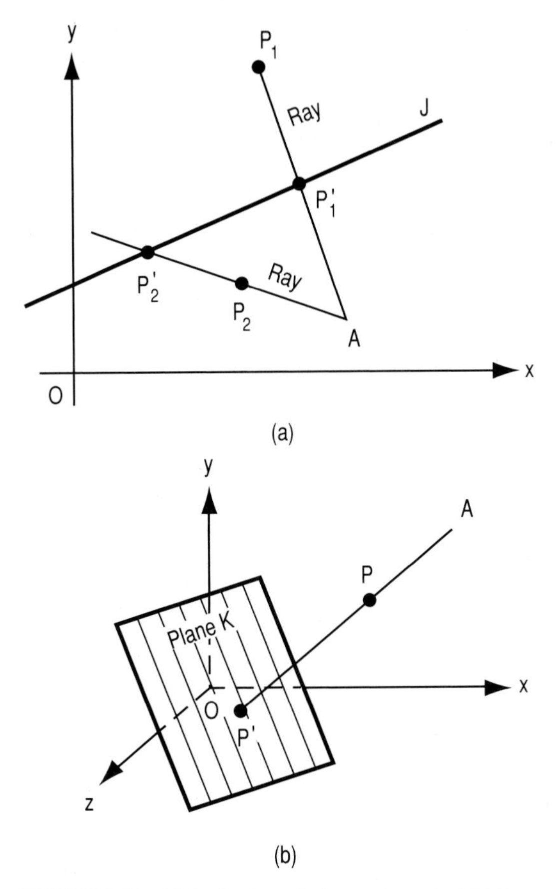

FIGURE 7.14 (*a*) Projection of plane onto a straight line; (*b*) projection of space onto a plane.

passes through A is termed a *ray*. Consider that the plane is transformed in such manner that a given point P is mapped onto the point P' at which the ray through P intersects J. Thus, P_1 and P_2 in the drawing are mapped onto P_1' and P_2', respectively. Since all image points lie on J, the plane has degenerated to this line. The plane is said to be *projected onto* J, and A and J are called the *center of projection* and *line of projection*, respectively. Point P_1' is the image not only of P_1 but of all points that lie on the ray through P_1.

Assume that A is placed at an infinite distance from J, causing all rays to be perpendicular to J. The projection is described as *orthographic*. Now consider that the plane is projected orthographically onto the line $y = d$. Then $x' = x$ and $y' = d$. Applying homogeneous coordinates with $w = 1$, we have

$$[x' \quad y' \quad 1] = [x \quad y \quad 1]\begin{bmatrix} 1 & 0 & 0 \\ 0 & 0 & 0 \\ 0 & d & 1 \end{bmatrix}$$

The projection of three-dimensional space is defined in an analogous manner. In Fig. 7.14*b*, point A is the center of projection and K is a fixed plane. Consider that the ray from A to a given point P is drawn and that P is mapped onto the point P' at which this ray intersects plane K. Space is said to be projected onto K, and K is termed the *plane of projection.*

Let L denote an arbitrary straight line in space and R denote the plane through L and the center of projection A. All rays from A to L lie in R, and the image of L is the line L' along which R intersects the plane of projection. We thus arrive at the following principle:

Theorem 7.12. When space is projected onto a plane, the image of a straight line is also a straight line.

Assume that the center of projection is placed at an infinite distance from the plane of projection, causing all rays to be parallel to one another. If the rays are perpendicular to the plane of projection, the projection is *orthographic*; otherwise, it is *oblique*. Orthographic projections are applied in the construction of engineering drawings.

In particular, consider that space is projected orthographically onto the plane $z = n$. Then $x' = x$, $y' = y$, and $z' = n$. By matrix multiplication,

$$[x' \quad y' \quad z' \quad 1] = [x \quad y \quad z \quad 1]\begin{bmatrix} 1 & 0 & 0 & 0 \\ 0 & 1 & 0 & 0 \\ 0 & 0 & 0 & 0 \\ 0 & 0 & n & 1 \end{bmatrix} \tag{7.49}$$

The transformation matrix is singular, as Theorem 7.1 requires. Now consider that space is projected orthographically onto the xy plane, making $z' = 0$. Then

$$[x' \quad y' \quad z'] = [x \quad y \quad z]\begin{bmatrix} 1 & 0 & 0 \\ 0 & 1 & 0 \\ 0 & 0 & 0 \end{bmatrix} \tag{7.49a}$$

7.3.2 Classification of Axonometric Projections

An *axonometric projection* of space is a composite transformation in which space is rotated about coordinate axes and then projected orthographically onto a plane. We shall take the xy plane as the plane of projection. In a particular instance, the rotation may be absent.

Axonometric projections can be divided into three categories: trimetric, dimetric, and isometric. These terms can be defined readily by referring to the unit cube in Fig. 7.15, which has a face on each coordinate plane. Under an axonometric projection, O is fixed, and points a, b, and c are transformed to a', b', and c', respectively. Consider the lengths Oa', Ob', and Oc'. If all three lengths differ, the projection is *trimetric*; if two and only two of these lengths are equal, the projection is *dimetric*; if all three lengths are equal, the drawing is *isometric*. We shall consider how dimetric and isometric projections can be obtained. The problem is simply to identify the rotation of space that will yield the desired result.

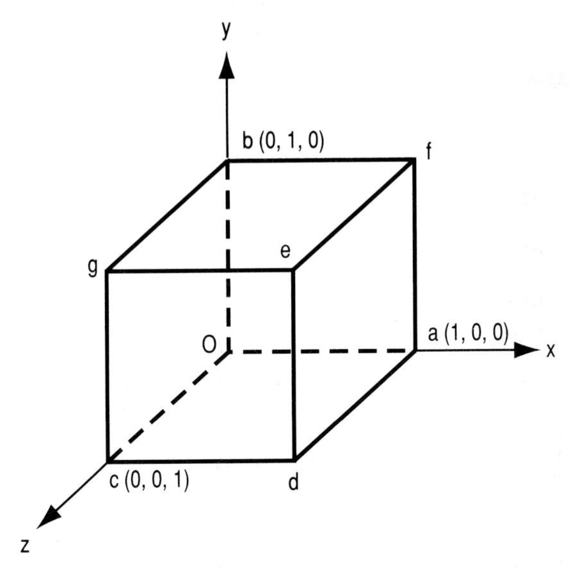

FIGURE 7.15 Location of unit cube.

7.3.3 Method of Identifying Hidden Lines

In forming an axonometric projection of a solid object, we are confronted with the basic problem of identifying visible and hidden lines and surfaces. For this purpose, we consider the projection of the solid to be the image of the object as perceived by an observer who is located at an infinite distance from the xy plane and views the plane from the positive side of the z axis.

To resolve this problem, let z_r denote the z coordinate of a point following the rotation of space but prior to the projection of space onto a plane. By computing the z_r values of points on the object, it becomes possible to ascertain which lines are visible and which lines are hidden.

It is also helpful to consider how the z_r value of an individual point can be obtained graphically by applying the orthographic projections of engineering. For illustrative purposes, we select the point $P(7, 8, 12)$. Consider that space is rotated about the y axis through 25°, mapping P onto P_1, and that space is then rotated about the x axis through 30°, mapping P_1 onto P_r. Figure 7.16*a* is a view parallel to the y axis (and normal to the xz plane), and Fig. 7.16*b* is a view parallel to the x axis (and normal to the yz plane). In Fig. 7.16*a*, we plot P and then locate P_1 in the indicated manner. We then locate P_1 in Fig. 7.16*b* by extending the ray from the first drawing and applying the given value of y. We then locate P_r in the indicated manner and scale its z coordinate, which is the z_r value of P.

7.3.4 Dimetric Projections

We shall formulate a dimetric projection of the cube in Fig. 7.15 in which $Oa' = Ob'$. Consider that the transformation of space consists of the following steps, in the specified order: a rotation about the y axis through an angle θ_y; a

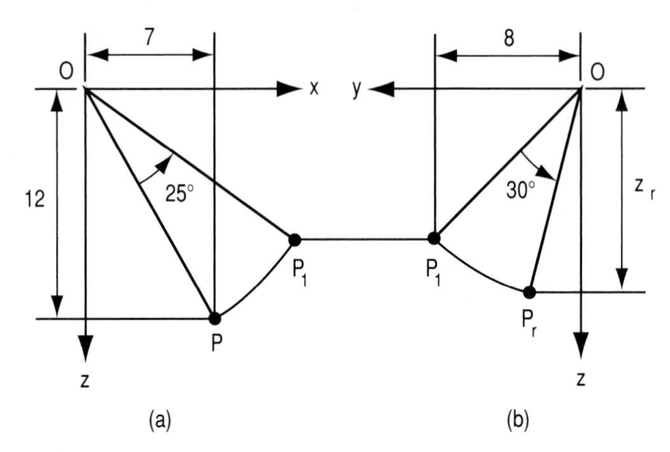

FIGURE 7.16 Transformation of P to P_r. (*a*) View parallel to y axis; (*b*) view parallel to x axis.

rotation about the x axis through an angle θ_x; orthographic projection onto the xy plane. Multiplying the transformation matrices in Eqs. (7.39), (7.38), and (7.49*a*), in that order, we obtain the following:

$$[x' \quad y' \quad z'] = [x \quad y \quad z] \begin{bmatrix} \cos\theta_y & \sin\theta_x \sin\theta_y & 0 \\ 0 & \cos\theta_x & 0 \\ \sin\theta_y & -\sin\theta_x \cos\theta_y & 0 \end{bmatrix} \tag{7.50}$$

The coordinates of points a, b, and c of the cube are recorded in Fig. 7.15. Applying Eq. (7.50), we obtain the following coordinates of their images: $a'(\cos\theta_y, \sin\theta_x \sin\theta_y)$; $b'(0, \cos\theta_x)$; $c'(\sin\theta_y, -\sin\theta_x \cos\theta_y)$. Then

$$\begin{aligned} (Oa')^2 &= \cos^2\theta_y + \sin^2\theta_x \sin^2\theta_y \\ (Ob')^2 &= \cos^2\theta_x \\ (Oc')^2 &= \sin^2\theta_y + \sin^2\theta_x \cos^2\theta_y \end{aligned} \tag{7.51}$$

By setting $Oa' = Ob'$, we obtain

$$\sin^2\theta_y = \tan^2\theta_x \tag{7.52}$$

The z_r value of a point is found by multiplying the transformation matrices in Eqs. (7.39) and (7.38), in that order, and we obtain this value:

$$z_r = -x \cos\theta_x \sin\theta_y + y \sin\theta_x + z \cos\theta_x \cos\theta_y \tag{7.53}$$

The values of θ_x and θ_y hinge on the value of Oc'. Let $Oc' = k$, where $0 < k < 1$. By combining the expression for Oc' with Eq. (7.52), we obtain

$$\sin^2\theta_x = \frac{k^2}{2} \tag{7.54}$$

Since we are at liberty to make the functions of these angles either positive or negative, there are multiple combinations of values of θ_x and θ_y that satisfy

Eqs. (7.52) and (7.54). We shall set $k = 0.5$, a value that is often applied in practice. One acceptable combination of values is the following:

$$\sin\theta_x = \sqrt{1/8} \qquad \cos\theta_x = \sqrt{7/8} \qquad \theta_x = 20.705^\circ$$

$$\sin\theta_y = \sqrt{1/7} \qquad \cos\theta_y = \sqrt{6/7} \qquad \theta_y = 22.208^\circ$$

The transformation matrix in Eq. (7.50) becomes

$$\begin{bmatrix} \sqrt{6/7} & \sqrt{1/56} & 0 \\ 0 & \sqrt{7/8} & 0 \\ \sqrt{1/7} & -\sqrt{3/28} & 0 \end{bmatrix}$$

The coordinates of the images of a, b, and c are as follows: $a'(\sqrt{6/7}, \sqrt{1/56})$; $b'(0, \sqrt{7/8})$; $c'(\sqrt{1/7}, -\sqrt{3/28})$. The lengths are $Oa' = Ob' = \sqrt{7/8}$ and $Oc' = 1/2$. These lengths satisfy our requirements.

By computing the coordinates of the images of the remaining vertices of the cube and plotting the points, we obtain the dimetric projection of the cube in Fig. 7.17*a*. We find that the z_r values of a and f are less than those of c, d, e, and g. Consequently, Oa', $a'd'$, and $a'f'$ are hidden lines. The face *bfeg* is visible; the face *Oadc* is hidden. (This conclusion can be reached more simply by visualizing the manner in which the cube rotates.)

Another acceptable combination of values is the following:

$$\sin\theta_x = -\sqrt{1/8} \qquad \cos\theta_x = -\sqrt{7/8} \qquad \theta_x = 200.705^\circ$$

$$\sin\theta_y = -\sqrt{1/7} \qquad \cos\theta_y = \sqrt{6/7} \qquad \theta_y = -22.208^\circ$$

Figure 7.17*b* is the corresponding dimetric projection of the cube. The face *Ocda* is visible; the face *bgef* is hidden.

We may modify the sequence of operations by rotating space about the x axis, then rotating it about the y axis, and then projecting it onto the xy plane. This sequence yields equations that correspond to Eqs. (7.52) and (7.54); the subscripts x and y are merely interchanged. In the dimetric projections based on this sequence of operations, point a' remains on the x axis and b' falls off the y axis.

7.3.5 Isometric Projections

We shall again apply the first transformation of space described in Art. 7.3.4: rotation about the y axis, rotation about the x axis, and projection onto the xy plane. We now extend the relationships in Art. 7.3.4 by setting $Oa' = Ob' = Oc'$. Since $(Ob')^2 = \cos^2\theta_x$ and $Oc' = k$, we replace k^2 in Eq. (7.54) with $\cos^2\theta_x$, and then apply Eq. (7.52). The results are these:

$$\sin^2\theta_x = 1/3 \tag{7.55}$$

$$\sin^2\theta_y = \cos^2\theta_y = 1/2 \tag{7.56}$$

If we make all functions positive, we obtain the following combination of values:

$$\sin\theta_x = \sqrt{1/3} \qquad \cos\theta_x = \sqrt{2/3} \qquad \theta_x = 35.2644^\circ$$

$$\sin\theta_y = \cos\theta_y = \sqrt{1/2} \qquad \theta_y = 45^\circ$$

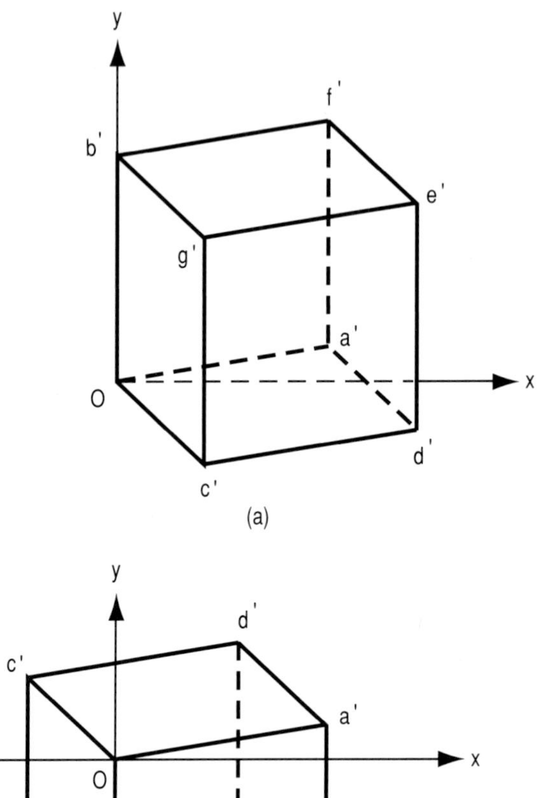

FIGURE 7.17 Dimetric projections of cube with $Oa' = Ob'$ and (*a*) $\theta_x = 20.7°$, $\theta_y = 22.2°$ and (*b*) $\theta_x = 200.7°$, $\theta_y = -22.2°$.

Figure 7.18*a* is the corresponding isometric projection of the cube in Fig. 7.15. The face *bfeg* is visible; the face *Oadc* is hidden. (The hidden lines have been omitted in the present instance for greater clarity.) In this drawing, $\alpha = \beta = 30°$. This fortunate characteristic makes an isometric projection simple to construct by use of a 30° triangle.

Alternative combinations of values may be used, such as the following:

$$\sin\theta_x = -\sqrt{1/3} \qquad \cos\theta_x = \sqrt{2/3} \qquad \theta_x = -35.2644°$$

$$\sin\theta_y = -\sqrt{1/2} \qquad \cos\theta_y = \sqrt{1/2} \qquad \theta_y = -45°$$

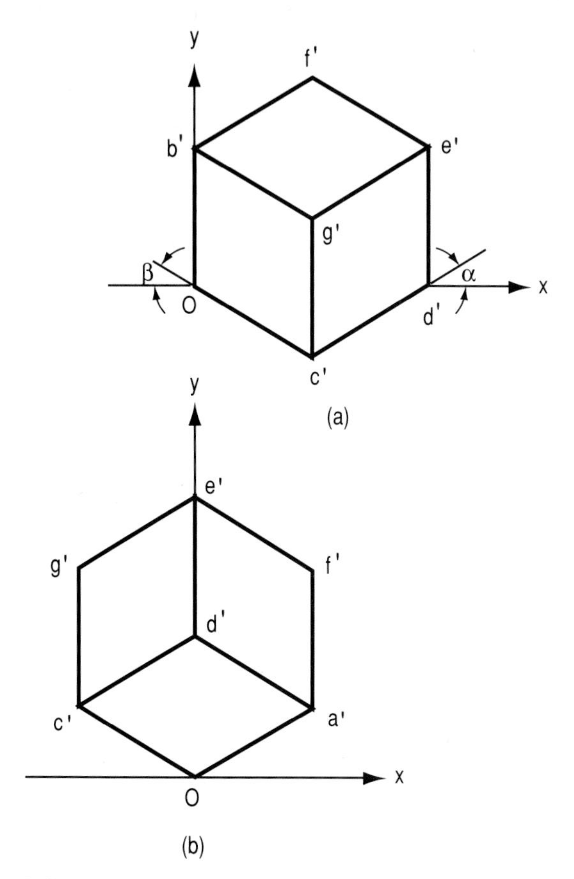

FIGURE 7.18 Isometric projections of cube with (*a*) $\theta_x = 35.3°$, $\theta_y = 45°$ and (*b*) $\theta_x = -45°$, $\theta_y = -45°$.

Figure 7.18*b* is the corresponding isometric projection of the cube. The face *Oadc* is visible; the face *bfeg* is hidden.

7.3.6 Terminology of Perspective Projections

In its general form, a *perspective projection* of space is a composite transformation in which translations and rotations are followed by the projection of space onto a plane, the center of projection being at a finite distance from the plane of projection. In a particular instance, the translations or rotations, or both, may be absent. The perspective projection of an object conforms closely with the image of the object as perceived by the human eye. For this reason, perspective projections are widely applied in architectural renderings of a building and in commercial illustrations.

Refer to Fig. 7.19. The observer of an object is assumed to view it from a single point. This is the center of projection, and it is now termed the *station point*. The horizontal plane on which the observer stands is called the *ground plane*. The rays extending from the station point to the object are called *visual rays*. The picture of

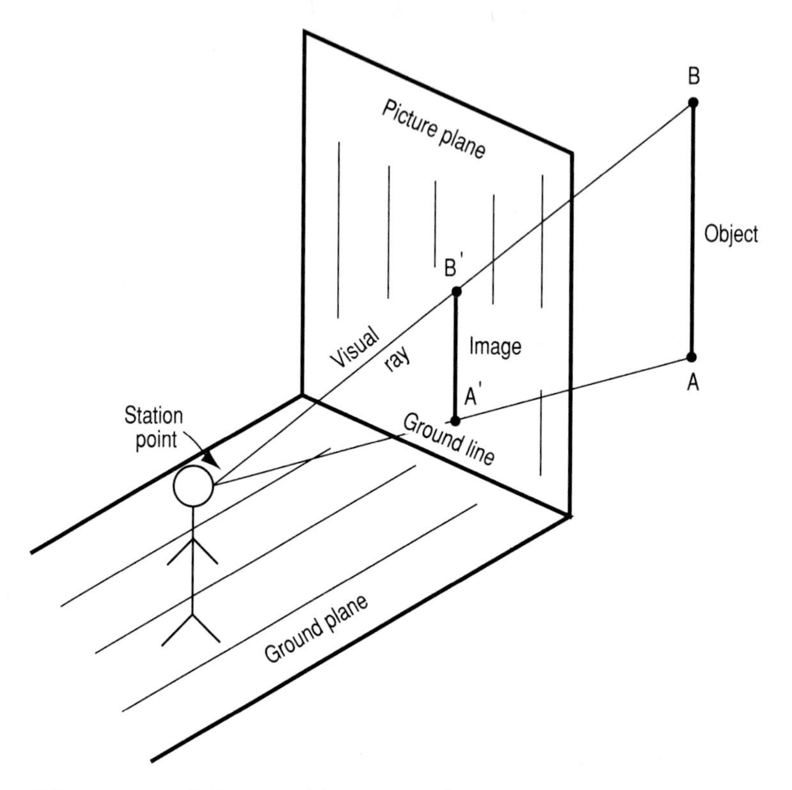

FIGURE 7.19 Definitions pertaining to perspective projections.

the object is formed on the plane of projection, which is now termed the *picture plane* or *viewing plane*. The line of intersection of the picture plane and the horizontal plane through the station point is the *horizon*; the line of intersection of the picture plane and the ground plane is the *ground line*.

When the picture plane is interposed between the observer and the object, the object is projected to a reduced size. Thus, in Fig. 7.19, the vertical line AB is projected as $A'B'$. Conversely, when the object lies between the picture plane and the station point, the object is projected to an enlarged size.

In Fig. 7.20, the station point is at A, L is an arbitrary straight line of infinite extent, P is a point on L, and P' is the image of P. We draw AB perpendicular to L and AC parallel to L, intersecting the picture plane at C. Consider that P moves along L to the right, receding indefinitely from the station point. As it does so, angle α diminishes and approaches 0 as a limit. Correspondingly, angle β increases and approaches 90° as a limit. Therefore, P' approaches C. Thus, although L is of infinite extent, its image terminates at C. Point C is called the *vanishing point* of L. (The vanishing point of a nonhorizontal line is sometimes called the *trace point*.) If we now draw a line M parallel to L, we find that C is also the vanishing point of M. Thus, all lines that are parallel to L appear to converge at C. A line N that is parallel to the picture plane has no vanishing point, and its image N' is parallel to N.

Generally, the station point is placed on the z axis, and the picture plane is made coincident with the xy plane. Since our basic objective is to have the perspective

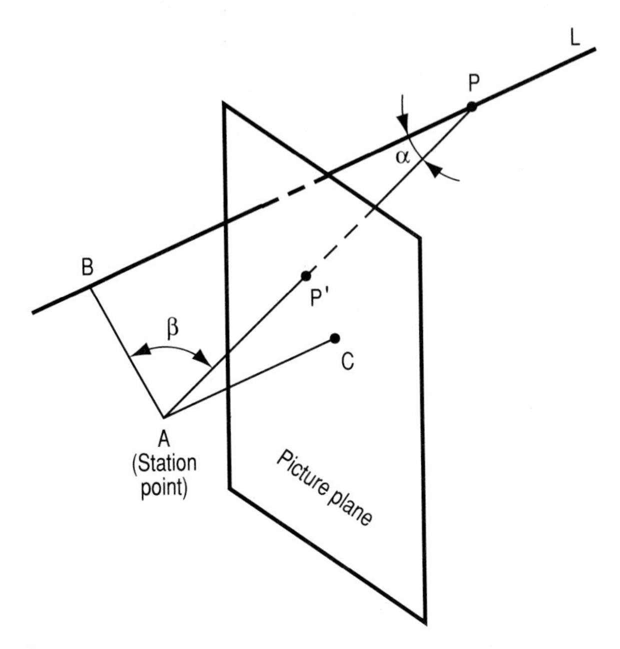

FIGURE 7.20 Location of vanishing point.

expose all the salient features of the object, it is imperative that we orient the object properly in relation to the observer. If the object is originally placed in some standard position that would produce a poor perspective, we can bring it into proper alignment by translating and rotating space before projecting space. Another objective in constructing a perspective is to avoid a gross distortion of the object. We can achieve this objective by confining the visual rays within a cone whose vertex angle does not exceed 30°.

7.3.7 Classification of Perspective Projections

Assume that an object contains lines that are parallel to the x, y, and z axes. In Fig. 7.21, we select an arbitrary point on or within the object and draw through this point axes t, u, and v parallel to the x, y, and z axes, respectively. The t, u, and v axes are the *principal axes* of the object. Unlike the coordinate axes, which are fixed, the principal axes change their position under a translation or rotation of space. Now consider that space is rotated about coordinate axes and then projected onto the xy plane, thus forming an image of the object. There are three types of perspective projections, and we shall now define them.

A *one-point* or *parallel perspective* is one in which two principal axes are parallel to the picture plane at the time space is projected. Specifically, if space is not rotated, or if it is rotated about the z axis prior to the projection, the t and u axes remain parallel to the xy plane. The image of the object then has a vanishing point corresponding to lines that are parallel to the v axis, but none corresponding to lines that are parallel to the t and u axes. (If the object contains other lines that are

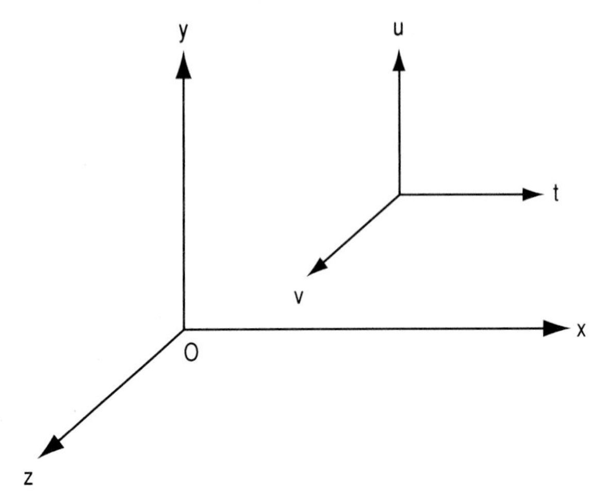

FIGURE 7.21 Location of principal axes of an object.

parallel to one another but not to a principal axis, the vanishing point corresponding to these lines is also significant.)

A *two-point* or *angular perspective* is one in which only one principal axis is parallel to the picture plane at the time space is projected. For example, consider that space is rotated about the x axis prior to the projection, the result being that t alone is parallel to the xy plane. The image of the object has a vanishing point corresponding to lines parallel to the u axis and another corresponding to lines parallel to the v axis.

A *three-point* or *oblique perspective* is one in which no principal axis is parallel to the picture plane at the time space is projected. This condition can arise if space is rotated about both the x and y axes prior to the projection. The image of the object has a vanishing point corresponding to lines parallel to each of the three principal axes.

7.3.8 Construction of Perspective Projections

Let $Q(a, b, c)$ denote the station point. Again let $P(x, y, z)$ denote an arbitrary point prior to projection and $P'(x', y', z')$ denote the image of P resulting from projection. Point P' lies on the visual ray QP, and Eq. (3.60) yields

$$\frac{x' - a}{x - a} = \frac{y' - b}{y - b} = \frac{z' - c}{z - c} \tag{7.57}$$

We now place the picture plane at the xy plane and the station point on the z axis at a distance of $1/r$ behind the picture plane. Thus, $a = b = 0$, $c = -1/r$, and $z' = 0$. Equation (7.57) yields

$$x' = \frac{x}{rz + 1} \qquad y' = \frac{y}{rz + 1} \tag{7.58}$$

To obtain these results by matrix multiplication, we proceed in two steps. First, applying homogeneous coordinates, we set

$$[X \quad Y \quad Z \quad W] = [x \quad y \quad z \quad 1]\begin{bmatrix} 1 & 0 & 0 & 0 \\ 0 & 1 & 0 & 0 \\ 0 & 0 & 0 & r \\ 0 & 0 & 0 & 1 \end{bmatrix} \tag{7.59a}$$

Then $\quad X = x \qquad Y = y \qquad Z = 0 \qquad W = rz + 1$

Now we set

$$[x' \quad y' \quad z' \quad w] = \left(\frac{1}{W}\right)[X \quad Y \quad Z \quad W] \tag{7.59b}$$

This operation yields the values of x' and y' as given by Eq. (7.58), and $w = 1$. The second step in this procedure is called *normalizing the coordinates.*

Consider that a point P moves along a line parallel to the z axis, receding indefinitely from the station point. Its x and y coordinates remain constant, but its z coordinate increases beyond bound. When the point is at infinity, Eq. (7.58) yields $x' = y' = 0$. Thus, the vanishing point of this line lies at the origin, and this mathematical result is in accord with the instruction for locating a vanishing point as presented in Art. 7.3.6.

EXAMPLE 7.10 The unit cube in Fig. 7.22 is positioned in the following manner: Face $EFGH$ lies on the xy plane, face $BFGC$ lies 0.6 units to the right of the yz plane, and face $DCGH$ lies 0.8 units above the xz plane. The station point lies at $Q(0, 0, 4)$, and the picture plane lies at the xy plane. Construct the image of this cube.

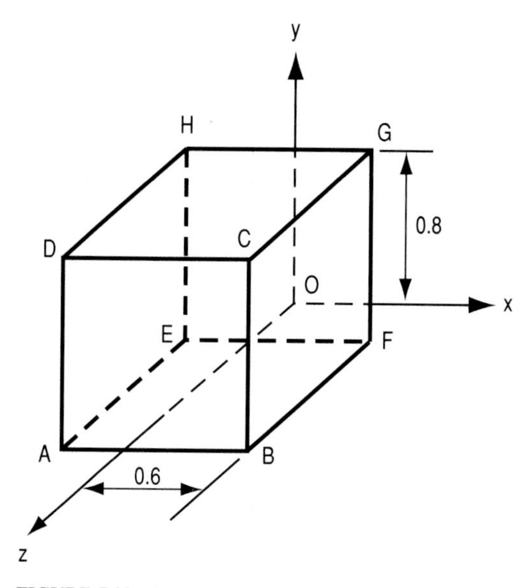

FIGURE 7.22 Location of unit cube.

TABLE 7.1 Coordinates of Vertices of Cube in Fig. 7.22

Vertex	Original coordinates			Transformed coordinates under projection	
	x	y	z	x'	y'
A	−0.4	−0.2	1	−0.5333	−0.2667
B	0.6	−0.2	1	0.8000	−0.2667
C	0.6	0.8	1	0.8000	1.0667
D	−0.4	0.8	1	−0.5333	1.0667
E	−0.4	−0.2	0	−0.4000	−0.2000
F	0.6	−0.2	0	0.6000	−0.2000
G	0.6	0.8	0	0.6000	0.8000
H	−0.4	0.8	0	−0.4000	0.8000

SOLUTION Since the object consists exclusively of lines that are parallel to the coordinate axes, the projection will be a one-point perspective, with a vanishing point corresponding to lines parallel to the z axis. Table 7.1 presents both the original coordinates of the vertices of the cube and the transformed coordinates as calculated by Eq. (7.58). We have $r = -1/4 = -0.25$. For $z = 0$, $rz + 1 = 1$. For $z = 1$, $rz + 1 = 0.75$. The projection of the cube appears in Fig. 7.23. The vanishing point for lines that were originally parallel to the z axis is at the origin. Thus, if lines $A'E'$, $B'F'$, $C'G'$, and $D'H'$ are prolonged, they intersect at the origin.

If space is rotated about the z axis before it is projected onto the xy plane, the effect is simply to rotate the image of the object through the same angle. The image is still a one-point perspective.

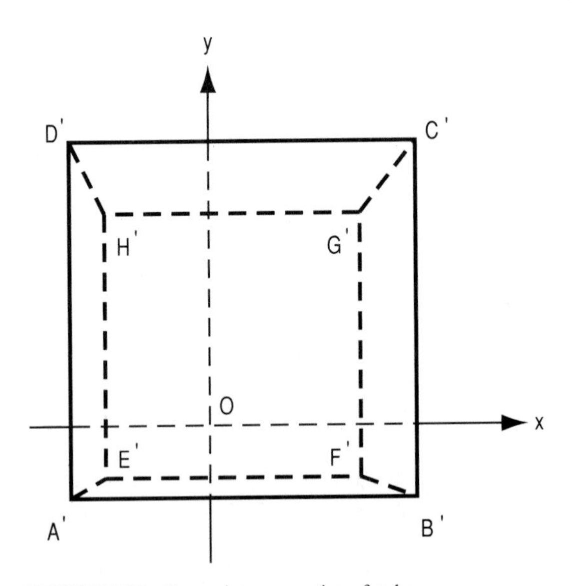

FIGURE 7.23 One-point perspective of cube.

Now consider that space is rotated about the x axis through an angle θ and is then projected onto the xy plane, the station point being at $Q(0, 0, -1/r)$. By expanding the transformation matrix in Eq. (7.38) to a fourth-order matrix and then postmultiplying it by the transformation matrix in Eq. (7.59a), we obtain the following matrix:

$$\mathbf{T}_c = \begin{bmatrix} 1 & 0 & 0 & 0 \\ 0 & \cos\theta & \sin\theta & 0 \\ 0 & -\sin\theta & \cos\theta & 0 \\ 0 & 0 & 0 & 1 \end{bmatrix} \begin{bmatrix} 1 & 0 & 0 & 0 \\ 0 & 1 & 0 & 0 \\ 0 & 0 & 0 & r \\ 0 & 0 & 0 & 1 \end{bmatrix}$$

$$= \begin{bmatrix} 1 & 0 & 0 & 0 \\ 0 & \cos\theta & 0 & r\sin\theta \\ 0 & -\sin\theta & 0 & r\cos\theta \\ 0 & 0 & 0 & 1 \end{bmatrix}$$

$$[X \quad Y \quad Z \quad W] = [x \quad y \quad z \quad 1]\mathbf{T}_c$$

$$X = x \qquad Y = y\cos\theta - z\sin\theta \qquad Z = 0$$

$$W = r(y\sin\theta + z\cos\theta) + 1$$

Normalizing the coordinates, we obtain

$$x' = \frac{x}{r(y\sin\theta + z\cos\theta) + 1} \qquad y' = \frac{y\cos\theta - z\sin\theta}{r(y\sin\theta + z\cos\theta) + 1} \tag{7.60}$$

These results can be obtained more simply by considering that P is transformed to $P_1(x_1, y_1, z_1)$ by the rotation about the x axis, and P_1 in turn is transformed to P' by the projection. Then

$$x_1 = x \qquad y_1 = y\cos\theta - z\sin\theta \qquad z_1 = y\sin\theta + z\cos\theta$$

We now apply Eq. (7.58), but replacing x, y, and z with x_1, y_1, and z_1, respectively.

If the angle of rotation about the x axis has any value other than 90° or a multiple thereof, neither the u nor v axis is parallel to the xy plane. Consequently, the image of the cube in Fig. 7.22 is a two-point perspective.

EXAMPLE 7.11 With reference to Example 7.10, consider that space is rotated about the x axis through a counterclockwise angle of 18° as viewed from the right of the yz plane. Space is then projected onto the xy plane from the specified station point. Construct the image of the cube, and locate the two vanishing points.

SOLUTION By our sign convention, θ is positive. For convenience, the original coordinates of the vertices of the cube are repeated in Table 7.2, and the transformed coordinates as calculated by Eq. (7.60) are then recorded. The image of the cube appears in Fig. 7.24.

The vanishing points are located in Fig. 7.25, which is a view parallel to the x axis (and normal to the yz plane). Point S denotes the station point, which lies on

TABLE 7.2

	Original coordinates			Transformed coordinates under rotation and projection	
Vertex	x	y	z	x'	y'
A	−0.4	−0.2	1	−0.5143	−0.6419
B	0.6	−0.2	1	0.7715	−0.6419
C	0.6	0.8	1	0.8566	0.6451
D	−0.4	0.8	1	−0.5711	0.6451
E	−0.4	−0.2	0	−0.3939	−0.1873
F	0.6	−0.2	0	0.5909	−0.1873
G	0.6	0.8	0	0.6395	0.8110
H	−0.4	0.8	0	−0.4263	0.8110

the z axis and at a distance k before the xy plane. Consider that space is rotated about the x axis through an angle θ, thereby causing axes u and v to assume the positions u' and v', respectively, as shown. Through S, we draw rays parallel to u' and v', intersecting the xy plane at Q and R, respectively. Points Q and R both lie on the y axis. Point Q is the vanishing point for lines that were originally parallel to the y axis, and R is the vanishing point for lines that were originally parallel to the z axis. Then $OQ = -k \cot \theta$ and $OR = k \tan \theta$. In the present case,

$$OQ = -4 \cot 18^\circ = -12.3107 \qquad OR = 4 \tan 18^\circ = 1.2997$$

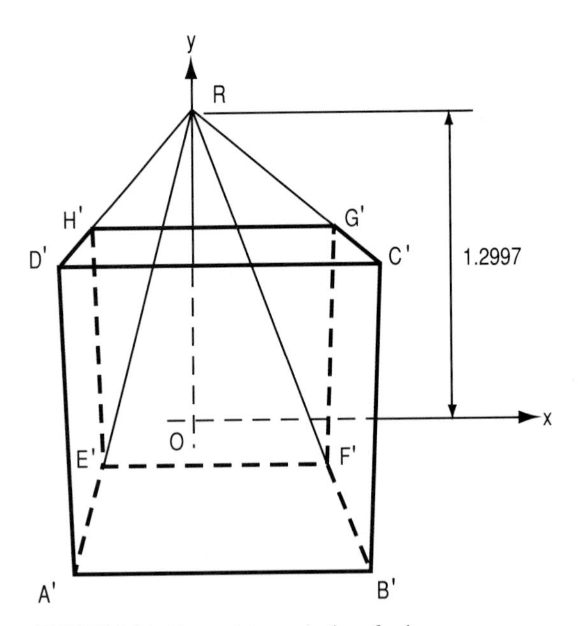

FIGURE 7.24 Two-point perspective of cube.

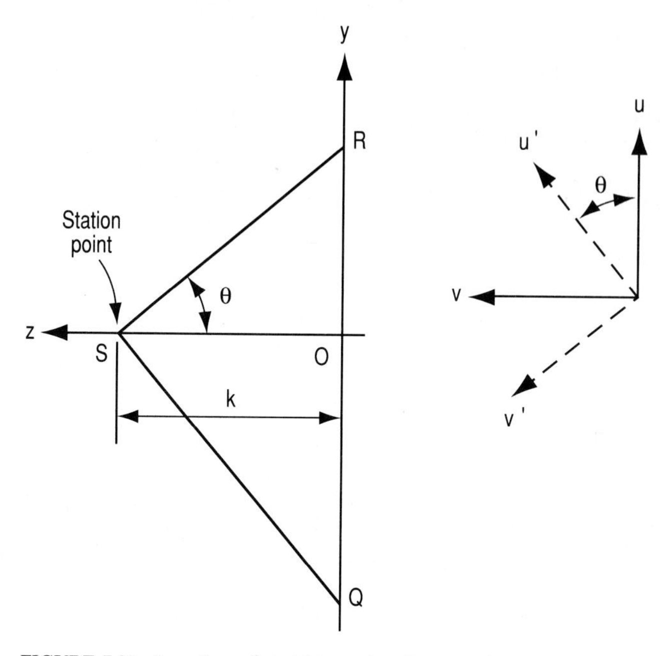

FIGURE 7.25 Location of vanishing points in two-point perspective.

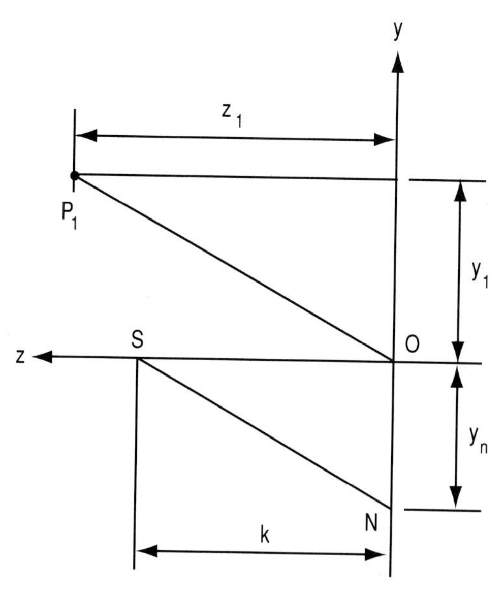

FIGURE 7.26 Location of a vanishing point in three-point perspective.

Point Q, which is the intersection point of lines $A'D'$, $B'C'$, $E'H'$, and $F'G'$ prolonged, falls outside the range of the drawing.

Finally, consider that space is rotated about both the x and y axes (and possibly the z axis) and is then projected onto the xy plane from a station point on the z axis. In general, neither the t, u, nor v axis is parallel to the xy plane, and the resulting image of the cube in Fig. 7.22 is a three-point perspective. The transformation matrix that corresponds to $\mathbf{T}_c$ has all nonzero elements in its fourth column.

Assume that we wish to locate the vanishing points in a three-point perspective for lines that were originally parallel to the coordinate axes. We shall present a general procedure for doing so. In Fig. 7.26, S is again the station point, and it has the indicated location on the z axis. Point $P_1(x_1, y_1, z_1)$ is an arbitrary point, and SN is a line that is parallel to OP_1 and intersects the xy plane at $N(x_n, y_n, 0)$. The coordinates of N are as follows:

$$x_n = -\frac{kx_1}{z_1} \qquad y_n = -\frac{ky_1}{z_1} \tag{7.61}$$

Now let P denote a point that was originally on a coordinate axis. Consider that space undergoes a series of rotations about the principal axes, and let P_1 be the image of P following these rotations. Point N in Fig. 7.26 is the vanishing point for lines that were originally parallel to the axis on which P was located, and Eq. (7.61) gives the location of N.

SECTION 8
BASIC STATISTICS

8.1 PRESENTATION OF STATISTICAL DATA

8.1.1 Definitions

If a variable assumes different values on different occasions, this set of values is referred to as *statistical data.* The value that the variable assumes on a given occasion is termed an *element* in the data. In the following discussion, the variable will be denoted by X.

If X can assume only discrete (isolated) values, it is referred to as a *discrete* or *step* variable. On the other hand, if the values that X can assume form a continuum, X is referred to as a *continuous* variable. For example, the number of neutrons in a potassium atom is a discrete variable (20, 21, or 22), and the temperature of a body at 1-h intervals is a continuous variable. In practice, continuous variables are generally treated as if they were discrete because the precision of our data must be restricted.

8.1.2 Frequencies and Frequency Distributions

Assume that the statistical data pertaining to a variable X have been compiled, and let n denote the number of elements in the data. Let f_i denote the number of times that X assumed the value X_i. Then f_i is the *frequency* of X_i, and f_i/n is the *relative frequency* of X_i, which is denoted by $f_{i,\text{rel}}$.

For example, assume that an experiment to determine the value of X was performed 20 times and that X assumed the value 4.7 on 3 occasions. The frequency of 4.7 is 3, and the relative frequency of this value is 3/20 or 15 percent. The sum of the relative frequencies of all values of X is 1.

A *cumulative frequency* is the sum of the frequencies of all values of X up to or beyond a specified value; it may or may not include that value. For example, assume that the values of X and their respective frequencies are as shown in Table 8.1. We have the following cumulative frequencies: the number of times X was less than 7 is $5+9+17=31$, the number of times X was at least 7 is $13+4+3=20$, and the number of times X was more than 7 is $4+3=7$. Cumulative frequency can also be expressed on a relative basis. For example, with reference to Table 8.1, the relative number of times that X was at least 7 is $20/51=39.2$ percent.

The values of X and the respective frequencies (or respective relative frequencies) of these values constitute the *frequency distribution* of X. Thus, Table 8.1 presents a frequency distribution.

TABLE 8.1

Value of X	Frequency
4	5
5	9
6	17
7	13
8	4
9	3
Total	51

8.1.3 Grouping of Data

Where the number of values of X is very large, a comprehensive recording of these values and their respective frequencies would be unwieldy. In this situation, the data are presented by grouping the values of X in *classes* (or *cells*) and recording the frequency of each class. The range of values associated with a given class is known as its *class interval*, and the end values of the interval are known as the *class limits*. The number of classes to be formed requires sound judgment, and it is deemed good practice to have this number lie between 5 and 20. Table 8.2 illustrates the method of grouping for a variable that is restricted to integral values. It shows, for example, that the number of times that $111 \leq X \leq 115$ is 22.

For analytic purposes, it is essential that there be no gaps between the classes; two successive classes must have a common boundary. Therefore, although the variable in Table 8.2 can assume only integral values, it is necessary to extend each class interval by one-half unit on each side to secure continuity. For example, the true interval of the second class is 105.5 to 110.5. The difference between the upper and lower limits of a class is called the *class width* or *class size*. Therefore, on the basis of the corrected class intervals, the class width in Table 8.2 is 5.

Grouping conceals the values of X that lie within a class. By convention, it is assumed that these values are uniformly distributed through the class interval. Consequently, each interval is divided into uniform *subintervals*. Moreover, the values are assumed to lie at the *upper ends* of the subintervals. For example, assume that a particular class has the true interval 80 to 85 and that its frequency is 20. The width of each subinterval is $5/20 = 0.25$. It is assumed that the first value within this

TABLE 8.2

Class interval	Frequency
101 to 105	6
106 to 110	13
111 to 115	22
116 to 120	19
121 to 125	11
126 to 130	5
131 to 135	2
Total	78

interval is 80.25, the second value is 80.50, etc. In general, with reference to a given class,

let L = lower class limit
c = class width
f = class frequency
X_r = rth value in class

Then
$$X_r = L + \frac{r}{f}c \tag{8.1}$$

EXAMPLE 8.1 A class having the true interval of 60 to 65 has a frequency of 13. Applying the conventional method, determine the ninth value in this class (to two decimal places).

SOLUTION By Eq. (8.1),

$$X_9 = 60 + (9/13)5 = 63.46$$

A class that lacks either a lower or upper limit is described as an *open class*. For example, a class having the designation "more than 700" is open because no upper limit exists. Since open classes are not amenable to statistical analysis, use of such classes should be eschewed.

The *midpoint* or *mark* of a class is the value of the variable that lies midway between the upper and lower limits. For example, the midpoint of the third class in Table 8.2 is 113.

If X is a continuous variable, it is imperative that the class limits be identified explicitly to avoid overlapping. For example, if one class has the interval 10 to 15 and the succeeding class has the interval 15 to 20, it is not clear to which class the value 15 should be assigned. This ambiguity can be resolved by making the intervals "10 to less than 15" and "15 to less than 20," thus definitively placing the value 15 in the second class.

In the subsequent material, it is understood that all classes are of uniform width.

8.1.4 Frequency-Distribution Diagrams

As an aid in visualizing and interpreting statistical data, it is advantageous to construct a diagram that exhibits the frequency distribution. In this diagram, values of the variable are plotted on the horizontal axis.

For ungrouped data, the diagram has the simple form illustrated by Fig. 8.1, which is a plotting of the data in Table 8.1. A vertical line is drawn above each value of X, the length of the line being equal to the frequency of that value.

Grouped data can be represented by a *histogram*, which is a diagram composed of rectangles. A rectangle is constructed above each class interval recorded on the horizontal axis, the area of the rectangle being equal to the frequency of that class. If the classes are of uniform width, the height of each rectangle is directly proportional to the frequency of that class. Thus, frequencies (or relative frequencies) can be plotted on the vertical axis. A histogram is illustrated by Fig. 8.2, which is a plotting of the data in Table 8.2.

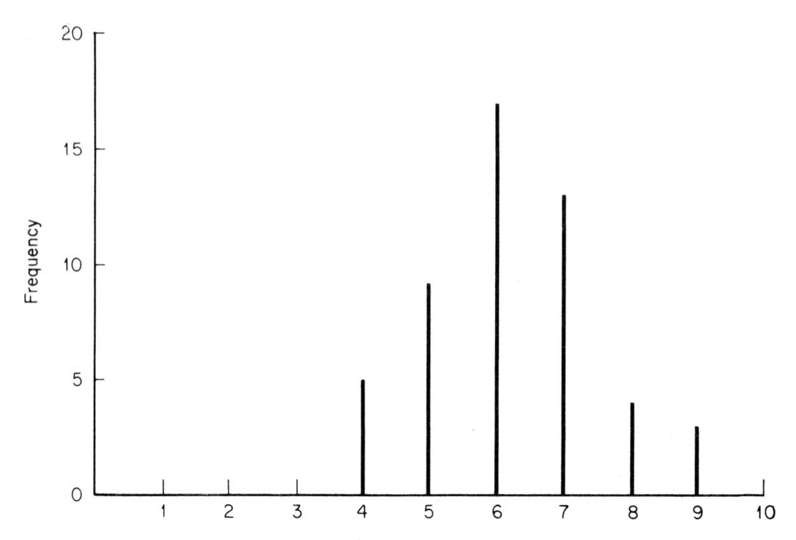

FIGURE 8.1 Frequency diagram. *(From Handbook of Engineering Economics; used with permission of McGraw-Hill, Inc.)*

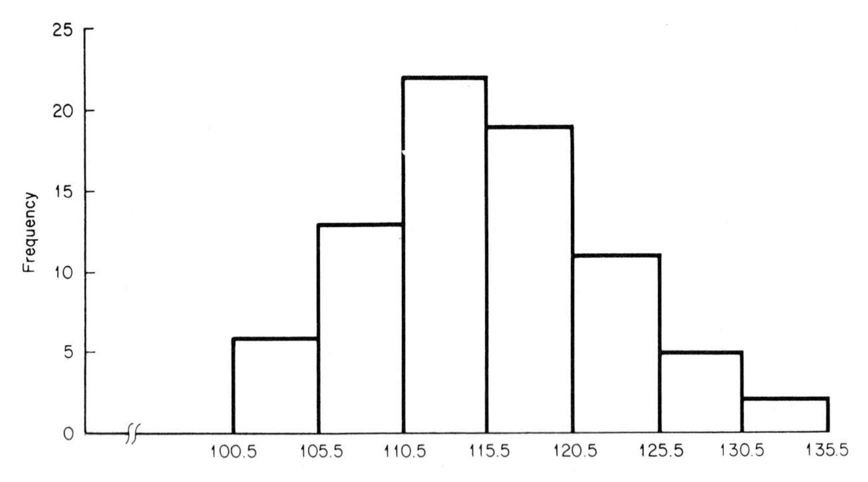

FIGURE 8.2 Histogram. *(From Handbook of Engineering Economics; used with permission of McGraw-Hill, Inc.)*

We shall now expand Table 8.2 by adding a class at each end. The class at the lower end has the true interval 95.5 to 100.5; its midpoint is 98 and its frequency is 0. The class at the upper end has the true interval 135.5 to 140.5; its midpoint is 138 and its frequency is 0. The data of the expanded table are plotted in Fig. 8.3*a*. The diagram is constructed by plotting points whose abscissas are the class midpoints and whose ordinates are the corresponding class frequencies, and the points are

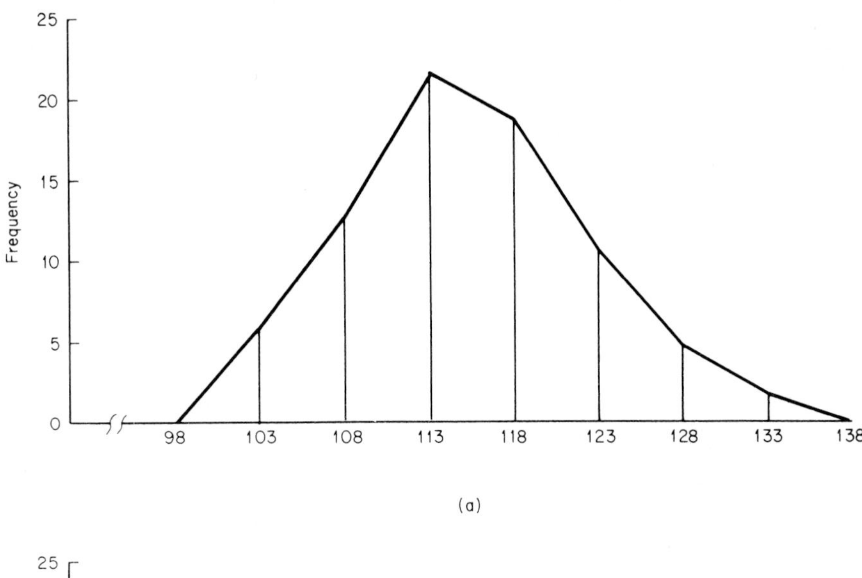

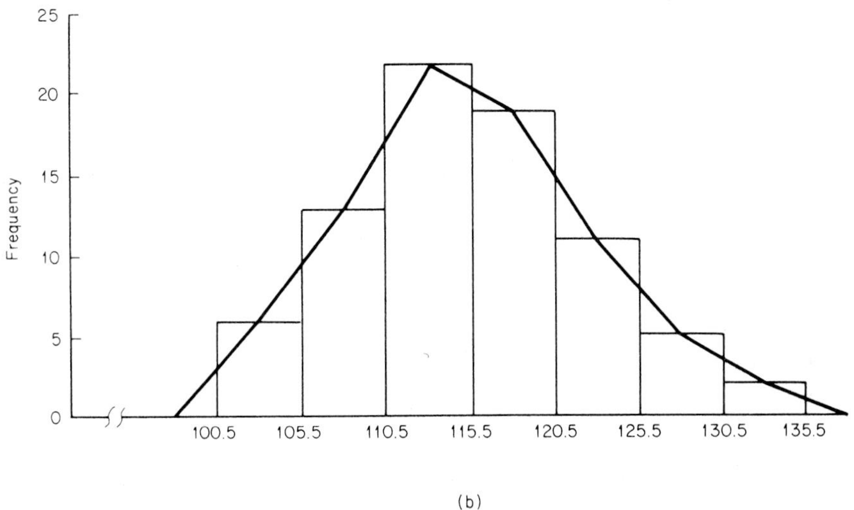

FIGURE 8.3 (*a*) Frequency polygon and (*b*) transformation of histogram to frequency polygon. (*From Handbook of Engineering Economics; used with permission of McGraw-Hill, Inc.*)

then connected by straight lines. This diagram is known as a *frequency polygon*. A histogram can be transformed to a frequency polygon by connecting the midpoints of the top lines of the rectangles with straight lines, as shown in Fig. 8.3*b*.

A diagram in which cumulative frequencies (or cumulative relative frequencies) are plotted against boundary values of X is termed an *ogive*. We shall illustrate the construction by a numerical example. Table 8.3 presents the income earned by members of an organization during a given period of time. The cumulative frequencies as computed on a "less-than" basis are in Table 8.4, and the

TABLE 8.3

Income, $ (class interval)	Number of members (frequency)
24,000 to less than 26,000	2
26,000 to less than 28,000	5
28,000 to less than 30,000	9
30,000 to less than 32,000	6
32,000 to less than 34,000	4
34,000 to less than 36,000	1
Total	27

TABLE 8.4

Income, $ (cumulative class interval)	Number of members (cumulative frequency)
Less than 24,000	0
Less than 26,000	2
Less than 28,000	7
Less than 30,000	16
Less than 32,000	22
Less than 34,000	26
Less than 36,000	27

cumulative frequencies as computed on an "or-more" basis are in Table 8.5. If the cumulative frequency on a given line of Table 8.4 is added to that on the corresponding line of Table 8.5, the sum is 27. The ogives corresponding to Tables 8.4 and 8.5 appear in Fig. 8.4*a* and *b*, respectively.

Assume that X is a continuous variable that can assume any value ranging from 2 to 16 and that we have compiled an infinite set of values of X. Consider that we first group the values of X in classes having a width of 2 and construct the

TABLE 8.5

Income, $ (cumulative class interval)	Number of members (cumulative frequency)
24,000 or more	27
26,000 or more	25
28,000 or more	20
30,000 or more	11
32,000 or more	5
34,000 or more	1
36,000 or more	0

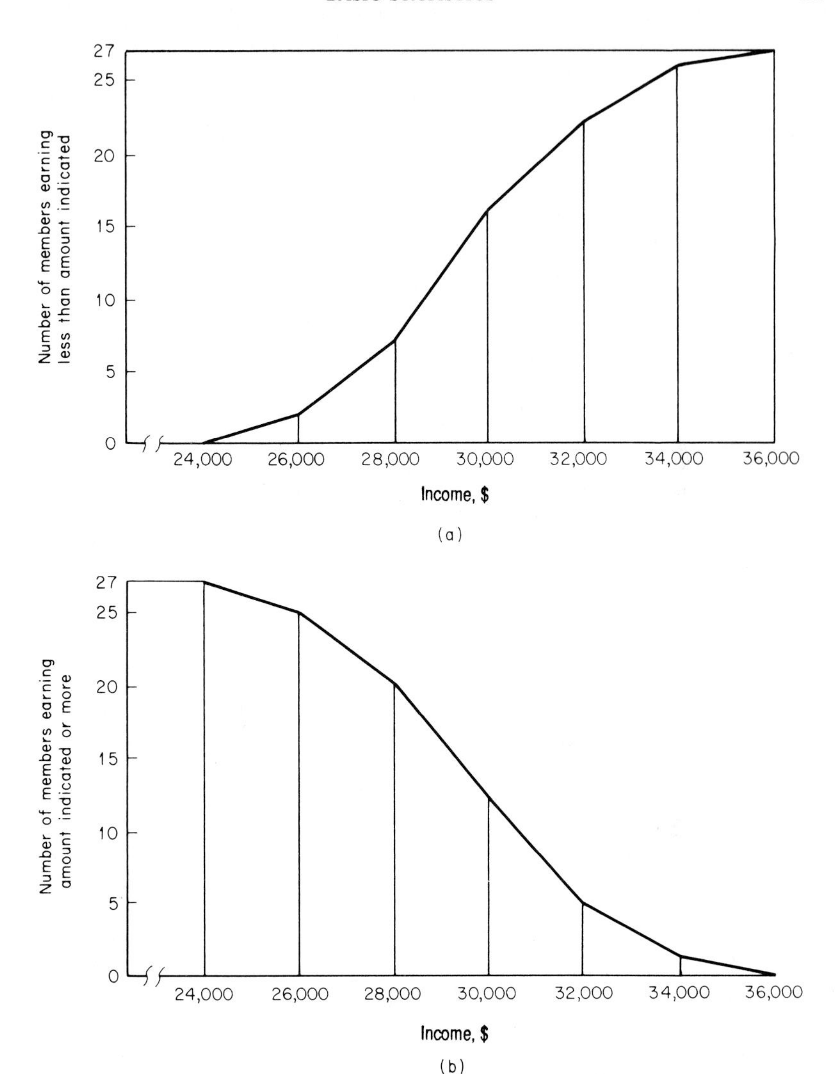

FIGURE 8.4 (*a*) "Less-than" ogive; (*b*) "or-more" ogive. (*From Handbook of Engineering Economics; used with permission of McGraw-Hill, Inc.*)

histogram shown in Fig. 8.5*a*, setting the area of each rectangle equal to the *relative* frequency of its respective class. Now consider that the class width is halved, thereby transforming the histogram in Fig. 8.5*a* to one containing 14 rectangles. If this process of halving the class width is continued indefinitely, the histogram approaches the smooth curve shown in Fig. 8.5*b*, which is referred to as a *frequency curve*.

The basic characteristic of a frequency curve is that the relative frequency corresponding to a given range of values equals the area bounded by the curve, the

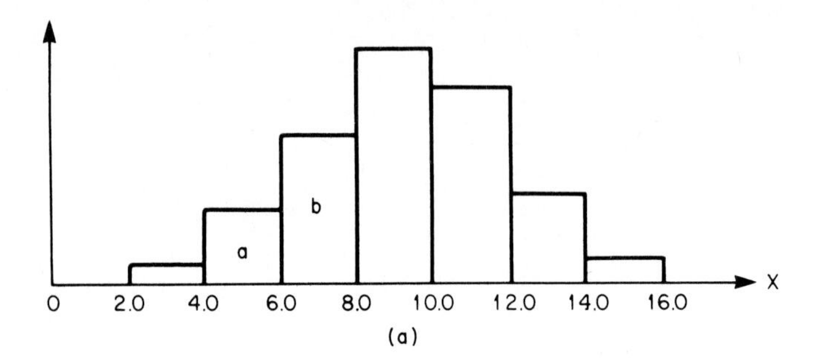

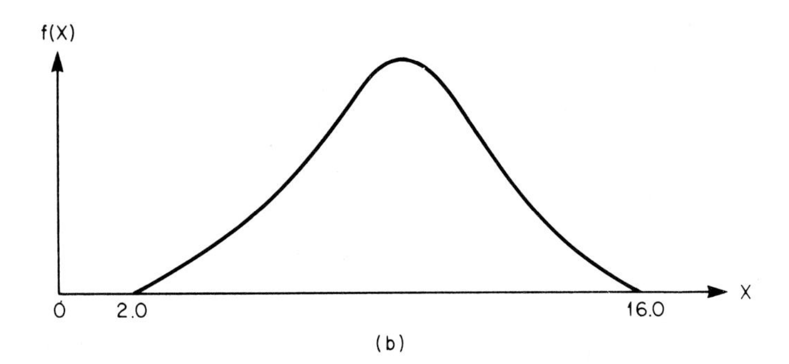

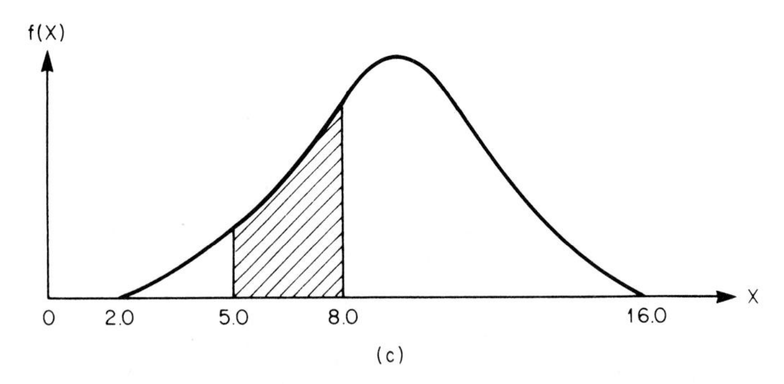

FIGURE 8.5 (*a*) Histograms based on grouped data; (*b*) smooth curve that evolves from successive reduction of class width; (*c*) determination of relative frequency by area. (*From Handbook of Engineering Economics; used with permission of McGraw-Hill, Inc.*)

horizontal axis, and vertical lines erected at the boundaries of the range. For example, with reference to Fig. 8.5*b*, the relative frequency of values of X ranging from 5.0 to 8.0 equals the shaded area in Fig. 8.5*c*. The ordinate of a frequency curve is denoted by $f(X)$, and it is called the *frequency-density function.*

Figure 8.6 illustrates three basic types of frequency curves. The bell-shaped

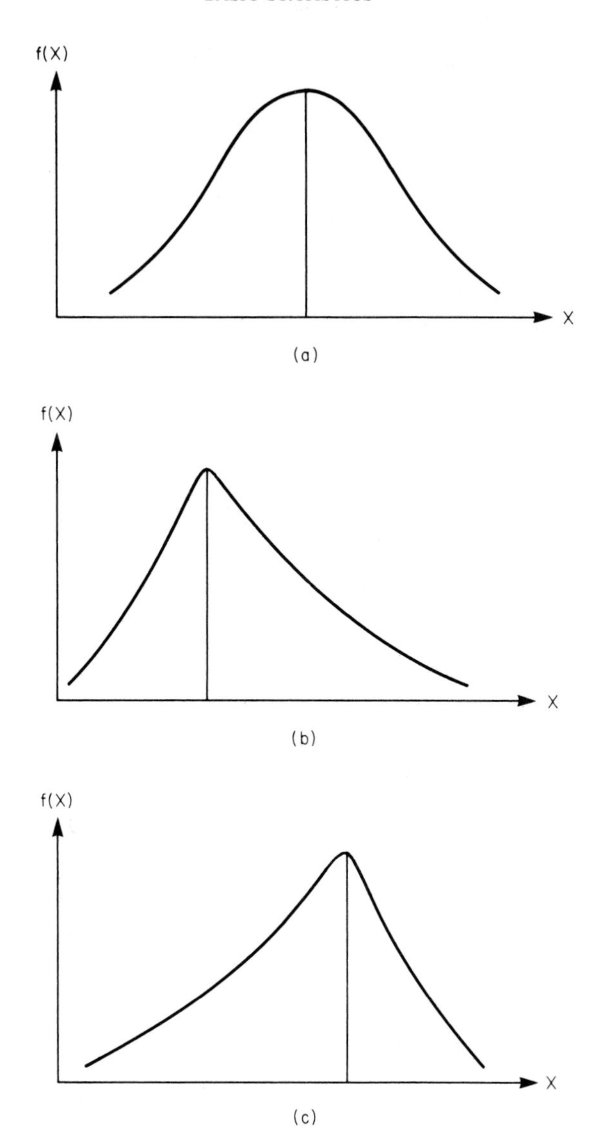

FIGURE 8.6 Classification of frequency curves. (*a*) Symmetrical curve; (*b*) curve skewed to the right; (*c*) curve skewed to the left. (*From Handbook of Engineering Economics; used with permission of McGraw-Hill, Inc.*)

curve in Fig. 8.6*a* is symmetrical about a vertical axis. The curves in Figs. 8.6*b* and *c* are unsymmetrical; therefore, they are described as *skewed.* If the longer tail of the curve lies to the right of the summit, as in Fig. 8.6*b*, the curve is *skewed to the right*; if the longer tail lies to the left of the summit, as in Fig. 8.6*c*, the curve is *skewed to the left.*

8.2 AVERAGES AND MEAN VALUES OF STATISTICAL DATA

8.2.1 Definition of Average and Mean

The first step that is generally required in analyzing statistical data is to establish some value of the variable X that may be regarded as representative of the entire set. This representative value is termed an *average*, and the most frequently applied type of average is the *mean*. We shall first present a general definition of the mean, and then we shall define specific types of means. To simplify the discussion, we shall initially confine our attention to discrete variables.

Assume that the given set of data consists of the numbers 12, 14, 17, and 23. This set of data has certain *properties:* a sum of 66, a product of 65,688, etc. Now consider that we replace each value of X with a constant without changing a particular property of the set of data. This constant is designated the *mean.* Specifically, if the property that remains unchanged is the sum, this constant is the *arithmetic mean*; if the property that remains unchanged is the product, this constant is the *geometric mean.* As an illustration, consider this set of data: 40, 72, 75. These numbers have a sum of 187 and a product of 216,000. The arithmetic mean of these numbers is 62.33 because $62.33 \times 3 = 187$. The geometric mean of these numbers is 60 because $60^3 = 216{,}000$. In addition to the arithmetic and geometric mean, there are a harmonic mean and a root mean square, which we shall presently define.

8.2.2 Arithmetic Mean

Assume that the statistical data are ungrouped. Let $X_1, X_2, \ldots, X_k$ denote the values assumed by X and $f_1, f_2, \ldots, f_k$ denote the respective frequencies of these values. Let $\bar{X}$ denote the arithmetic mean. Again let n denote the number of elements in the data; this number equals the sum of the frequencies. Then

$$\bar{X} = \frac{f_1 X_1 + f_2 X_2 + \cdots + f_k X_k}{n}$$

or in sigma notation

$$\bar{X} = \frac{\Sigma fX}{n} \tag{8.2}$$

This equation is also applicable to grouped data if $X_1, X_2, \ldots, X_k$ equal the class midpoints and $f_1, f_2, \ldots, f_k$ equal the respective frequencies of the classes.

EXAMPLE 8.2 With reference to Table 8.3, compute the arithmetic mean of the income of members of the organization for the given period (to three significant figures).

SOLUTION Applying the class midpoints, we have the following:

$$\Sigma fX = 2 \times 25{,}000 + 5 \times 27{,}000 + 9 \times 29{,}000 + 6 \times 31{,}000 + 4 \times 33{,}000 + 35{,}000 = \$799{,}000$$

$$\bar{X} = 799{,}000/27 = \$29{,}600$$

The arithmetic mean can also be expressed in terms of relative frequencies, in this manner:

$$\bar{X} = f_{1,\text{rel}}X_1 + f_{2,\text{rel}}X_2 + \cdots + f_{k,\text{rel}}X_k$$

or

$$\bar{X} = \Sigma\, f_{\text{rel}}X \tag{8.3}$$

Let $d_{m,i} = X_i - \bar{X}$. The quantity $d_{m,i}$ is termed the *deviation* of X_i from $\bar{X}$. The arithmetic mean has the following important property:

$$\Sigma\, fd_m = 0 \tag{8.4}$$

Expressed verbally, the algebraic sum of the deviations from the arithmetic mean is zero. The proof is as follows:

$$\Sigma\, fd_m = \Sigma\, f(X - \bar{X}) = \Sigma\, fX - n\bar{X} = \Sigma\, fX - n\frac{\Sigma\, fX}{n} = 0$$

As stated in Art. 8.1.4, the frequency distribution of a continuous variable X is represented by a frequency curve. The arithmetic mean of the X values lies at the vertical centroidal axis of the area lying between the frequency curve and the horizontal axis, as indicated in Fig. 8.7. (The centroidal axis of an area is defined in Art. 12.9.2.)

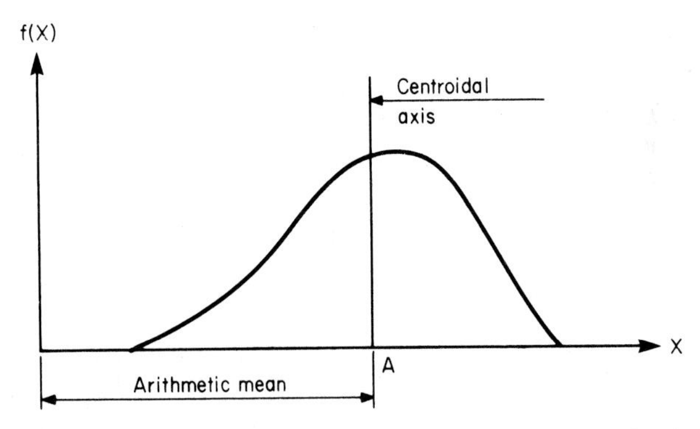

FIGURE 8.7 (*From Handbook of Engineering Economics; used with permission of McGraw-Hill, Inc.*)

8.2.3 Weighted Arithmetic Mean

It is often necessary to compute the arithmetic mean of a set of values of X while recognizing that these values are of varying importance. For example, assume that the values were obtained by measurement but certain measurements are known to be more accurate than others. In this situation, we can assign *weights* to the X values on the basis of their presumed reliability and then compute a *weighted* arithmetic mean of these values.

In general, let $w_1, w_2, \ldots, w_k$ denote the weights assigned to the values

$X_1, X_2, \ldots, X_k$, respectively. Let $\bar{X}_w$ denote the weighted arithmetic mean. Then

$$\bar{X}_w = \frac{w_1 f_1 X_1 + w_2 f_2 X_2 + \cdots + w_k f_k X_k}{w_1 f_1 + w_2 f_2 + \cdots + w_k f_k}$$

or

$$\bar{X}_w = \frac{\Sigma\, wfX}{\Sigma\, wf} \tag{8.5}$$

EXAMPLE 8.3 A student received the examination grades recorded in column 2 of Table 8.6. To arrive at the term grade, the instructor assigned to the examinations the weights shown in column 3. Determine the term grade of this student.

SOLUTION Multiply the grades by their respective weights to obtain the products shown in column 4. Total columns 3 and 4. Then

$$\text{Term grade} = \frac{488.75}{6.25} = 78.2$$

The simple (unweighted) arithmetic mean of the grades is 76. Since the student fared better in the examinations carrying greater weight, it was to be anticipated that the weighted grade would exceed 76.

Weighted arithmetic means abound in engineering and science. For example, the specific heat of a mixture of perfect gases is the weighted arithmetic mean of the specific heats of the component gases, the statistical weights being the masses of the component gases.

TABLE 8.6

Examination number (1)	Grade (2)	Weight (3)	Product (4)
1	61	1.00	61.00
2	75	1.50	112.50
3	83	1.75	145.25
4	85	2.00	170.00
Total		6.25	488.75

8.2.4 The Median

As a representative value of a set of data, the arithmetic mean has the following serious limitation: If an extreme value occurs, it strongly influences the arithmetic mean and thereby distorts its significance. Therefore, it is advantageous to devise other forms of average that may be applied in conjunction with the arithmetic mean. One such average is the *median*. Consider that all values of X are arranged in ascending order of magnitude; this arrangement is called the *ascending array* of X values. If the number of elements is odd, the median is the value that occupies the central position in the array. If the number of elements is even, the median is equated to the arithmetic mean of the two values that occupy the central position. In either case, the sum of frequencies of values below the median equals the sum of frequencies of values above the median.

To illustrate the definition, consider first the following set of X values:

Value	12	13	15	16	21	32
Frequency	1	4	2	2	1	1

The ascending array of this set of values is as follows:

12 13 13 13 13 15 15 16 16 21 32

Since there are 11 elements, the median is determined by the sixth element in the array, and it is 15. (The arithmetic mean of these values is 16.27.)

Now consider the following set of X values:

Value	5	7	22	23	24	25	26
Frequency	1	1	8	7	10	5	2

Since there are 34 elements, the seventeenth and eighteenth elements jointly determine the median. These elements are 23 and 24, and the median is 23.5. (The arithmetic mean is 22.53.)

In establishing the median of grouped data, we apply the convention described in Art. 8.1.3, which assumes the following: The values within an interval divide that interval into subintervals of uniform width, and each element lies at the upper end of its respective subinterval. Therefore, the ascending array of X values contains a total of n subintervals (but of nonuniform width). The rth element in the ascending array lies at the end of the rth subinterval, and it is followed by $n - r$ subintervals. We now select the element that lies at the center of the ascending array and make this the median. Thus, the median is the $(n/2)$th element in the ascending array of X values. This element is real if n is even, and it is hypothetical if n is odd. The $(n/2)$th element is called the *median element*, and the class in which it lies is called the *median class*.

EXAMPLE 8.4 Determine the median of the set of X values in columns 1 and 2 of Table 8.7.

SOLUTION Since there are 25 elements, the median lies midway between the twelfth and thirteenth elements. Record the cumulative frequencies in column 3 of the table. These values disclose that the median lies in the third class and that it lies

TABLE 8.7

Class interval (1)	Frequency (2)	Cumulative frequency (3)
30 to less than 32	2	2
32 to less than 34	8	10
34 to less than 36	9	19
36 to less than 38	6	25
Total	25	

midway between the second and third elements in that class. By Eq. (8.1),

$$\text{Median} = 34 + (2.5/9)2 = 34.56$$

Figure 8.8 is the frequency curve of a continuous variable, and the vertical line L bisects the area between the frequency curve and the horizontal axis. Since the area under a frequency curve equals the total frequency within the corresponding range of X values, it follows that the sum of frequencies below A equals the sum of frequencies above A. Therefore, point A lies at the median.

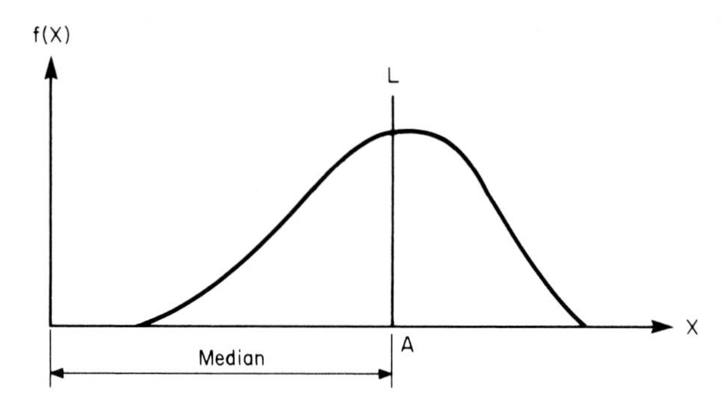

FIGURE 8.8 (*From Handbook of Engineering Economics; used with permission of McGraw-Hill, Inc.*)

8.2.5 The Mode

Another form of average that is often applied in conjunction with the arithmetic mean is the *mode*. Assume that X is a discrete variable and that its values are presented in ungrouped form. The mode is the X value that has the highest frequency. As an illustration, consider the following data:

Value	7	8	9	10	11	12
Frequency	2	3	12	15	9	5

The mode is 10.

Now consider the following data:

Value	10	11	12	13	14	15
Frequency	2	6	9	9	7	4

This set of data has two modes, 12 and 13, and therefore it is described as *bimodal*.

The foregoing definition of the mode can be extended to grouped data, and the extended definition applies to both discrete and continuous variables. The class that has the highest frequency is termed the *modal class*, and it becomes necessary to select some value within the interval of this class as the mode. The method of selection is largely a matter of convention, and two methods are widely applied.

In the *crude method*, the mode is placed at the midpoint of the modal class. In the *interpolation method*, the mode is computed by applying the frequencies of the two adjacent classes. The procedure can best be explained graphically. Refer to Fig. 8.9*a*, which is the partial histogram of a variable. The frequencies are 12 for the modal class, 10 for the preceding class, and 7 for the succeeding class, as indicated. The mode is placed at a point that divides the class interval into segments that are directly proportional to the *differences* between successive frequencies. Then

$$\text{Mode} = 150 + \left(\frac{2}{2+5}\right)10 = 152.9$$

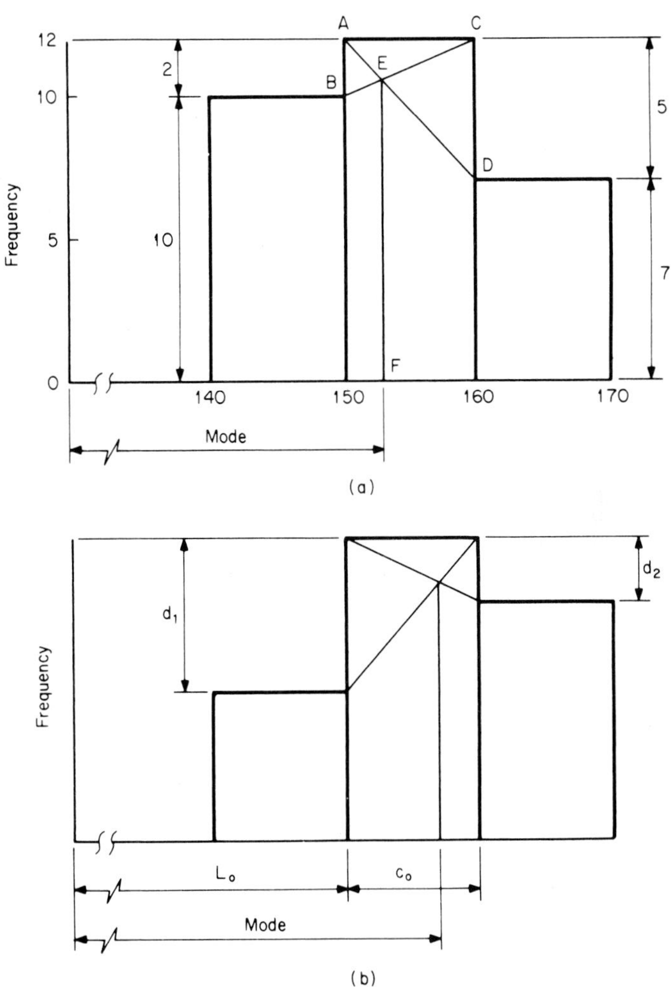

FIGURE 8.9 Determination of mode under interpolation method by use of a histogram. (*a*) Partial histogram in given case; (*b*) partial histogram in general case. (*From Handbook of Engineering Economics; used with permission of McGraw-Hill, Inc.*)

This result can be obtained graphically by drawing lines *AD* and *BC*, which intersect at *E*, and then drawing line *EF*. Point *F* lies at the mode.

The result of the interpolation method can now be generalized. With reference to Fig. 8.9*b*,

let L_o = lower limit of modal class
c_o = width of modal class
d_1 = difference between frequency of modal class and that of preceding class
d_2 = difference between frequency of modal class and that of succeeding class

Then
$$\text{Mode} = L_o + \frac{d_1}{d_1 + d_2} c_o \tag{8.6}$$

8.2.6 Relationship among Mean, Median, and Mode

The general relationship that exists among the arithmetic mean, the median, and the mode can be visualized most effectively by referring to a frequency curve, such as that in Fig. 8.10, which is skewed to the right. The mode lies at the vertical line passing through the summit, the median lies at the vertical line that bisects the area between the curve and the horizontal axis, and the arithmetic mean lies at the vertical centroidal axis of this area.

Consider that we start with a symmetrical frequency curve and transform it to an unsymmetrical one by adding area at one side. Both the median and the arithmetic mean are displaced to the side where area is added, and the displacement of the arithmetic mean exceeds that of the median. This condition can be explained by the fact that the arithmetic mean is more sensitive to extreme values than is the

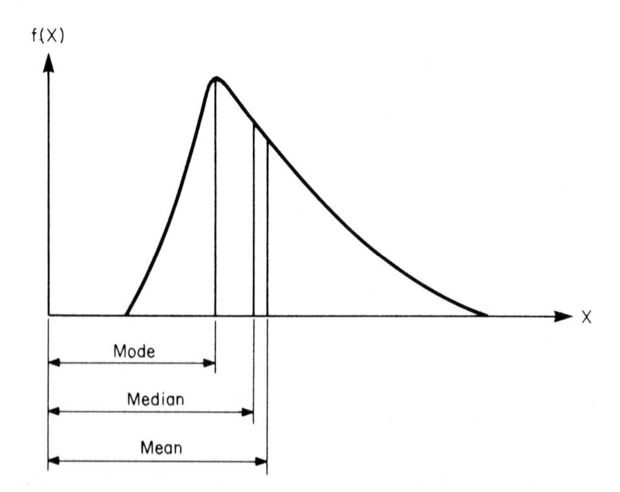

FIGURE 8.10 (*From Handbook of Engineering Economics; used with permission of McGraw-Hill, Inc.*)

median. Consequently, the median always lies between the mode and the arithmetic mean. It follows that the arithmetic mean exceeds the median if the curve is skewed to the right, as in Fig. 8.10, and the reverse is true if the curve is skewed to the left.

8.2.7 Geometric Mean and Rates of Change

The geometric mean is defined in Art. 8.2.1, and it follows from the definition that the geometric mean of a set of n numbers equals the nth root of their product. Let $X_1, X_2, \ldots, X_n$ denote the numbers and G denote their geometric mean. Then

$$G = (X_1 X_2 \cdots X_n)^{1/n} \tag{8.7}$$

Equation (1.8) yields

$$\log G = \frac{\log X_1 + \log X_2 + \cdots + \log X_n}{n}$$

or

$$\log G = \frac{\Sigma\ (\log X)}{n} \tag{8.8}$$

Thus, the log of G is the arithmetic mean of the logarithms of the X values.

As an illustration, consider the following set of numbers: 23, 28, 31, 40, 41, 44, 52, 55. The sum of their common logarithms is 12.6149. Then

$$\log G = 12.6149/8 = 1.5769 \qquad G = 37.75$$

In calculating the geometric mean of grouped data, all values within a class are equated to the midpoint of that class. Let X denote the class midpoint and f the class frequency. Equation (8.8) assumes this form:

$$\log G = \frac{\Sigma\ (f \log X)}{n} \tag{8.8a}$$

The geometric mean is of utility in dealing with rates of change. The *rate of increase* of a variable across a given interval of time or space equals the increase in value divided by the value at the beginning of the interval. For example, if $X = 100$ at the beginning of an interval and $X = 107$ at the end of the interval, the rate of increase across the interval is $7/100 = 0.07$. Assume that we are given the rate of increase of a variable X for consecutive time intervals (which are not necessarily uniform) and that we wish to find the mean rate of increase during the entire time. By definition, the mean rate is a constant rate that can replace each rate in the set without changing the final value of the variable.

Let n = number of time intervals
r_i = rate of increase of X during ith time interval
r^* = mean rate of increase of X during the n intervals
X_0 = initial value of X
X_n = value of X at end of nth interval

Then

$$X_n = X_0(1 + r_1)(1 + r_2) \cdots (1 + r_n)$$

and
$$X_n = X_0(1 + r^*)^n \tag{8.9a}$$

Solving for r^*,

$$r^* = \left(\frac{X_n}{X_0}\right)^{1/n} - 1 \tag{8.9b}$$

Then
$$r^* = [(1 + r_1)(1 + r_2) \cdots (1 + r_n)]^{1/n} - 1 \tag{8.10}$$

Thus, $1 + r^*$ is the geometric mean of the $1 + r$ values.

EXAMPLE 8.5 A body was placed in a rapidly changing environment, and its temperature was read at 1-h intervals. The changes in temperature during the first 5 h were as follows: an increase of 23 percent, an increase of 15 percent, a decrease of 12 percent, an increase of 28 percent, a decrease of 9 percent. What was the mean rate of increase in temperature during this period (expressed as a percent)?

SOLUTION

$$r^* = [(1.23)(1.15)(0.88)(1.28)(0.91)]^{1/5} - 1 = 7.7128 \text{ percent}$$

EXAMPLE 8.6 The bacterial count in a culture was found to be 10,300 at a certain time. If it is assumed that the number of bacteria increased at the constant rate of 38 percent per day, what was the bacterial count 4 days earlier?

SOLUTION Let X denote the bacterial count. By Eq. (8.9a),

$$X_{-4} = X_0(1 + r^*)^{-4} = 10{,}300(1.38)^{-4} = 2840$$

8.2.8 Harmonic Mean

The harmonic mean of a set of numbers is obtained by the following formula: Form the reciprocal of each number; compute the arithmetic mean of the reciprocals; take the reciprocal of this arithmetic mean. The result is the harmonic mean of the given set of numbers.

As an illustration, consider the following set of numbers: 4, 5, 8, 10, 20. Then

$$\text{Sum of reciprocals} = 0.250 + 0.200 + 0.125 + 0.100 + 0.050 = 0.725$$

$$\text{Arithmetic mean of reciprocals} = 0.725/5 = 0.145$$

$$\text{Harmonic mean of given set of numbers} = 1/0.145 = 6.8966$$

Let H denote the harmonic mean. Continuing the previous notation, we have

$$H = \frac{n}{\Sigma f(1/X)} \tag{8.11}$$

The harmonic mean has numerous applications. For example, if equal distances are traversed at varying rates of speed, the mean speed at which the total distance is traversed is the harmonic mean of the given speeds.

EXAMPLE 8.7 A vehicle must travel from A to B and then back to A, making several round trips in the course of a day. On five successive round trips, the mean speeds were 78, 89, 93, 76, and 103 km/h (kilometers per hour). Find the mean speed s^* corresponding to these five round trips.

SOLUTION Applying Eq. (8.11) but replacing H with s^*, we obtain

$$s^* = \frac{5}{1/78 + 1/89 + 1/93 + 1/76 + 1/103} = 86.69 \text{ km/h}$$

The harmonic mean also arises frequently in engineering and science. As an illustration, refer to Fig. 8.11*a*, which shows an electric network with resistances R_1, R_2, and R_3 in parallel. Consider that the network is transformed to an equivalent one by replacing each resistance with a new resistance R^*, as shown in Fig. 8.11*b*. Then R^* is the harmonic mean of R_1, R_2, and R_3.

In calculating the harmonic mean of grouped data, all values within a class are again equated to the midpoint of the class.

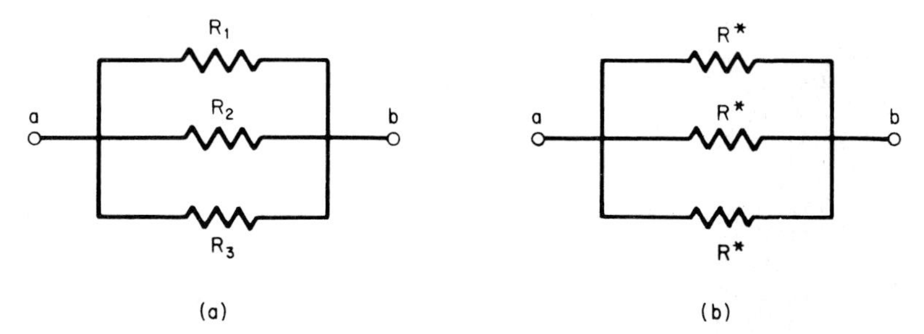

FIGURE 8.11 Equivalent electric networks. (*a*) True network; (*b*) equivalent network with three identical resistances. (*From Handbook of Engineering Economics; used with permission of McGraw-Hill, Inc.*)

8.2.9 Root Mean Square

Another type of mean that arises frequently in engineering and science is the *root mean square* (rms), also known as the *quadratic mean*. The definition is as follows:

$$\text{rms} = \sqrt{\frac{\Sigma fX^2}{n}} \tag{8.12}$$

Thus, the expression "root mean square" is a contraction of the expression "square root of the arithmetic mean of the squares."

As a simple illustration of the significance of the rms, let L denote the length of a line and let P_x, P_y, and P_z denote its length as projected onto the x, y, and z axes, respectively. Now consider that the line is rotated in such manner that all three projected lengths have the identical value P^*. Then

$$L = \sqrt{P_x^2 + P_y^2 + P_z^2} = P^*\sqrt{3}$$

It follows that P^* is the rms of the original projected lengths.

8.2.10 Relationship among the Means

Assume that all X values are positive. By applying the definitions of the various means, we find that they have this relationship:

$$\text{rms} > \bar{X} > G > H \tag{8.13}$$

As an illustration, consider this set of X values: 7, 8, 10, 13, 15, 19. Performing the calculations, we obtain the following:

$$\bar{X} = 12 \qquad G = 11.29 \qquad H = 10.64 \qquad \text{rms} = 12.70$$

These values conform with the foregoing inequality.

8.2.11 Arithmetic Mean and Root Mean Square of a Continuous Variable

Equations (8.2) and (8.12) pertain to a discrete variable, and we shall now convert them to their counterparts for a continuous variable. Consider that Y is a function of X, that X in turn is a function of the continuous variable U, and that U ranges from a to b, where $b > a$.

To achieve the conversion, it is simply necessary to replace ΣfY with $\int Y\,dU$ and to replace n with $b - a$. The defining equations of the means assume these forms:

Arithmetic mean:

$$\bar{X} = \frac{\int X\,dU}{b - a} \tag{8.14}$$

Root mean square:

$$\text{rms} = \sqrt{\frac{\int X^2\,dU}{b - a}} \tag{8.15}$$

EXAMPLE 8.8 A variable X has the equation $X = A \sin \theta$, where A is a constant. Find the arithmetic mean of X when θ varies from 0 to π.

SOLUTION Applying Eq. (8.14) and Integral (150) in Table A.2 in the appendix, we obtain

$$\bar{X} = A\left(-\frac{\cos \pi - \cos 0}{\pi}\right) = \frac{2A}{\pi} = 0.6366A$$

As an approximate check, assume that X varied *linearly* from 0 to A and then back to 0. The arithmetic mean of X would be $0.5A$. From the shape of the sine curve, we deduce that the true arithmetic mean must exceed this value by a small amount, and it does.

The range of θ could have been taken as 0 to $\pi/2$ because the X values then repeat themselves (in reverse order).

EXAMPLE 8.9 Find the root mean square of the variable in Example 8.8 when θ varies from 0 to 2π.

SOLUTION Applying Eq. (8.15) and Integral (151), we obtain the following:

$$\int X^2\, d\theta = A^2\left[\frac{\theta}{2} - \frac{\sin 2\theta}{4}\right]_0^{2\pi} = A^2\left(\frac{2\pi}{2}\right) = A^2\pi$$

$$\text{rms} = \sqrt{\frac{A^2\pi}{2\pi}} = \frac{A}{\sqrt{2}} = 0.7071A$$

The range of θ could have been taken as 0 to π because the values of X^2 for this interval repeat themselves cyclically. In electrical engineering, the rms of a sinusoidally varying voltage or current is termed its *effective value*.

8.2.12 Other Means of a Continuous Variable

There are myriad situations in engineering and science where we must compute the mean value of a continuous variable X and the mean falls outside the categories previously enumerated. In these situations, our interest centers on the *effect* that X produces: acceleration of a body, resistance to heat transfer, etc. The mean of X is found by answering this question: If X remained constant while producing the same effect as the true values of X, what would be this constant value?

As an illustration, consider that a force F that varies in magnitude and direction acts on a body during a time interval T. If our interest centers on the acceleration of the body, the mean value of F is a hypothetical force that is constant in magnitude and direction and that would produce the same acceleration of the body if it acted on the body for the same time interval T.

In Example 8.10, we investigate heat transfer through a wall.

Let R = resistance of wall to heat flow
A = area of wall normal to direction of heat flow
t = thickness of wall in direction of heat flow
k = thermal conductivity of wall

Then

$$R = \frac{t}{kA} \qquad (a)$$

EXAMPLE 8.10 Heat flows through a cylindrical pipe having an internal area A_i and external area A_e. If our interest centers on the resistance of the pipe to heat flow, what is the mean area A_m of the pipe?

SOLUTION In the pipe, heat flows in the radial direction. However, we visualize a wall that is straight, has the same length and thickness as the pipe, is composed of the same material as the pipe, and offers the same resistance to heat flow as does the pipe. The area of this hypothetical wall is the mean area of the pipe.

We may consider that the pipe is composed of concentric cylindrical rings, as shown in Fig. 8.12. The resistance of the pipe is the sum of the resistances of

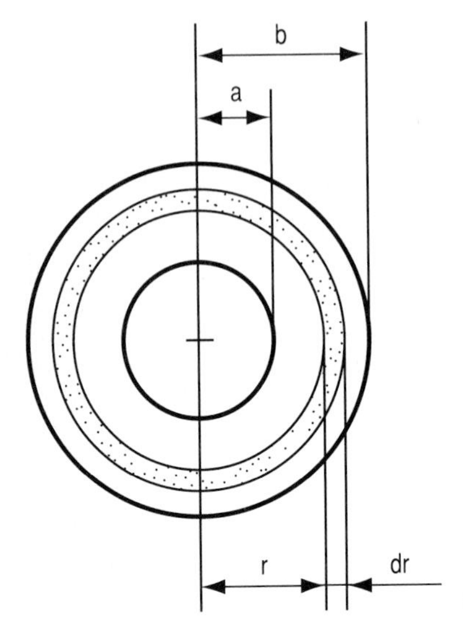

FIGURE 8.12 Division of pipe into cylindrical rings.

these rings.

Let r = internal radius of ring
dr = thickness of ring
a = internal radius of pipe
b = external radius of pipe
L = length of pipe

For a ring, $t = dr$, and we may set $A = 2\pi rL$. Applying Eq. (a) and Integral (5) in Table A.2, we obtain

$$R = \frac{t}{kA_m} = \int_a^b \frac{dr}{kA} = \frac{1}{2\pi kL}\int_a^b \frac{dr}{r} = \frac{\ln b - \ln a}{2\pi kL} = \frac{\ln (b/a)}{2\pi kL}$$

$$A_m = \frac{2\pi Lt}{\ln (b/a)} = \frac{2\pi L(b-a)}{\ln (b/a)} = \frac{A_e - A_i}{\ln (A_e/A_i)} = \frac{A_e - A_i}{\ln A_e - \ln A_i}$$

For example, if $A_i = 7\ \text{m}^2$ and $A_e = 11\ \text{m}^2$, then $A_m = 8.85\ \text{m}^2$. In practice, A_m is referred to as the *logarithmic mean area* to distinguish it from the arithmetic mean of A_i and A_e.

8.3 ANALYSIS OF DISTRIBUTION OF DATA

8.3.1 Significance of Dispersion

Consider the following sets of data:

Set A:	23	24	25	26	27
Set B:	2	6	20	44	53

The two sets of data are alike in having 25 as their arithmetic mean, but they differ drastically in their basic structure. The values in set A cluster within a narrow range while those in set B are widely scattered. The extent to which the values are scattered is termed the *dispersion* of the data.

A knowledge of the dispersion of data is important for numerous reasons. For example, we may wish to know whether a given quantity, such as the temperature of a body made of a new material, varies widely or only slightly. The dispersion of a set of temperatures supplies the answer. Similarly, dispersion can reveal whether the given data are reliable. For example, assume that an experiment is performed repeatedly to establish some numerical constant. If there is considerable variation among the results obtained, the validity of the experiment becomes questionable.

The importance of dispersion makes it imperative that we devise some means of gauging dispersion. Several indices of dispersion are available, and we shall study each one in turn.

8.3.2 The Range

The simplest index of dispersion is the *range* of the data. This is the difference between the highest and lowest values. For example, with reference to the two sets of data recorded in Art. 8.3.1, the range of set A is $27 - 23 = 4$, and the range of set B is $53 - 2 = 51$.

Since the range is governed solely by extreme values, it is a very crude method of gauging dispersion, and there is scant justification for its use. Nevertheless, the concept is a useful one, and we shall apply it in Art. 8.3.5.

8.3.3 Mean Absolute Deviation

In Art. 8.2.2, we defined the deviation $d_{m,i}$ of a given value X_i from the arithmetic mean $\bar{X}$ as $d_{m,i} = X_i - \bar{X}$. Deviations from the arithmetic mean reflect the extent to which the values in the set are scattered, and therefore they are useful in gauging dispersion. However, since the algebraic sum of these deviations is zero, it is impossible to use the "raw" deviations directly.

One method of resolving the problem is to apply the *absolute* values of the deviations from the arithmetic mean. The arithmetic mean of the absolute values of d_m is called the *mean absolute deviation* (m.a.d.). Again let f denote the frequency of a given value and n denote the total number of elements in the set of data. Then

$$\text{m.a.d.} = \frac{\Sigma\,(f|d_m|)}{n} \tag{8.16}$$

EXAMPLE 8.11 Compute the m.a.d. of the set B data in Art. 8.3.1.

SOLUTION The sum of the absolute deviations is

$$-(2-25)-(6-25)-(20-25)+(44-25)+(53-25)=94$$

$$\text{m.a.d.} = 94/5 = 18.8$$

Since the sum of the d_m values is zero, the task of finding the sum of the $|d_m|$ values can be simplified by taking solely the values of X that lie above or below $\bar{X}$, finding the sum of their $|d_m|$ values, and doubling the result.

8.3.4 Standard Deviation and Variance

The m.a.d. is the most accurate index of dispersion available, but it has a severe weakness: Since it is composed of absolute values rather than true algebraic values, it is not amenable to mathematical analysis.

The most widely applied index of dispersion is the *standard deviation*, which is denoted by s. The standard deviation is the root mean square of the deviations from the arithmetic mean. Then

$$s = \sqrt{\frac{\Sigma f d_m^2}{n}} = \sqrt{\frac{\Sigma f(X - \bar{X})^2}{n}} \tag{8.17}$$

EXAMPLE 8.12 Find the standard deviation of the data in Table 8.8.

SOLUTION We total the frequencies and then compute the values recorded in column 3. Then $\bar{X} = 255/17 = 15$. We now record the values in columns 4 and 5, and

$$s = \sqrt{\frac{174}{17}} = 3.1993$$

Some calculators have a function that supplies the standard deviation of a set of data.

By expanding $(X - \bar{X})^2$ in Eq. (8.17), we obtain this alternative form of the equation:

$$s = \sqrt{\frac{\Sigma f X^2}{n} - \bar{X}^2} = \sqrt{(\text{rms})^2 - \bar{X}^2} \tag{8.17a}$$

Equation (8.17) can also be recast in terms of *relative frequencies*, giving

$$s = \sqrt{\Sigma f_{\text{rel}} d_m^2} = \sqrt{\Sigma f_{\text{rel}}(X - \bar{X})^2} \tag{8.18}$$

The calculation of standard deviation can often be simplified by selecting an arbitrary value A, calculating the deviation $d_{A,i} = X_i - A$, and applying the following relationship:

$$s = \sqrt{\frac{\Sigma f d_A^2}{n} - \left(\frac{\Sigma f d_A}{n}\right)^2} \tag{8.19}$$

TABLE 8.8

X (1)	Frequency, f (2)	fX (3)	$d_m = X - 15$ (4)	fd_m^2 (5)
9	1	9	−6	36
12	3	36	−3	27
13	4	52	−2	16
16	5	80	1	5
18	2	36	3	18
21	2	42	6	72
Total	17	255		174

The advantage of Eq. (8.19) is that it obviates the need to calculate the arithmetic mean if our sole objective is to find the standard deviation. It also emphasizes that the standard deviation is purely an index of the variability of the data and is not directly related to the arithmetic mean. Thus, if all values of X increase by an amount h, the arithmetic mean increases by h, but the standard deviation is unchanged.

The square of the standard deviation is termed the *variance* and it is denoted simply by s^2.

Now consider that X is a continuous variable, that it is a function of the independent variable U, and that U ranges from a to b, where $b > a$. The standard deviation of X is

$$s = \sqrt{\frac{\int (X - \bar{X})^2 \, dU}{b - a}} \tag{8.20}$$

The analog of the first form of Eq. (8.17a) is

$$s = \sqrt{\frac{\int X^2 \, dU}{b - a} - (\bar{X})^2} \tag{8.20a}$$

EXAMPLE 8.13 Find the standard deviation of the variable X in Example 8.8 when θ varies from 0 to π.

SOLUTION We shall apply the second form of Eq. (8.17a). From Examples 8.8 and 8.9, we have the following values for the given range of θ:

$$\bar{X} = \frac{2A}{\pi} \qquad \text{rms} = \frac{A}{\sqrt{2}}$$

Then

$$s = A\sqrt{\frac{1}{2} - \frac{4}{\pi^2}} = 0.3078A$$

The standard deviation of a continuous variable is analogous to the radius of gyration of an area, which is defined in Art. 12.9.3.

8.3.5 Quartile Deviation

As stated in Art. 8.3.2, the range of a set of data does not offer a valid measure of its dispersion because the range is governed solely by the two extreme values. We shall now consider a method of gauging dispersion that retains the concept of the range but is considerably more refined. The method can be described most effectively by use of a drawing.

Again let n denote the number of values of X. We divide a straight line into n segments of unit length, starting at point O. We then place the X values at the *midpoints* of these segments, in ascending order of magnitude. Thus, the rth value in ascending order lies at a distance of $r - 0.5$ from O. The *first quartile* Q_1 is the value that lies at a distance of $n/4$ from O, and the *third quartile* Q_3 is the value that lies at a distance of $3n/4$ from O. These quartiles are true X values if n is 6, 10, 14, or any other number that has the form $4m + 2$, where m is an arbitrary integer. In all other instances, these quartiles are hypothetical X values that are obtained by interpolation on the basis of their positions on the scale, as we shall illustrate.

The difference $Q_3 - Q_1$ is termed the *interquartile range*, and one-half this difference is termed the *quartile deviation* (qd). Then

$$\mathrm{qd} = \frac{Q_3 - Q_1}{2} \tag{8.21}$$

EXAMPLE 8.14 Find the quartile deviation of the following set of numbers: 12, 14, 19, 20, 25, 27, 30, 34, 37.

SOLUTION Refer to Fig. 8.13. We have $n = 9$, $n/4 = 2.25$, and $3n/4 = 6.75$. Therefore, the position of Q_1 is three-fourths of the way from 14 to 19. By interpolation,

$$Q_1 = 14 + 0.75(19 - 14) = 17.75$$

Similarly, the position of Q_3 is one-fourth of the way from 30 to 34. By interpolation,

$$Q_3 = 30 + 0.25(34 - 30) = 31.00$$

$$\mathrm{qd} = (31.00 - 17.75)/2 = 6.625$$

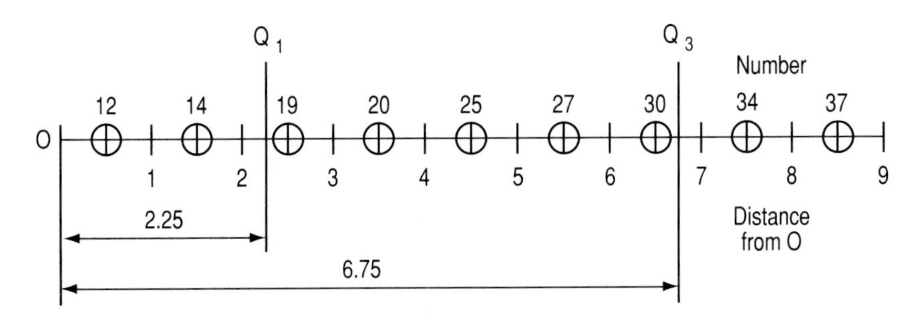

FIGURE 8.13 Location of first and third quartiles.

8.3.6 Relative Dispersion

In many instances, the variability of one set of data must be compared with that of another. A direct comparison of their standard deviations would lack significance if the data pertain to dissimilar items or if the data are based on different units. Therefore, for comparative purposes, it is necessary to formulate indices of *relative* dispersion. We shall define three indices that are frequently applied.

Let V_s = coefficient of variation
$V_{\mathrm{m.a.d.}}$ = coefficient of mean absolute deviation
V_{qd} = coefficient of quartile deviation

The defining equations are as follows:

$$V_s = \frac{s}{\bar{X}} \tag{8.22}$$

$$V_{\text{m.a.d.}} = \frac{\text{m.a.d.}}{\bar{X}} \tag{8.23}$$

$$V_{\text{qd}} = \frac{Q_3 - Q_1}{Q_3 + Q_1} \tag{8.24}$$

EXAMPLE 8.15 Two aptitude tests, one in mathematics and one in engineering, were given to a group of 12 students. The students were assigned grades corresponding to the number of problems that they solved correctly. The grades were as follows:

Mathematics:	21, 22, 22, 24, 25, 25, 25, 26, 26, 27, 28, 29
Engineering:	33, 35, 36, 38, 40, 44, 49, 51, 54, 59, 61, 64

Applying the coefficient of variation, determine which set of grades has the greater variability.

SOLUTION Performing the necessary calculations, we obtain the following results: For the mathematics test,

$$\bar{X} = 25 \qquad s = 2.3452 \qquad V_s = \frac{2.3452}{25} = 0.0938$$

For the engineering test,

$$\bar{X} = 47 \qquad s = 10.4003 \qquad V_s = \frac{10.4003}{47} = 0.2213$$

The grades in engineering have far greater variability than those in mathematics. Thus, on a relative basis, the students exhibited much greater diversity in their engineering aptitude than in their mathematical aptitude.

8.3.7 Standardized Variables and Standard Units

The deviation of a given value X_i from the arithmetic mean $\bar{X}$ must often be expressed on a *relative* basis for the result to be meaningful. A relative value can be obtained by dividing the deviation by the standard deviation, and the result is termed a *standardized variable*. Let z_i denote this quantity. Then

$$z_i = \frac{d_{m,i}}{s} = \frac{X_i - \bar{X}}{s} \tag{8.25a}$$

The quantity z_i, which is a pure number, represents the number of standard deviations contained in the given deviation. Therefore, it is said to be expressed in *standard units*.

EXAMPLE 8.16 A class was given an examination in astronomy and in chemistry. In the astronomy examination, the mean grade was 66 and the standard deviation was 3. In the chemistry examination, the mean grade was 78 and the standard deviation was 8. A student received a grade of 70 in astronomy and 85 in chemistry. In which examination was his performance more satisfactory?

SOLUTION The relative superiority of this student is as follows:

$$\text{Astronomy:} \quad z = \frac{70 - 66}{3} = 1.333$$

$$\text{Chemistry:} \quad z = \frac{85 - 78}{8} = 0.875$$

On a relative basis, the student fared better in the astronomy examination than in the chemistry examination.

In many instances, z_i is known and X_i must be determined. Rearranging Eq. (8.25*a*), we obtain

$$X_i = z_i s + \bar{X} \tag{8.25b}$$

8.3.8 Moments

In Arts. 12.9.1 and 12.9.3, we define the statical moment (or first moment) and moment of inertia (or second moment) of an area with respect to a given axis. The term *moments* is also applied in statistics, but with a modified meaning.

Let $m_{r,A}$ denote the rth moment of a set of data with respect to a given number A. The definition of this moment is as follows:

$$m_{r,A} = \frac{\Sigma f(X - A)^r}{n} \tag{8.26}$$

It is seen at once that variance is the second moment of the set of data with respect to the arithmetic mean. For grouped data, X is the class midpoint.

EXAMPLE 8.17 Given the following set of numbers: 14, 11, 20, 17, 28, compute the third moment of this set of numbers with respect to the number 15.

SOLUTION The numerator in Eq. (8.26) is

$$(14 - 15)^3 + (11 - 15)^3 + (20 - 15)^3 + (17 - 15)^3 + (28 - 15)^3 = 2265$$

$$m_{3,15} = \frac{2265}{5} = 453$$

The rth moment of a set of data with respect to the arithmetic mean will be denoted by $m_{r,m}$.

8.3.9 Skewness

A frequency distribution that is unsymmetrical about the mode is said to be *skewed*, and the degree of asymmetry is termed the *skewness* of the data. Figures 8.6*b* and *c* show frequency curves that are skewed to the right and to the left, respectively. By convention, the skewness is considered positive if the frequency curve or histogram is skewed to the right and negative if it is skewed to the left. Manifestly, skewness is a major characteristic of a set of data, and therefore it becomes necessary to devise a means of measuring it. We shall present the measure of skewness that is most widely applied.

A set of data that has zero skewness is symmetrical about its arithmetic mean. Every positive deviation from the mean is matched by a corresponding negative deviation. Thus, skewness is determined by the deviations of the X values from their arithmetic mean. However, if we take the algebraic sum of these deviations, we obtain zero, and if we take the sum of the squares of the deviations, we lose their algebraic signs. Both these obstacles can be overcome if we apply the third moment of the data with respect to the arithmetic mean. The algebraic sign of the skewness agrees with that of the third moment. To obtain a *relative measure* of skewness, we divide this moment by the cube of the standard deviation, and the result is a pure number.

Let a_3 denote the skewness. Then

$$a_3 = \frac{m_{3,m}}{s^3} = \frac{m_{3,m}}{m_{2,m}^{1.5}} \tag{8.27}$$

EXAMPLE 8.18 Calculate the skewness of the data in Example 8.17.

SOLUTION

$$\bar{X} = \frac{14 + 11 + 20 + 17 + 28}{5} = 18$$

$$m_{2,m} = \frac{(-4)^2 + (-7)^2 + 2^2 + (-1)^2 + 10^2}{5} = 34$$

Similarly, $m_{3,m} = 600/5 = 120$. Then

$$a_3 = \frac{120}{34^{1.5}} = 0.6053$$

The data are positively skewed.

By convention, the distribution of a set of data is classified on the basis of the absolute value of its skewness, in the manner shown in Table 8.9.

TABLE 8.9

Classification	Condition
Fairly symmetrical	$\lvert a_3 \rvert < 0.5$
Moderately skewed	$0.5 < \lvert a_3 \rvert < 1$
Highly skewed	$\lvert a_3 \rvert > 1$

8.3.10 Kurtosis

Figure 8.14 presents two frequency curves that have two attributes in common: They are both symmetrical, and they have the same range of values. However, they differ strikingly in their structure, for one rises to a peak far more sharply than the other. The degree to which its frequency curve or histogram is peaked is called the *kurtosis* of a frequency distribution, and this term applies to both symmetrical and unsymmetrical unimodal distributions.

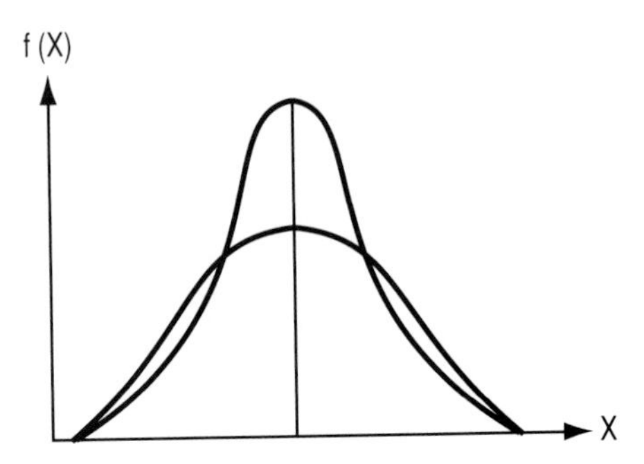

FIGURE 8.14 (*From Handbook of Engineering Economics; used with permission of McGraw-Hill, Inc.*)

Kurtosis is generally measured by applying the fourth moment of the data with respect to its arithmetic mean. Let a_4 denote the kurtosis. Then

$$a_4 = \frac{m_{4,m}}{s^4} = \frac{m_{4,m}}{m_{2,m}^2} \tag{8.28}$$

This equation expresses kurtosis on a relative basis and yields a pure number.

The rationale underlying Eq. (8.28) is as follows: Consider that we start with a frequency curve that is only slightly peaked and then transfer area from the outer regions to the region where the peak occurs. During this process, both the numerator and denominator of the fraction in Eq. (8.28) decrease, but the denominator decreases more rapidly than the numerator. As a result, a_4 increases as the area becomes more concentrated.

The normal probability curve, which is discussed in Art. 9.3.3, has a kurtosis of 3, and this value is often applied as a standard in assessing the kurtosis of a given frequency distribution.

SECTION 9
PROBABILITY

9.1 CALCULATION OF PROBABILITY

9.1.1 Definitions

If the value that a variable will assume on a given occasion cannot be predetermined because it is influenced by chance, this quantity is referred to as a *random* or *stochastic variable*. For example, the number of organisms of a specified type contained in a 1-cm^3 specimen of water drawn from a lake is a random variable.

A process that yields a value of the random variable is known as a *trial* or *experiment*, and the value the variable assumes in a given trial is called the *outcome*. As an illustration, consider the standard type of cubical die, which has dots on each of its six faces. The number of dots on a face ranges from 1 to 6. When the die is tossed, the manner in which it lands is governed by chance, and therefore the number of dots on the face that lands on top is a variable. In this situation, the process of tossing the die is the trial, and the number of dots on the top face is the outcome. There are six possible outcomes, namely, the integers from 1 to 6, inclusive.

The foregoing definition of a trial can be extended to occurrences that do not yield numbers by assigning code numbers to the possible occurrences. For example, assume that a projectile is fired at a target area. The outcome may be labeled 1 if the projectile lands in the target area and 0 if it fails to do so. Thus, the act of firing a projectile becomes a trial in accordance with our definition.

Assume that a trial has n possible outcomes, designated $O_1, O_2, O_3, \ldots, O_n$, and that one outcome is just as likely as any other. The *probability* of a particular outcome is defined as $1/n$. Let $P(O_i)$ denote the probability of O_i. Then

$$P(O_i) = \frac{1}{n} \tag{9.1}$$

A specified outcome or set of outcomes is termed an *event*. For example, with reference to tossing a die, we may define the following events: The outcome is 4; the outcome is even (which comprises the outcomes 2, 4, and 6); the outcome is at least 3 (which comprises the outcomes 3, 4, 5, and 6).

Two events are said to be *mutually exclusive* or *disjoint* if it is impossible for both to result from a single trial. Thus, with reference to tossing a die, the following events are mutually exclusive: The outcome is even; the outcome is 3 or 5.

Two events are said to be *overlapping* if there is one or more outcomes that will satisfy both events. For example, with reference to tossing a die, consider the following pair of events: The outcome is even; the outcome is less than 4. There is one outcome (namely, 2) that satisfies both events, and therefore they are overlapping.

A set of events is said to be *exhaustive* if it includes all possible outcomes. If two events are mutually exclusive and constitute an exhaustive set, the events are said to be *complementary* to each other. For example, assume that a box contains both spheres and cubes and that an object is to be drawn at random from the box. If the given event consists of drawing a sphere, the complementary event consists of drawing a cube. The notation for complementary events is as follows: If E denotes the given event, $\bar{E}$ denotes the complementary event.

9.1.2 Venn Diagrams

A given situation involving probability can be visualized more readily by constructing a *Venn diagram.* In this diagram, each possible outcome of a trial is represented by a unique point on a plane, which is termed a *sample point.* All sample points corresponding to a given trial are uniformly distributed on the plane, and the region of the plane occupied by the sample points is termed the *sample space* of the trial.

With reference to Fig. 9.1*a*, consider that the sample points corresponding to a given trial lie within the rectangle *abcd.* This rectangle is the sample space of the trial. Assume that an event E is satisfied solely by outcomes having sample points within the circle indicated. This circle is said to *represent* event E. With reference to Fig. 9.1*b*, the events E_1 and E_2 represented by the circles indicated are mutually exclusive because there are no outcomes that are common to both events. On the

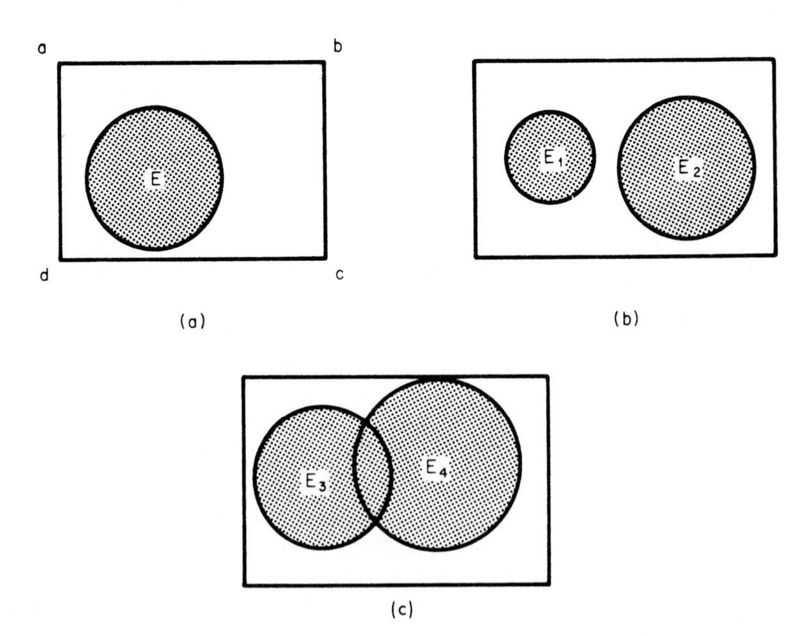

FIGURE 9.1 Venn diagrams. (*a*) Sample space; (*b*) mutually exclusive events; (*c*) overlapping events. (*From Handbook of Engineering Economics; used with permission of McGraw-Hill, Inc.*)

other hand, with reference to Fig. 9.1*c*, events E_3 and E_4 are overlapping because such outcomes do exist.

9.1.3 Laws of Probability

There are certain simple laws underlying the determination of probability, and we shall develop several of these at this point. The probability of an event E is denoted by $P(E)$.

Theorem 9.1. Assume that a trial has n possible outcomes of equal probability. If any one of r outcomes will produce an event E, the probability that E will occur is $P(E) = r/n$.

For example, with reference to tossing a die, consider this event: The outcome is odd and more than 2. This event is satisfied by two of the six possible outcomes (namely, 3 and 5). Therefore, the probability that this event will occur is 2/6 or 1/3.

It follows as a corollary of Theorem 9.1 that an impossible event has a probability of 0 and an event that is certain to occur has a probability of 1. Therefore, the probability of an event can range from 0 to 1. The value can be expressed as an ordinary fraction, a decimal fraction, or a percent.

Theorem 9.1 can be expressed in this alternative manner: Assume that a task can be performed by n alternative methods, that the actual method by which it will be performed will be determined by chance, and that all n methods have equal likelihood of being the one followed. If r of these methods cause an event E to occur, the probability of E is $P(E) = r/n$.

To calculate the probability of an event, it is often necessary to calculate the number of possible permutations or combinations. Therefore, we shall apply Eqs. (4.1) and (4.5) extensively in the present material.

EXAMPLE 9.1 A four-place number will be formed by selecting integers from 1 to 7, inclusive, at random, each digit appearing only once in the number. What is the probability that the number formed will contain the digits 2, 3, and 7, without reference to their order?

SOLUTION Since we are concerned solely with the identity of the digits that form the number, each selection of digits is a *combination* of seven digits taken four at a time. The number of possible combinations is ${}_7C_4 = 7!/(4!3!) = 35$. The digits 2, 3, and 7 can be combined with any of the remaining 4 digits; therefore, the number of satisfactory combinations is 4. Thus, the probability of the specified event is $4/35 = 0.1143$.

EXAMPLE 9.2 A bowl contains nine chips that are numbered consecutively from 1 to 9. Six chips will be drawn at random from the bowl and placed in a row. What is the probability that the chips marked 7 and 3 will occupy, respectively, the first and fifth positions in the row?

SOLUTION The number of ways in which the row can be formed is ${}_9P_6 = 9!/3! = 60{,}480$. Consider that the chips marked 7 and 3 are placed in the designated positions. Four positions remain to be filled, and seven chips are now available. Therefore, the number of ways the row can be completed is ${}_7P_4 = 7!/3! = 840$. Thus, the probability of the specified event is $840/60{,}480 = 0.0139$.

EXAMPLE 9.3 A committee consisting of six members of equal rank is to be formed. There are 15 individuals, including Smith and Jones, available for appointment to the committee. Because these individuals are all equally qualified, the committee members will be selected randomly. What is the probability that the committee will include Jones but not Smith?

SOLUTION Since the committee members will be of equal rank, each possible committee represents a combination of 15 individuals taken 6 at a time. The number of possible committees if no restrictions are imposed is ${}_{15}C_6 = 15!/(6!9!) = 5005$. Consider that Jones has been appointed to the committee but Smith is excluded from consideration. The 5 remaining committee members can be selected from 13 individuals, and the number of ways in which the committee can be completed is ${}_{13}C_5 = 13!/(5!8!) = 1287$. Therefore, the probability that the committee formed will meet the specifications is $1287/5005 = 0.2571$.

Theorem 9.2. If two events E_1 and E_2 are mutually exclusive, the probability that either E_1 or E_2 will occur is the sum of their respective probabilities. Expressed symbolically,

$$P(E_1 \text{ or } E_2) = P(E_1) + P(E_2)$$

This relationship becomes self-evident when we refer to Fig. 9.1*b*.

Consider that all events in a set are mutually exclusive and that the set is exhaustive. Since it is certain that one and only one event in the set will occur and the probability of certainty is 1, it follows that the sum of the respective probabilities of these events is 1. In particular, if the set consists of two events, these events are complementary to each other, and therefore $P(\bar{E}) = 1 - P(E)$. For example, if an object is to be drawn at random from a box and the probability of drawing a red object is 0.68, the probability of drawing an object of some other color is 0.32.

EXAMPLE 9.4 With reference to Example 9.3, what is the probability that the committee will include either Smith or Jones, or both?

SOLUTION Let E denote the given event. Then $\bar{E}$ denotes the complementary event: neither Smith nor Jones will be appointed. If both these individuals are excluded from consideration, the number of possible committees is ${}_{13}C_6 = 13!/(6!7!) = 1716$. Then $P(\bar{E}) = 1716/5005$, and $P(E) = 1 - 1716/5005 = 0.6571$.

EXAMPLE 9.5 In a game of chance, seven individuals compose a team. Each individual selects an integer from 1 to 9, inclusive, in privacy. The team wins an award if one integer is selected three times, another integer is selected twice, and the two remaining integers are distinct. For example, the selections 1, 4, 4, 5, 1, 4, 2 and 8, 3, 2, 7, 8, 2, 8 are satisfactory. Compute the probability that the team will win the award.

SOLUTION Since each integer that is selected is associated with a specific player, the selection of integers by the team is a *permutation* of nine numbers taken seven at a time, with duplications permitted. To demonstrate this fact, assume for convenience that the players are arranged in a row. We may consider that the integer selected by the rth player in the row is the rth integer selected. Moreover, in accordance with the definition in Art. 4.1.11, the selection of integers is a *repeated* permutation. By Eq. (4.10), the number of *possible* permutations is

$9^7 = 4{,}782{,}969$. We must now establish the number of *satisfactory* permutations, and we shall do this by two distinct methods.

Method 1. We first consider how many satisfactory *sets of integers* are available, without reference to their order. Let A and B denote the integers that appear three times and twice, respectively, and C and D denote the integers that appear once. Since A and B are of different rank, the number of ways in which they can be selected from nine integers is ${}_9P_2 = 72$. Since C and D are of equal rank, the number of ways in which they can be selected from the remaining seven integers is ${}_7C_2 = 21$. By Theorem 4.1, the number of available sets of integers is $72 \times 21 = 1512$.

A satisfactory selection of integers by the players is a permutation of seven numbers in which three numbers are identical and two other numbers are identical. By Eq. (4.2), the quantity of permutations that can be formed is $7!/(3!2!) = 420$. By Theorem 4.1, the number of possible satisfactory permutations is $1512 \times 420 = 635{,}040$.

In computing the number of satisfactory sets of integers, the sequence in which we select the integers is immaterial; that is, we may select A and B first and then C and D, or vice versa. Mathematically, we have ${}_9P_2 \times {}_7C_2 = {}_9C_2 \times {}_7P_2 = 9!/(2!5!) = 1512$.

Method 2. An alternative approach consists of matching players with integers. We first select three of the seven players and assign one of the nine integers to them. The number of possible ways of doing this is ${}_7C_3 \times 9 = 35 \times 9 = 315$. We now select two of the remaining four players and assign one of the remaining eight integers to them. The number of possible ways of doing this is ${}_4C_2 \times 8 = 6 \times 8 = 48$. Finally, we assign one of the remaining seven integers to one player and then one of the remaining six integers to the final player. The number of possible ways of doing this is $7 \times 6 = 42$.

Each assignment of integers to players constitutes a permutation, and the number of satisfactory permutations is $315 \times 48 \times 42 = 635{,}040$. This result coincides with that obtained by Method 1.

Since all permutations have equal likelihood of materializing, the probability that the team will win the award equals the number of satisfactory permutations divided by the number of possible permutations. Thus, the probability is $635{,}040/4{,}782{,}969 = 0.1328$.

Two trials are said to be *independent* of each other if the outcome of one trial has no bearing on the outcome of the other trial. For example, assume that a die will be tossed twice. Since the outcome of the first toss does not influence the outcome of the second toss, the two trials are independent of each other.

Theorem 9.3. Assume that a trial T_1 has n_1 possible outcomes of equal probability and that a trial T_2, independent of T_1, has n_2 possible outcomes of equal probability. The two trials may be performed simultaneously or in sequence.

1. The number of possible combined outcomes of the two trials is $n_1 n_2$, all of equal probability.
2. If an event E_1 can be produced by any one of r_1 outcomes of T_1 and an event E_2 can be produced by any one of r_2 outcomes of T_2, then *both* E_1 *and* E_2 can be produced by any one of $r_1 r_2$ combined outcomes.

Both statements stem directly from Theorem 4.1.

Theorem 9.4. Assume that two independent trials will be performed. The probability that the first trial will produce an event E_1 and the second trial will produce an event E_2 is the product of their respective probabilities. Expressed symbolically,

$$P(E_1 \text{ and } E_2) = P(E_1) \times P(E_2)$$

This statement stems directly from Theorem 9.3, and it can be extended to include any number of independent trials.

Consider that a set of events is specified. The probability that all events in the set will occur is known as the *joint probability* of the set. Thus, Theorem 9.4 is a statement of joint probability where the events are independent of one another.

EXAMPLE 9.6 A die is to be tossed twice. What is the probability of obtaining a number less than 3 on the first toss and an odd number on the second toss? Verify the answer by recording the satisfactory combined outcomes.

SOLUTION The first event is satisfied by the outcomes 1 and 2, and the second event is satisfied by the outcomes 1, 3, and 5. Then

$$P(E_1) = 2/6 \qquad P(E_2) = 3/6$$

$$P(E_1 \text{ and } E_2) = (2/6)(3/6) = 1/6$$

The combined outcomes that cause both E_1 and E_2 to occur are as follows: 1 and 1; 1 and 3; 1 and 5; 2 and 1; 2 and 3; 2 and 5. Thus, there are $2 \times 3 = 6$ satisfactory combined outcomes. The number of possible combined outcomes is $6 \times 6 = 36$. Then

$$P(E_1 \text{ and } E_2) = 6/36 = 1/6$$

EXAMPLE 9.7 The probability that a certain event will occur on each of the next three days is as follows: Monday, 0.35; Tuesday, 0.45; Wednesday, 0.60. If X denotes the number of days on which the event occurs, what is the probability that $X = 2$?

SOLUTION We must consider each possible combination of days on which the event can occur. The probabilities that the event will occur on the specified days are as follows:

$$\text{Mon. and Tues., not Wed.: } (0.35)(0.45)(1 - 0.60) = 0.0630$$

$$\text{Mon. and Wed., not Tues.: } (0.35)(1 - 0.45)(0.60) = 0.1155$$

$$\text{Tues. and Wed., not Mon.: } (1 - 0.35)(0.45)(0.60) = 0.1755$$

Since these combined outcomes are mutually exclusive, we have

$$P(X = 2) = 0.0630 + 0.1155 + 0.1755 = 0.3540$$

Theorem 9.5 a. Let E_1 and E_2 denote two overlapping events. If an event E results from the occurrence of E_1 and E_2, or both, the probability of E is

$$P(E) = P(E_1) + P(E_2) - P(E_1 \text{ and } E_2)$$

The validity of this statement can be demonstrated by means of the Venn diagram in Fig. 9.2*a*, where

$$P(E) = a + b + c \qquad P(E_1) = a + b \qquad P(E_2) = b + c$$

$$P(E_1 \text{ and } E_2) = b$$

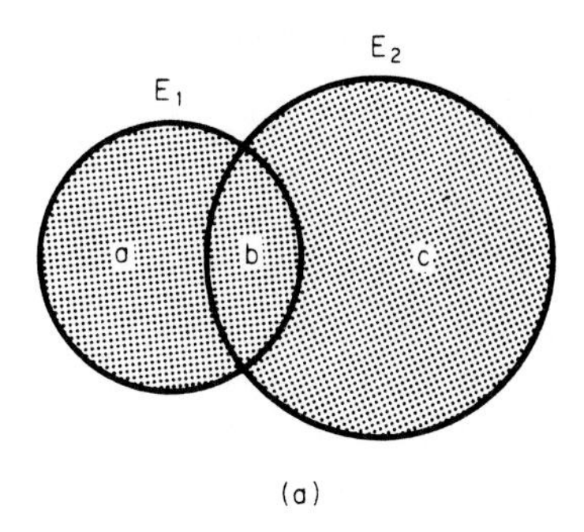

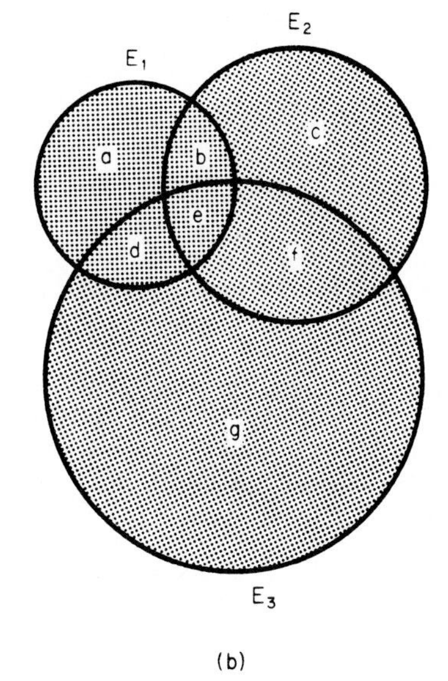

FIGURE 9.2 Probability of event resulting from overlapping events. (*a*) Venn diagram for two overlapping events; (*b*) Venn diagram for three overlapping events. (*From Handbook of Engineering Economics; used with permission of McGraw-Hill, Inc.*)

When these expressions are substituted, the foregoing equation is found to be correct.

EXAMPLE 9.8 With reference to Example 9.7, what is the probability that the event will occur on two successive days?

SOLUTION Two satisfactory combinations exist: The event occurs on Monday and Tuesday, and the event occurs on Tuesday and Wednesday. However, these combinations encompass a third combination: the event occurs on all three days. Applying Theorem 9.5*a*, we obtain this probability:

$$(0.35)(0.45) + (0.45)(0.60) - (0.35)(0.45)(0.60) = 0.3330$$

Theorem 9.5 b. Let E_1, E_2, and E_3 denote three overlapping events. If an event E results from the occurrence of E_1, E_2, or E_3, or any combination of them, the probability of E is

$$P(E) = P(E_1) + P(E_2) + P(E_3) - P(E_1 \text{ and } E_2) - P(E_1 \text{ and } E_3) - P(E_2 \text{ and } E_3) + P(E_1, E_2, \text{ and } E_3)$$

The validity of this statement can be demonstrated by means of the Venn diagram in Fig. 9.2*b*, where

$$P(E) = a + b + c + d + e + f + g$$

The proof is similar to that of Theorem 9.5*a*.

Theorem 9.6. Let $E_1, E_2, E_3, \ldots, E_n$ denote n independent events. If an event E results from the occurrence of any event in this set, or any combination of these events, the probability of E is

$$P(E) = 1 - [1 - P(E_1)][1 - P(E_2)][1 - P(E_3)] \cdots [1 - P(E_n)]$$

The proof of this statement lies in the fact that the complementary event $\bar{E}$ occurs if none of the specified events occurs, and $P(E) = 1 - P(\bar{E})$.

EXAMPLE 9.9 With reference to the electric circuit in Fig. 9.3, all five relays function independently, and the probability of any relay being closed is p. What is the probability that a current exists between a and b?

SOLUTION There are three alternative paths from a to b, and these are assigned the numbers shown in circles. A current exists if all relays along any path are closed. By Theorem 9.4, the probability of complete closure along a path is p^r, where r is the number of relays along the path. Therefore, the probability of complete closure along a path has the following values: path 1, p^2; path 2, p; path 3, p^2. Applying Theorem 9.6, we have

$$P(\text{current}) = 1 - (1 - p^2)(1 - p)(1 - p^2)$$
$$= p + 2p^2 - 2p^3 - p^4 + p^5$$

EXAMPLE 9.10 To obtain a passing grade, a student must pass all five parts of an examination. The probability of failure in any part is as follows: part A, 13 percent;

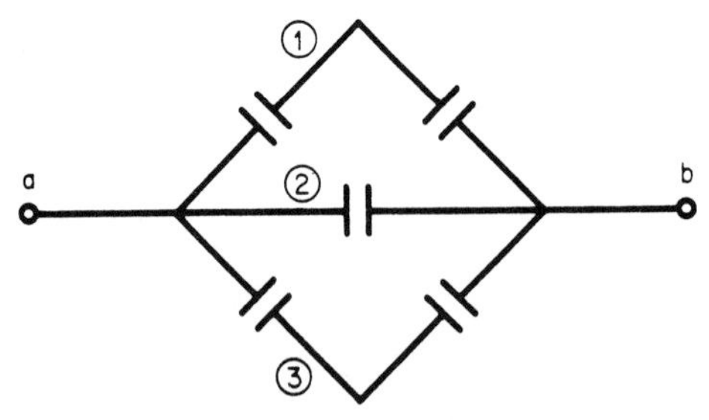

FIGURE 9.3 Electric circuit with relays. (*From Handbook of Engineering Economics; used with permission of McGraw-Hill, Inc.*)

part B, 18 percent; part C, 6 percent; part D, 11 percent; part E, 9 percent. What is the probability that the student will fail the examination?

SOLUTION

$$P(\text{failure}) = 1 - (1 - 0.13)(1 - 0.18)(1 - 0.06)(1 - 0.11)(1 - 0.09)$$
$$= 0.4569 = 45.69 \text{ percent}$$

In the subsequent material, it is to be understood that if a set of items exists and one item is to be drawn at random, all items in the set have equal likelihood of being drawn.

9.1.4 Dependent Trials and Conditional Probability

Consider that two trials, T_1 and T_2, will be performed in the sequence indicated. If the outcome of T_1 influences the outcome of T_2, then T_2 is *dependent* on T_1.

Let E_1 and E_2 denote events that may arise from T_1 and T_2, respectively. The probability of E_2 cannot be determined definitively until the outcome of T_1 is known. If the probability of E_2 is calculated on the premise that E_1 has occurred, the result is called the *conditional probability* of E_2. Events E_1 and E_2 are said to constitute a *compound event*.

Theorem 9.7. Assume that there will be two trials, the second dependent on the first. The probability that the first trial will yield an event E_1 and the second trial will yield an event E_2 is the product of their respective probabilities, where the probability of E_2 is calculated on the premise that E_1 has occurred.

Let $P(E_2|E_1)$ denote the probability that E_2 will occur, given that E_1 has occurred. Theorem 9.7 may be expressed symbolically in this manner:

$$P(E_1 \text{ and } E_2) = P(E_1) \times P(E_2|E_1)$$

This principle can be extended to compound events consisting of any number of individual events. The probability of the rth event in the chain is based on the premise that the preceding $r - 1$ events that were specified have in fact occurred.

EXAMPLE 9.11 A particle is ejected from a rotating mechanism, and it then enters either of two passages: A or B. The probability is 0.58 that it will enter A and 0.42 that it will enter B. In the passage, the particle is subjected to a force that determines whether it emerges from the passage or comes to rest within it. The probability that the particle will emerge from the passage is as follows: if it entered A, 0.76; if it entered B, 0.68. Compute the probability corresponding to every possible compound event.

SOLUTION

$$P(\text{enter A, emerge}) = (0.58)(0.76) = 0.4408$$
$$P(\text{enter A, not emerge}) = (0.58)(0.24) = 0.1392$$
$$P(\text{enter B, emerge}) = (0.42)(0.68) = 0.2856$$
$$P(\text{enter B, not emerge}) = (0.42)(0.32) = 0.1344$$

These probabilities total 1, as they must.

EXAMPLE 9.12 A state lottery operates in this manner: A player selects three sets of numbers. Each set contains five integers that can range from 1 to 52, inclusive. The player wins if any of his or her set of numbers is the one drawn by the lottery commission. What is the probability that the player will win?

SOLUTION

Method 1. Assume that the player selects a total of 15 distinct numbers. Each set of numbers is a combination of 52 integers taken 5 at a time. The number of possible sets is ${}_{52}C_5 = 52!/(5!47!) = 2{,}598{,}960$. Therefore, the probability that the player will win is $3/2{,}598{,}960 = 1/866{,}320$.

Method 2. We shall compute the probability that a particular set of numbers is the one drawn by the commission. The probability that any number in this set will be drawn is 5/52. Assume that this event has occurred. The pool now contains 51 numbers, and the quantity of numbers remaining in the set is 4. Therefore, the probability that another number in the set will be drawn is 4/51. Continuing in this manner and applying Theorem 9.7, we find that the probability that a particular set of numbers is drawn is $(5/52)(4/51)(3/50)(2/49)(1/48) = 1/2{,}598{,}960$. The probability that the player will win is three times this amount, or 1/866,320.

EXAMPLE 9.13 A case contains three type A units and eight type B units. However, since the units were not properly labeled, they can be identified only by examination. A unit will be drawn at random from the case and examined, the process being continued until a type A unit has been found. What is the probability that a type A unit will be found on the fourth drawing?

SOLUTION Each drawing reduces the number of units remaining in the case. The given event requires that four units be drawn and that they be drawn in the sequence B-B-B-A. By Theorem 9.7, the probability of obtaining this sequence is

$$(8/11)(7/10)(6/9)(3/8) = 7/55 = 0.1273$$

9.1.5 Bayes' Theorem

Assume that a trial is to be performed, and the probability that a given event will occur is calculated. The trial is then performed, and partial information concerning the outcome becomes available. The probability that the given event has occurred is then calculated on the basis of this partial information. The value of probability as originally calculated is known as the *prior probability*, and the value as subsequently calculated is known as the *posterior probability*.

With reference to the Venn diagram in Fig. 9.4*a*, the sample space of a trial is the rectangle *adeh*. Let E_1, E_2, and E_3 denote three events that are mutually exclusive and constitute an exhaustive set. Now let F denote an event that overlaps E_1, E_2, and E_3 and is represented by the rectangle *ilmp*. The prior probability of E_3 is

$$P(E_3) = \frac{\text{area } cdef}{\text{area } adeh}$$

and the probability of both E_3 and F is

$$P(E_3 \text{ and } F) = \frac{\text{area } klmn}{\text{area } adeh}$$

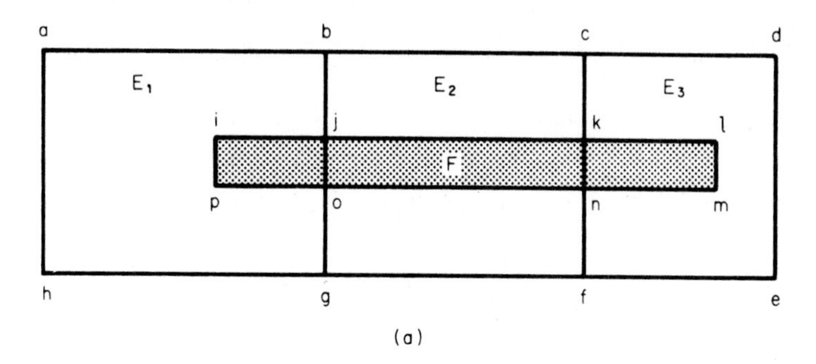

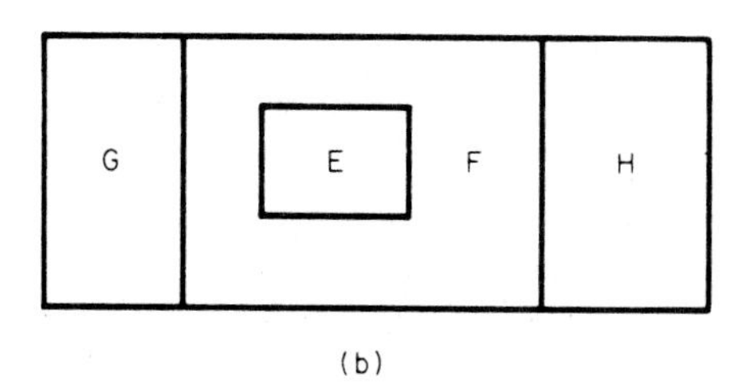

FIGURE 9.4 Venn diagrams to develop Bayes' theorem. (*a*) General case: (*b*) special case where given event lies within a single higher event. (*From Handbook of Engineering Economics; used with permission of McGraw-Hill, Inc.*)

Consider that event F is known to have occurred. What is the probability that E_3 has occurred?

$$P(E_3|F) = \frac{\text{area } klmn}{\text{area } ilmp} = \frac{(\text{area } klmn)/(\text{area } adeh)}{(\text{area } ilmp)/(\text{area } adeh)}$$

$$= \frac{P(E_3 \text{ and } F)}{P(F)}$$

In general, we may write

$$P(E_r|F) = \frac{P(E_r \text{ and } F)}{P(F)} \tag{9.2}$$

This relationship is known as *Bayes' theorem*.

EXAMPLE 9.14 A firm purchases cables for use in its operations. They are procured from three suppliers, in these proportions: 40 percent from Company A, 35 percent from Company B, and 25 percent from Company C. The probability that a cable procured from a particular supplier is defective is as follows: Company A, 5 percent; Company B, 9 percent; Company C, 10 percent. If a cable is found to be defective, what is the probability that it was supplied by Company A?

SOLUTION If a cable is selected at random, the probabilities concerning its supplier and its condition are as follows:

$$P(\text{A}) = 0.40 \qquad P(\text{B}) = 0.35 \qquad P(\text{C}) = 0.25$$

$$P(\text{A and defective}) = (0.40)(0.05) = 0.0200$$

$$P(\text{B and defective}) = (0.35)(0.09) = 0.0315$$

$$P(\text{C and defective}) = (0.25)(0.10) = 0.0250$$

$$P(\text{defective}) = 0.0200 + 0.0315 + 0.0250 = 0.0765$$

In the terminology of the previous discussion, the trial consists of drawing a cable at random, and the partial information is that it is defective. By Bayes' theorem,

$$P(\text{A}|\text{defective}) = \frac{P(\text{A and defective})}{P(\text{defective})} = \frac{0.0200}{0.0765} = 0.2614$$

A special case of Bayes' theorem arises where an event E can occur only if an event F occurs. This type of situation is represented by the Venn diagram in Fig. 9.4*b*, where the sample space of E lies wholly within the sample space of F. Then

$$P(E|F) = \frac{P(E)}{P(F)} \tag{9.2a}$$

This relationship is a special case of Eq. (9.2), for in the present situation the occurrence of E means the occurrence of both E and F.

9.1.6 Relationship between Probability and Relative Frequency

Assume that n independent trials of identical type were performed and that an event E occurred r times. In accordance with the definition in Art. 8.1.2, the *relative frequency* f_{rel} of E is r/n.

Where a relatively small number of independent trials of identical type is performed, the relative frequency of a given event may differ somewhat from its probability. However, it is logical to assume that as the number of trials n increases indefinitely, the relative frequency of the event approaches its probability as a limiting value.

In general, let $f_{rel}(E)$ denote the relative frequency of a given event E. Then

$$\lim_{n \to \infty} f_{rel}(E) = P(E) \tag{9.3}$$

Expressed verbally, probability represents relative frequency *in the long run*.

This relationship has two extremely important applications. First, consider a situation where the probability of a given event cannot be predetermined because the mathematical complexity is formidable. The probability can be established empirically (within acceptable limits of precision) by performing a vast number of trials or by drawing on the results of past trials. We simply equate probability to relative frequency. Second, this relationship affords a relatively simple and practical method of calculating probability without invoking esoteric laws of probability. We simply assume that a large number of trials has been performed and equate the relative frequency of the event to its probability.

EXAMPLE 9.15 Solve Example 9.14 by equating the probability that a cable is defective to the actual incidence of defectives in the long run.

SOLUTION Assume that 10,000 cables are purchased.

$$\text{No. defectives procured from A} = 10{,}000(0.40)(0.05) = 200$$

$$\text{No. defectives procured from B} = 10{,}000(0.35)(0.09) = 315$$

$$\text{No. defectives procured from C} = 10{,}000(0.25)(0.10) = \underline{250}$$

$$\text{Total no. defectives} = 765$$

$$\text{Probability that defective came from A} = 200/765 = 0.2614$$

9.1.7 Definition of a Probability Distribution

Let X denote a discrete random variable. The entire set of possible values of X and their respective probabilities constitute the *probability distribution* of X. Since it is a certainty that X will assume one particular value in this set, it follows from Theorem 9.2 that the sum of the probabilities is 1.

EXAMPLE 9.16 A box contains 13 objects. Of these, 9 are yellow and 4 are purple. Seven objects will be drawn from the box at random, without being replaced. If X denotes the number of yellow objects that are drawn, establish the probability distribution of X.

SOLUTION Since there are only 4 purple objects, the minimum value of X is 3.

The 7 objects that are drawn constitute a *combination* of 13 objects taken 7 at a time. By Eq. (4.5), the number of possible combinations is ${}_{13}C_7 = 1716$. The X yellow objects that are drawn constitute a combination of 9 objects taken X at a time, and the number of possible combinations of yellow objects is ${}_9C_X$. The $7 - X$ purple objects that are drawn constitute a combination of 4 objects taken $7 - X$ at a time, and the number of possible combinations of purple objects is ${}_4C_{7-X}$. In accordance with Theorem 4.1, the number of ways in which the 7 objects can be drawn is obtained by multiplying the number of combinations of yellow objects by the number of combinations of purple objects.

Since all possible combinations have equal likelihood of being drawn, we obtain the following results:

$$P(X=3) = \frac{{}_9C_3 \times {}_4C_4}{1716} = \frac{84 \times 1}{1716} = 0.0490$$

$$P(X=4) = \frac{{}_9C_4 \times {}_4C_3}{1716} = \frac{126 \times 4}{1716} = 0.2937$$

$$P(X=5) = \frac{{}_9C_5 \times {}_4C_2}{1716} = \frac{126 \times 6}{1716} = 0.4406$$

$$P(X=6) = \frac{{}_9C_6 \times {}_4C_1}{1716} = \frac{84 \times 4}{1716} = 0.1958$$

$$P(X=7) = \frac{{}_9C_7 \times {}_4C_0}{1716} = \frac{36 \times 1}{1716} = 0.0210$$

These probabilities total 1, as they must.

EXAMPLE 9.17 With reference to Example 9.13, let X denote the number of units that must be drawn to obtain a type A unit. Establish the probability distribution of X.

SOLUTION Proceeding as in Example 9.13, we obtain the following results:

$$P(X = 1) = 3/11 = 0.2727 \qquad P(X = 2) = (8/11)(3/10) = 0.2182$$

$$P(X = 3) = (8/11)(7/10)(3/9) = 0.1697$$

Continuing these calculations, we obtain

$$P(X = 4) = 0.1273 \qquad P(X = 5) = 0.0909 \qquad P(X = 6) = 0.0606$$

$$P(X = 7) = 0.0364 \qquad P(X = 8) = 0.0182 \qquad P(X = 9) = 0.0061$$

Again, these probabilities total 1, as they must.

The probability distribution of a discrete variable gives the probability that the variable will assume a *particular value* on a given occasion. Similarly, the probability distribution of a continuous variable gives the probability that the variable will assume a value that lies within a *particular interval* on a given occasion.

The probability distribution of a continuous variable can readily be visualized by constructing a *probability curve*, which has this property: The area bounded by the curve, the X axis, and vertical lines erected at $X = a$ and at $X = b$ equals the probability that $a \leq X \leq b$. For example, with reference to Fig. 9.5, the shaded area is the probability that $1.40 \leq X \leq 2.05$. The total area between the probability curve and the X axis is 1.

Let $f(X)$ denote the ordinate of the probability curve. Then

$$P(a \leq X \leq b) = \int_a^b f(X)\, dX = \text{area under probability curve between } a \text{ and } b$$

The ordinate $f(X)$ is referred to as the *probability-density function*.

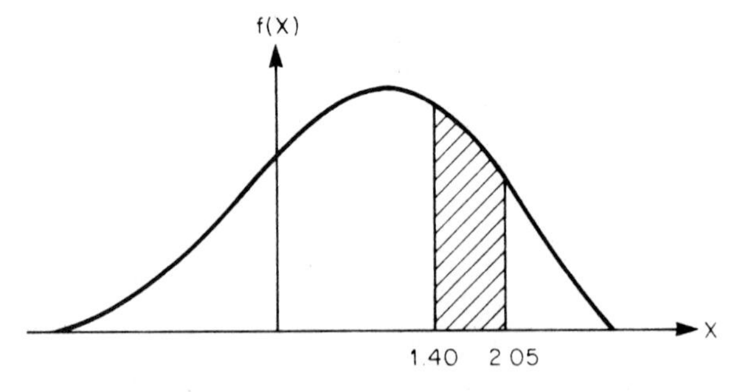

FIGURE 9.5 Probability curve. (*From Handbook of Engineering Economics; used with permission of McGraw-Hill, Inc.*)

9.1.8 Properties of a Probability Distribution

Consider that a trial is performed repeatedly to yield m values of a random variable X. This set of values has an arithmetic mean $\bar{X}$ and a standard deviation s, these terms being defined in Arts. 8.2.2 and 8.3.4, respectively. Now consider that the number of values m increases beyond bound. The limiting values approached by $\bar{X}$ and s are termed the arithmetic mean and standard deviation, respectively, of the *probability distribution* of X, and they are denoted by μ and σ, respectively.

Equations (8.3) and (8.18) present, respectively, the arithmetic mean and standard deviation of a discrete variable in terms of relative frequency. Since the relative frequency of an X value approaches the probability of that value as m becomes infinite, these equations can be adapted to the present case by replacing relative frequency with probability. Then

$$\mu = \Sigma\, X[P(X)] \tag{9.4}$$

$$\sigma = \sqrt{\Sigma\,(X-\mu)^2[P(X)]} \tag{9.5}$$

The arithmetic mean of a probability distribution is also called the *expected value* of the variable, and it is then denoted by $E(X)$.

EXAMPLE 9.18 With reference to Example 9.16, compute the expected value of X.

SOLUTION

$$E(X) = 3(0.0490) + 4(0.2937) + 5(0.4406) + 6(0.1958) + 7(0.0210) = 4.8466$$

EXAMPLE 9.19 An individual has purchased a ticket that entitles her to participate in a game of chance. The sums that may be won and their respective probabilities are recorded in Table 9.1. What is the player's expected winning?

SOLUTION

$$\text{Expected winning} = 10(0.15) + 25(0.04) + 50(0.02) + 75(0.01)$$
$$= \$4.25$$

Thus, if the number of times this game is played increases beyond bound, the average amount won in a single play approaches \$4.25 as a limit.

EXAMPLE 9.20 An experiment will be performed repeatedly until a required result is obtained. Let X denote the number of times it will be necessary to perform the

TABLE 9.1

Prospective winning, \$	Probability
0	0.78
10	0.15
25	0.04
50	0.02
75	0.01
Total	1.00

experiment. The probability distribution of X is as follows:

$$P(X = 1) = 0.23 \qquad P(X = 2) = 0.28 \qquad P(X = 3) = 0.21$$

$$P(X = 4) = 0.16 \qquad P(X = 5) = 0.09 \qquad P(X = 6) = 0.03$$

The cost of performing the experiment is estimated to be \$8 for the first experiment, \$6 for the second experiment, and \$5 for each subsequent experiment. What is the expected cost of the entire set of experiments?

SOLUTION Let C denote this cost. The calculations are performed in Table 9.2. First, we record the value of C corresponding to every possible value of X. For example, if $X = 4$, $C = 8 + 6 + 2 \times 5 = \$24$. The value of C can range from \$8 (when $X = 1$) to \$34 (when $X = 6$). We then record the probability of every possible value of C (which equals the probability of the corresponding X value), form the products $C[P(C)]$, and sum the products. The result is $E(C) = \$17.22$.

TABLE 9.2

X	C, \$	$P(C)$	$C[P(C)]$, \$
1	8	0.23	1.84
2	8 + 6 = 14	0.28	3.92
3	14 + 5 = 19	0.21	3.99
4	19 + 5 = 24	0.16	3.84
5	24 + 5 = 29	0.09	2.61
6	29 + 5 = 34	0.03	1.02
Total			17.22

9.2 DISCRETE PROBABILITY DISTRIBUTIONS

Probability distributions tend to follow certain clearly defined patterns, and we shall now discuss the most important ones. We start with those cases where the random variable is discrete. For convenience of reference, Table 9.3 presents the definitions of the four basic distributions.

TABLE 9.3 Probability Distributions

Type of distribution	Process	Definition of X
Binomial	Perform n independent trials	Number of times given event occurs
Geometric	Perform independent trials until given event occurs	Number of trials required
Pascal	Perform independent trials until given event occurs k times	Number of trials required
Hypergeometric	Draw items from a set, without replacement	Number of times item of given type is drawn

9.2.1 Binomial Distribution

The possible outcomes of a trial can be divided into two categories: those that yield a given event, and those that fail to do so. Consider that n independent trials are performed, and let X denote the number of times the given event occurs. For a reason that will soon become apparent, the probability distribution of X is called *binomial*. We shall first consider a specific case of a binomial distribution.

EXAMPLE 9.21 A case contains 5 type A and 9 type B units. A unit will be drawn at random from the case and replaced with one of identical type, the process being repeated until 8 units have been drawn. Let X denote the number of type A units that are drawn. Determine the probability that $X = 3$.

SOLUTION Each drawing has two possible outcomes: A and B. As a result of the replacement, the 8 drawings are independent of one another. Therefore, the probability distribution of X is binomial. The probabilities associated with each drawing are these: $P(\text{A}) = 5/14$; $P(\text{B}) = 9/14$.

Consider this set of outcomes; A-A-A-B-B-B-B-B. The probability of this set is $(5/14)^3(9/14)^5$. We can rearrange the 3 A's and 5 B's to form another acceptable set of outcomes having the same probability. By Eq. (4.2), the number of possible arrangements is $8!/(3!5!)$. Alternatively, we may consider that the positions occupied by the A's constitute a combination of 8 positions taken 3 at a time, and the number of possible combinations is ${}_8C_3 = 8!/(3!5!)$. Then

$$P(X = 3) = \frac{8!}{3!5!}\left(\frac{5}{14}\right)^3\left(\frac{9}{14}\right)^5 = 0.2801$$

We shall now generalize on the basis of Example 9.21. Let

P = probability given event will occur on single trial

Q = probability given event will fail to occur on single trial $= 1 - P$

n = number of independent trials

X = number of times given event occurs

Applying Eq. (4.5) and the notation of Art. 1.22, we have

$$P(X) = {}_nC_X P^X Q^{n-X} = \frac{n!}{X!(n-X)!} P^X Q^{n-X} = \binom{n}{X} P^X Q^{n-X} \tag{9.6}$$

Reference to Eq. (1.34) reveals that the expression for $P(X)$ is the $(X + 1)$th term in the binomial expansion $(Q + P)^n$, and it is for this reason that the probability distribution of X is termed *binomial*. Assigning to X all possible values from 0 to n, we obtain

$$P(0) + P(1) + P(2) + \cdots + P(n) = (Q + P)^n = 1^n = 1$$

EXAMPLE 9.22 A firm manufactures a standard commodity, and 35 percent of the units are produced by machine A. If 7 units are selected at random, what is the probability that not more than 3 of these units came from machine A?

SOLUTION Let X denote the number of units drawn that came from machine A. We assume that the number of units in stock is vast, and therefore the reduction in stock caused by each drawing is negligible. As a result, the 7 drawings may be considered to be independent for all practical purposes, and the probability distribution of X is binomial. Applying Eq. (9.6), we obtain the following results:

$$P(0) = (0.65)^7 = 0.0490 \qquad P(1) = 7(0.35)(0.65)^6 = 0.1848$$

$$P(2) = 21(0.35)^2(0.65)^5 = 0.2985$$

$$P(3) = 35(0.35)^3(0.65)^4 = 0.2679$$

Summing these results, we obtain $P(X \leq 3) = 0.8002$.

The arithmetic mean and standard deviation of the binomial probability distribution are as follows:

$$\mu = nP \tag{9.7}$$

$$\sigma = \sqrt{nPQ} = \sqrt{\mu Q} \tag{9.8}$$

EXAMPLE 9.23 The incidence of defective units produced by a machine is found to average 4.8 percent. If 600 units are produced per week, determine the mean number of defective units produced per week and the standard deviation from this mean.

SOLUTION Since the production of each unit represents an independent trial, the probability distribution of the number of defectives is binomial.

$$\mu = 600(0.048) = 28.8 \qquad \sigma = \sqrt{28.8(0.952)} = 5.24$$

EXAMPLE 9.24 The probability that an event will occur in a single trial is 0.62. Let X denote the number of times the event occurs in 4 independent trials. Compute the arithmetic mean and standard deviation of X without applying Eqs. (9.7) and (9.8). Then test the results by applying these equations.

SOLUTION Applying Eq. (9.6), we obtain the following results:

$$P(0) = 0.0209 \qquad P(1) = 0.1361 \qquad P(2) = 0.3330$$

$$P(3) = 0.3623 \qquad P(4) = 0.1478$$

$$\mu = 1(0.1361) + 2(0.3330) + 3(0.3623) + 4(0.1478) = 2.48$$

$$\begin{aligned}\sigma^2 &= (0 - 2.48)^2(0.0209) + (1 - 2.48)^2(0.1361) + (2 - 2.48)^2(0.3330) \\ &\quad + (3 - 2.48)^2(0.3623) + (4 - 2.48)^2(0.1478) \\ &= 0.9428\end{aligned}$$

$$\sigma = 0.9710$$

Equations (9.7) and (9.8) yield the following values:

$$\mu = 4(0.62) = 2.48 \qquad \sigma = \sqrt{4(0.62)(0.38)} = 0.9708$$

The slight discrepancy in the value of σ stems from rounding effects.

9.2.2 Geometric Distribution

Consider that a trial is performed repeatedly, each trial being independent of the preceding trials, until a given event occurs. Let X denote the number of trials that are required. For example, if the given event occurs on the fifth trial, $X = 5$. The probability distribution of X is said to be *geometric*. In theory, X can assume any positive integral value from 1 to infinity.

It is to be emphasized that the trials must be independent of one another to produce a geometric probability distribution. Thus, the variable in Example 9.13 does not have this distribution because the outcome of one drawing influences the outcome of the next. Similarly, assume that a series of experiments will be performed until a desired result is obtained. Realistically, the experimenter is likely to gain proficiency, and the probability of success improves as the experiments continue.

EXAMPLE 9.25 Projectiles will be fired in succession until one strikes a target, and the probability that any projectile will strike the target is 0.18. If X denotes the number of firings that are required, what is the probability that $X = 5$?

SOLUTION Since each firing is independent of the preceding ones, X has a geometric probability distribution. The variable assumes the value 5 if the first 4 firings are unsuccessful and the fifth one is successful. By Theorem 9.4, the probability of the specified sequence of outcomes is

$$P(5) = (0.82)^4(0.18) = 0.0814$$

In general, let

P = probability given event will occur on single trial

Q = probability given event will fail to occur on single trial $= 1 - P$

X = number of trials required to obtain given event

$$P(X) = Q^{X-1}P \tag{9.9}$$

Let r denote a positive integer. If the first r trials are unsuccessful, $X > r$. Then

$$P(X > r) = Q^r \tag{9.10}$$

EXAMPLE 9.26 With reference to Example 9.25, what is the probability that at least 4 firings will be required?

SOLUTION

$$P(X > 3) = (0.82)^3 = 0.5514$$

The mean and standard deviation of the geometric probability distribution are as follows:

$$\mu = \frac{1}{P} \tag{9.11}$$

$$\sigma = \frac{\sqrt{Q}}{P} \tag{9.12}$$

Equation (9.11) is in accord with simple logic. For example, if the probability that a trial will yield a given event is 1/8, it is plausible to consider that on the average it will require 8 trials to produce the event.

EXAMPLE 9.27 The probability that an experiment will be successful is 0.08, regardless of the number of preceding experiments. The experiment will be performed repeatedly until success is achieved. The first experiment will cost \$15, the second will cost \$12, and each subsequent experiment will cost \$10. What is the expected cost of the set of experiments?

SOLUTION The expected number of experiments is $\mu = 1/0.08 = 12.5$. Therefore,

$$\text{Expected cost} = 15 + 12 + 10.5 \times 10 = \$132$$

9.2.3 Pascal Distribution

Again consider that a trial is performed repeatedly, each trial being independent of all preceding ones, until a given event has occurred k times, where k is a positive integer. Let X denote the number of trials that are required. This variable is said to have a *Pascal probability distribution.*

As an illustration, assume the following: A trial has two possible outcomes, A and B, the event is defined as the occurrence of B, and we set $k = 3$. An acceptable sequence of outcomes is A-B-A-A-A-B-A-B. In this case, $X = 8$. The geometric probability distribution is a special form of the Pascal distribution in which $k = 1$.

EXAMPLE 9.28 Objects are ejected randomly from a rotating mechanism, and the probability that an object will enter a stationary receptacle after leaving the mechanism is 0.28. The process of ejecting objects will continue until four objects have entered the receptacle. If X denotes the number of objects that must be ejected, what is the probability that $X = 10$?

SOLUTION Let S and F denote, respectively, success and failure of an ejection. Then $P(\mathrm{S}) = 0.28$ and $P(\mathrm{F}) = 0.72$. In the present instance, $k = 4$ and $X = 10$. Consider the following set of outcomes that meets the requirements:

F-S-F-F-F-S-F-F-S-S

The probability of this set of outcomes is $(0.28)^4(0.72)^6$. However, this set of outcomes can be transformed to another satisfactory set having the same probability by holding the fourth S in the final position but assigning the first 3 S's to other positions. Each assignment represents a combination of nine positions taken three at a time. Therefore, the number of satisfactory sets is ${}_9C_3 = 84$. Then

$$P(10) = 84(0.28)^4(0.72)^6 = 0.0719$$

In general, let

P = probability given event will occur on a single trial

$Q = 1 - P$

k = number of times given event is to occur

X = number of trials required

By generalizing on the basis of Example 9.28, we obtain the following:

$$P(X) = {}_{X-1}C_{k-1}P^kQ^{X-k} = \frac{(X-1)!}{(k-1)!(X-k)!}P^kQ^{X-k} \tag{9.13}$$

The mean and standard deviation of the Pascal probability distribution are as follows:

$$\mu = \frac{k}{P} \tag{9.14}$$

$$\sigma = \frac{\sqrt{kQ}}{P} \tag{9.15}$$

A comparison of Eq. (9.13) with Eq. (1.34) reveals that the expression for $P(X)$ in the Pascal distribution equals P^k multiplied by the kth term in the expansion of $(Q+1)^{X-1}$. The Pascal distribution is sometimes referred to as the *negative-binomial distribution.*

9.2.4 Hypergeometric Distribution

Consider that there is a set of N items, several of which are of a given type. Now consider that n items are drawn at random from the set, where $n \leq N$, and that these items are discarded without being replaced. The number of items remaining in the set diminishes by 1 with each drawing. If X denotes the number of times an item of a given type is drawn, the probability distribution of X is termed *hypergeometric.*

EXAMPLE 9.29 A case contains 7 type A units and 9 type B units. Eleven units will be drawn from the case at random, without being replaced. If X denotes the number of type A units that are drawn, what is the probability that $X = 5$?

SOLUTION Since the sequence in which the units are drawn is immaterial, the units that are drawn constitute a *combination* of 16 units taken 11 at a time, and the number of possible combinations is ${}_{16}C_{11}$.

Set $X = 5$. The units that are drawn may be divided into two groups: 5 type A and 6 type B units. The first group represents a combination of 7 units taken 5 at a time, and the second group represents a combination of 9 units taken 6 at a time. Therefore, the total number of ways in which the 11 units can be drawn is ${}_7C_5 \times {}_9C_6$. Then

$$P(5) = \frac{{}_7C_5 \times {}_9C_6}{{}_{16}C_{11}} = \frac{21 \times 84}{4368} = 0.4038$$

In general, let

N = number of items in original set

n = number of items drawn

a = number of items of given type in original set

X = number of items of given type drawn

P = probability that first item drawn is of given type = a/N

$Q = 1 - P$

Generalizing on the basis of Example 9.29, we obtain

$$P(X) = \frac{{}_aC_X \times {}_{N-a}C_{n-X}}{{}_NC_n} \tag{9.16}$$

The mean and standard deviation of the hypergeometric distribution are as follows:

$$\mu = nP \tag{9.17}$$

$$\sigma = \sqrt{\frac{nPQ(N-n)}{N-1}} \tag{9.18}$$

If N is large, the calculation of hypergeometric probability by means of Eq. (9.16) becomes arduous, but fortunately a published set of tables is available for this purpose.[1] If N is extremely large in relation to n, a binomial distribution may be applied as a reasonable approximation, as was done in Example 9.22.

9.2.5 Multinomial Distribution

In the binomial probability distribution, n independent trials are performed and each trial has two possible outcomes. This concept can be generalized. Consider that n independent trials are performed and that each trial has k possible outcomes, labeled $O_1, O_2, O_3, \ldots, O_k$. Let X_i denote the number of times O_i occurs. The set of numbers $X_1, X_2, X_3, \ldots, X_k$ is said to have a *multinomial* probability distribution.

EXAMPLE 9.30 Twelve projectiles will be fired in succession, and each will land in area A, B, or C. The probability corresponding to each landing is as follows: A, 0.35; B, 0.41; C, 0.24. What is the probability that 4 projectiles will land in A, 5 will land in B, and 3 will land in C?

SOLUTION Consider the following series of outcomes that satisfies the requirement:

A-A-A-A-B-B-B-B-B-C-C-C

The probability of this series is

$$(0.35)^4(0.41)^5(0.24)^3$$

Other satisfactory series of outcomes can be devised by forming other arrangements of the letters, and they all have the same probability. How many such arrangements are possible? The 4 positions occupied by the A's represent a combination of 12 positions taken 4 at a time; the 5 positions occupied by the B's represent a combination of the 8 remaining positions taken 5 at a time. Therefore, the number of possible arrangements is

$${}_{12}C_4 \times {}_8C_5 = 495 \times 56 = 27{,}720$$

---

[1] Gerald J. Lieberman and Donald B. Owen, *Tables of the Hypergeometric Probability Distribution*, Stanford University Press, Stanford, Calif., 1961.

The probability of the given event is

$$27{,}720(0.35)^4(0.41)^5(0.24)^3 = 0.0666$$

9.2.6 Time and Space Events

An event that occurs repeatedly is termed a *recurrent event*. Thus far, we have considered solely recurrent events for which the interval between successive occurrences is a discrete random variable. For example, assume that a case contains type A and type B units, and that units will be drawn at random, either with or without replacement. We define the given event as drawing a type A unit. The interval between successive occurrences is the number of trials required to produce the event after the preceding event has occurred.

We shall now consider two important types of recurrent events for which the interval between successive occurrences is a continuous random variable. The following are illustrations:

1. The event is defined as the arrival of a vehicle at the entrance to a bridge. The interval between successive occurrences is the time intervening between successive arrivals. This event is known as a *time event*.
2. A botanist surveys a field in search of a particular type of plant, and the event is defined as her discovering such a plant. The interval between successive occurrences is the distance between two such plants that were found in succession. This event is known as a *space* event.

Assume that a given time or space event occurs randomly but has a uniform probability of occurrence across the interval of time or space under consideration. For example, assume that the probability that a continuously operating machine will break down during a given month is the same for every month of its service life, and assume that the probability that a typographical error will occur in one column of newsprint is the same for all columns. In the subsequent material, we shall assume that a time or space event has a uniform probability of occurrence.

Now assume the following: A given time event has been found to occur randomly and to occur 5 times per day on the average. The *expected* number of occurrences is as follows: in 1 day, 5; in 3 days, 15; in 1 h, 5/24.

9.2.7 Poisson Distribution

We shall now investigate the probability distribution of a time event that occurs randomly but has a uniform probability of occurrence, and analogous statements will apply to a space event of this type. Each instant of time is a trial in the sense that the given event either will or will not occur at that instant. However, since the number of trials in a given time interval is infinite, the problem must be approached in an indirect manner.

We start by dividing the time interval under consideration into subintervals of definite and uniform length, and we view each subinterval as an independent trial. Thus, we assume that the number of times the event will occur in a subinterval is either 0 or 1. In this manner, we start with a finite number of trials. We now approach reality by making the length of each subinterval progressively smaller, continuing the process indefinitely. The length of each subinterval becomes infinitesimally small and the number of trials becomes infinitely large. With reference to the initial conditions, let

T = time interval under consideration

n = number of time subintervals contained in T

P = probability given event will occur in one subinterval

$Q = 1 - P$

m = expected number of events in interval T

X = true number of events in interval T

Under the initial conditions, X has a binomial probability distribution, and we tentatively set $P = m/n$. By Eq. (9.6), the probability that the event will occur in precisely X subintervals is

$$P(X) = \frac{n!}{X!(n-X)!} P^X Q^{n-X}$$

Now consider that the length of the subinterval is successively halved while T and X remain constant. The value of n is successively doubled, the value of P is successively halved, and the value of m remains constant. As the process continues indefinitely, $P(X)$ approaches the following limit:

$$P(X) = \frac{m^X}{X!} e^{-m} \tag{9.19}$$

where e is the quantity defined in Art. 1.28. In the special case where $X = 0$, Eq. (9.19) reduces to

$$P(0) = e^{-m} \tag{9.19a}$$

Equation (9.19) was first derived by the French mathematician Poisson, and a time or space event to which this equation applies is said to have a Poisson probability distribution. It should be emphasized, however, that the statement "A given event has a Poisson distribution" is actually a contraction of the statement "The number of times a given event occurs in a given interval of time or space has a Poisson distribution."

EXAMPLE 9.31 A firm manufactures rolls of tape and then cuts the rolls into lengths of 360 m (1181 ft). Extensive measurements have disclosed that defects in the roll occur at random and the average distance between defects is 120 m (393.7 ft). If a Poisson distribution is assumed, what is the probability (to four decimal places) that a tape has more than four defects?

SOLUTION Let X denote the number of defects in a 360-m tape.

$$m = 360/120 = 3 \qquad e^{-3} = 0.049787$$

By Eq. (9.19),

$$P(0) = e^{-3} = 0.04979$$

$$P(1) = (3/1!)e^{-3} = (3/1)(0.049787) = 0.14936$$

$$P(2) = (3^2/2!)e^{-3} = (9/2)(0.049787) = 0.22404$$

$$P(3) = (3^3/3!)e^{-3} = (27/6)(0.049787) = 0.22404$$

$$P(4) = (3^4/4!)e^{-3} = (81/24)(0.049787) = 0.16803$$

$$P(X \leq 4) = P(0) + P(1) + P(2) + P(3) + P(4) = 0.8153$$

$$P(X > 4) = 1 - 0.8153 = 0.1847$$

EXAMPLE 9.32 A radioactive substance emits particles randomly, and it has been ascertained that the average interval between emissions is 12 s. Assume that the number of particles emitted during a given time interval has a Poisson distribution.

a. What is the probability that the substance will emit more than three particles in a 30-s interval?

b. What is the probability that the substance will emit more than three particles exactly twice in five consecutive 30-s intervals?

SOLUTION Let X denote the number of particles emitted in a 30-s interval.

$$m = 30/12 = 2.5 \qquad e^{-2.5} = 0.082085$$

Applying Eq. (9.19), we obtain these results:

$$P(0) = 0.08209 \qquad P(1) = 0.20521$$

$$P(2) = 0.25652 \qquad P(3) = 0.21376$$

Part a. By summation,

$$P(X \leq 3) = 0.7576 \qquad P(X > 3) = 1 - 0.7576 = 0.2424$$

Part b. Each 30-s interval is an independent trial in the sense that the number of particles emitted in one interval has no influence on the number of particles emitted in any subsequent interval. Therefore, the number of times that X exceeds 3 has a binomial probability distribution. Let Y denote the number of times that X exceeds 3. Applying Eq. (9.6) and setting ${}_5C_2 = 10$, we obtain

$$P(Y = 2) = 10(0.2424)^2(0.7576)^3 = 0.2555$$

EXAMPLE 9.33 With reference to Example 9.32, a counting device is used to establish the number of particles emitted. The probability that the device will actually count an emission is 0.92. If C denotes the number of emissions counted in a 30-s interval, what is the probability that $C = 3$?

SOLUTION The device increases the count by 1 if two events occur: a particle is emitted and the device functions properly. Therefore, the expected number of emissions counted equals the expected number of emissions times the reliability of the device.

$$m = (2.5)(0.92) = 2.3$$

Applying Eq. (9.19), we obtain $P(C = 3) = 0.2033$.

When a time or space event has a Poisson probability distribution, interest often centers on the interval between successive occurrences. Since this interval is a continuous random variable, we can only specify the probability that this value lies within a given range. Let a denote the expected number of occurrences in 1 time unit and U denote the interval between successive occurrences. The expected number of occurrences in K time units is aK. By Eq. (9.19a), the probability that U exceeds K is

$$P(U > K) = e^{-aK} \tag{9.20}$$

With reference to a time event, the interval between successive occurrences is often referred to as the *waiting time* of the event. Waiting time constitutes the basis of queuing theory.

EXAMPLE 9.34 Experience indicates that a certain machine breaks down on an average of once every 18 operating days. If a Poisson distribution is assumed, what is the probability that the time between successive breakdowns will exceed 21 operating days?

SOLUTION

$$P(U > 21) = e^{-(1/18)21} = 0.3114$$

The arithmetic mean and standard deviation of the Poisson probability distribution are as follows:

$$\mu = m \tag{9.21}$$

$$\sigma = \sqrt{m} = \sqrt{\mu} \tag{9.22}$$

Equation (9.19) for Poisson probability was developed by starting with binomial probability and allowing the number of trials to increase beyond bound. Thus, Poisson probability represents a limiting case of binomial probability. Therefore, Poisson probability can be applied as a reasonable approximation to binomial probability if the number of trials is extremely large and P is close to either 0 or 1. This approximation vastly simplies the calculations.

EXAMPLE 9.35 It has been found that the incidence of defective parts produced by a machine averages 1 in every 180. Applying a Poisson approximation, calculate the probability that a case containing 225 parts has more than 2 defective parts.

SOLUTION Let X denote the number of defectives in the case. The expected value of X is $225/180 = 1.25$. Applying Eq. (9.19) with $m = 1.25$, we obtain

$$P(0) = 0.2865 \qquad P(1) = 0.3581 \qquad P(2) = 0.2238$$

By summation,

$$P(X \leq 2) = 0.8684 \qquad P(X > 2) = 1 - 0.8684 = 0.1316$$

9.3 CONTINUOUS PROBABILITY DISTRIBUTIONS

9.3.1 Characteristics of a Continuous Distribution

In Art. 9.1.7, we defined the probability-density function $f(X)$ as the ordinate of the probability curve. Since the total area under the curve is 1, this function must satisfy the following equation:

$$\int_{-\infty}^{+\infty} f(X)\, dX = 1 \tag{9.23}$$

The mean and variance of the probability distribution are defined by the following equations:

$$\mu = \int_{-\infty}^{+\infty} [f(X)]X\, dX \tag{9.24}$$

$$\sigma^2 = \int_{-\infty}^{+\infty} [f(X)](X-\mu)^2\, dX \tag{9.25}$$

However, Eq. (9.25) can be transformed to the following equation, which is simpler to apply:

$$\sigma^2 = \int_{-\infty}^{+\infty} [f(X)]X^2\, dX - \mu^2 \tag{9.26}$$

Expressed verbally, Eq. (9.26) states

$$\text{Variance} = \text{mean of } X^2 \text{ values} - (\text{mean of } X)^2$$

EXAMPLE 9.36 A continuous random variable X has the following probability-density function:

$$f(X) = 0.024X^2 \qquad \text{if } 0 < X \le 5$$

$$f(X) = 0 \qquad \text{elsewhere}$$

a. Prove that the equation for $f(X)$ is valid.

b. Compute the arithmetic mean and standard deviation of this distribution.

c. Compute the probability that, on a given occasion, X will assume a value lying between 2 and 4.

SOLUTION

Part a

$$\int_0^5 f(X)\, dX = \frac{(0.024)5^3}{3} = 1$$

Since Eq. (9.23) is satisfied, the equation for $f(X)$ is valid.

Part b. By Eqs. (9.24) and (9.26),

$$\mu = 0.024 \int_0^5 X^3\, dX = \frac{(0.024)5^4}{4} = 3.75$$

$$\sigma^2 = 0.024 \int_0^5 X^4\,dX - 3.75^2 = \frac{(0.024)\,5^5}{5} - 3.75^2 = 0.9375$$

$$\sigma = \sqrt{0.9375} = 0.968$$

Part c

$$P(2 \le X \le 4) = \int_2^4 f(X)\,dX = \frac{(0.024)(4^3 - 2^3)}{3} = 0.448$$

9.3.2 Uniform Distribution

A uniform probability distribution is one in which the probability-density function is constant. Assume that X can range from a to b, inclusive, where a and b are constants and $b > a$. Then

$$f(X) = \frac{1}{b-a} \qquad \text{if } a \le X \le b$$
$$f(X) = 0 \qquad \text{elsewhere} \tag{9.27}$$

In this special case, the probability "curve" is a straight line, as shown in Fig. 9.6.

EXAMPLE 9.37 It is known that a signal will be received sometime between 6 and 11 a.m., and it is just as likely to be received at one instant within that interval as any other. What is the probability that the signal will be received between 7:05 and 9:20 a.m.?

SOLUTION The total interval is 5 h, and the restricted interval is 2.25 h. Therefore, the probability of the specified event is $2.25/5 = 0.45$.

The arithmetic mean and standard deviation of the uniform distribution are as follows:

$$\mu = \frac{a+b}{2} \tag{9.28}$$

$$\sigma = \frac{b-a}{\sqrt{12}} \tag{9.29}$$

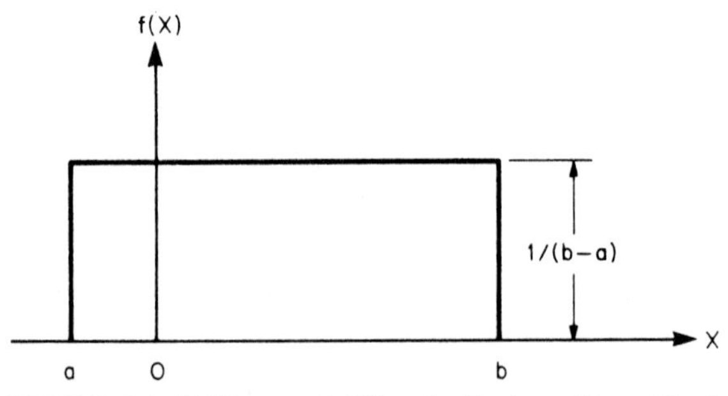

FIGURE 9.6 Uniform probability distribution. (*From Handbook of Engineering Economics; used with permission of McGraw-Hill, Inc.*)

9.3.3 Normal Distribution

A continuous random variable is said to have a *normal* or *gaussian probability distribution* if the range of its possible values is infinite and its probability-density function has this form:

$$f(X) = \frac{1}{b\sqrt{2\pi}} e^{-K}$$

where

$$K = \frac{(X-a)^2}{2b^2} \qquad (9.30)$$

In this equation, e is the quantity defined in Art. 1.28, and a and b are constants.

Equation (9.30) was first formulated by DeMoivre in solving a problem in gambling, but it was later discovered that this equation also describes a vast number of random variables that appear in natural phenomena.

Figure 9.7*a* is the normal probability curve for assumed values of a and b. This bell-shaped curve has a summit at $X = a$, and it is symmetrical about the vertical line through the summit. Thus, the constant a in Eq. (9.30) is the arithmetic mean

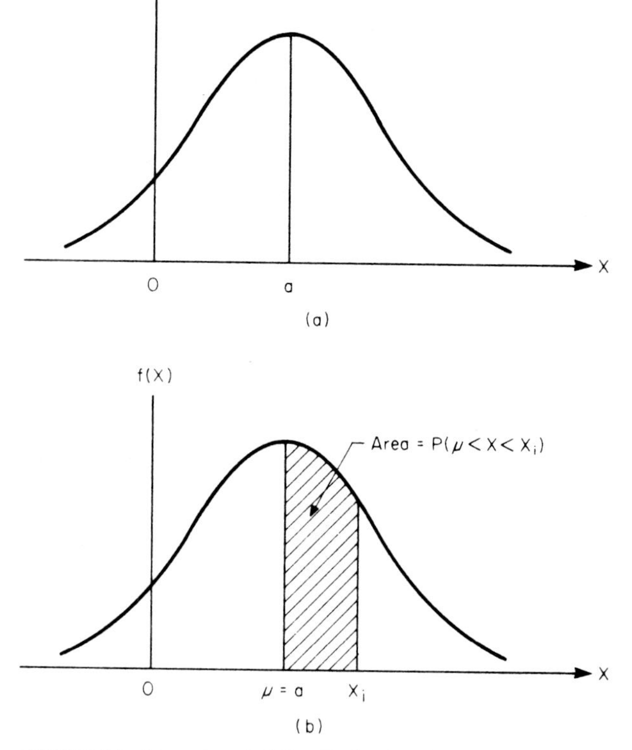

FIGURE 9.7 Normal probability distribution. (*a*) Probability curve; (*b*) method of measuring area under curve. (*From Handbook of Engineering Economics; used with permission of McGraw-Hill, Inc.*)

of the normal probability distribution. It can also be demonstrated that the constant b is the standard deviation. Expressed symbolically,

$$\mu = a \qquad \sigma = b$$

It is convenient to express the width of an interval extending from μ to a given value X_i in *standard units* by dividing this interval by the standard deviation. The width is then denoted by z_i. Thus,

$$z_i = \frac{X_i - \mu}{\sigma} \tag{9.31}$$

Table 9.4 presents the area of an interval under the normal curve, where one boundary of the interval lies at the centerline, as shown in Fig. 9.7*b*, and the width of the interval is expressed in standard units as given by Eq. (9.31). Since the curve is symmetrical about its centerline, corresponding positive and negative values of z_i have equal areas. Since the total area under the curve is 1, the area lying on either side of the centerline is 0.5.

EXAMPLE 9.38 A continuous random variable X having a normal probability distribution is known to have an arithmetic mean of 14 and a standard deviation of 2.5. What is the probability that, on a given occasion, (*a*) X lies between 14 and 17, (*b*) X lies between 12 and 16.2, and (*c*) X is less than 10?

SOLUTION Refer to Fig. 9.8.

$$\mu = 14 \qquad \sigma = 2.5$$

Part a

$$z = \frac{17 - 14}{2.5} = 1.20$$

From Table 9.4, the corresponding area is 0.38493. Then

$$P(14 \leq X \leq 17) = 0.38493$$

Part b. Resolve the interval into two parts by cutting it at $X = \mu = 14$. For the first part,

$$z = \frac{12 - 14}{2.5} = -0.8 \qquad \text{Area} = 0.28814$$

For the second part,

$$z = \frac{16.2 - 14}{2.5} = 0.88 \qquad \text{Area} = 0.31057$$

$$\therefore\ P(12 \leq X \leq 16.2) = 0.28814 + 0.31057 = 0.59871$$

Part c. First take the interval from $X = 10$ to $X = 14$.

$$z = \frac{10 - 14}{2.5} = -1.6 \qquad \text{Area} = 0.44520$$

$$\therefore\ P(10 \leq X \leq 14) = 0.44520$$

$$P(X < 10) = 0.5 - 0.44520 = 0.05480$$

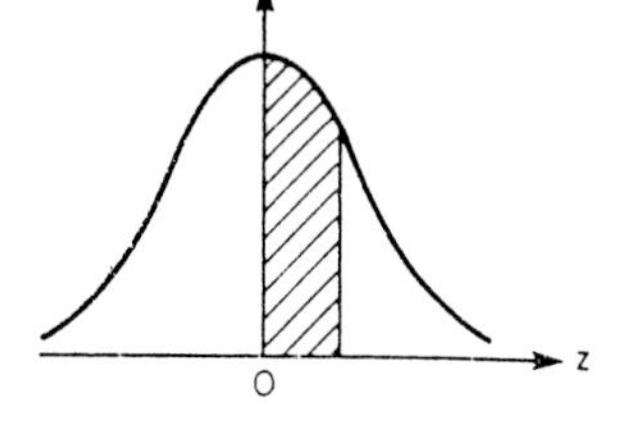

TABLE 9.4 Area under the Standard Normal Curve

Table presents value of area from $z = 0$ to indicated value of z.

z	.00	.01	.02	.03	.04	.05	.06	.07	.08	.09
0.0	.00000	.00399	.00798	.01197	.01595	.01994	.02392	.02790	.03188	.03586
0.1	.03983	.04380	.04776	.05172	.05567	.05962	.06356	.06749	.07142	.07535
0.2	.07926	.08317	.08706	.09095	.09483	.09871	.10257	.10642	.11026	.11409
0.3	.11791	.12172	.12552	.12930	.13307	.13683	.14058	.14431	.14803	.15173
0.4	.15542	.15910	.16276	.16640	.17003	.17364	.17724	.18082	.18439	.18793
0.5	.19146	.19497	.19847	.20194	.20540	.20884	.21226	.21566	.21904	.22240
0.6	.22575	.22907	.23237	.23565	.23891	.24215	.24537	.24857	.25175	.25490
0.7	.25804	.26115	.26424	.26730	.27035	.27337	.27637	.27935	.28230	.28524
0.8	.28814	.29103	.29389	.29673	.29955	.30234	.30511	.30785	.31057	.31327
0.9	.31594	.31859	.32121	.32381	.32639	.32894	.33147	.33398	.33646	.33891
1.0	.34134	.34375	.34614	.34850	.35083	.35314	.35543	.35769	.35993	.36214
1.1	.36433	.36650	.36864	.37076	.37286	.37493	.37698	.37900	.38100	.38298
1.2	.38493	.38686	.38877	.39065	.39251	.39435	.39617	.39796	.39973	.40147
1.3	.40320	.40490	.40658	.40824	.40988	.41149	.41309	.41466	.41621	.41774
1.4	.41924	.42073	.42220	.42364	.42507	.42647	.42786	.42922	.43056	.43189
1.5	.43319	.43448	.43574	.43699	.43822	.43943	.44062	.44179	.44295	.44408
1.6	.44520	.44630	.44738	.44845	.44950	.45053	.45154	.45254	.45352	.45449
1.7	.45543	.45637	.45728	.45818	.45907	.45994	.46080	.46164	.46246	.46327
1.8	.46407	.46485	.46562	.46638	.46712	.46784	.46856	.46926	.46995	.47062
1.9	.47128	.47193	.47257	.47320	.47381	.47441	.47500	.47558	.47615	.47670
2.0	.47725	.47778	.47831	.47882	.47932	.47982	.48030	.48077	.48124	.48169
2.1	.48214	.48257	.48300	.48341	.48382	.48422	.48461	.48500	.48537	.48574
2.2	.48610	.48645	.48679	.48713	.48745	.48778	.48809	.48840	.48870	.48899
2.3	.48928	.48956	.48983	.49010	.49036	.49061	.49086	.49111	.49134	.49158
2.4	.49180	.49202	.49224	.49245	.49266	.49286	.49305	.49324	.49343	.49361
2.5	.49379	.49396	.49413	.49430	.49446	.49461	.49477	.49492	.49506	.49520
2.6	.49534	.49547	.49560	.49573	.49585	.49598	.49609	.49621	.49632	.49643
2.7	.49653	.49664	.49674	.49683	.49693	.49702	.49711	.49720	.49728	.49736
2.8	.49744	.49752	.49760	.49767	.49774	.49781	.49788	.49795	.49801	.49807
2.9	.49813	.49819	.49825	.49831	.49386	.49841	.49846	.49851	.49856	.49861
3.0	.49865	.49869	.49874	.49878	.49882	.49886	.49889	.49893	.49897	.49900
3.1	.49903	.49906	.49910	.49913	.49916	.49918	.49921	.49924	.49926	.49929
3.2	.49931	.49934	.49936	.49938	.49940	.49942	.49944	.49946	.49948	.49950
3.3	.49952	.49953	.49955	.49957	.49958	.49960	.49961	.49962	.49964	.49965
3.4	.49966	.49968	.49969	.49970	.49971	.49972	.49973	.49974	.49975	.49976
3.5	.49977	.49978	.49978	.49979	.49980	.49981	.49981	.49982	.49983	.49983
3.6	.49984	.49985	.49985	.49986	.49986	.49987	.49987	.49988	.49988	.49989
3.7	.49989	.49990	.49990	.49990	.49991	.49991	.49992	.49992	.49992	.49992
3.8	.49993	.49993	.49993	.49994	.49994	.49994	.49994	.49995	.49995	.49995
3.9	.49995	.49995	.49996	.49996	.49996	.49996	.49996	.49996	.49997	.49997
4.0	.49997									

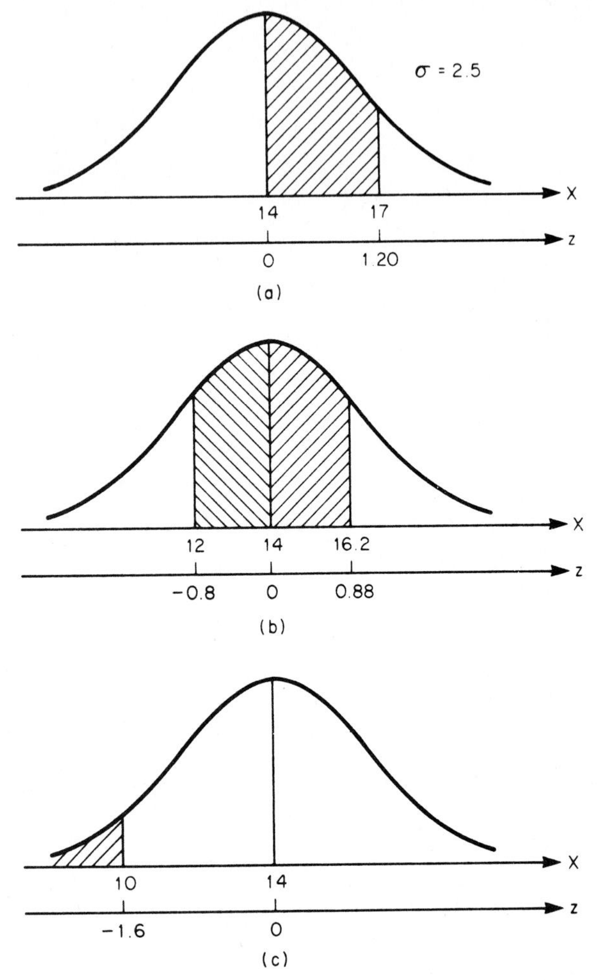

FIGURE 9.8 Determination of probability that (*a*) *X* lies between 14 and 17, (*b*) *X* lies between 12 and 16.2, and (*c*) *X* is less than 10. (*From Handbook of Engineering Economics; used with permission of McGraw-Hill, Inc.*)

In theory, a random variable can have a normal distribution only if its range of allowable values is unrestricted. In practice, however, many variables having a restricted range of allowable values are also considered to have a normal distribution.

EXAMPLE 9.39 A firm manufactures cylindrical machine parts. The diameter of the part is assumed to be normally distributed, with a mean of 8.350 cm (3.287 in) and a standard deviation of 0.093 cm (0.0366 in). A part is considered satisfactory if its diameter lies between 8.205 cm (3.230 in) and 8.490 cm (3.343 in). What is the proportion of defective parts?

SOLUTION Let X denote the diameter, and set $X = 8.205$ cm. Then

$$z = \frac{8.205 - 8.350}{0.093} = -1.559 \qquad \text{Area} = 0.4405$$

$$P(X < 8.205) = 0.5 - 0.4405 = 0.0595$$

Now set $X = 8.490$ cm. Then

$$z = \frac{8.490 - 8.350}{0.093} = 1.505 \qquad \text{Area} = 0.4338$$

$$P(X > 8.490) = 0.5 - 0.4338 = 0.0662$$

By summation, the probability that a part selected at random is defective is

$$P(\text{defective}) = 0.0595 + 0.0662 = 0.1257$$

Since probability represents relative frequency in the long run, the proportion of defective parts is 12.57 percent.

The binomial probability distribution can be approximated by the normal distribution where the number of trials n is large and the value of P is not close to 0 or 1. The rule generally followed is to apply this approximation if both nP and nQ exceed 5.

EXAMPLE 9.40 When a projectile is fired from a given location, the probability that it lands in a target area is 0.30. If 210 projectiles are fired, what is the probability (to three decimal places) that more than 66 land in the target area?

SOLUTION Let X denote the number of projectiles that land in the target area. Since all firings are independent of one another, the probability distribution of X is binomial. However, since an exact appraisal of probability would be prohibitively arduous, a normal approximation will be applied. For this purpose, it is necessary to transform X from a discrete to a continuous variable. This is accomplished by replacing each integral value by an *interval* that extends one-half unit below and above that value. For example, the value $X = 66$ is replaced by $65.5 \leq X \leq 66.5$. By Eqs. (9.7) and (9.8),

$$\mu = 210(0.30) = 63 \qquad \sigma = \sqrt{63(0.70)} = 6.64$$

Since we require the probability of values of X greater than 66, we set $X = 66.5$.

$$z = \frac{66.5 - 63}{6.64} = 0.527 \qquad \text{Area} = 0.201$$

$$P(X > 66.5) = 0.5 - 0.201 = 0.299$$

9.3.4 Negative-Exponential Distribution

A random variable X is said to have a *negative-exponential* (or simply *exponential*) probability distribution if its probability-density function $f(X)$ is of this form:

$$\begin{aligned} f(X) &= ae^{-aX} && \text{if } X \geq 0 \\ f(X) &= 0 && \text{if } X < 0 \end{aligned} \tag{9.32}$$

where a is a positive constant and e is the quantity defined in Art. 1.28. The probability curve of X appears in Fig. 9.9a. Many mechanical and electronic devices have negative-exponential life spans, and the quantity of a radioactive substance that remains at a given time has an equation of the same form as Eq. (9.32).

Let K denote a positive constant. By integrating $f(X)\,dX$ between the limits of 0 and K, we obtain

$$P(X \leq K) = 1 - e^{-aK} \tag{9.33a}$$

$$\therefore\ P(X > K) = e^{-aK} \tag{9.33b}$$

Equation (9.33a) expresses *cumulative probability*, and the graph of this equation appears in Fig. 9.9b.

The arithmetic mean and standard deviation of the negative-exponential distribution are as follows:

$$\mu = \sigma = \frac{1}{a} \tag{9.34}$$

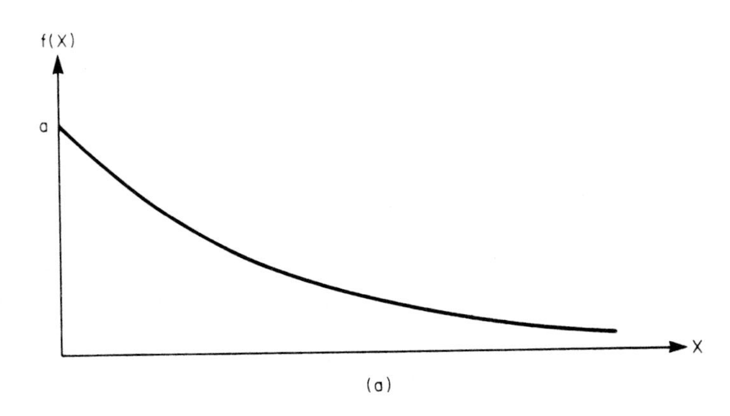

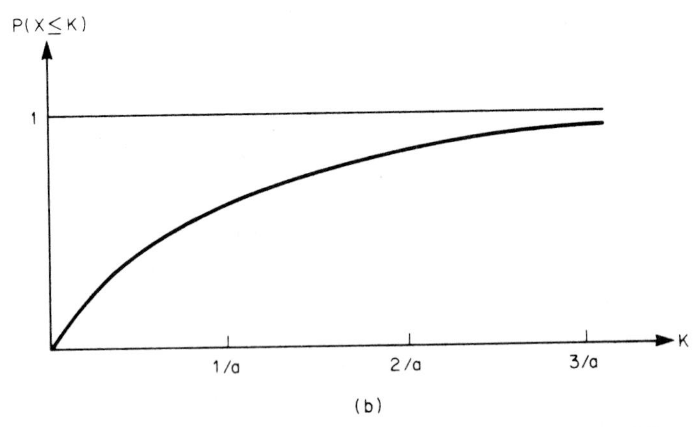

FIGURE 9.9 Negative-exponential probability distribution. (a) Probability curve: (b) cumulative-probability curve. (*From Handbook of Engineering Economics; used with permission of McGraw-Hill, Inc.*)

EXAMPLE 9.41 The life span of a mechanism that operates continuously has a negative-exponential distribution, and the mean life span is eight days.

a. What is the probability that the life span will exceed nine days?

b. If the mechanism has been in operation for the past five days, what is the probability that it will still be operating nine days hence?

SOLUTION Let X denote the life span in days. By Eq. (9.34), $a = 1/8 = 0.125$.
Part a

$$P(X > 9) = e^{-(0.125)9} = 0.3247$$

Part b. We shall recast the question in this form: Given that $X > 5$, what is the probability that $X > 14$? Since the life span must exceed 5 days if it is to exceed 14 days, we apply the special form of Bayes' theorem as embodied in Eq. (9.2*a*).

$$P(X > 14 | X > 5) = \frac{P(X > 14)}{P(X > 5)} = \frac{e^{-(0.125)14}}{e^{-(0.125)5}} = e^{-(0.125)9}$$
$$= 0.3247$$

Alternatively, this result can be obtained by equating probability to relative frequency and applying this reasoning: Assume that a vast number of such mechanisms are activated simultaneously. The proportion that survive the first 5 days is $e^{-(0.125)5} = 0.535261$, and the proportion that survive the first 14 days is $e^{-(0.125)14} = 0.173774$. Now consider the mechanisms that have survived the first 5 days. The proportion of these that will survive the following 9 days is $0.173774/0.535261 = 0.3247$, and this is the required probability.

The values obtained in Parts *a* and *b* of Example 9.41 are equal, and it is apparent that this equality is general. If a mechanism has a negative-exponential life span, the probability that it will survive the next m time units is independent of its present life. Thus, the device does not "age" as it operates.

Assume that a time or space event has a Poisson probability distribution and let U denote the interval between successive events. The quantity U is a continuous random variable, and the probability that U will exceed a given value K is given by Eq. (9.20). A comparison of this equation with Eq. (9.33*b*) reveals that U has a negative-exponential distribution, with the parameter a in Eq. (9.32) equal to the expected number of events in 1 time unit.

9.3.5 Gamma Distribution

A random variable X is said to have a *gamma* probability distribution if its probability-density function $f(X)$ is of this form:

$$f(X) = \frac{a^b X^{b-1} e^{-aX}}{(b-1)!} \qquad \text{if } X \geq 0$$
$$f(X) = 0 \qquad \text{if } X < 0 \tag{9.35}$$

where a and b are positive constants. If $b = 1$, the first equation reduces to $f(X) = ae^{-aX}$, and therefore the negative-exponential distribution is a special case of the gamma distribution. The arithmetic mean and standard deviation of the gamma distribution are as follows:

$$\mu = \frac{b}{a} \qquad \sigma = \frac{\sqrt{b}}{a} \tag{9.36}$$

It follows that

$$a = \frac{\mu}{\sigma^2} \quad \text{and} \quad b = \left(\frac{\mu}{\sigma}\right)^2 \tag{9.36a}$$

Figure 9.10 is the probability curve of a random variable that has a gamma distribution with $b = 3$. Unlike the normal curve, this gamma curve is unsymmetrical, being skewed to the right.

If b is an integer, the cumulative probability of X can be found by applying the following equation:

$$P(X \leq K) = 1 - e^{-aK} \sum_{n=0}^{b-1} \frac{(aK)^n}{n!} \tag{9.37}$$

where K is positive. If b has a nonintegral value, cumulative probabilities are obtained by referring to a table.

EXAMPLE 9.42 A random variable X has a gamma distribution with a mean of 20 and standard deviation of 10. Compute the probability that X lies between 12 and 18.

SOLUTION By Eq. (9.36*a*),

$$a = 20/100 = 0.2 \qquad b = (20/10)^2 = 4$$

FIGURE 9.10 Curve of gamma distribution with $b = 3$. *(From Handbook of Engineering Economics; used with permission of McGraw-Hill, Inc.)*

When $K = 12$, $aK = 2.4$, and Eq. (9.37) yields

$$P(X \leq 12) = 1 - e^{-2.4}\left[1 + 2.4 + \frac{(2.4)^2}{2} + \frac{(2.4)^3}{6}\right] = 0.2213$$

When $K = 18$, $aK = 3.6$, and

$$P(X \leq 18) = 1 - e^{-3.6}\left[1 + 3.6 + \frac{(3.6)^2}{2} + \frac{(3.6)^3}{6}\right] = 0.4848$$

$$\therefore\ P(12 \leq X \leq 18) = 0.4848 - 0.2213 = 0.2635$$

For simplicity, we initially defined the gamma distribution by assuming that the lower bound of X is 0, but the definition can be broadened. In general, let c denote the lower bound of X, where c is a constant. By replacing X with $X - c$ on the right side of Eq. (9.35), we extend the definition of the gamma distribution to a variable X that can range from c to infinity. The arithmetic mean μ of the distribution increases by c, but the standard deviation is not affected.

9.4 MARKOV PROBABILITY

9.4.1 Description of a Markov Process

Assume the following: A given trial will be performed repeatedly, the outcome of each trial is directly influenced by the N preceding outcomes, and the character of this influence remains unchanged as the number of trials increases. A trial having this characteristic is called a *Markov process*, and it is of the Nth order. A set of consecutive outcomes resulting from a Markov process is termed a *Markov chain*. We shall confine our investigation to a Markov process of the first order, where each outcome is directly influenced solely by its immediate predecessor (although *indirectly* it is influenced by *all* preceding outcomes).

As an illustration, assume the following: A Markov process has two possible outcomes, A and B. If the $(n-1)$th outcome is A, the probability that the nth outcome will also be A is 0.72; if the $(n-1)$th outcome is B, the probability that the nth outcome will also be B is 0.59. We shall apply this notation: A_p and B_q denote, respectively, that the pth outcome is A and the qth outcome is B. Similarly, $P(A_p)$ denotes the probability that the pth outcome will be A. Applying the notation for conditional probability presented in Art. 9.1.4, we may record the given probabilities in this manner:

$$P(A_n|A_{n-1}) = 0.72 \qquad P(B_n|B_{n-1}) = 0.59$$

These conditional probabilities are termed *transition probabilities*. It is convenient to record them in the form of a matrix, termed a *transition matrix*, in which each element is the probability associated with the indicated pair of successive outcomes. Table 9.5 is the transition matrix for the present system.

A Markov chain is denoted by use of hyphens. For example, the notation B_1-E_2-A_3 denotes that the first outcome is B, the second outcome is E, and the third outcome is A.

During the interval of time between successive trials, the system that generates an outcome is said to be in a *state* corresponding to the most recent outcome. For example, if the possible outcomes are red, green, and blue and the most recent

TABLE 9.5 Transition Matrix

$(n-1)$th outcome	nth outcome	
	A_n	B_n
A_{n-1}	0.72	0.28
B_{n-1}	0.41	0.59

outcome was blue, the system is in the state blue until the next trial occurs. Let m denote the number of trials performed in the past. While m has a finite value, the system is in a *transient state*; as m increases without limit, the system approaches its *steady state*. As we shall find, each system has steady-state probabilities that are governed solely by the transition probabilities.

9.4.2 Transient Conditions with Two Possible Outcomes

We undertake our study of first-order Markov processes by considering the simplest type of process: one that has only two possible outcomes.

EXAMPLE 9.43 A mechanism contains two sounding devices, A and B. The mechanism has been constructed in a manner that causes one (and only one) device to emit a sound at the end of every 1-min interval. However, whether it is A or B that emits the sound on a given occasion is determined partly by random factors. An analysis of past performance reveals the following: The probability that a sound from A will be followed immediately by another sound from A is 0.75; the probability that a sound from B will be followed immediately by another sound from B is 0.60. Calculate the probabilities corresponding to the next four sounds.

SOLUTION The outcome of a trial is the emission of a sound, and there are two possible outcomes, A and B. Table 9.6 is the transition matrix. The outcomes will be numbered consecutively, starting with 0 for the most recent outcome and 1 for the first outcome in the future.

Method 1. From Table 9.6, we have the following:

$$P(A_n) = 0.75P(A_{n-1}) + 0.40P(B_{n-1})$$

Replacing $P(B_{n-1})$ with $1 - P(A_{n-1})$ and simplifying, we obtain

$$P(A_n) = 0.35P(A_{n-1}) + 0.40 \qquad (a)$$

Setting $n = 1$ and applying Eq. (a), we obtain

$$P(A_1) = 0.35P(A_0) + 0.40 \qquad (b)$$

TABLE 9.6 Transition Matrix

$(n-1)$th outcome	nth outcome	
	A_n	B_n
A_{n-1}	0.75	0.25
B_{n-1}	0.40	0.60

Now setting $n = 2$, applying Eq. (a), and replacing $P(\text{A}_1)$ with its expression in Eq. (b), we obtain

$$P(\text{A}_2) = 0.1225P(\text{A}_0) + 0.54 \qquad (c)$$

Continuing this process, we obtain

$$P(\text{A}_3) = 0.0429P(\text{A}_0) + 0.589 \qquad (d)$$

$$P(\text{A}_4) = 0.0150P(\text{A}_0) + 0.6062 \qquad (e)$$

Assume that the most recent outcome was A. Setting $P(\text{A}_0) = 1$ and substituting in the foregoing equations, we obtain three results:

$$P(\text{A}_1) = 0.75 \qquad P(\text{A}_2) = 0.6625$$

$$P(\text{A}_3) = 0.6319 \qquad P(\text{A}_4) = 0.6212$$

Now assume that the most recent outcome was B. Setting $P(\text{A}_0) = 0$ and substituting in the foregoing equations, we obtain these results:

$$P(\text{A}_1) = 0.40 \qquad P(\text{A}_2) = 0.54$$

$$P(\text{A}_3) = 0.589 \qquad P(\text{A}_4) = 0.6062$$

The points representing these probabilities are plotted in Fig. 9.11, and they are

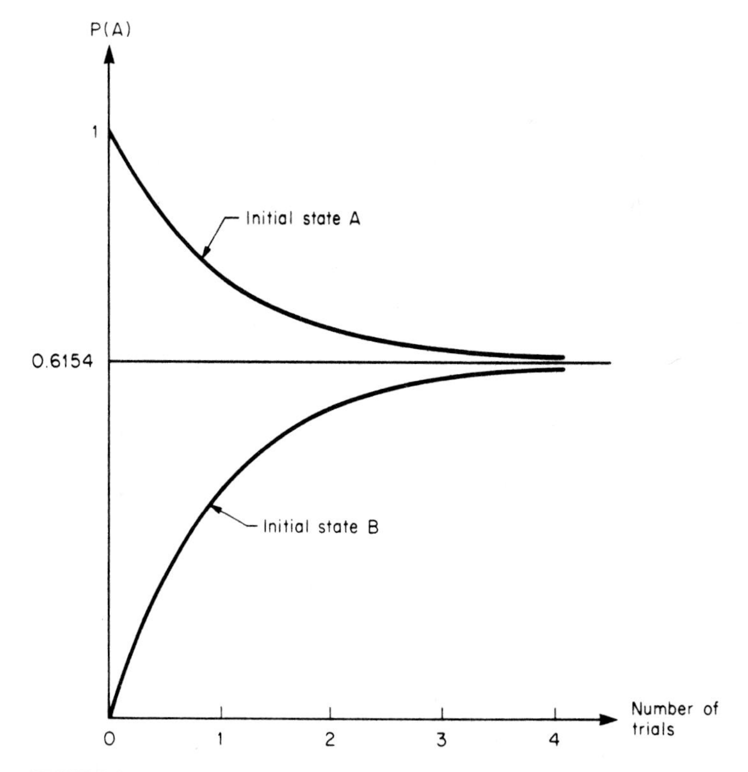

FIGURE 9.11 Probabilities of outcome A in Markov chain. (*From Handbook of Engineering Economics; used with permission of McGraw-Hill, Inc.*)

then connected with smooth curves. We shall demonstrate in Art. 9.4.3 that the two sets of values converge to the limiting probability of 0.6154.

Method 2. The probabilities concerning future outcomes can be obtained by constructing a *probability tree*, which exhibits every possible chain of outcomes. Refer to Fig. 9.12, which is the probability tree for the first three outcomes in the future as based on the assumption that the most recent outcome was **B**. An outcome is represented by a node, and two successive outcomes are connected by a branch. The probability that the first outcome will be followed by the second outcome is recorded directly above the branch.

The probability of obtaining a given chain of outcomes is found by multiplying all probabilities along that chain, in accordance with Theorem 9.7. For example,

$$P(\mathrm{B}_0\text{-}\mathrm{A}_1\text{-}\mathrm{B}_2\text{-}\mathrm{B}_3) = (0.40)(0.25)(0.60) = 0.060$$

In the probability tree, the probability associated with each chain is recorded directly to the right of the last outcome in the chain. By summing the probabilities

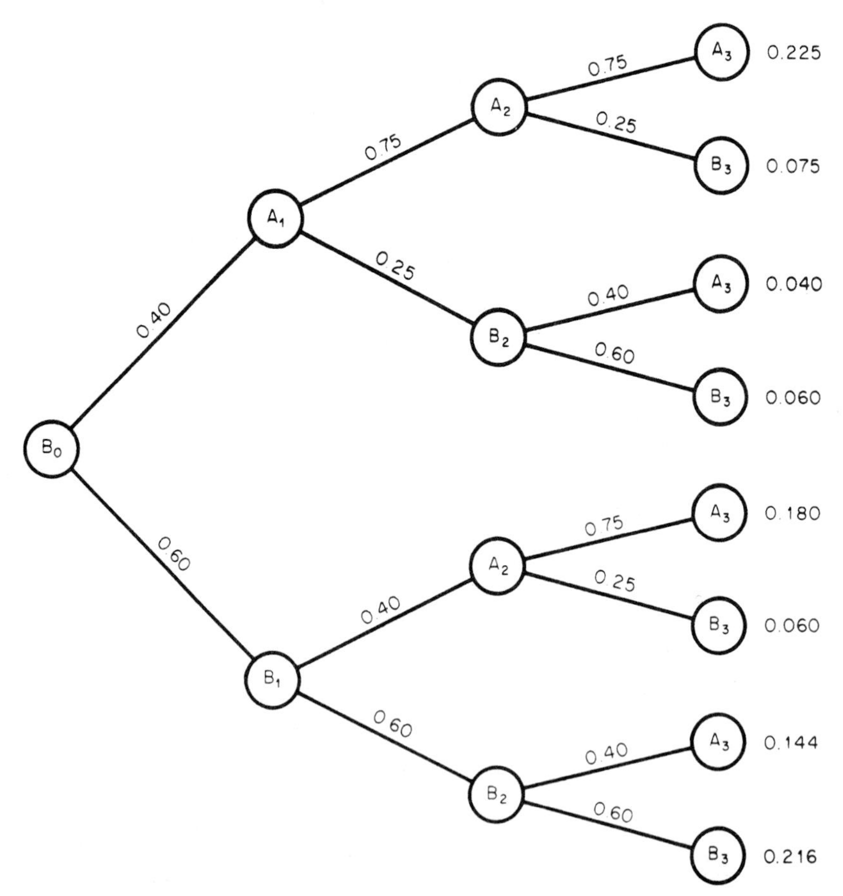

FIGURE 9.12 Probability tree for Markov chain with initial outcome B. (*From Handbook of Engineering Economics; used with permission of McGraw-Hill, Inc.*)

associated with all chains culminating in A_3, we obtain

$$P(A_3|B_0) = 0.225 + 0.040 + 0.180 + 0.144 = 0.589$$

This result agrees with that obtained by Method 1. We can now construct a probability tree corresponding to A_0.

Figure 9.13 may be constructed as an alternative to the probability tree in Fig. 9.12. Each possible future outcome is represented by a box, and the probability that that outcome will materialize is recorded directly in the box. Each possible chain of outcomes is connected by rays, and the probability that a particular pair of successive outcomes will materialize is recorded directly above the corresponding ray.

Assume again that the most recent outcome was B. Then $P(B_0\text{-}A_1) = 0.40$ and $P(B_0\text{-}B_1) = 0.60$. These values are recorded in the manner shown, and $P(A_1)$ and $P(B_1)$ are recorded in their respective boxes. Assume that A_1 materializes. Then $P(A_1\text{-}A_2) = (0.40)(0.75) = 0.30$ and $P(A_1\text{-}B_2) = (0.40)(0.25) = 0.10$. The probabilities of $B_1\text{-}A_2$ and $B_1\text{-}B_2$ are found in a similar manner. We now have $P(A_2) = 0.30 + 0.24 = 0.54$ and $P(B_2) = 0.10 + 0.36 = 0.46$. As a check on our calculations to this point, we have $P(A_2) + P(B_2) = 0.54 + 0.46 = 1$. Continuing in this manner, we obtain the probabilities pertaining to the third and fourth outcomes, and the results agree with those previously obtained. The principal advantage of a diagram such as that in Fig. 9.13, as compared with a probability tree, is that the former remains compact in the vertical direction.

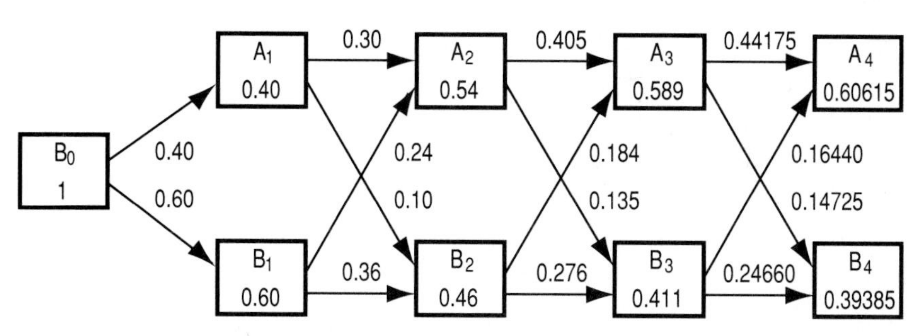

FIGURE 9.13 Probabilities corresponding to Markov chain with initial outcome B.

9.4.3 Steady-State Conditions with Two Possible Outcomes

Assume again that a Markov process has only two possible outcomes, A and B, and that the transition probabilities remain constant as the trial is performed repeatedly. We shall demonstrate that the probabilities concerning a particular outcome approach limiting values as the number of trials increases beyond bound. Let

$$P(A_n|A_{n-1}) = a \qquad \text{and} \qquad P(A_n|B_{n-1}) = b$$

Then

$$P(A_n) = aP(A_{n-1}) + bP(B_{n-1})$$
$$= aP(A_{n-1}) + b[1 - P(A_{n-1})]$$

$$P(A_n) = (a - b)P(A_{n-1}) + b \tag{9.38}$$

Proceeding as in Example 9.43, we obtain the following:

$$P(\mathrm{A}_n) = (a-b)^n P(\mathrm{A}_0) + \frac{b[1-(a-b)^n]}{1-(a-b)}$$

Now consider that n becomes infinitely large. Since $|a-b| < 1$, it follows that $(a-b)^n$ vanishes. Let $P^*(\mathrm{A})$ denote the limiting value of $P(\mathrm{A}_n)$ as n becomes infinitely large. Then

$$P^*(\mathrm{A}) = \frac{b}{1-a+b} \tag{9.39}$$

This result reveals that the limiting value of $P(\mathrm{A}_n)$ is governed solely by the inherent characteristics of the system, which we assume remain constant. It is independent of the present state of the system, which is ephemeral.

Alternatively, Eq. (9.39) can be obtained directly from Eq. (9.38). Since the difference between $P(\mathrm{A}_n)$ and $P(\mathrm{A}_{n-1})$ vanishes in the long run, we may replace $P(\mathrm{A}_{n-1})$ in Eq. (9.38) with $P(\mathrm{A}_n)$. When the resulting equation is solved for $P(\mathrm{A}_n)$, Eq. (9.39) emerges.

EXAMPLE 9.44 With reference to Example 9.43, find the limiting values of $P(\mathrm{A}_n)$ and $P(\mathrm{B}_n)$.

SOLUTION We apply Eq. (9.39) with $a = 0.75$ and $b = 0.40$. Then

$$P^*(\mathrm{A}) = \frac{0.40}{1-0.75+0.40} = 0.6154$$

$$P^*(\mathrm{B}) = 1 - 0.6154 = 0.3846$$

The steady-state probabilities can also be found by constructing a *recurrent chain*. As an illustration, consider this Markov chain:

A-A-A-B-B-A-B-B-B-A-A-A-A

Assume that this chain is typical and will therefore recur endlessly. Thus, the final outcome in this chain will be followed by A. By counting the successors of these 13 outcomes, we arrive at these results: A is followed by A six times, A is followed by B two times, B is followed by A two times, and B is followed by B three times. Therefore, the transition probabilities are as follows:

$$P(\mathrm{A}_n|\mathrm{A}_{n-1}) = \frac{6}{6+2} = 0.75$$

$$P(\mathrm{A}_n|\mathrm{B}_{n-1}) = \frac{2}{2+3} = 0.40$$

These transition probabilities coincide with those in Example 9.43.

We may now compute the probability of A_n on the basis of relative frequency, without reference to any outcome in the past. The recurrent chain contains eight A's and five B's. Therefore, the probability of A_n is $8/13 = 0.6154$. This result agrees with that obtained by applying Eq. (9.39).

9.4.4 Transient Conditions for the General Markov Process

We shall now investigate the general Markov process, where the number of possible outcomes can have any integral value greater than 1. Our study will involve the use of matrix algebra, which is presented in Sec. 5.

With reference to Table 9.6, consider that we replace $n-1$ with 0 and replace n with 1; we now label this transition matrix **P**. If we calculate all probabilities corresponding to outcome 2 and scrutinize the calculations, we find that the probabilities for outcome 2 can be obtained by multiplying **P** by itself. This conclusion is entirely general, and it leads to the following principle:

Theorem 9.8. If **P** denotes the transition matrix for a Markov process, the probabilities corresponding to the nth outcome in the future are the elements of the matrix $\mathbf{P}^n$.

EXAMPLE 9.45 A Markov process has four possible outcomes, A, B, C, and D, and the transition matrix appears in Table 9.7. Determine the probabilities corresponding to the third outcome in the future.

SOLUTION Let **P** denote the matrix in Table 9.7 but with the subscripts $n-1$ and n replaced with 0 and 1, respectively. In recording a matrix, we shall include its row and column designations. Then

$$\mathbf{P}^2 = \begin{array}{c} \\ A_0 \\ B_0 \\ C_0 \\ D_0 \end{array} \begin{array}{c} \begin{array}{cccc} A_2 & B_2 & C_2 & D_2 \end{array} \\ \begin{bmatrix} 0.3675 & 0.3250 & 0.1700 & 0.1375 \\ 0.3000 & 0.3200 & 0.1375 & 0.2425 \\ 0.2850 & 0.3400 & 0.1775 & 0.1975 \\ 0.5775 & 0.2250 & 0.0900 & 0.1075 \end{bmatrix} \end{array}$$

$$\mathbf{P}^3 = \begin{array}{c} \\ A_0 \\ B_0 \\ C_0 \\ D_0 \end{array} \begin{array}{c} \begin{array}{cccc} A_3 & B_3 & C_3 & D_3 \end{array} \\ \begin{bmatrix} 0.3945 & 0.2955 & 0.1358 & 0.1743 \\ 0.3596 & 0.3208 & 0.1619 & 0.1578 \\ 0.3926 & 0.3038 & 0.1459 & 0.1578 \\ 0.3225 & 0.3175 & 0.1468 & 0.2133 \end{bmatrix} \end{array}$$

The probabilities corresponding to the third outcome are given by $\mathbf{P}^3$. For example, if the initial state of the system is B, then $P(C_3) = 0.1619$; if the initial state of the system is D, then $\mathbf{P}(A_3) = 0.3225$. In matrices $\mathbf{P}^2$ and $\mathbf{P}^3$, the elements in each row have a sum of 1, as they must.

TABLE 9.7 Transition Matrix

	A_n	B_n	C_n	D_n
A_{n-1}	0.20	0.35	0.15	0.30
B_{n-1}	0.70	0.15	0.10	0.05
C_{n-1}	0.55	0.25	0	0.20
D_{n-1}	0	0.55	0.35	0.10

9.4.5 Steady-State Conditions for the General Markov Process

We now consider the problem of evaluating the steady-state probabilities of a Markov process where the number of possible outcomes can have any value whatever. To illustrate the procedure, we shall apply the data in Example 9.45, where there are four possible outcomes.

Continuing the previous notation, let $P^*(\mathrm{A})$ denote the probability of outcome A in the long run. Expanding the notation in Art. 9.4.3, let

$$P(\mathrm{A}_n|\mathrm{A}_{n-1}) = a \qquad P(\mathrm{A}_n|\mathrm{B}_{n-1}) = b$$

$$P(\mathrm{A}_n|\mathrm{C}_{n-1}) = c \qquad P(\mathrm{A}_n|\mathrm{D}_{n-1}) = d$$

We now have

$$P(\mathrm{A}_n) = aP(\mathrm{A}_{n-1}) + bP(\mathrm{B}_{n-1}) + cP(\mathrm{C}_{n-1}) + dP(\mathrm{D}_{n-1})$$

As n becomes infinitely large, the difference between $P(\mathrm{A}_n)$ and $P(\mathrm{A}_{n-1})$ vanishes, and $P(\mathrm{A}_n)$ approaches $P^*(\mathrm{A})$. Therefore, we may replace all probabilities in the foregoing equation with the corresponding steady-state probabilities, and the result is

$$P^*(\mathrm{A}) = aP^*(\mathrm{A}) + bP^*(\mathrm{B}) + cP^*(\mathrm{C}) + dP^*(\mathrm{D})$$

Transposing, we have

$$-(1-a)P^*(\mathrm{A}) + bP^*(\mathrm{B}) + cP^*(\mathrm{C}) + dP^*(\mathrm{D}) = 0$$

Applying the data in Example 9.45, we have

$$-0.80P^*(\mathrm{A}) + 0.70P^*(\mathrm{B}) + 0.55P^*(\mathrm{C}) = 0 \qquad (f)$$

Similarly, by setting up the expressions for $P^*(\mathrm{B})$ and $P^*(\mathrm{C})$, we obtain the following:

$$0.35P^*(\mathrm{A}) - 0.85P^*(\mathrm{B}) + 0.25P^*(\mathrm{C}) + 0.55P^*(\mathrm{D}) = 0 \qquad (g)$$

$$0.15P^*(\mathrm{A}) + 0.10P^*(\mathrm{B}) - P^*(\mathrm{C}) + 0.35P^*(\mathrm{D}) = 0 \qquad (h)$$

We also have

$$P^*(\mathrm{A}) + P^*(\mathrm{B}) + P^*(\mathrm{C}) + P^*(\mathrm{D}) = 1 \qquad (i)$$

The system consisting of Eqs. (f), (g), (h), and (i) has this solution:

$$P^*(\mathrm{A}) = 0.3710 \qquad P^*(\mathrm{B}) = 0.3083$$

$$P^*(\mathrm{C}) = 0.1472 \qquad P^*(\mathrm{D}) = 0.1735$$

9.4.6 Definition of an Absorbing Chain

Assume that a Markov process has three possible outcomes, A, B, and C. Also assume that A and B both have A, B, and C as possible successors, but C can be followed only by C. Thus, when the system that generates the Markov process reaches state C, it loses the capacity to change, and the Markov chain is considered to have terminated.

The termination of a Markov chain is referred to as *absorption*. Therefore, a chain that has the capacity to terminate itself is called an *absorbing chain*, and an outcome that causes absorption is called an *absorbing outcome*. Thus, in our illustrative case, C is an absorbing outcome while A and B are nonabsorbing outcomes.

We shall describe the system as *active* if it has the capacity to change its state and *inactive* if it has lost that capacity. Thus, in our illustrative case, the system is active if the most recent outcome was A or B, and it is inactive if the most recent outcome was C.

We shall investigate systems that have only one absorbing outcome. However, our methods of investigation can readily be extended to systems that have multiple absorbing outcomes.

9.4.7 Probabilities Pertaining to an Absorbing Chain

An absorbing chain will end eventually, but the length of the chain is a random variable. It is often necessary to compute probabilities pertaining to this variable.

EXAMPLE 9.46 A Markov process that is performed at 1-min intervals has the transition matrix appearing in Table 9.8. Assume that the process was just performed and the outcome was A, B, or D. What are the probabilities that the system will remain active for at least 4 min?

SOLUTION If the system is to remain active for at least 4 min, it is necessary that each of the first three outcomes be A, B, or D. We shall compute the probability of obtaining a satisfactory chain. For compactness, we shall delete the row and column pertaining to C; this deletion does not affect our present calculations. The matrix that remains is the following:

$$\mathbf{N} = \begin{matrix} \\ A_0 \\ B_0 \\ D_0 \end{matrix} \begin{matrix} \begin{matrix} A_1 & B_1 & D_1 \end{matrix} \\ \begin{bmatrix} 0.10 & 0.40 & 0.45 \\ 0.30 & 0.20 & 0.25 \\ 0.15 & 0.05 & 0.20 \end{bmatrix} \end{matrix}$$

In accordance with Theorem 9.8, the probability matrices corresponding to the second and third outcomes are $\mathbf{N}^2$ and $\mathbf{N}^3$, respectively, and they are as follows:

$$\mathbf{N}^2 = \begin{matrix} \\ A_0 \\ B_0 \\ D_0 \end{matrix} \begin{matrix} \begin{matrix} A_2 & B_2 & D_2 \end{matrix} \\ \begin{bmatrix} 0.1975 & 0.1425 & 0.2350 \\ 0.1275 & 0.1725 & 0.2350 \\ 0.0600 & 0.0800 & 0.1200 \end{bmatrix} \end{matrix}$$

TABLE 9.8 Transition Matrix

	A_n	B_n	C_n	D_n
A_{n-1}	0.10	0.40	0.05	0.45
B_{n-1}	0.30	0.20	0.25	0.25
C_{n-1}	0	0	1	0
D_{n-1}	0.15	0.05	0.60	0.20

$$N^3 = \begin{array}{c} \\ A_0 \\ B_0 \\ D_0 \end{array} \begin{array}{c} \begin{array}{ccc} A_3 & B_3 & D_3 \end{array} \\ \begin{bmatrix} 0.09775 & 0.11925 & 0.17150 \\ 0.09975 & 0.09725 & 0.14750 \\ 0.04800 & 0.04600 & 0.07100 \end{bmatrix} \end{array}$$

Assume that the most recent outcome was A. The probability that the third outcome in the future will be A, B, or D is the sum of the elements in the first row of $\mathbf{N}^3$. Let X denote the time in minutes that the system remains active. Then

$$P(X \geq 4|A_0) = 0.09775 + 0.11925 + 0.17150 = 0.3885$$

$$P(X \geq 4|B_0) = 0.09975 + 0.09725 + 0.14750 = 0.3445$$

$$P(X \geq 4|D_0) = 0.04800 + 0.04600 + 0.07100 = 0.1650$$

In matrix **N**, the elements in each row and in each column have a sum less than 1. Therefore, as n becomes infinitely large, $\mathbf{N}^n$ approaches the null matrix, and the probability that $X \geq n$ approaches 0. This conclusion is intuitively self-evident, for the system will ultimately succumb to absorption.

EXAMPLE 9.47 With reference to Example 9.46, what are the probabilities that the system will remain active for precisely 3 min?

SOLUTION

Method 1. Assume the most recent outcome was A. The following relationship applies:

$$P(X = 3|A_0) = P(X \geq 3|A_0) - P(X \geq 4|A_0)$$

The first probability at the right is obtained by summing the elements in the first row of $\mathbf{N}^2$, and the second probability at the right was obtained in Example 9.46. Then

$$P(X = 3|A_0) = 0.5750 - 0.3885 = 0.1865$$

Similarly, for the remaining possibilities, we have

$$P(X = 3|B_0) = 0.5350 - 0.3445 = 0.1905$$

$$P(X = 3|D_0) = 0.2600 - 0.1650 = 0.0950$$

Method 2. The system will remain active for precisely 3 min if the second outcome in the future is A, B, or D, and the third outcome is C. Then

$$P(X = 3) = P(A_2\text{-}C_3) + P(B_2\text{-}C_3) + P(D_2\text{-}C_3)$$

Assume the most recent outcome was B_0. Applying the probabilities appearing in $\mathbf{N}^2$ and Table 9.8, we have the following:

$$P(X = 3|B_0) = (0.1275)(0.05) + (0.1725)(0.25) + (0.2350)(0.60) = 0.1905$$

Under Method 2, it is unnecessary to construct $\mathbf{N}^3$.

9.4.8 Expected Life of an Active System

Article 9.1.8 presents the formula for calculating the *expected value* of a random variable, which is the mean value of the variable in the long run. Where a Markov chain is absorbing, it is often necessary to determine the expected length of time that the system will remain active.

Assume that a Markov process is performed regularly at the beginning of a time unit, such as 1 h, 1 day, etc. Thus, the active life of the Markov system is restricted to an integral number of time units. Let X denote the active life, and as before let $P(n)$ denote the probability that $X = n$. By Eq. (9.4), we have

$$E(X) = P(1) + 2P(2) + 3P(3) + \cdots \tag{j}$$

We now set

$$P(n) = P(X \geq n) - P(X \geq n + 1) \tag{k}$$

By replacing each term in Eq. (j) with its expression in Eq. (k), we obtain the following:

$$E(X) = P(X \geq 1) + P(X \geq 2) + P(X \geq 3) + \cdots \tag{9.40}$$

With reference to Example 9.46, assume that the initial outcome is A. Then

$$P(X \geq 1) = 1 = \text{sum of elements in first row of } \mathbf{I} \text{ (unit matrix)}$$

$$P(X \geq 2) = \text{sum of elements in first row of } \mathbf{N}$$

$$P(X \geq 3) = \text{sum of elements in first row of } \mathbf{N}^2$$

and so on. Equation (9.40) reveals that $E(X)$ can be obtained by this procedure:

1. Form the matrix $\mathbf{S}$, where $\mathbf{S} = \mathbf{I} + \mathbf{N} + \mathbf{N}^2 + \mathbf{N}^3 + \cdots$. It can be demonstrated that $\mathbf{S}$ is the inverse of $\mathbf{I} - \mathbf{N}$.
2. Set $E(X)$ equal to the sum of the elements in the first row of $\mathbf{S}$.

EXAMPLE 9.48 With reference to Example 9.46, assume that the initial outcome is A, B, or D. Compute the expected number of minutes the system will remain active.

SOLUTION

$$\mathbf{I} - \mathbf{N} = \begin{bmatrix} 1 & 0 & 0 \\ 0 & 1 & 0 \\ 0 & 0 & 1 \end{bmatrix} - \begin{bmatrix} 0.10 & 0.40 & 0.45 \\ 0.30 & 0.20 & 0.25 \\ 0.15 & 0.05 & 0.20 \end{bmatrix}$$

$$= \begin{bmatrix} 0.90 & -0.40 & -0.45 \\ -0.30 & 0.80 & -0.25 \\ -0.15 & -0.05 & 0.80 \end{bmatrix}$$

Inverting this matrix by the method presented in Art. 5.6.2, we obtain

$$\mathbf{S} = (\mathbf{I} - \mathbf{N})^{-1} = \begin{bmatrix} 1.5968 & 0.8716 & 1.1704 \\ 0.7062 & 1.6604 & 0.9160 \\ 0.3437 & 0.2673 & 1.5267 \end{bmatrix}$$

If the initial outcome is A,

$$E(X) = 1.5968 + 0.8716 + 1.1704 = 3.64 \text{ min}$$

If the initial outcome is B,

$$E(X) = 0.7062 + 1.6604 + 0.9160 = 3.28 \text{ min}$$

If the initial outcome is D,

$$E(X) = 0.3437 + 0.2673 + 1.5267 = 2.14 \text{ min}$$

Each element of matrix **S** gives the expected length of time the system will be in a particular state during its active life. For example, if the initial outcome is A, the expected length of time the system will be in state B is 0.8716 min; if the initial outcome is D, the expected length of time the system will be at state D is 1.5267 min. In general, of course, the system will be at a particular state intermittently rather than continuously.

The procedure we have formulated for finding the expected life of a system that has one absorbing outcome has an important application with reference to a nonabsorbing system. Assume that we are given the initial state of the system and that we must determine the expected number of trials that will be required to bring the system to some specified state different from the initial one. We may view the outcome that produces the specified state as an absorbing outcome and then apply the foregoing procedure to find the expected life of the system.

SECTION 10

STATISTICAL INFERENCE AND REGRESSION ANALYSIS

In our present investigation, we shall apply the terminology and principles of statistics and probability that are presented in Secs. 8 and 9, respectively.

10.1 SAMPLING DISTRIBUTIONS

10.1.1 Basic Definitions and Notation

Consider that we have a set of items and that there is a distinctive numerical value X associated with each item. For example, X may be the length, molecular weight, or temperature of an item. The set of X values has a property R. Thus, R may be the harmonic mean, median, or standard deviation of the X values. The set of items is termed the *population*, and R is termed a *parameter*.

Now assume that it is necessary to evaluate R. An exact evaluation would require that we examine each item to establish its X value and then perform the necessary calculations. However, this procedure is often impractical or impossible to undertake, for several reasons. First, the population is frequently so vast as to preclude an examination of each item. Second, the mere act of evaluating X may entail destruction of the item. For example, the only way to determine the breaking strength of a cable is to stress it to its breaking point, and the only way to evaluate the life span of an electronic unit is to operate it until it expires. Third, the population may exist only as a prospect rather than a reality. As an illustration, assume that we are exploring the possibility of manufacturing a commodity by a novel method. All units that will be produced by this method if it is adopted constitute the population.

In these situations, we must content ourselves with *estimating* R by examining a limited number of items to establish their X values and then computing their property S that corresponds to R. For example, if R is the arithmetic mean of the X values of the population, S is the arithmetic mean of the X values that are actually obtained. We then apply S as an estimate of R. The items that are examined constitute a *sample*, and S is termed a *statistic*.

The process of estimating a parameter R by use of a statistic S is referred to as *statistical inference*. The number of items included in the sample is called the sample *size*. The difference between S and R is termed the *sampling error*.

We shall illustrate the definitions and relationships of statistical inference by means of highly simplified examples.

EXAMPLE 10.1 A set of items has the X values 14, 16, 19, 21, 23, 30, 34, 39, 40, 44. The sample consists of the items having the X values 16, 30, 44. If we are concerned with the arithmetic mean of the population, what is the sampling error?

SOLUTION The population contains 10 items, and the sum of their X values is 280. The sample contains 3 items, and the sum of their X values is 90. Then

$$R = \frac{280}{10} = 28 \qquad S = \frac{90}{3} = 30$$

$$\text{Sampling error} = 30 - 28 = 2$$

In several situations, it is essential that the statistical analyst exercise some measure of control in drawing the sample to ensure that the sample represents the population with reasonable accuracy. However, we shall assume in our investigation that the sample is drawn in a purely random manner. Consequently, all prospective samples have equal likelihood of being drawn, and it follows that the statistic is a random variable.

The notational system is as follows:

Let $\bar{X}$ = arithmetic mean of sample
s = standard deviation of sample
n = number of items in sample (i.e., sample size)
μ = arithmetic mean of population
σ = standard deviation of population
N = number of items in population

The sample represents a combination of N items taken n at a time. Therefore, by Eq. (4.5),

$$\text{Number of possible samples} = {}_NC_n = \frac{N!}{n!(N-n)!}$$

For simplicity, the terms *population* and *sample* are often applied in referring to the set of X values of the population and of the sample, respectively.

10.1.2 Definition and Properties of a Sampling Distribution

Since we are applying a statistic S to estimate a parameter R, we are confronted with this question: How reliable is the estimate? In seeking an answer, we must invert reality by following this procedure: Take a population that is known rather than unknown, form all possible samples of a specified size, compute the S value of each sample, and then compare these S values with the known value of R. Manifestly, the greater the variability of the S values, the less reliable is the estimate. The set of S values is known as the *sampling distribution* of S.

EXAMPLE 10.2 The population consists of the numbers 13, 16, 18, 22, and the sample size is 2. Establish the sampling distribution of the following statistics: the arithmetic mean, the geometric mean, and the variance.

SOLUTION The number of possible samples is $4!/(2!2!) = 6$. Refer to Table 10.1. Each possible sample is recorded in the column at the left, and the required statistics of that sample are recorded in the remaining columns. The calculations for the first sample are presented for illustrative purposes.

$$\bar{X} = (13 + 16)/2 = 14.5 \qquad G = \sqrt{13 \times 16} = 14.42$$

$$s^2 = [(13 - 14.5)^2 + (16 - 14.5)^2]/2 = 2.25$$

The sampling distributions are as follows:

Arithmetic mean: 14.5, 15.5, 17.0, 17.5, 19.0, 20.0

Geometric mean: 14.42, 15.30, 16.91, 16.97, 18.76, 19.90

Variance: 1.00, 2.25, 4.00, 6.25, 9.00, 20.25

In the subsequent material, it is understood that the term *mean* refers to the arithmetic mean exclusively.

As previously stated, a statistic S is a random variable. A sampling distribution presents all possible values of S and, where duplications occur, the frequency of each value. If the frequency is divided by the number of samples, the result is a relative frequency, and the relative frequency of a particular value is the probability that S will assume that value when the true sample is drawn. Therefore, a sampling distribution is completely analogous to a probability distribution, the sole difference being that the sampling distribution presents relative frequencies rather than probabilities.

Since it is a set of numbers, a sampling distribution has a mean and standard deviation. These are denoted by μ and σ, respectively, with a subscript to identify the statistic. For example, $\mu_{\bar{X}}$ and μ_{s^2} denote, respectively, the mean of the sampling distribution of the mean and the mean of the sampling distribution of the variance. Similarly, $\sigma_{\bar{X}}$ and σ_s^2 denote, respectively, the standard deviation of the sampling distribution of the mean and the variance of the sampling distribution of the standard deviation. Table 10.2 summarizes the notational system presented thus far.

EXAMPLE 10.3 With reference to Example 10.2, find the following: the mean of the sampling distribution of the mean, the mean of the sampling distribution of the variance, and the standard deviation of the sampling distribution of the variance.

TABLE 10.1 Sampling Distributions

Sample	Sample arithmetic mean, $\bar{X}$	Sample geometric mean, G	Sample variance, s^2
13, 16	14.5	14.42	2.25
13, 18	15.5	15.30	6.25
13, 22	17.5	16.91	20.25
16, 18	17.0	16.97	1.00
16, 22	19.0	18.76	9.00
18, 22	20.0	19.90	4.00

TABLE 10.2 Notation

Property	Sample	Population	Sampling distribution of statistic S
Mean	$\bar{X}$	μ	μ_S
Standard deviation	s	σ	σ_S
Number of items	n	N	

SOLUTION

$$\mu_{\bar{X}} = \frac{14.5 + 15.5 + 17.0 + 17.5 + 19.0 + 20.0}{6} = 17.25$$

$$\mu_{s^2} = \frac{1.00 + 2.25 + 4.00 + 6.25 + 9.00 + 20.25}{6} = 7.125$$

The calculations for the third value are as follows:

$$(1.00 - 7.125)^2 + (2.25 - 7.125)^2 + (4.00 - 7.125)^2 + (6.25 - 7.125)^2 + (9.00 - 7.125)^2 + (20.25 - 7.125)^2 = 247.5938$$

$$\sigma_{s^2} = \sqrt{\frac{247.5938}{3}} = 9.08$$

The standard deviation of the sampling distribution of a statistic is known as the *standard error* of that statistic. For example, $\sigma_{\bar{X}}$ is the standard error of the mean.

We shall now consider the properties of several specific types of sampling distributions.

10.1.3 Sampling Distribution of the Mean

The sampling distribution of the mean is of major importance in statistical analysis. It can be demonstrated that the properties of the sampling distribution and those of the population are related in this manner:

$$\mu_{\bar{X}} = \mu \tag{10.1}$$

$$\sigma_{\bar{X}} = \sigma \sqrt{\frac{N - n}{n(N - 1)}} \tag{10.2}$$

If the population is infinite, Eq. (10.2) reduces to

$$\sigma_{\bar{X}} = \frac{\sigma}{\sqrt{n}} \tag{10.2a}$$

EXAMPLE 10.4 The population consists of the numbers 11, 14, 15, 20, 22, 24, 27, and the sample size is 2. Find the mean and the standard deviation of the sampling distribution of the mean (*a*) without recourse to Eqs. (10.1) and (10.2) and (*b*) by applying these equations.

SOLUTION

Part a. The population contains seven items and the sample contains two items. Therefore, the number of possible samples is $7!/(2!5!) = 21$. Refer to Table 10.3, where each possible sample is recorded in column 1. The mean of the sample is recorded in column 2, and the set of numbers in this column is the sampling distribution of the mean. Totaling these numbers, we obtain

$$\mu_{\bar{X}} = \frac{399.0}{21} = 19$$

(It can be demonstrated that the sum of the sample means should be three times the sum of the numbers in the population, and it is.)

The deviations of the sample means from 19 are recorded in column 3. For example, for the sample consisting of 11 and 24, we have $d = 17.5 - 19 = -1.5$. The squares of the deviations are recorded in column 4. Then

$$\sigma_{\bar{X}} = \sqrt{\frac{255.00}{21}} = 3.485$$

Part b. The mean of the population is

$$\mu = \frac{11 + 14 + 15 + 20 + 22 + 24 + 27}{7} = 19$$

TABLE 10.3

Sample (1)	Sample mean, $\bar{X}$ (2)	Deviation, $d = \bar{X} - 19$ (3)	d^2 (4)
11, 14	12.5	−6.5	42.25
11, 15	13.0	−6.0	36.00
11, 20	15.5	−3.5	12.25
11, 22	16.5	−2.5	6.25
11, 24	17.5	−1.5	2.25
11, 27	19.0	0	0
14, 15	14.5	−4.5	20.25
14, 20	17.0	−2.0	4.00
14, 22	18.0	−1.0	1.00
14, 24	19.0	0	0
14, 27	20.5	1.5	2.25
15, 20	17.5	−1.5	2.25
15, 22	18.5	−0.5	0.25
15, 24	19.5	0.5	0.25
15, 27	21.0	2.0	4.00
20, 22	21.0	2.0	4.00
20, 24	22.0	3.0	9.00
20, 27	23.5	4.5	20.25
22, 24	23.0	4.0	16.00
22, 27	24.5	5.5	30.25
24, 27	25.5	6.5	42.25
Total	399.0	0	255.00

The variance of the population is found in this manner:

$$(11-19)^2+(14-19)^2+(15-19)^2+(20-19)^2 + (22-19)^2+(24-19)^2+(27-19)^2=204$$

$$\sigma^2=\frac{204}{7}$$

By Eqs. (10.1) and (10.2),

$$\mu_{\bar{X}}=\mu=19$$

$$\sigma_{\bar{X}}=\sqrt{\frac{204}{7}\times\frac{7-2}{2(7-1)}}=\sqrt{\frac{255}{21}}=3.485$$

Table 10.3 clearly reveals why $\mu_{\bar{X}}$ is coincident with μ. Each item in the population appears in six samples. Consequently, all items in the population have the same effect on the value of $\mu_{\bar{X}}$, as they do on the value of μ.

Equation (9.30) describes the probability curve of a variable that has a normal probability distribution. If the equation of the relative-frequency curve of a continuous variable has the same form as Eq. (9.30), that variable is said to be *normally distributed.*

Consider again that all possible samples of a given size have been drawn. The *central-limit theorem* states the following:

1. If the population is extremely large and normally distributed, the sample means are also normally distributed.
2. If the population is extremely large but not normally distributed, the distribution of the sample means is approximately normal if the sample size is relatively large. (The sample size is considered to be relatively large if $n \geq 30$.)

EXAMPLE 10.5 A firm manufactured 1000 rods having a mean length of 2.108 m and a standard deviation of 0.056 m. Assuming that the lengths are normally distributed, find the probability that a sample of 50 rods drawn at random has a mean length less than 2.101 m.

SOLUTION In accordance with the central-limit theorem, the sample means are normally distributed. By Eqs. (10.1) and (10.2), the mean and standard deviation of this normal distribution are as follows:

$$\mu_{\bar{X}}=2.108 \text{ m}$$

$$\sigma_{\bar{X}}=0.056\sqrt{\frac{1000-50}{50\times 999}}=0.00772 \text{ m}$$

Refer to Fig. 10.1 and Table 9.4. By Eq. (9.31), when $\bar{X}=2.101$ m,

$$z=\frac{2.101-2.108}{0.00772}=-0.907 \qquad \text{Area}=0.318$$

$$P(\bar{X}<2.101)=0.5-0.318=0.182$$

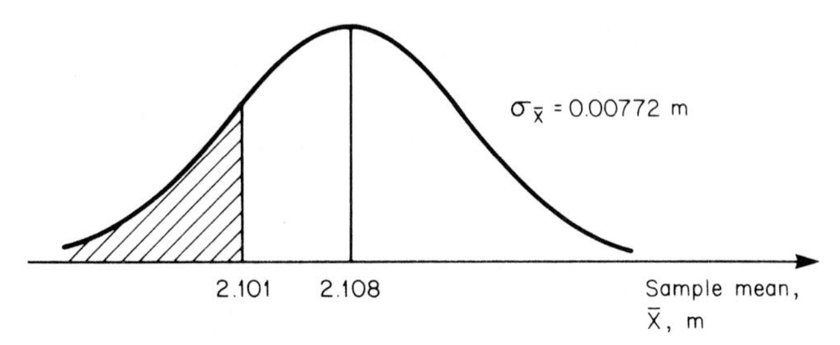

FIGURE 10.1 Sampling distribution of the mean. (*From Handbook of Engineering Economics; used with permission of McGraw-Hill, Inc.*)

10.1.4 Sampling Distribution of the Variance

It can be demonstrated that the sampling distribution of the variance has the following value:

$$\mu_{s^2} = \frac{N(n-1)}{(N-1)n}\sigma^2 \tag{10.3}$$

If the population is infinite, this equation reduces to

$$\mu_{s^2} = \frac{n-1}{n}\sigma^2 \tag{10.3a}$$

EXAMPLE 10.6 The population consists of the numbers 21, 24, 33, and 42, and the sample size is 3. Compute the mean of the sampling distribution of the variance (*a*) without recourse to Eq. (10.3) and (*b*) by applying Eq. (10.3).

SOLUTION

Part a. The number of possible samples is 4!/(3!)(1!) = 4. Refer to Table 10.4. The calculations for the first sample are presented for illustrative purposes.

$$\bar{X} = \frac{21 + 24 + 33}{3} = 26$$

$$s^2 = \frac{(21-26)^2 + (24-26)^2 + (33-26)^2}{3} = \frac{78}{3} = 26$$

The numbers in the last column of the table constitute the sampling distribution

TABLE 10.4

Sample	Sample mean, $\bar{X}$	Sample variance, s^2
21, 24, 33	26	26
21, 24, 42	29	86
21, 33, 42	32	74
24, 33, 42	33	54
Total		240

of the variance. The mean of this sampling distribution is

$$\mu_{s^2} = \frac{240}{4} = 60$$

Part b. For the population,

$$\mu = \frac{21 + 24 + 33 + 42}{4} = 30$$

$$\sigma^2 = \frac{(21-30)^2 + (24-30)^2 + (33-30)^2 + (42-30)^2}{4}$$

$$= 67.5$$

Applying Eq. (10.3) with $N = 4$ and $n = 3$, we obtain

$$\mu_{s^2} = \frac{4 \times 2 \times 67.5}{3 \times 3} = 60$$

10.1.5 Sampling Distribution of the Proportion

Thus far, we have investigated sets of items in which each item has some numerical value, such as length or temperature. We shall now investigate a set of items in which each item can be classified on the basis of its *type*. For example, the alternative types may be acceptable and unacceptable; red, green, and yellow; visible and invisible; etc. The ratio of the number of items of a given type to the total number of items in the set is termed the *proportion* of those items. Thus, if the set contains 9 cylinders and 15 spheres, the proportion of cylinders is 9/24 or 0.375.

Assume that we must determine the proportion of items of a given type in a set. If the set is extremely large, the only practical procedure consists of drawing a sample and applying the proportion of the sample to estimate the proportion of the population. Our notational system is as follows:

P = proportion of items of given type in population

Q = proportion of items of all other types in population $= 1 - P$

p = proportion of items of given type in sample

As an illustration, assume that the population consists of the numbers 23, 25, 30, 31, 33, 38, 40, 45 and the sample consists of the numbers 30, 38, and 45. With reference to even numbers, $P = 3/8$, $Q = 5/8$, and $p = 2/3$.

Consider that all possible samples of a given size have been formed, thereby generating a set of p values. This set of values is termed the *sampling distribution of the proportion*. Conforming to our general notational system, let μ_p and σ_p denote the mean and standard deviation, respectively, of this sampling distribution. It can be demonstrated that

$$\mu_p = P \tag{10.4}$$

$$\sigma_p = \sqrt{\frac{PQ(N-n)}{n(N-1)}} \tag{10.5}$$

If the population is infinite, Eq. (10.5) reduces to

$$\sigma_p = \sqrt{\frac{PQ}{n}} \tag{10.5a}$$

EXAMPLE 10.7 The population consists of the labels A1, A2, A3, B1, B2, and the sample size is 2. Our interest centers on the labels that begin with A. Find the mean and standard deviation of the sampling distribution of the proportion of these labels (*a*) without recourse to Eqs. (10.4) and (10.5) and (*b*) by applying these equations.

SOLUTION

Part a. The number of possible samples is $5!/(2!3!) = 10$. Refer to Table 10.5, where each sample is recorded in column 1. The proportion of labels that begin with A is recorded in column 2, and the numbers in this column constitute the sampling distribution of the proportion. Then

$$\mu_p = \frac{6.0}{10} = 0.6$$

The deviations of the sample proportions from 0.6 are recorded in column 3, and the squares of these deviations are recorded in column 4. Then

$$\sigma_p = \sqrt{\frac{0.90}{10}} = 0.3$$

Part b. In the population, the proportion of labels that begin with A is $P = 3/5 = 0.6$. By Eqs. (10.4) and (10.5),

$$\mu_p = 0.6 \qquad \sigma_p = \sqrt{\frac{(0.6)(0.4)3}{2 \times 4}} = 0.3$$

If the sample size is equal to or greater than 30, the sampling distribution of the proportion is approximately normal. However, if we are to take advantage of this relationship, we must transform the number of items of a given type from a discrete

TABLE 10.5

Sample (1)	Proportion, p (2)	Deviation, $d = p - 0.6$ (3)	d^2 (4)
A1, A2	1.0	0.4	0.16
A1, A3	1.0	0.4	0.16
A1, B1	0.5	−0.1	0.01
A1, B2	0.5	−0.1	0.01
A2, A3	1.0	0.4	0.16
A2, B1	0.5	−0.1	0.01
A2, B2	0.5	−0.1	0.01
A3, B1	0.5	−0.1	0.01
A3, B2	0.5	−0.1	0.01
B1, B2	0.0	−0.6	0.36
Total	6.0	0	0.90

to a continuous variable. This transformation is achieved by replacing an integral value with an *interval* that extends one-half unit above and below the integer.

EXAMPLE 10.8 A firm produces several types of units, and 30 percent of all the units are of type A. If a sample of 120 units is drawn at random, what is the probability that the proportion of type A units is (*a*) more than 35 percent or (*b*) less than 28 percent?

SOLUTION The population is considered to be infinite. By Eqs. (10.4) and (10.5*a*),

$$\mu_p = P = 0.30 \qquad \sigma_p = \sqrt{\frac{(0.30)(0.70)}{120}} = 0.04183$$

Thus, the sampling distribution of the proportion is approximately a normal distribution having a mean of 0.30 and a standard deviation of 0.04183. Let X denote the number of type A units in the sample.

Part a. Refer to Fig. 10.2*a* and Table 9.4. Set $p = 0.35$. Then

$$X = 120(0.35) = 42$$

We replace this integral value with the interval 41.5 to 42.5. Since we are concerned

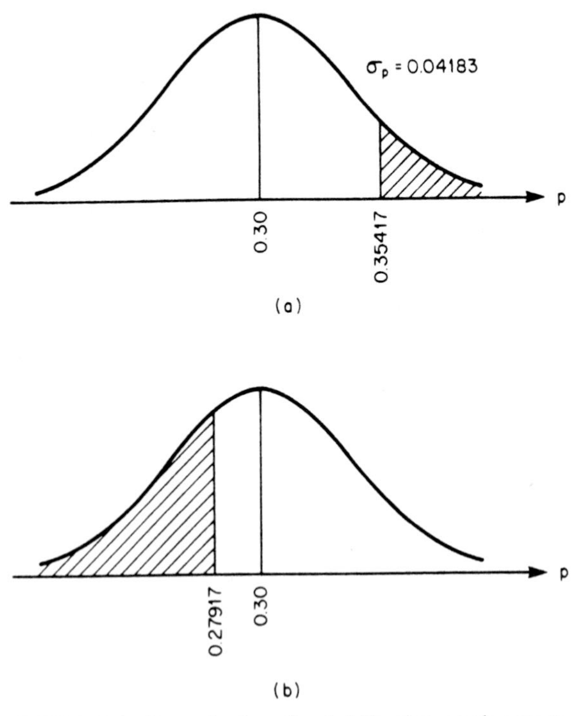

FIGURE 10.2 Determination of probability that sample proportion is (*a*) more than 35 percent and (*b*) less than 28 percent. (*From Handbook of Engineering Economics; used with permission of McGraw-Hill, Inc.*)

with values *above* the interval, we set $X = 42.5$. The adjusted value of the proportion is

$$p = \frac{42.5}{120} = 0.35417$$

$$z = \frac{0.35417 - 0.30}{0.04183} = 1.295 \qquad \text{Area} = 0.402$$

$$\text{Probability} = 0.5 - 0.402 = 0.098$$

Part b. Refer to Fig. 10.2*b*.

$$X = 120(0.28) = 33.6$$

We replace this value with the interval 33.5 to 34.5. Since we are concerned with values *below* this interval, we set $X = 33.5$. Then

$$p = \frac{33.5}{120} = 0.27917$$

$$z = \frac{0.27917 - 0.30}{0.04183} = -0.498 \qquad \text{Area} = 0.191$$

$$\text{Probability} = 0.5 - 0.191 = 0.309$$

10.2 ESTIMATION OF PARAMETERS

10.2.1 Types of Estimates and Estimators

As stated in Art. 10.1.1, statistical inference is the process of estimating a parameter, which is a property of the population, by means of a statistic, which is a property of the sample that has been drawn from the population. Having defined and analyzed sampling distributions, it now becomes possible to apply these distributions in statistical inference.

There are two types of estimates: point and interval. A *point estimate* expresses a specific value, and an *interval estimate* expresses a range of values. For example, if we estimate that the mean temperature of a group of bodies is 23°C, we have a point estimate; if we estimate that the mean diameter of a group of spherical bodies lies between 32.05 and 32.13 cm, we have an interval estimate. In the subsequent material, only interval estimates will be considered, and each statement of an estimate will be coupled with a statement of its reliability, this being the probability that the estimate is correct.

A statistic that is applied to estimate a parameter is called an *estimator*. If the mean of the sampling distribution of that statistic equals the corresponding parameter, the statistic is classified as an *unbiased* estimator. If this equality is lacking, the statistic is classified as a *biased* estimator.

Equation (10.1) discloses that the sample mean $\bar{X}$ is an unbiased estimator of the population mean μ, and Eq. (10.4) discloses that the sample proportion p is an unbiased estimator of the population proportion P. On the other hand, Eq. (10.3) discloses that the sample variance s^2 is a biased estimator of the population variance σ^2. However, if the population is vast and the sample size is relatively large, the fraction in Eq. (10.3) has a value close to 1, and the degree of bias is relatively small. For this reason, s^2 and s are generally applied as estimators of σ^2 and σ, respectively, when these conditions exist.

10.2.2 Confidence Intervals and Confidence Levels

At this point, we shall develop a simple principle pertaining to inequalities that lies at the core of statistical inference. Let a, b, and c denote constants that are related in this manner:

$$a - b < c < a + b$$

If we take $a - b < c$ and transpose b, we obtain $a < c + b$. If we now take $c < a + b$ and transpose b, we obtain $c - b < a$. Combining these relationships, we have

$$c - b < a < c + b$$

It is helpful to express these results verbally, in the following manner:

Theorem 10.1. If c lies between $a - b$ and $a + b$, then a lies between $c - b$ and $c + b$, and vice versa.

Now assume that a statistic S is continuous and has a normal sampling distribution, as shown in Fig. 10.3*a*. Again let μ_S and σ_S denote the mean and standard deviation, respectively, of the sampling distribution of S. Consider the interval $\mu_S - \sigma_S$ to $\mu_S + \sigma_S$, which is shown shaded in Fig. 10.3*a*. Let M denote the proportion of samples that have S values that lie within this interval. When a sample is drawn at random, the probability that its S value lies within this interval is M. At the boundaries of this interval, $z = \pm 1$. From Table 9.4, the area under the normal curve from the centerline to a boundary is 0.34134. Then $M = 2(0.34134) = 0.6827$, and the following probability exists:

$$P(\mu_S - \sigma_S < S < \mu_S + \sigma_S) = 0.6827$$

We now invert the situation by considering that S is known and μ_S is to be determined. In accordance with Theorem 10.1, if S lies within the interval $\mu_S - \sigma_S$ to $\mu_S + \sigma_S$, then μ_S lies within the interval $S - \sigma_S$ to $S + \sigma_S$. It follows that

$$P(S - \sigma_S < \mu_S < S + \sigma_S) = 0.6827 \qquad (a)$$

For example, if a sample yields the value S_1 shown in Fig. 10.3*b*, there is a probability of 68.27 percent that μ_S lies within the interval AB in that drawing.

In statistical inference, the term *confidence* is applied in a general sense to denote the reliability of an estimate. With reference to Fig. 10.3*b*, the interval $S_1 - \sigma_S$ to $S_1 + \sigma_S$ is called the *confidence interval*, and the probability of 68.27 percent that μ_S lies within this interval is called the *confidence level* or *confidence coefficient*. The boundaries of this interval, $S_1 - \sigma_S$ and $S_1 + \sigma_S$, are called the *confidence limits*.

Now consider the interval $\mu_S - 2\sigma_S$ to $\mu_S + 2\sigma_S$. At the boundaries, $z = \pm 2$. The area under the normal curve across this interval is $2(0.47725) = 0.9545$. Proceeding as before, we obtain

$$P(S - 2\sigma_S < \mu_S < S + 2\sigma_S) = 0.9545 \qquad (b)$$

Thus, a z value of 2 (in absolute value) corresponds to a confidence level of 95.45 percent, and vice versa.

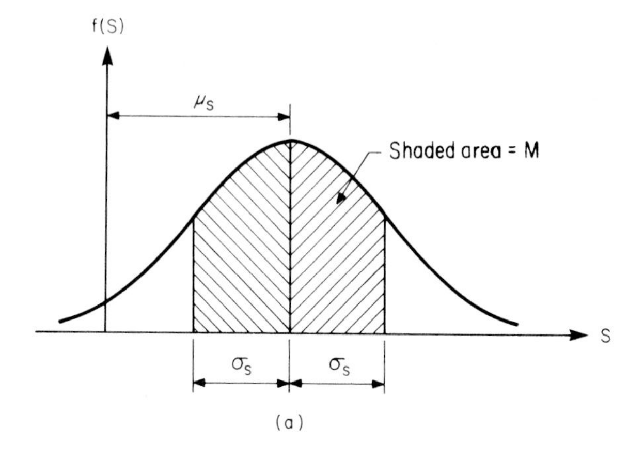

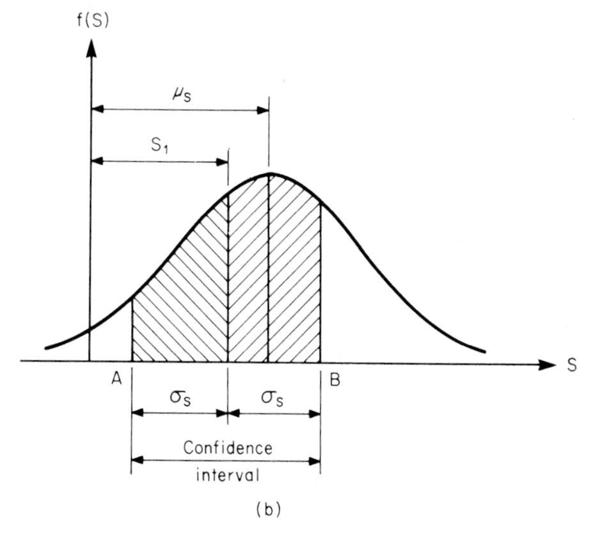

FIGURE 10.3 Probabilistic relationship between a statistic and the mean of the sampling distribution of the statistic where (*a*) mean is known or (*b*) statistic is known. (*From Handbook of Engineering Economics; used with permission of McGraw-Hill, Inc.*)

Continuing these calculations, we obtain the results recorded in Table 10.6. In general, the confidence interval corresponding to a given value of z is denoted by $S \pm z\sigma_S$, where S denotes the statistic under consideration.

Assume that μ_S is an unbiased estimator of the parameter that corresponds to S; in accordance with our definition, μ_S is equal to this parameter. It is then possible to replace μ_S with the parameter in Eqs. (*a*) and (*b*). It now follows that the confidence level corresponding to a given confidence interval expresses the probability that the *parameter* lies within that interval. For example, if p denotes the

TABLE 10.6

Confidence level, %	z
50	0.6745
68.27	1.00
90	1.645
95	1.96
95.45	2.00
99	2.58
99.73	3.00

proportion of a particular sample, Eq. (*a*) becomes

$$P(p - \sigma_p < P < p + \sigma_p) = 0.6827$$

Expressed verbally, the probability that the population proportion P lies within the interval $p \pm \sigma_S$ is 68.27 percent.

The foregoing discussion provides a basis for estimating a parameter by use of a sample when the following conditions exist:

1. The sampling distribution of the corresponding statistic S is normal or approximately normal. We shall assume that such is always the case.
2. The mean of the sampling distribution of S is equal to the given parameter.

However, this method of estimation has a serious obstacle: It requires a knowledge of σ_S. Since in reality the population is unknown, the value of σ_S cannot be determined. This obstacle can be circumvented by expressing σ_S in terms of σ, the standard deviation of the population, and estimating σ by applying the sample data. As stated in Art. 10.2.1, the sample standard deviation s may be applied as an estimator of σ if the population is vast and the sample size is relatively large.

10.2.3 Estimation of Population Mean

The statistic that corresponds to the population mean μ is the sample mean $\bar{X}$. In accordance with the central-limit theorem, the sampling distribution of $\bar{X}$ is normal or approximately normal if $n \geq 30$. By Eq. (10.1), $\mu_{\bar{X}} = \mu$. It follows that the population mean can be estimated by the method formulated in Art. 10.2.2 whenever the sample has the specified size.

Let CI denote the confidence interval corresponding to a given value of z. Then $\text{CI} = \bar{X} \pm z\sigma_{\bar{X}}$. For a finite population, $\sigma_{\bar{X}}$ is given by Eq. (10.2). Applying s as an estimator of σ, we obtain the following equation for the confidence interval with respect to the mean:

$$\text{CI} = \bar{X} \pm zs\sqrt{\frac{N-n}{n(N-1)}} \tag{10.6}$$

For a population that may be considered infinite, Eq. (10.2*a*) applies, and

$$\text{CI} = \bar{X} \pm \frac{zs}{\sqrt{n}} \tag{10.6a}$$

EXAMPLE 10.9 A firm manufactures rods of a standard length. It selected 80 rods at random and found that these rods had a mean length of 1.723 m and a standard deviation of 0.069 m.

a. Estimate the mean length of all rods manufactured by this firm, using a 95 percent confidence level.

b. What is the probability that the mean length of all rods falls within the interval 1.723 ± 0.008 m?

c. What is the probability that the mean length of all rods falls within the interval 1.723 ± 0.006 m?

SOLUTION Since the population is extremely large, Eq. (10.6*a*) applies.

$$n = 80 \qquad \bar{X} = 1.723 \text{ m} \qquad s = 0.069 \text{ m}$$

Part a. From Table 10.6, $z = 1.96$ when the confidence level is 95 percent. Substituting in Eq. (10.6*a*), we obtain

$$\text{CI} = 1.723 \pm \frac{1.96(0.069)}{\sqrt{80}} = 1.723 \pm 0.015$$

The confidence limits are $1.723 - 0.015 = 1.708$ m and $1.723 + 0.015 = 1.738$ m. Thus, there is a 95 percent probability that the mean length of all rods manufactured by this firm lies between 1.708 and 1.738 m.

Part b. Substituting in Eq. (10.6*a*) and then solving for z, we obtain the following:

$$\text{CI} = 1.723 \pm \frac{z(0.069)}{\sqrt{80}} = 1.723 \pm 0.008 \qquad z = 1.037$$

From Table 9.4,

$$\text{Area} = 0.3501 \qquad M = 2(0.3501) = 0.7002$$

Thus, there is a 70.02 percent probability that the mean length of all rods lies between 1.715 and 1.731 m.

Part c. Proceeding as before, we obtain $z = 0.778$ and $M = 2(0.2817) = 0.5634$. Thus, there is a 56.34 percent probability that the mean length of all rods lies between 1.717 and 1.729 m.

As we narrow the width of the interval from 0.016 m in Part *b* to 0.012 m in Part *c*, the probability that the population mean lies within that interval drops from 70.02 percent to 56.34 percent.

EXAMPLE 10.10 Solve Example 10.9, Part *a* if the sample size is 130 and the sample results are the same.

SOLUTION

$$\text{CI} = 1.723 \pm \frac{1.96(0.069)}{\sqrt{130}} = 1.723 \pm 0.012$$

Thus, there is a 95 percent probability that the mean length of all rods lies between 1.711 and 1.735 m.

An increase in the sample size enables us to place the population mean within a narrower interval while maintaining the same degree of confidence.

10.2.4 Estimation of Population Proportion

Again let P and p denote the proportion of items of a given type in the population and in the sample, respectively. If $n \geq 30$, the population proportion can be estimated by the method formulated in Art. 10.2.2. Following the same procedure as in Art. 10.2.3, we find that the confidence interval of P corresponding to a given value of z is as follows: For a finite population,

$$\text{CI} = p \pm z\sqrt{\frac{p(1-p)(N-n)}{n(N-1)}} \tag{10.7}$$

For an infinite population,

$$\text{CI} = p \pm z\sqrt{\frac{p(1-p)}{n}} \tag{10.7a}$$

EXAMPLE 10.11 It is known that a culture contains several thousand organisms of varying types. An examination of 75 organisms revealed that 12 of them were of type A. Estimate the proportion of type A organisms in the culture, using a 90 percent confidence level.

SOLUTION Since the population is extremely large, Eq. (10.7a) applies

$$n = 75 \qquad p = 12/75 = 0.16$$

From Table 10.6, $z = 1.645$ when the confidence level is 90 percent. Then

$$\text{CI} = 0.16 \pm 1.645\sqrt{\frac{(0.16)(0.84)}{75}} = 0.16 \pm 0.070$$

The confidence limits are $0.16 - 0.070 = 0.090$ and $0.16 + 0.070 = 0.230$. Thus, there is a 90 percent probability that the proportion of type A organisms in the culture lies between 9.0 percent and 23.0 percent.

10.2.5 Decision Making on Basis of an Estimate

We are continually confronted with the need to make decisions. Generally, making a correct decision requires that we know some parameter of a population: the mean life span of a set of motors, the proportion of defective machine parts in a shipment, the proportion of engineering students who would develop structural visualization more effectively by some novel approach, etc. For the reasons previously discussed, we must often rely on a sample to allow us to estimate the parameter. Since a sample cannot reflect a population with complete precision, the decision must be based on probability rather than certainty.

Many decision-making processes require that we make some assumption concerning a population in order to obtain a basis for investigation. This assumption is termed a *hypothesis*. It is then necessary to determine whether the sample that is drawn confirms or disproves the hypothesis. This investigation is called a *test* of the hypothesis. Thus, the decision-making process requires statistical inference. We shall illustrate the process.

EXAMPLE 10.12 The time required to produce 1 unit of a standard commodity is known to have a mean value of 3.50 h and a standard deviation of 0.64 h. An industrial engineer claims that a revised method of production will substantially reduce production time. His proposed method was tested on 40 units, and it was found that the mean production time was 3.37 h per unit. Management has decided that it will adopt the proposed method only if there is a probability of 95 percent or more that the engineer's claim is valid. What is your recommendation?

SOLUTION The population consists of all units that will be produced under the revised method if it is adopted, and the sample consists of the 40 units that were actually produced under this method. We place the burden of proof on the industrial engineer and formulate the hypothesis that production time under the revised method is coincident with that under the present method. Therefore, according to our hypothesis, the population parameters are as follows:

$$\mu = 3.50 \text{ h} \qquad \sigma = 0.64 \text{ h}$$

If our hypothesis is correct, the difference between the population mean of 3.50 h and the sample mean of 3.37 h is ascribable to sampling error; thus, it is the result of chance. We now test our hypothesis.

We assume that the sampling distribution of the mean is normal, and the resulting sampling-distribution diagram appears in Fig. 10.4. By Eqs. (10.1) and (10.2*a*), the mean of the sampling distribution of the mean is also 3.50 h and the standard deviation of the sampling distribution is

$$\sigma_{\bar{X}} = \frac{\sigma}{\sqrt{n}} = \frac{0.64}{\sqrt{40}} = 0.101 \text{ h}$$

There is a 95 percent probability that the engineer's claim is valid and our hypothesis is false if the true sample mean of 3.37 h is less than the mean corresponding to 95 percent of all possible samples of 40 units. Let $X_{\min}$ denote the minimum sample mean at which our hypothesis is tenable, and let B in Fig. 10.4 denote the point corresponding to $X_{\min}$. The area to the right of B is 0.95. Therefore, for the interval AB,

$$\text{Area} = 0.95 - 0.50 = 0.45 \qquad z = -1.645 \qquad \text{(from Table 9.4)}$$

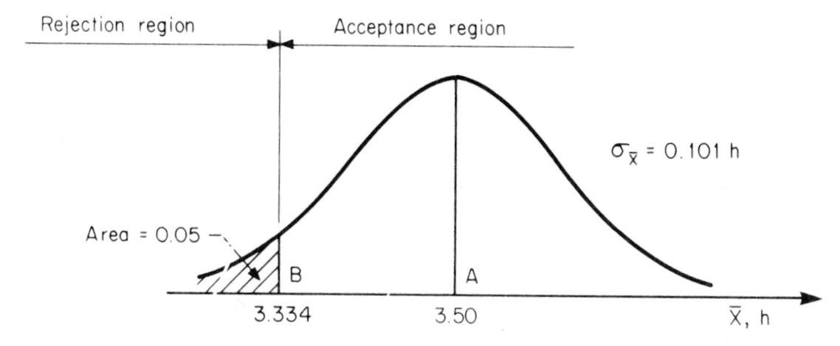

FIGURE 10.4 Acceptance and rejection regions of sampling distribution. *(From Handbook of Engineering Economics; used with permission of McGraw-Hill, Inc.)*

By Eq. (9.31),

$$X_{min} = 3.50 + (-1.645)(0.101) = 3.334 \text{ h}$$

Since the sample mean of 3.37 h exceeds the limiting value, our hypothesis stands. We must therefore recommend that the present production method be retained. In making this recommendation, we are not stating conclusively that the engineer's claim is invalid; we are merely saying that it has failed to demonstrate its superiority to the extent demanded by management. The requirement of 95 percent probability is based on the fact that the firm would incur considerable expense by changing its production method, and the sample result failed to offer sufficient evidence that the expense is warranted. The apparent reduction in production time can be explained by chance.

In Example 10.12, we formulated the hypothesis that production time under the revised method has a mean value of 3.50 h and a standard deviation of 0.64 h. This hypothesis was to be accepted or rejected according to whether the sample mean had a value greater than or less than 3.334 h, respectively. Therefore, with reference to Fig. 10.4, the region of the diagram to the right of B is termed the *acceptance region* corresponding to our hypothesis, and that to the left of B is termed the *rejection region*, as indicated.

10.3 REGRESSION ANALYSIS

10.3.1 Basic Problem in Regression Analysis

Consider that the value assumed by a variable Y is determined primarily by a nonrandom variable X and secondarily by one or more random variables. As an illustration, assume that a projectile is fired from a fixed point with a fixed angle of inclination. Let X denote the velocity with which the projectile is fired and Y denote the horizontal distance from the point of firing to the point at which the projectile strikes the ground. The value of Y is determined primarily by X, but it is also influenced by wind velocity.

Now assume that we are given several pairs of simultaneous values of X and Y and that we must establish the relationship between them. Although the random variables exert only a minor effect on Y, they obscure the relationship that we seek. Therefore, we must apply the known values and establish the relationship that most closely "fits" these values.

In our present investigation, the term *regression* is applied to signify a *relationship*, and the problem of fashioning a relationship to fit a given set of paired values of X and Y is called *regression analysis*.[1] Since the relationship between X and Y must be established on the basis of a limited number of paired values, regression analysis falls within the context of statistical inference.

The definition of regression analysis can be extended. If Y is primarily a function of a single nonrandom variable X, the regression is *simple*. On the other hand, if Y

[1] Use of the term regression in the present context originated with the treatise "Regression Towards Mediocrity in Hereditary Stature" by Francis Galton in 1885. Galton maintained that the children of exceptionally tall or exceptionally short parents tend to be closer to the population mean in stature, and he termed this tendency *regression to the mean*.

is primarily a function of several nonrandom variables $X_1, X_2, \ldots, X_k$, the regression is *multiple*. We shall confine our study to simple regression.

10.3.2 Definition of Regression Line

Consider again that Y is a function of X and also consider that the following pairs of values have been obtained:

X	1	2	3	4	5	6
Y	44	53	54	85	92	117

The points representing these pairs of values are located in Fig. 10.5, where they are labeled A, B, C, D, E, and F. For example, C has an abscissa of 3 and ordinate of 54. These points constitute the *scatter diagram* of X and Y.

The random variables produce an irregularity in the known values of X and Y, and this irregularity manifests itself in the failure of the points in the scatter diagram to fall on a smooth curve or straight line. However, it is possible to draw a curve, such as curve Q, that comes reasonably close to these points, and we therefore say that curve Q *fits* the scatter diagram fairly well. Therefore, the problem of finding the relationship that best fits the known values of X and Y may be described geometrically as that of finding the curve or straight line that best fits the scatter diagram. This curve or straight line is termed the *regression line* or *line of best fit*, and its equation is termed the *regression equation*. It now becomes necessary to formulate a precise definition of the expression *best fit*.

Figure 10.6 shows a curve Q' and a point A that lies off the curve. The vertical distance from A to Q' is termed the *deviation* of A from Q' and is denoted by d. The deviation of a point from Q' is positive or negative according to whether the point lies above or below Q', respectively. Thus, in Fig. 10.6, the deviation of A is positive and that of B is negative.

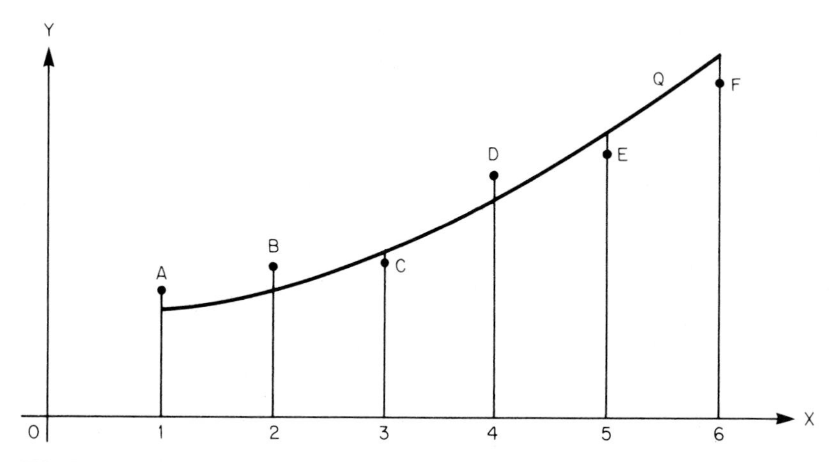

FIGURE 10.5 Scatter diagram and line of reasonable fit. (*From Handbook of Engineering Economics; used with permission of McGraw-Hill, Inc.*)

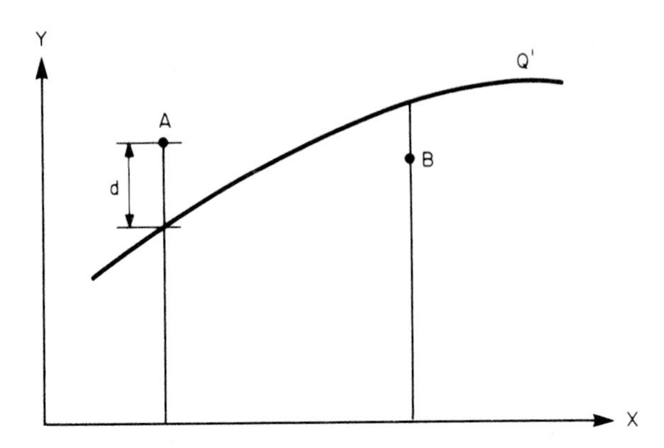

FIGURE 10.6 Deviation of a point from a line. (*From Handbook of Engineering Economics; used with permission of McGraw-Hill, Inc.*)

In Fig. 10.5, some points in the scatter diagram have positive deviations from Q while other points have negative deviations. If we square all deviations, we obtain positive results exclusively. By convention, the regression line is considered to be that for which the sum of the squared deviations (Σd^2) is minimum. Consequently, the regression line is also known as the *least-squares line.*

If the shape of the regression line is unknown, it may be necessary to experiment with several shapes until the most satisfactory one is found. We shall consider three shapes: a straight line, a parabolic arc, and a Gompertz curve. However, the procedure for identifying the regression line is general, and it may be applied to a line of any shape whatever.

In presenting the regression equation, we do not claim that it expresses the exact relationship between X and Y, for that is impossible to identify. We merely state that this equation expresses the *most probable* relationship between the variables.

10.3.3 Equations for Linear Regression

Assume that the relationship between X and Y is known or assumed to be linear. For a given value of X, let Y denote the true value of the dependent variable and Y_c denote the ordinate of the point on the regression line. Then

$$Y_c = a + bX \tag{10.8}$$

where a and b are constants that must be evaluated. These constants are called the *regression coefficients* (or *regression parameters*). Their significance is this: b is the slope of the regression line, and a is its Y intercept (i.e., the distance from the origin to the point where the line intersects the Y axis). The deviation of a point in the scatter diagram from the regression line is

$$d = Y - Y_c = Y - a - bX$$

Let n denote the number of pairs of X and Y values. The sum of the squared deviations for the n pairs of values is

$$\Sigma d^2 = \Sigma Y^2 + na^2 + b^2 \Sigma X^2 - 2a \Sigma Y - 2b \Sigma XY + 2ab \Sigma X$$

To establish the values of a and b at which Σd^2 is minimum, we form the partial derivative of Σd^2 with respect to both a and b and then set each partial derivative equal to zero. This procedure yields the following equations:

$$\Sigma Y = na + b \Sigma X \tag{10.9a}$$

$$\Sigma XY = a \Sigma X + b \Sigma X^2 \tag{10.9b}$$

Solving this system of simultaneous equations for b, we obtain

$$b = \frac{n \Sigma XY - (\Sigma X)(\Sigma Y)}{n \Sigma X^2 - (\Sigma X)^2} \tag{10.10}$$

From Eq. (10.9a),

$$a = \frac{\Sigma Y - b \Sigma X}{n} \tag{10.11}$$

Let $\bar{X}$ and $\bar{Y}$ denote the arithmetic mean of the X and Y values, respectively. Equation (10.11) can be recast in this form:

$$a = \bar{Y} - b\bar{X} \tag{10.11a}$$

EXAMPLE 10.13 The relationship between X and Y is known to be linear, and Table 10.7 presents the observed values of these variables. Establish the regression equation. Then compare the true values of Y with those given by the regression line.

SOLUTION Table 10.7 is completed in the manner shown. Equations (10.10) and (10.11) yield the following:

$$b = \frac{6(60.915) - (23.4)(17.94)}{6(104.84) - (23.4)^2} = -0.6665$$

$$a = \frac{17.94 - (-0.6665)(23.4)}{6} = 5.5894$$

Then

$$Y_c = 5.5894 - 0.6665X$$

The given values of Y and the corresponding values of Y_c are recorded in Table 10.8, and the deviations d are then computed. We find that $\Sigma d = 0$ (with allowance for rounding effects), and the reason for this condition will soon be discussed.

TABLE 10.7

Observation	X	Y	X^2	XY
1	1.7	4.31	2.89	7.327
2	2.5	4.10	6.25	10.250
3	3.6	3.02	12.96	10.872
4	4.3	2.83	18.49	12.169
5	5.2	2.39	27.04	12.428
6	6.1	1.29	37.21	7.869
Total	23.4	17.94	104.84	60.915

TABLE 10.8

Observation	X	Y	Y_c	$d = Y - Y_c$
1	1.7	4.31	4.46	−0.15
2	2.5	4.10	3.92	0.18
3	3.6	3.02	3.19	−0.17
4	4.3	2.83	2.72	0.11
5	5.2	2.39	2.12	0.27
6	6.1	1.29	1.52	−0.23
Total				0.01

As an interesting hypothetical case, assume that we are given only two pairs of X and Y values. By applying Eqs. (10.10) and (10.11), we find that the regression line contains the two points in the scatter diagram, as it logically should.

Equation (10.9) is referred to as the *normal equation* of the straight regression line. The individual equations can readily be memorized by applying this formula: In Eq. (10.8), replace Y_c with Y and then sum each term for the n values. The result is Eq. (10.9a). Returning to Eq. (10.8), again replace Y_c with Y, multiply both sides of the resulting equation by X, and again sum each term. The result is Eq. (10.9b).

10.3.4 Special Relationships in Linear Regression

Let M denote the point in the plane having the coordinates $X = \bar{X}$, $Y = \bar{Y}$. The straight regression line has the following important properties:

1. The algebraic sum of the deviations of the points in the scatter diagram from the regression line is zero.
2. The regression line contains point M.

The proof of property 1 is as follows: Rearranging Eq. (10.9a), we have

$$\Sigma\, Y - na - b\, \Sigma\, X = 0$$

As before,

$$d = Y - Y_c = Y - a - bX$$

Summing the values, we obtain

$$\Sigma\, d = \Sigma\, Y - na - b\, \Sigma\, X = 0$$

The proof of property 2 is as follows: Rearranging Eq. (10.11a), we have

$$\bar{Y} = a + b\bar{X}$$

This equation has the same form as Eq. (10.8), and it follows that point M lies on the regression line.

It is to be emphasized that property 1 is not unique to the regression line, for it applies to all straight lines through point M. Therefore, in Example 10.13, the fact that $\Sigma\, d = 0$ is not conclusive proof that the regression equation is correct.

Let $\Sigma\, Y_c$ and $\bar{Y}_c$ denote the sum and arithmetic mean, respectively, of the ordinates of the regression line corresponding to the given values of X. Applying property 1 of the regression line, we obtain the following:

$$\Sigma d = \Sigma Y - \Sigma Y_c = 0$$

$$\Sigma Y_c = \Sigma Y \qquad \bar{Y}_c = \bar{Y} \tag{10.12}$$

Squaring both sides of Eq. (10.8) and summing, we obtain

$$\Sigma Y_c^2 = na^2 + 2ab\, \Sigma X + b^2\, \Sigma X^2$$

We now perform these operations: Multiply Eq. (10.8) by Y and sum, multiply Eq. (10.9a) by a, multiply Eq. (10.9b) by b, and combine the resulting set of equations. The result is

$$\Sigma Y_c^2 = \Sigma YY_c \tag{10.13}$$

It then follows that

$$\Sigma d^2 = \Sigma Y^2 - \Sigma YY_c = \Sigma Y^2 - \Sigma Y_c^2 \tag{10.14}$$

Since $\Sigma d^2 \geq 0$, we have

$$\Sigma Y^2 \geq \Sigma Y_c^2 \tag{10.15}$$

10.3.5 Simplified Calculations for Linear Regression

The calculations for obtaining the slope of a straight regression line can be simplified considerably, as we shall now demonstrate. In Fig. 10.7, L is the regression line. Its slope is b, its Y intercept is a, and it contains the point M that has the coordinates $X = \bar{X}$, $Y = \bar{Y}$. We displace the origin from O to M.

Let x and y denote the new horizontal and vertical axes, respectively. For any point in the plane, the relationship between the new and original coordinates is as follows:

$$x = X - \bar{X} \qquad y = Y - \bar{Y}$$

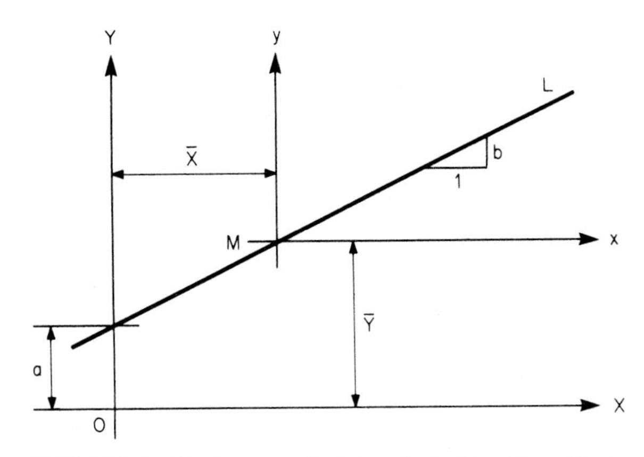

FIGURE 10.7 Displacement of origin to $M(\bar{X}, \bar{Y})$. (*From Handbook of Engineering Economics; used with permission of McGraw-Hill, Inc.*)

Thus, x and y are the deviations of X and Y, respectively, from their arithmetic means. As stated in Art. 8.2.2, the algebraic sum of these deviations is zero; therefore, $\Sigma x = 0$ and $\Sigma y = 0$. Equation (10.10) thus reduces to

$$b = \frac{\Sigma xy}{\Sigma x^2} \tag{10.16}$$

The value of a is then obtained by applying Eq. (10.11*a*). Equation (10.16) reveals, as it must, that the slope of the line is positive when X and Y vary in the same direction and negative when X and Y vary in the opposite direction.

EXAMPLE 10.14 With reference to Example 10.13, formulate the regression equation by applying deviations from the arithmetic mean.

SOLUTION The given values of X and Y are repeated in Table 10.9, and we have the following:

$$\bar{X} = \frac{23.4}{6} = 3.9 \qquad \bar{Y} = \frac{17.94}{6} = 2.99$$

$$x = X - 3.9 \qquad y = Y - 2.99$$

The x and y values are recorded in Table 10.9, as well as the xy and x^2 values. By Eqs. (10.16) and (10.11*a*),

$$b = \frac{-9.051}{13.58} = -0.6665$$

$$a = 2.99 - (-0.6665)(3.9) = 5.5894$$

The regression equation is identical with that in Example 10.13.

TABLE 10.9

Observation	X	Y	x	y	x^2	xy
1	1.7	4.31	−2.2	1.32	4.84	−2.904
2	2.5	4.10	−1.4	1.11	1.96	−1.554
3	3.6	3.02	−0.3	0.03	0.09	−0.009
4	4.3	2.83	0.4	−0.16	0.16	−0.064
5	5.2	2.39	1.3	−0.60	1.69	−0.780
6	6.1	1.29	2.2	−1.70	4.84	−3.740
Total	23.4	17.94	0	0	13.58	−9.051

10.3.6 Transformation of Nonlinear Relationships to Linear Form

By the complete or partial use of logarithms, it is possible to transform two major classes of nonlinear relationships to linear ones. First, consider the relationship

$$y = cx^b \tag{c}$$

where b is a constant, c is a positive constant, and x can assume only positive values. The variable y is restricted to positive values. By Eqs. (1.8) and (1.10), we may recast Eq. (c) in the logarithmic form

$$\log y = \log c + b \log x$$

If we set $Y = \log y$, $X = \log x$, and $a = \log c$, we obtain the linear equation

$$Y = a + bX$$

Thus, $\log y$ is a linear function of $\log x$. Therefore, if the constants c and b must be established by regression analysis, the analysis is vastly simplified by the use of logarithms because the relationship is then transformed to a linear one.

As an aid in visualizing the relationship between x and y, it is helpful to construct the scatter diagram of x and y values in the form of a *log-log drawing*. In a drawing of this type, distances from the origin to points on the x and y axes equal the logarithms of the quantities indicated at these points. Thus, a graph in the plane is a plotting of $\log y$ versus $\log x$, and the graph of Eq. (c) is a straight line. Since $\log 1 = 0$, it follows that $x = y = 1$ at the origin.

As an illustration, refer to Fig. 10.8, where line L is a plotting of the equation

$$y = 0.62x^{1.3}$$

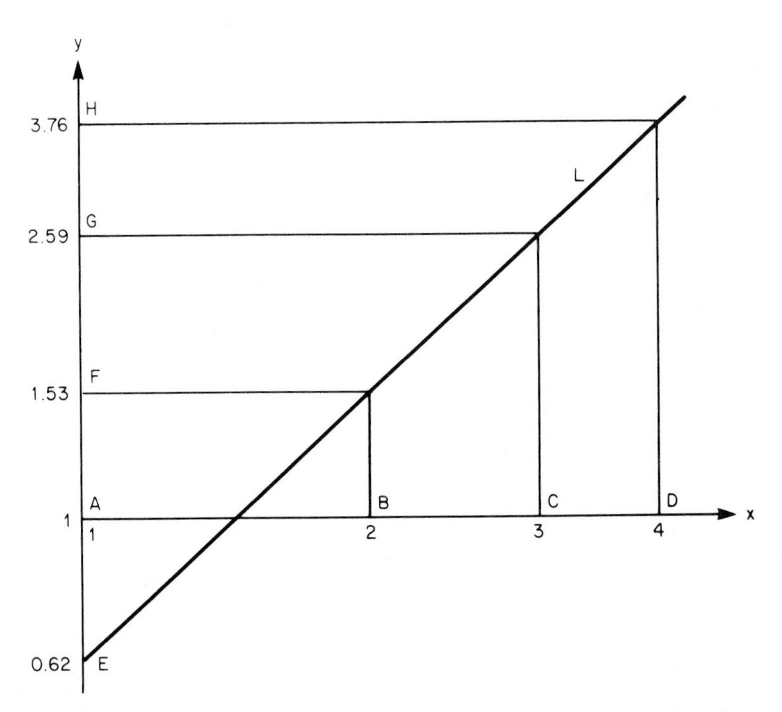

FIGURE 10.8 Graph of $y = 0.62x^{1.3}$ on log-log paper. *Note:* Both scales logarithmic. (*From Handbook of Engineering Economics; used with permission of McGraw-Hill, Inc.*)

We have the following pairs of simultaneous values:

x	1	2	3	4
y	0.62	1.53	2.59	3.76

In Fig. 10.8, $AB = \log 2$, $AC = \log 3$, $AE = \log 0.62$, $AH = \log 3.76$, etc.

As an illustration of a relationship having the form of Eq. (*c*), consider that a perfect gas is subjected to a polytropic process in which the pressure and volume vary. The quantities are related in this manner:

$$pV^n = C \tag{10.17}$$

where p and V denote pressure and volume, respectively, C is a constant, and n is a constant that is characteristic of the particular process. When simultaneous values of p and V are obtained, the precise relationship between these variables may be obscured by some imprecision in the readings. Therefore, n and C must be evaluated by regression analysis, and the analysis can be performed most easily by the use of logarithms. In Example 10.15, we have intentionally exaggerated the imprecision in the readings for illustrative purposes.

EXAMPLE 10.15 During a polytopic expansion of a gas, readings of pressure and volume were taken at various times, and the results are recorded in Table 10.10. Applying regression analysis, estimate the pressure when the volume will be 0.38 m².

SOLUTION In logarithmic form, Eq. (10.17) becomes

$$\log p + n \log V = \log C$$

or

$$\log p = \log C - n \log V$$

Let $X = -\log V$ and $Y = \log p$. The foregoing equation becomes

$$Y = \log C + nX$$

By Eq. (1.9), $X = \log(1/V)$. The values of X and Y as based on common logarithms are recorded in Table 10.10, and the table is then completed in the manner shown.

TABLE 10.10

Pressure (p), kPa	Volume (V), m²	X	Y	X^2	XY
551.6	0.0852	1.0696	2.7416	1.1440	2.9324
479.2	0.0979	1.0092	2.6805	1.0185	2.7052
425.8	0.1025	0.9893	2.6292	0.9787	2.6011
355.3	0.1189	0.9248	2.5506	0.8553	2.3588
291.7	0.1340	0.8729	2.4649	0.7620	2.1516
228.3	0.1725	0.7632	2.3585	0.5825	1.8000
165.9	0.2145	0.6686	2.2198	0.4470	1.4842
103.4	0.3110	0.5072	2.0145	0.2573	1.0218
Total		6.8048	19.6596	6.0453	17.0551

In the present case, the regression coefficients in Eq. (10.8) are $a = \log C$ and $b = n$. Applying Eqs. (10.10) and (10.11), we obtain the following values:

$$n = \frac{8(17.0551) - (6.8048)(19.6596)}{8(6.0453) - (6.8048)^2} = 1.294$$

$$\log C = \frac{19.6596 - (1.294)(6.8048)}{8} = 1.3568 \qquad C = 22.74$$

Thus, the regression equation in the specified units is

$$pV^{1.294} = 22.74$$

When $V = 0.38\ \text{m}^2$, the estimated pressure is

$$p = \frac{22.74}{(0.38)^{1.294}} = 79.5\ \text{kPa}$$

Now consider the relationship

$$y = df^{gx} \tag{d}$$

where g is a constant and d and f are positive constants. The variable y is restricted to positive values. By Eq. (1.2), we may rewrite Eq. (d) in this form:

$$y = d(f^g)^x$$

Then

$$\log y = \log d + (\log f^g)x$$

If we set $Y = \log y$, $a = \log d$, and $b = \log f^g$, we obtain the linear equation

$$Y = a + bx$$

Thus, $\log y$ is a linear function of x. Therefore, if d and f^g must be found by regression analysis, the analysis is vastly simplified by applying the given x values and the logarithms of the given y values.

The graph of Eq. (d) is a straight line when plotted in the form of a *semilogarithmic drawing*. In a drawing of this type, the standard scale is used along the x axis and the logarithmic scale is used along the vertical axis. Thus, at the origin, $x = 0$ and $y = 1$.

As an illustration, refer to Fig. 10.9, where line L' is a plotting of the equation

$$y = 3 \times 8^{-0.4x}$$

We have the following pairs of simultaneous values:

x	0	1	2	3	4
y	3	1.31	0.57	0.25	0.11

In Fig. 10.9, $AB = 1$, $AC = 2$, $AF = \log 3$, $AG = \log 1.31$, etc.

Equation (d) is the general form of a vast number of equations that arise in engineering and science. For example, it encompasses Eq. (d) in Art. 12.4.2, which is an expression of the law of natural growth or decay.

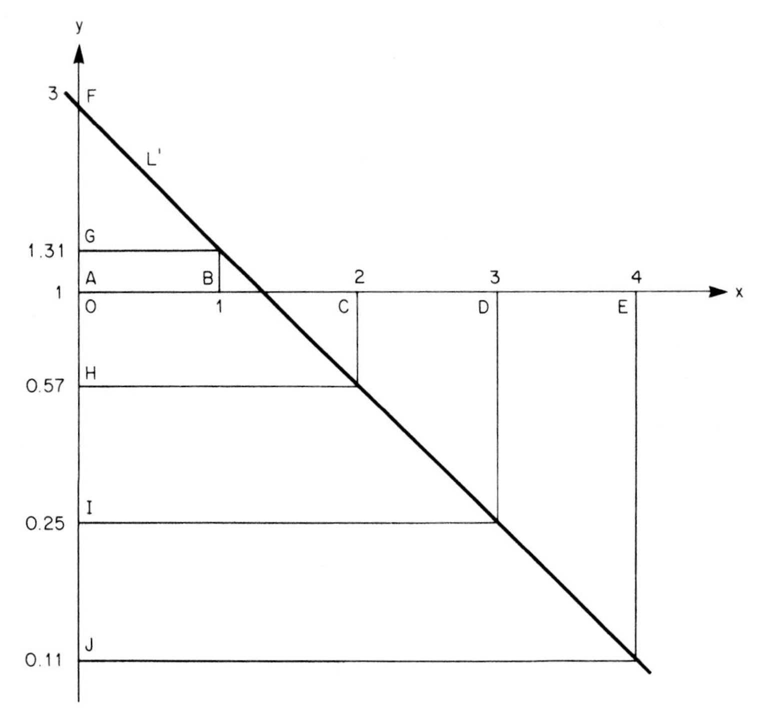

FIGURE 10.9 Graph of $y = 3 \times 8^{-0.4x}$ on semilogarithmic paper. *Note:* x scale standard; y scale logarithmic. (*From Handbook of Engineering Economics; used with permission of McGraw-Hill, Inc.*)

10.3.7 Parabolic Regression

Again consider that Y is primarily a function of X and now assume that the relationship is known or believed to be of the second degree in X. The regression line is a parabolic arc, and it becomes necessary to formulate the equation of this arc.

As before, let Y_c denote the ordinate of the point on the regression line corresponding to a given value of X. Then

$$Y_c = a + bX + cX^2 \tag{10.18}$$

where a, b, and c are constants that must be determined. The deviation d of a point in the scatter diagram from the regression line is

$$d = Y - Y_c = Y - a - bX - cX^2$$

Again, the regression line is considered to be that for which $\Sigma\, d^2$ is minimum. To evaluate the regression coefficients, we form the partial derivative of $\Sigma\, d^2$ with respect to a, b, and c and then set each partial derivative equal to zero. The following equations emerge:

$$\Sigma\, Y = na + b\,\Sigma\, X + c\,\Sigma\, X^2 \tag{10.19a}$$

$$\Sigma\, XY = a\,\Sigma\, X + b\,\Sigma\, X^2 + c\,\Sigma\, X^3 \tag{10.19b}$$

$$\Sigma\, X^2 Y = a\,\Sigma\, X^2 + b\,\Sigma\, X^3 + c\,\Sigma\, X^4 \tag{10.19c}$$

The foregoing are referred to as the *normal equations* of the parabolic regression line. By solving this system of simultaneous equations, we obtain the values of a, b, and c. In Art. 10.3.3, we presented a simple formula for memorizing the normal equations of linear regression. An analogous formula that is based on Eq. (10.18) enables us to memorize Eq. (10.19) as well.

Equation (10.19a) yields $\Sigma\, d = 0$. Following a procedure similar to that in Art. 10.3.4, we find that Eqs. (10.13), (10.14), and (10.15) apply to parabolic as well as linear regression. However, the two forms of regression have this important difference: Whereas the straight regression line contains the point $M(\bar{X}, \bar{Y})$, the parabolic regression line does not.

EXAMPLE 10.16 Table 10.11 presents observed values of X and Y. Assuming that Y is a second-degree function of X, formulate the regression equation. Verify the result by demonstrating that Eq. (10.14) is satisfied.

SOLUTION Since the X values form an arithmetic progression, we can simplify the calculations by introducing the variable $x = X - \bar{X}$. In this manner, $\Sigma\, x$ and $\Sigma\, x^3$ are reduced to zero. Then

$$\bar{X} = \frac{\Sigma\, X}{n} = \frac{36}{9} = 4 \qquad \text{and} \qquad x = X - 4$$

Table 10.11 is completed in the manner shown. Substitution in Eq. (10.19) yields the following:

$$68.7 = 9a + 60c$$

$$34.2 = 60b$$

$$412.8 = 60a + 708c$$

TABLE 10.11

	X	Y	x	x^2	x^4	xY	x^2Y
	0	4.0	−4	16	256	−16.0	64.0
	1	5.7	−3	9	81	−17.1	51.3
	2	6.6	−2	4	16	−13.2	26.4
	3	8.0	−1	1	1	−8.0	8.0
	4	8.6	0	0	0	0	0
	5	9.1	1	1	1	9.1	9.1
	6	9.3	2	4	16	18.6	37.2
	7	8.8	3	9	81	26.4	79.2
	8	8.6	4	16	256	34.4	137.6
Total	36	68.7		60	708	34.2	412.8

The solution of this system of simultaneous equations is

$$a = 8.6117 \qquad b = 0.5700 \qquad c = -0.14675$$

Then
$$Y_c = 8.6117 + 0.5700x - 0.14675x^2$$
$$= 8.6117 + (0.5700)(X - 4) - (0.14675)(X - 4)^2$$
$$Y_c = 3.9837 + 1.7440X - 0.14675X^2$$

In Table 10.12, the given values of X and Y are repeated, and the calculated values of Y_c are recorded. The table is completed in the manner shown. We find that $\Sigma\, YY_c = \Sigma\, Y_c^2$ (with allowance for rounding effects), and the regression equation is thus confirmed.

TABLE 10.12

X		Y	Y_c	YY_c	Y_c^2
0		4.0	3.98	15.920	15.8404
1		5.7	5.58	31.806	31.1364
2		6.6	6.88	45.408	47.3344
3		8.0	7.89	63.120	62.2521
4		8.6	8.61	74.046	74.1321
5		9.1	9.03	82.173	81.5409
6		9.3	9.16	85.188	83.9056
7		8.8	9.00	79.200	81.0000
8		8.6	8.54	73.444	72.9316
	Total	68.7	68.67	550.305	550.0735

10.3.8 Gompertz Curve as Regression Line

There are many situations in engineering, science, and economics where the line that best fits the scatter diagram is a Gompertz curve. The equation of this type of curve is as follows:

$$Y = ka^{b^X} \tag{10.20}$$

where a, b, and k are positive constants and a and b have values other than 1. Then Y is positive, and the equation can be recast in the logarithmic form

$$\log Y = \log k + b^X \log a \tag{10.20a}$$

The Gompertz curve has four possible forms. These forms and their corresponding characteristics are as follows:

Form 1A:	$0 < a < 1$	$0 < b < 1$
Form 1B:	$0 < a < 1$	$b > 1$
Form 2A:	$a > 1$	$0 < b < 1$
Form 2B:	$a > 1$	$b > 1$

Figure 10.10 presents curves having the forms 1A and 2A.

We shall now apply the following relationship:

$$b^X = \left(\frac{1}{b}\right)^{-X}$$

Therefore, if we start with a given point on a Gompertz curve, replace b in Eq. (10.20) with its reciprocal and replace X with $-X$, we obtain another point that

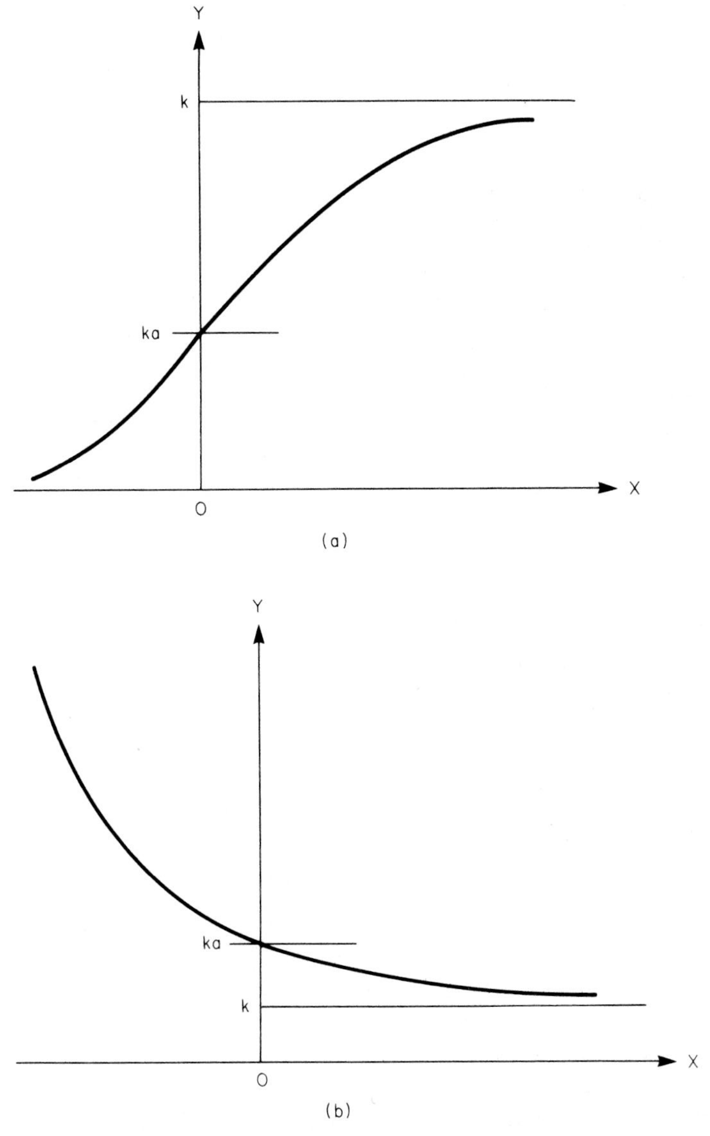

FIGURE 10.10 Gompertz curves. (*a*) Form 1A; (*b*) form 2A. (*From Handbook of Engineering Economics; used with permission of McGraw-Hill, Inc.*)

has the same ordinate and is symmetrically located with respect to the Y axis. It follows that the replacement of b with its reciprocal transforms a 1A curve to a 1B curve (or vice versa), and it transforms a 2A curve to a 2B curve (or vice versa). The transformed curve is a mirror image of the original curve with respect to the Y axis.

We shall now analyze the shape of the Gompertz curve, and this can be done most easily by assigning numerical values to a and b. For simplicity, we set $c = b^X$, and the equation of the curve becomes $Y = ka^c$. When $X = 0$, $c = 1$, and $Y = ka$. Now consider the following cases:

Case 1 (form 1A curve):

$$Y = k(0.6)^c \qquad \text{where } c = (0.2)^X$$

Case 2 (form 2A curve):

$$Y = k(1.5)^c \qquad \text{where } c = (0.2)^X$$

When X increases, c decreases. Therefore, Y increases in Case 1 and decreases in Case 2. As X becomes infinitely large, c approaches 0. Therefore, Y approaches k in both cases. As X decreases without limit, c becomes infinitely large. Therefore, Y approaches 0 in Case 1 and Y becomes infinitely large in Case 2. As Fig. 10.10 shows, the 1A curve has double curvature because Y is bounded at both ends, and the 2A curve has single curvature because Y is unbounded at one end.

Let X denote elapsed time, and consider solely the part of the Gompertz curve that lies to the right of the Y axis. The 1A curve describes a variable that increases consistently as time elapses but approaches a specific limit. Similarly, the 2A curve describes a variable that decreases consistently as time elapses but approaches a specific nonzero limit.

The procedure for devising the Gompertz regression equation is relatively simple where the following conditions exist: The initial value of X is 0; X assumes consecutive integral values; and the number of pairs of known X and Y values is a multiple of 3. The procedure is as follows: Let n denote the number of pairs of values, and let $p = n/3$. Taking the Y values in sequence, divide them into three groups, each containing p values. Let $\Sigma_1 \log Y$, $\Sigma_2 \log Y$, and $\Sigma_3 \log Y$ denote the sum of the logarithms in the first, second, and third groups, respectively. The parameters of the regression equation have these values:

$$b^p = \frac{\Sigma_3 \log Y - \Sigma_2 \log Y}{\Sigma_2 \log Y - \Sigma_1 \log Y} \tag{10.21a}$$

$$\log a = (\Sigma_2 \log Y - \Sigma_1 \log Y)\frac{b-1}{(b^p-1)^2} \tag{10.21b}$$

$$\log k = \frac{1}{p}\left(\Sigma_1 \log Y - \frac{b^p-1}{b-1}\log a\right) \tag{10.21c}$$

In applying these equations, we may use either common or natural logarithms. In Example 10.17, we have again intentionally exaggerated the imprecision in our readings.

EXAMPLE 10.17 The mass of a certain substance that is present in a solution diminishes as time elapses but approaches a nonzero limit. Readings of the mass were taken at uniform intervals, and the results are recorded in columns 2 and 3 of Table 10.13, where t denotes elapsed time and m denotes the mass. It may be

TABLE 10.13

Group (1)	t (2)	m (3)	m_c (4)
1	0	32.12	32.15
	1	29.59	29.54
	2	27.54	27.38
	3	25.43	25.58
2	4	24.21	24.07
	5	23.10	22.80
	6	22.04	21.71
	7	20.09	20.78
3	8	19.98	19.98
	9	19.27	19.29
	10	18.59	18.70
	11	18.31	18.18

assumed that the graph of mass versus time is a Gompertz curve. Applying regression analysis, formulate the equation of the curve.

SOLUTION We replace X and Y in Eq. (10.20) with t and m, respectively. Let m_c denote the mass that is present as given by the regression equation. Divide the given set of values into the three groups shown.

$$n = 12 \qquad p = 12/3 = 4$$

Taking the common logarithms of the m values and totaling them for each group, we obtain the following results:

$$\Sigma_1 \log m = 5.82323 \qquad \Sigma_2 \log m = 5.39380 \qquad \Sigma_3 \log m = 5.11744$$

Equation (10.21) yields the following:

$$b^4 = \frac{5.11744 - 5.39380}{5.39380 - 5.82323} = 0.6436 \qquad b = 0.8957$$

$$\log a = (-0.42943)\frac{-0.1043}{(-0.3564)^2} = 0.35262 \qquad a = 2.2523$$

$$\log k = \frac{1}{4}\left[5.82323 - \frac{-0.3564}{-0.1043}(0.35262)\right] \qquad k = 14.2750$$

The regression equation is

$$m_c = 14.2750(2.2523)^{0.8957^t}$$

For comparison purposes, the values of m_c are recorded in column 4 of Table 10.13.

10.3.9 Reversal of Functional Relationship

Again consider that two quantities, X and Y, vary jointly and that we must establish the relationship between them by regression analysis. It is necessary to assign the role of independent variable to one quantity and the role of dependent variable to the other. In most situations, the roles are apparent. However, there are other situations where the roles can be assigned in an arbitrary manner. For example, where the pressure and volume of a gas both vary during a thermodynamic process, either quantity may be considered the independent variable. In Example 10.15, our task was to estimate the pressure corresponding to a given volume, and it was logical to view pressure as a function of volume.

Regardless of the manner in which the roles are assigned, the X and Y axes are always drawn in a horizontal and vertical position, respectively. However, as can readily be demonstrated, a reversal of the roles causes a change in the regression line. If we consider Y to be a function of X, the regression line that results is called the *regression line of Y on X*; if we consider X to be a function of Y, this line is called the *regression line of X on Y*.

Our study of regression analysis has been based on the assumption that Y is a function of X. If we consider the reverse to be true, it becomes necessary to adjust the preceding equations by interchanging X and Y. As an illustration, consider that X is a linear function of Y, and let X_c denote the abscissa of a point on the regression line corresponding to a given value of Y. Equation (10.8) becomes

$$X_c = a + bY$$

and Eq. (10.10) becomes

$$b = \frac{n \Sigma XY - (\Sigma X)(\Sigma Y)}{n \Sigma Y^2 - (\Sigma Y)^2}$$

10.4 CORRELATION AND ACCURACY OF ESTIMATES

10.4.1 Definitions and Basic Problems

If there is a discernible relationship between the manner in which X varies and the manner in which Y varies, these two variables are said to be *correlated*, and the strength of the relationship is termed the *correlation* between X and Y. For example, there is a correlation between the position of the moon relative to the earth and the height of the water surface at a given point in the ocean, and there is a correlation between the demand for a commodity and the price of the commodity.

If two quantities are correlated, it does not necessarily follow that one is a function of the other. For example, assume that X is a function of U and Y is a function of both U and W. A change in U produces related changes in X and Y; therefore, X and Y are correlated. If the correlation is high, we may conclude that Y is influenced strongly by U and only marginally by W.

When two quantities are correlated, a regression line can be drawn to exhibit their relationship. The correlation is said to be *linear* or *parabolic* if the regression line is a straight line or a parabolic arc, respectively. The correlation is described as *positive* or *direct* if the regression line has a positive slope at all points and *negative* or *inverse* if this line has a negative slope at all points.

In the preceding analysis of regression, we dealt with the problem of formulating the regression equation for two quantities that are known or assumed to be correlated. However, it is now apparent that a more fundamental problem exists: to determine whether two quantitites are truly correlated, and if they are, to appraise the correlation. Thus, it becomes necessary to devise an effective measure of correlation.

Another problem that exists is the following: Assume that we apply the regression equation to estimate the values of one quantity corresponding to given values of the other quantity. By what amount can we reasonably expect the true values of the first quantity to differ from the estimated values? Thus, we must appraise the *accuracy* of our estimates.

The two problems we have presented will be considered in reverse order. To simplify our subsequent discussion, we shall assume that Y is a function of X. For convenience, we repeat our notational system. For a given value of X, Y and Y_c denote, respectively, the true value of the function and its value as given by the regression equation; d denotes the deviation $Y - Y_c$; and n denotes the number of pairs of known values of X and Y.

10.4.2 Standard Deviation of Regression

Assume that we have applied the regression equation to obtain estimated values of Y corresponding to assigned values of X. What magnitude of error is possible? In an individual case, the error is $Y - Y_c = d$. Therefore, we must compute some average value of the deviations to secure an answer to the question we have posed.

We have found that $\Sigma\, d = 0$ for both linear and parabolic regression; therefore, the arithmetic mean of the deviations is useless for our present purpose. However, we can square the deviations, find the arithmetic mean of the d^2 values, and then take the square root of the mean. The result is known as the *standard deviation of regression* or *standard error of estimate* of Y on X, and it is denoted by s_{YX}. Then

$$s_{YX} = \sqrt{\frac{\Sigma\, d^2}{n}} = \sqrt{\frac{\Sigma\,(Y - Y_c)^2}{n}} \tag{10.22}$$

This quantity is the root mean square of the deviations, and it is completely analogous to the standard deviation of a variable from its arithmetic mean. The square of s_{YX} is known as the *variance of regression.*

If the regression is linear or parabolic, Eq. (10.14) enables us to express s_{YX} directly in terms of the regression coefficients. For linear regression, we have

$$s_{YX} = \sqrt{\frac{\Sigma\, Y^2 - a\,\Sigma\, Y - b\,\Sigma\, XY}{n}} \tag{10.23a}$$

For parabolic regression,

$$s_{YX} = \sqrt{\frac{\Sigma\, Y^2 - a\,\Sigma\, Y - b\,\Sigma\, XY - c\,\Sigma\, X^2Y}{n}} \tag{10.23b}$$

10.4.3 Coefficient of Determination

The standard deviation of regression is of no value as an aid in appraising correlation because it expresses probable error on an absolute rather than a relative basis. For example, assume that $Y - Y_c = 1$ in an individual case. If $Y_c = 1000$, the error of estimate is 0.1 percent. On the other hand, if $Y_c = 5$, this error is 20 percent.

Manifestly, an effective measure of correlation must be based on *relative* values, and we shall now proceed to devise such a measure. It rests on the following relationship that applies to both linear and parabolic regression:

$$\Sigma (Y - \bar{Y})^2 = \Sigma (Y_c - \bar{Y})^2 + \Sigma (Y - Y_c)^2 \tag{10.24}$$

This relationship is proved by expanding each term and applying Eqs. (10.12) and (10.13). Each term in Eq. (10.24) has a nonnegative value, and the last term is zero in the hypothetical case where all points in the scatter diagram lie on the regression line. Therefore,

$$\Sigma (Y_c - \bar{Y})^2 \leq \Sigma (Y - \bar{Y})^2 \tag{10.25}$$

It can be demonstrated that this relationship is true in general.

Figure 10.11 shows a straight regression line L. This line contains the point $M(\bar{X}, \bar{Y})$. Let Y denote the ordinate of an arbitrary point A in the scatter diagram.

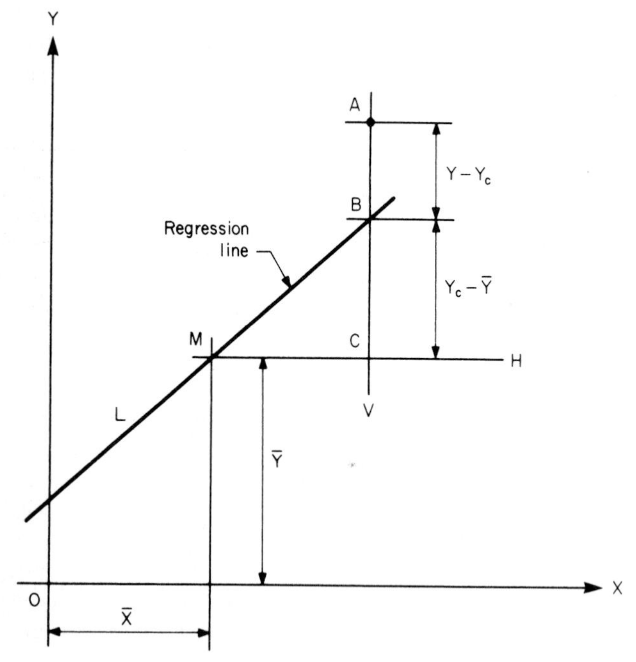

FIGURE 10.11 Deviation of point in scatter diagram from mean value. CB = explained deviation; BA = unexplained deviation. (*From Haddbook of Engineering Economics; used with permission of McGraw-Hill, Inc.*)

Through M, draw line H parallel to the X axis; through A, draw line V parallel to the Y axis. Lines L and V intersect at B; lines H and V intersect at C. The distance $CA = Y - \bar{Y}$ is termed the *deviation* of Y from $\bar{Y}$, and it may be resolved into two parts: $CB = Y_c - \bar{Y}$, and $BA = Y - Y_c$. The part CB is ascribable to the fact that the X value of A differs from $\bar{X}$. Thus, this part reflects the influence of X on Y, and it is accordingly termed the *explained deviation*. The part BA reflects the influence of the random variables on Y, and it is accordingly termed the *unexplained deviation*.

Both the explained and unexplained deviations may be positive or negative. To obtain positive values exclusively, we square the deviations, and we then sum the squared deviations for all given values of X. We now introduce the following definitions: $\Sigma (Y - \bar{Y})^2$ is the *total variation* of the given data with reference to $\bar{Y}$, $\Sigma (Y_c - \bar{Y})^2$ is the *explained variation*, and $\Sigma (Y - Y_c)^2$ is the *unexplained variation*. These definitions enable us to express Eq. (10.24) verbally, in this manner: For both linear and parabolic regression,

$$\text{Total variation} = \text{explained variation} + \text{unexplained variation}$$

It is now apparent that the correlation between X and Y can be measured by the following quantity, which is known as the *coefficient of determination*:

$$r^2 = \frac{\text{explained variation}}{\text{total variation}} = \frac{\Sigma (Y_c - \bar{Y})^2}{\Sigma (Y - \bar{Y})^2} \tag{10.26}$$

Although this definition was formulated by referring to a straight regression line, it is to be emphasized that the definition applies to both linear and nonlinear regression.

The coefficient of determination is perforce positive, and Eq. (10.25) discloses that it has an upper limit of 1. This upper limit corresponds to the hypothetical situation where all points on the scatter diagram lie on the regression line. In this situation, Y is governed exclusively by X, and the correlation is perfect. At the other extreme, if Y is governed predominantly by random variables and only marginally by X, the points in the scatter diagram deviate very widely from the regression line, and the coefficient of determination is extremely small.

EXAMPLE 10.18 It is known that Y is a linear function of X, and the known values of X and Y are recorded in Table 10.14. Compute the coefficient of determination.

TABLE 10.14

X		Y	Y_c	$(Y - \bar{Y})^2$	$(Y_c - \bar{Y})^2$
5		21.42	22.32	19.3600	28.0900
6		20.96	20.80	15.5236	14.2884
7		20.06	19.29	9.2416	5.1529
8		18.43	17.78	1.9881	0.5776
9		15.46	16.27	2.4336	0.5625
10		15.02	14.75	4.0000	5.1529
11		13.95	13.24	9.4249	14.2884
12		10.86	11.73	37.9456	27.9841
	Total	136.16		99.9174	96.0968

SOLUTION

$$\bar{Y} = \frac{\Sigma Y}{n} = \frac{136.16}{8} = 17.02$$

Applying Eqs. (10.10) and (10.11), we obtain the regression equation

$$Y_c = 29.88 - 1.5126X$$

Table 10.14 is completed in the manner shown. By Eq. (10.26),

$$r^2 = \frac{96.0968}{99.9174} = 0.9618$$

The coefficient of determination can be expressed in alternative forms. First, by applying Eq. (10.24) for linear and parabolic regression, we can transform Eq. (10.26) to the following:

$$r^2 = \frac{\Sigma (Y - \bar{Y})^2 - \Sigma (Y - Y_c)^2}{\Sigma (Y - \bar{Y})^2}$$

or

$$r^2 = 1 - \frac{\Sigma (Y - Y_c)^2}{\Sigma (Y - \bar{Y})^2} \tag{10.27}$$

or

$$r^2 = 1 - \frac{\text{unexplained variation}}{\text{total variation}}$$

The fraction in Eq. (10.27) is known as the *coefficient of nondetermination* and its square root is known as the *coefficient of alienation*. Now

let $s_{Y,c}$ = standard deviation of Y_c values from $\bar{Y}$
s_Y = standard deviation of Y values from $\bar{Y}$

If we divide numerator and denominator of the fraction in Eq. (10.26) by n, we obtain

$$r^2 = \frac{s_{Y,c}^2}{s_Y^2} \tag{10.28}$$

Similarly, Eq. (10.27) becomes

$$r^2 = 1 - \frac{s_{YX}^2}{s_Y^2} \tag{10.29}$$

10.4.4 Coefficient of Correlation

Since the coefficient of determination is perforce positive, it fails to reveal whether the correlation between the two variables is positive or negative. To incorporate this information, we take the square root of the coefficient of determination, which we denote by r, and assign to it a plus or minus sign according to whether the correlation is positive or negative, respectively. The quantity r is termed the *coefficient of correlation*. Thus,

$$r = \pm\sqrt{\frac{\Sigma (Y_c - \bar{Y})^2}{\Sigma (Y - \bar{Y})^2}} \tag{10.30}$$

From Eq. (10.28),

$$r = \pm \frac{s_{Y,c}}{s_Y} \tag{10.31}$$

Since r^2 has an upper limit of 1, r can range from $+1$ (for perfect positive correlation) to -1 (for perfect negative correlation).

If the regression is linear, an alternative expression for r is available. Consider that we displace the origin to the point having the original coordinates $\bar{X}$ and $\bar{Y}$, as shown in Fig. 10.7. Equation (10.30) can be recast in the notation of Art. 10.3.5, giving

$$r = \pm \sqrt{\frac{\Sigma\, y_c^2}{\Sigma\, y^2}}$$

where $y_c = Y_c - \bar{Y}$. By Eq. (10.16),

$$y_c = bx = \frac{\Sigma\, xy}{\Sigma\, x^2} x$$

Substituting above and omitting the $\pm$ sign, we obtain

$$r = \frac{\Sigma\, xy}{\sqrt{(\Sigma\, x^2)(\Sigma\, y^2)}} \tag{10.32}$$

Equation (10.32) is referred to as the *product-moment formula* for the coefficient of linear correlation. It has the following important properties:

1. The term $\Sigma\, xy$ is positive or negative according to whether X and Y tend to vary in the same or opposite direction, respectively. Expressed in our present terminology, $\Sigma\, xy$ is positive or negative according to whether the correlation between X and Y is positive or negative, respectively. Therefore, if we consider the denominator in Eq. (10.32) to be positive, we obtain the proper algebraic sign of r directly from the numerical values. It is for this reason that the $\pm$ sign is omitted.
2. It obviates the need for formulating the regression equation when our sole objective is to evaluate the correlation.
3. An interchange of x and y does not affect the value of r. Thus, it is immaterial whether we consider Y to be a function of X or vice versa. Since we stated that the concept of correlation does not require or imply any functional relationship, it was to be anticipated that the expression for correlation would allow an interchange of the variables, and it does.

EXAMPLE 10.19 With reference to Example 10.18, compute the coefficient of correlation of the two variables by applying the product-moment formula.

SOLUTION The X and Y values are repeated in Table 10.15. We found that $\bar{Y} = 17.02$, and we have $\bar{X} = 68/8 = 8.5$. Then

$$x = X - \bar{X} = X - 8.5 \qquad y = Y - \bar{Y} = Y - 17.02$$

The values of y^2 appear in Table 10.14 under the heading $(Y - \bar{Y})^2$, and $\Sigma\, y^2 = 99.9174$. The remaining values that are required in our present calculation

TABLE 10.15

X	Y	x	y	x^2	xy
5	21.42	−3.5	4.40	12.25	−15.400
6	20.96	−2.5	3.94	6.25	−9.850
7	20.06	−1.5	3.04	2.25	−4.560
8	18.43	−0.5	1.41	0.25	−0.705
9	15.46	0.5	−1.56	0.25	−0.780
10	15.02	1.5	−2.00	2.25	−3.000
11	13.95	2.5	−3.07	6.25	−7.675
12	10.86	3.5	−6.16	12.25	−21.560
Total				42.00	−63.530

are recorded in Table 10.15. Equation (10.32) yields

$$r = \frac{-63.530}{\sqrt{42(99.9174)}} = -0.9807$$

If we apply the coefficient of determination obtained in Example 10.18, we obtain

$$r = -\sqrt{0.9618} = -0.9807$$

Thus, our present result is consistent with that in Example 10.18.

10.5 REFERENCES

Guenther, William C.: *Concepts of Statistical Inference*, 2d ed., McGraw-Hill, New York, 1973.

Kurtz, Max: *Handbook of Engineering Economics*, McGraw-Hill, New York, 1984 (Secs. 7, 9, and 12).

Lapin, Lawrence L.: *Probability and Statistics for Modern Engineering*, 2d ed., PWS-Kent, Boston, 1990.

McPherson, Glen: *Statistics in Scientific Investigation*, Springer-Verlag, New York, 1990.

Wadsworth, Harison M., Jr.: *Handbook of Statistical Methods for Engineers and Scientists*, McGraw-Hill, New York, 1990.

Wall, Francis J.: *Statistical Data Analysis Handbook*, McGraw-Hill, New York, 1986.

SECTION 11

INTRODUCTION TO CYBERNETICS

In cybernetics, the term *machine* or *dynamic system* is applied in a very broad sense to refer to any entity that is capable of undergoing a change of state. The entity may be an electronic mechanism, an industrial system, the body of an organism, or a satellite in space. The change of state is assumed to occur discretely. This premise does not conflict with the fact that many changes occur continuously. By making the interval between successive changes progressively smaller, we approach continuity, and that is the technique applied in calculus.

Cybernetics is essentially a system of logic. Because it studies the behavior of machines in general, its results are applicable to diverse areas of engineering and science. Cybernetics provides a unifying thread, enabling us to see analogies and similarities among problems that arise in seemingly unrelated subjects. Two individuals, Norbert Wiener and C. E. Shannon, pioneered in the development of cybernetics.

11.1 TRANSFORMATIONS

11.1.1 Definitions and Notation

If an object has a characteristic that is capable of changing, that characteristic is a *variable*. The variable can be the color of a traffic light, the phase of the moon, or the temperature of a body. A specific quality or numerical value of the variable is termed the *state* of the object. For example, if the variable is the direction in which a pendulum is swinging, there are two possible states: clockwise and counterclockwise. In a general discussion, the state of an object is denoted by a letter, in either uppercase or lowercase form.

Consider that state a is converted to state b by an agent Z. State a is the *operand*, state b is the *transform*, agent Z is the *operator*, and the conversion of a to b is a *transition*. As an illustration, assume that heat is applied to convert ice to water. The variable is the phase (solid, liquid, or vapor) of the substance. The solid phase (ice) is the operand, the liquid phase (water) is the transform, the applied heat is the operator, and the conversion of ice to water is the transition.

Usually, the effect of the operator is to change the state of a *set of objects* rather than that of an individual object. For example, a gust of wind may cause leaves to fall, papers to blow, and a boat to capsize. The entire set of transitions is called a *transformation*, and it is denoted by an uppercase letter.

If the transformation is relatively simple, it can be exhibited by recording the set of operands in a row and placing the transforms directly below their corresponding operands. An arrow is then drawn from the upper to the lower row. As an illustration, consider that the entire set of integers from 0 to 9, inclusive, is to be transformed by replacing each integer with its successor, with 9 replaced with 0. This transformation, which we shall label T, can be represented in this manner:

$$T: \quad \downarrow \begin{array}{cccccccc} 0 & 1 & 2 & 3 & 4 & \cdots & 8 & 9 \\ 1 & 2 & 3 & 4 & 5 & \cdots & 9 & 0 \end{array}$$

If the state of an object is a numerical quantity and the transformation is expressible by means of an equation, the transformation is described as *algebraic*. As an illustration of an algebraic transformation, assume that all integers are to be increased by 6. In this transformation, every integer is an operand. Let x denote an operand and x' denote its transform. The equation that expresses the transformation is this:

$$x' = x + 6$$

In this illustrative case, the set of operands is of infinite extent.

A transformation is *closed* if each transform is present in the set of operands; otherwise, it is *open*. As an illustration, consider the following transformations:

$$\downarrow \begin{array}{ccccc} 5 & 6 & 7 & 8 & 9 \\ 8 & 7 & 9 & 6 & 5 \end{array} \qquad \downarrow \begin{array}{ccccc} 1 & 2 & 3 & 4 & 5 \\ 4 & 5 & 6 & 7 & 8 \end{array}$$

The first transformation is closed, but the second is open because the transforms 6, 7, and 8 are not present in the set of operands.

A *single-valued transformation* is one in which each operand has only one possible transform; a *multivalued transformation* is one in which at least one operand has several possible transforms. At present, we shall deal exclusively with single-valued transformations.

A *one-one transformation* occurs when each transform corresponds to a unique operand, and a *many-one transformation* occurs when at least one transform corresponds to multiple operands. The following transformations serve as illustrations:

$$\downarrow \begin{array}{ccccc} A & B & C & D & E \\ C & D & E & B & A \end{array} \qquad \downarrow \begin{array}{ccccc} A & B & C & D & E \\ C & D & D & E & A \end{array}$$

The first transformation is one-one, but the second is many-one because the transform D corresponds to two operands: B and C.

An *identity transformation* is one in which each transform coincides with its operand. Thus, the effect of the transformation is null. We shall denote the identity transformation by I.

It is often necessary to determine whether a given transformation is reversible or irreversible. Manifestly, a transformation is reversible only if it is closed and one-one.

In many instances, it is desirable to express a transformation in matrix form. The construction is as follows: Assign a *column* to each operand and a *row* to each possible transform. Label the columns and rows. Place a 1 in a cell if its operand and transform correspond; otherwise, place a 0 in the cell. For example, the closed transformation

$$\downarrow \begin{array}{ccccc} A & B & C & D & E \\ C & E & D & E & A \end{array}$$

is represented in this manner:

$$\begin{array}{c|ccccc|} \downarrow & A & B & C & D & E \\ A & 0 & 0 & 0 & 0 & 1 \\ B & 0 & 0 & 0 & 0 & 0 \\ C & 1 & 0 & 0 & 0 & 0 \\ D & 0 & 0 & 1 & 0 & 0 \\ E & 0 & 1 & 0 & 1 & 0 \end{array}$$

Since A changes to C, we place a 1 at the intersection of the column for A and the row for C. Examination of this matrix discloses that B is not a transform and that the transformation is many-one because two operands (B and D) have a common transform (E).

11.1.2 Product of Multiple Transformations

Consider that a set of variables is subjected to multiple transformations in a definite sequence. The transform of one transformation becomes the operand of the succeeding transformation.

Now consider that a set of variables is subjected to a closed transformation T and then a closed transformation U. Let V denote a single transformation that will produce the same effect as T and U combined. Thus, V is *equivalent* to T and U, taken in that sequence. The relationship is expressed symbolically as

$$V = UT$$

and V is called the *product* or *composition* of T and U. As an illustration, let the transformations be as follows:

$$T: \quad \downarrow \begin{array}{cccc} A & B & C & D \\ D & B & A & C \end{array} \qquad U: \quad \downarrow \begin{array}{cccc} A & B & C & D \\ B & D & D & A \end{array}$$

Applying the transformations in the indicated sequence, we arrive at these results: A becomes D and then reverts to A; B remains B and then becomes D; C becomes A and then B; D becomes C and then reverts to D. Therefore, V is as follows:

$$V: \quad \downarrow \begin{array}{cccc} A & B & C & D \\ A & D & B & D \end{array}$$

It is to be emphasized that the sequence in which transformations are applied is crucial. Thus, UT and TU have entirely disparate meanings. To illustrate this point, let $W = TU$. Applying the transformation U and then T, we obtain the following:

$$W: \quad \downarrow \begin{array}{cccc} A & B & C & D \\ B & C & C & D \end{array}$$

Thus, V and W differ.

The product of several transformations can also be obtained by multiplying their corresponding matrices. (Article 5.2.4 presents matrix multiplication.) If we apply transformation T and then transformation U, the matrix for T is the postfactor and the matrix for U is the prefactor in the matrix multiplication. We obtain the following:

$$V: \quad \begin{array}{c|cccc|} \downarrow & A & B & C & D \\ \hline A & 0 & 0 & 0 & 1 \\ B & 1 & 0 & 0 & 0 \\ C & 0 & 0 & 0 & 0 \\ D & 0 & 1 & 1 & 0 \end{array}\; \overset{U}{} \quad \begin{array}{cccc} A & B & C & D \\ \hline 0 & 0 & 1 & 0 \\ 0 & 1 & 0 & 0 \\ 0 & 0 & 0 & 1 \\ 1 & 0 & 0 & 0 \end{array}\; \overset{T}{}$$

$$= \begin{array}{c|cccc|} \downarrow & A & B & C & D \\ \hline A & 1 & 0 & 0 & 0 \\ B & 0 & 0 & 1 & 0 \\ C & 0 & 0 & 0 & 0 \\ D & 0 & 1 & 0 & 1 \end{array}$$

Thus, under transformation V, A remains A, B changes to D, C changes to B, and D remains D. These results agree with those previously obtained.

11.1.3 Powers of a Transformation

Now consider that a closed transformation T is applied n times. The product is denoted by T^n, and it is called the *nth power* of T. As an illustration, consider the transformation

$$T: \quad \downarrow \begin{array}{ccccc} A & B & C & D & E \\ A & E & D & A & C \end{array}$$

Applying this transformation repeatedly, we obtain these results:

$$T^2: \quad \downarrow \begin{array}{ccccc} A & B & C & D & E \\ A & C & A & A & D \end{array}$$

$$T^3: \quad \downarrow \begin{array}{ccccc} A & B & C & D & E \\ A & D & A & A & A \end{array}$$

$$T^4: \quad \downarrow \begin{array}{ccccc} A & B & C & D & E \\ A & A & A & A & A \end{array}$$

For example, the variable at state B is first transformed to E, then to C, then to D, and then to A. Therefore, T^4 transforms B to A. With the fourth transformation, all variables become A, and no further change is possible. Thus, T^n is identical with T^4 if $n > 4$.

We shall now investigate the effect of applying an algebraic transformation repeatedly. Again let x denote a numerical operand and x' denote its transform under transformation T. The transform of x under T^n is denoted by $T^n(x)$. Thus, $x' = T(x)$. Now consider the transformation

$$T: \quad x' = 3x + 7$$

We obtain the following:

$$T^2(x) = 3x' + 7 = 3(3x + 7) + 7 = 3^2x + (3 + 1)7 = 9x + 28$$

$$T^3(x) = 3[3T^2(x) + 7] + 7 = 3[3^2x + (3 + 1)7] + 7$$
$$= 3^3x + (3^2 + 3 + 1)7 = 27x + 91$$

The pattern is now clear. Applying Eq. (1.17) for the sum of a geometric series, we arrive at this result:

$$T^n(x) = 3^n x + \left(\frac{3^n - 1}{2}\right)7 = 3^n x + 3.5(3^n - 1) \qquad (a)$$

All algebraic transformations can be analyzed by this recursive procedure.

The validity of Eq. (*a*) can be demonstrated by mathematical induction, which is presented in Art. 1.15. Applying the formula for x', we obtain

$$\begin{aligned} T^{n+1}(x) &= 3T^n(x) + 7 = 3[3^n x + 3.5(3^n - 1)] + 7 \\ &= 3^{n+1}x + 3.5(3^{n+1} - 1) \end{aligned}$$

and this result is consistent with Eq. (*a*).

The pattern governing $T^n(x)$ can often be readily discerned by recording the calculations in the form of a nested series. (Refer to Art. 1.18.) For example, with reference to the foregoing transformation T, we have

$$\begin{aligned} T^3(x) = 3(3(3x + 7) + 7) + 7 &= 3^3 x + 7(1 + 3 + 3^2) \\ &= 27x + 91 \end{aligned}$$

In many instances, $T^n(x)$ approaches a limiting value as n becomes infinite. As an illustration, consider the following transformation:

$$T: \qquad x' = \frac{x}{3} + 7 \qquad (b)$$

Proceeding as before, we obtain

$$T^n(x) = \frac{x}{3^n} + 10.5\left(1 - \frac{1}{3^n}\right)$$

As n becomes infinite, the expression at the right approaches 10.5 as a limit. The specific value of x is immaterial in this respect.

When it is known that $T^n(x)$ approaches a limiting value as n becomes infinite, this limit can also be found by the simple device of equating x' to x and solving for x. This value is the limit of $T^n(x)$. For example, Eq. (*b*) yields

$$x = \frac{x}{3} + 7 \qquad \text{and} \qquad x = 10.5$$

The rationale underlying this method is the fact that $x' - x$ approaches 0 as $T^n(x)$ approaches its limit.

11.1.4 Inverse Transformations

Let T and U denote closed transformations whose product is the identity transformation I. Transformation U is the *inverse* of T, and it can be denoted by T^{-1}. The following transformations are illustrative:

$$T: \quad \downarrow \begin{matrix} A & B & C & D & E \\ D & A & E & C & B \end{matrix} \qquad T^{-1}: \quad \downarrow \begin{matrix} A & B & C & D & E \\ B & E & D & A & C \end{matrix}$$

That the second transformation is the inverse of the first can be demonstrated by applying the transformations in turn. Thus, A changes to D and then reverts to A; B changes to A and then reverts to B; etc. Transformation T^{-1} is obtained from T by inverting the operand-transform relationship, and the matrix for T^{-1} is derived from that for T merely by interchanging the rows and columns. Manifestly, only a one-one transformation can have an inverse.

The inverse relationship is reflexive; that is, if U is the inverse of T, then T is the inverse of U. Moreover, the sequence in which they are applied is immaterial, and we have $T^{-1}T = TT^{-1} = I$.

EXAMPLE 11.1 Two variables, x and y, undergo the following transformation T:

$$x' = 2x + 12y$$

$$y' = 3x + 16y$$

Identify the inverse of T.

SOLUTION If we solve the foregoing system of equations for x and y, we obtain the following:

$$x = -4x' + 3y'$$

$$y = 0.75x' - 0.5y'$$

To obtain T^{-1}, we simply make the following interchanges in the second system of equations: x and x', y and y'. The result is the following:

$$x' = -4x + 3y$$

$$y' = 0.75x - 0.5y$$

These are the equations of T^{-1}.

As an illustration, we shall set $x = 12$, $y = 10$. Applying T, we obtain $x' = 144$, $y' = 196$. We now set $x = 144$, $y = 196$ and apply T^{-1}. The result is $x' = 12$, $y' = 10$. Thus, under T and T^{-1}, x changed from 12 to 144 and then reverted to 12, and y changed from 10 to 196 and then reverted to 10.

11.1.5 Basins

Consider again that a closed transformation T is performed repeatedly. Let O_i denote the ith element in the original set of operands; then $T^n(O_i)$ denotes the transform of O_i under T^n. In Art. 11.1.3, we encountered a transformation where $T^n(O_i)$ is A for all values of i when $n \geq 4$. Thus, no further transformation is possible. In Art. 11.1.6, we shall encounter transformations where $T^n(O_i)$ goes through a recurrent cycle of states after n attains a certain value. The term *basin* is applied to denote either a state at which transformation ceases or a cycle of states that recurs indefinitely as n assumes successively higher values. As we shall find, every closed transformation must produce at least one basin.

11.1.6 Kinematic Graphs

When a closed transformation is applied repeatedly, the successive changes that occur can be traced readily by constructing a diagram that exhibits all operand-transform relationships. This diagram is known as a *kinematic graph*. In this diagram, each operand is followed by its transform, and an arrow leads from the operand to the transform. Alternative geometric arrangements are often possible; the sole requirement is that the operand-transform relationship be shown correctly. A kinematic graph is particularly helpful in identifying basins.

We shall illustrate a kinematic graph by starting with the transformation T in Art. 11.1.3, which we repeat for convenience.

$$T: \quad \downarrow \begin{matrix} A & B & C & D & E \\ A & E & D & A & C \end{matrix}$$

The kinematic graph of T appears in Fig. 11.1*a*. Since A remains A when T is applied, there is no arrow leading from A. Consider that we start with the operand C. This changes to D under the first transformation and then to A under the second transformation. Thus, $T^2(C)$ is A. We find that all variables ultimately become A, and therefore A is a basin. The longest path is that leading from B to A, and this requires four steps. It follows that all variables are A under T^4, and this conclusion is in accord with our finding in Art. 11.1.3.

Now consider the following closed transformation:

$$U: \quad \downarrow \begin{matrix} A & B & C & D & E & F & G \\ G & D & A & C & A & B & D \end{matrix}$$

B → E → C → D → A

(a)

(b)

(c)

(d)

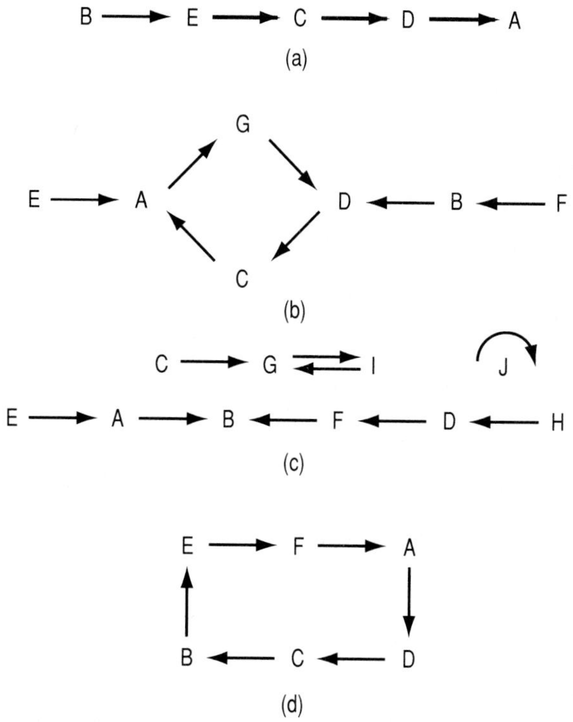

FIGURE 11.1 Kinematic graph of (*a*) T, (*b*) U, (*c*) V, and (*d*) W.

The kinematic graph of U appears in Fig. 11.1*b*. Inspection of this graph reveals that the cycle D-C-A-G-D is recurrent, and this cycle is a basin. For example, $T^2(F)$ is D, $T^3(F)$ is C, $T^4(F)$ is A, $T^5(F)$ is G, $T^6(F)$ is D, and the specified cycle recurs.

Next, consider the transformation

$$V: \downarrow \begin{matrix} A & B & C & D & E & F & G & H & I & J \\ B & B & G & F & A & B & I & D & G & J \end{matrix}$$

The kinematic graph of V appears in Fig. 11.1*c*. The operand J is isolated from all other operands, and this condition is signified by placing a semicircular arrow above the mark. Transformation V has three basins: B, J, and the cycle G-I-G. When T is applied the third time, the operands A, D, E, F, and H have all been transformed to B, and B never changes.

Finally, consider the transformation

$$W: \downarrow \begin{matrix} A & B & C & D & E & F \\ D & E & B & C & F & A \end{matrix}$$

The kinematic graph in Fig. 11.1*d* reveals that the basin is a recurring cycle that encompasses all variables.

Kinematic graphs have many other advantages. For example, they enable us to recognize inverse transformations instantly. Thus, Fig. 11.2 presents the kinematic graphs of the transformations T and T^{-1} in Art. 11.1.4. One graph is the mirror image of the other.

FIGURE 11.2 Kinematic graphs of inverse transformations.

11.2 BEHAVIOR OF DETERMINATE MACHINES

11.2.1 Definitions and Notation

As previously stated, the term *machine* or *dynamic system* is applied in cybernetics in referring to any entity that is capable of undergoing a change of state. A machine is *determinate* if its behavior under a given set of conditions is fully predictable, and we shall deal at present exclusively with determinate machines. The successive states assumed by a machine under repeated transformations constitute its *trajectory* or *line of behavior*. The trajectory of a machine is expressed by recording the successive states in the order of their occurrence. Thus, if the state of the machine is initially *d*, then *b*, then *f*, and then *e*, the trajectory is *dbfe*. We shall assume at present that only one transformation T can be applied to the machine and that it is applied repeatedly.

If a machine is made up of parts that are capable of undergoing a change of state, it is a *compound machine*; otherwise, it is a *simple machine*. Each part of a compound machine is at a particular state at a given time. These individual states constitute a *vector*, and this vector defines the state of the machine. The state of a part is a *component* of the vector. The notation for a vector is as follows: Let a_i denote the state of the ith part of a machine that has n parts. The vector of the machine is represented by $(a_1, a_2, \ldots, a_n)$.

To illustrate these definitions, let the machine be the position of a point moving in a plane. The state of the machine is the composite of the x and y coordinates of the point, and the vector is (x, y). (Alternatively, the components may be the polar coordinates of the point.) Now let the machine consist of three lights arranged in series, and assume that the color of each light changes periodically. If at a given instant the first light is red, the second yellow, and the third blue, the vector of the machine is (red, yellow, blue).

11.2.2 Transformation of Vectors

The transformation of a vector is a composite of the transformation of its components. Therefore, to describe how the vector is transformed, we must specify how each component is transformed.

EXAMPLE 11.2 A compound machine contains two variables, x and y. At regular intervals, the machine is subjected to the following transformation:

$$T: \quad \begin{aligned} x' &= 2x - 0.4y \\ y' &= 0.8x + 1.5y \end{aligned}$$

Initially, $x = 40$ and $y = 10$. Establish the vectors that result from the first three transformations.

SOLUTION

$$T(x) = x' = 2(40) - 0.4(10) = 76$$

$$T(y) = y' = 0.8(40) + 1.5(10) = 47$$

$$T^2(x) = 2(76) - 0.4(47) = 133.2$$

$$T^2(y) = 0.8(76) + 1.5(47) = 131.3$$

$$T^3(x) = 2(133.2) - 0.4(131.3) = 213.88$$

$$T^3(y) = 0.8(133.2) + 1.5(131.3) = 303.51$$

Thus, the vector, which was originally (40, 10), becomes the following in succession: (76, 47), (133.2, 131.3), and (213.88, 303.51).

When a closed transformation is applied several times, we shall number the successive states in this manner: The initial state of the machine is state 0, and the state of the machine following the mth transformation is state m.

11.2.3 Transient and Steady States

If a machine is subjected to a closed transformation T repeatedly, its behavior will ultimately become repetitive, for every closed transformation contains at least one basin. The machine is at a *transient state* up to the point where the first repetition occurs; beyond that point, it is at a *steady state.*

As an illustration, assume that a simple machine starts at state A, is subjected to T repeatedly, and has the trajectory $ACGDEGDE\ldots$, with the cycle GDE recurring indefinitely. The machine is at a transient state during the phase $ACGDE$, and it enters its steady state when G occurs a second time. Thus, the fifth application of T brings the machine to its steady state. Similarly, assume that the machine starts at state D, is subjected to a transformation U repeatedly, and has the trajectory $DGMCFAAA\ldots$, with the state A recurring indefinitely. The machine is at a transient state during the phase $DGMCFA$. The sixth application of U brings the machine to its steady state.

A machine may have multiple potential steady states. For example, consider this transformation:

$$\downarrow \begin{matrix} A & B & C & D & E \\ D & C & B & A & E \end{matrix}$$

Three basins are present. One consists of A and D, another consists of B and C, and the third is E. Which steady state materializes depends of course on the initial state of the system.

11.2.4 Equilibrium

Assume that, under repeated application of a closed transformation T, a machine eventually reaches a state A and then remains at that state. The recurrent state A is termed a *state of equilibrium.* Manifestly, for A to be a state of equilibrium, the requirement is $T(A) = A$.

A machine may have multiple potential states of equilibrium. For example, under the transformation

$$\downarrow \begin{matrix} A & B & C & D & E & F \\ C & B & D & F & E & F \end{matrix}$$

there are three potential states of equilibrium (B, E, and F). Similarly, a machine may have a potential state of equilibrium without ever attaining that state. Thus, consider the transformation

$$\downarrow \begin{matrix} A & B & C & D \\ B & A & C & C \end{matrix}$$

State C is a state of equilibrium, but the machine will attain that state only if it starts at state C or D.

Where the transformation of a machine is algebraic, the presence of states of equilibrium can be investigated by this procedure: Let x denote a variable that has an equilibrium value. When x attains this value, the succeeding transformation fails to change its value. Therefore, we may set $x' = x$ and solve for the equilibrium value of x.

EXAMPLE 11.3 A compound machine has the variables x and y, and these change by this formula:

$$x' = 3x - 5y - 41$$

$$y' = 7x + 4y$$

Determine whether this machine has any state of equilibrium under the given transformation.

SOLUTION When we replace x' and y' in these equations with x and y, respectively, and rearrange the resulting equations, we obtain the following:

$$2x - 5y = 41$$

$$7x + 3y = 0$$

This system of simultaneous equations has the solution $x = 3$, $y = -7$. Thus, $(3, -7)$ is a state of equilibrium. However, the machine must start at this state in order to remain there.

A state of equilibrium may exist as a *limiting state*, one that the machine approaches asymptotically. We encountered such a situation in Art. 11.1.3, and we offer the following transformation as a second illustration:

$$x' = 0.8x + 1.4$$

Replacing x' with x and solving the resulting equation, we obtain $x = 7$ as an equilibrium value. However, this value of x will recur only if x is initially at 7. Assume that x has some initial value x_0 other than 7. Proceeding as in Art. 11.1.3, we find that

$$T^n(x) = (0.8)^n x_0 + 1.4\left[\frac{1-(0.8)^n}{0.2}\right]$$

$$\lim_{n \to \infty} T^n(x) = \frac{1.4}{0.2} = 7$$

Thus, $x = 7$ is a state of equilibrium that the machine approaches but theoretically never reaches.

Geometrically, the foregoing result may be viewed in this manner: The transforming equation may be recast as

$$x' = x + 0.2(7 - x)$$

In Fig. 11.3, $OQ = 7$ and P is a moving point such that OP_i is the value of x at state i. The foregoing equation discloses that we advance one-fifth of the distance between P and Q in each step. This process brings us ever closer to Q, and Q is our limiting position.

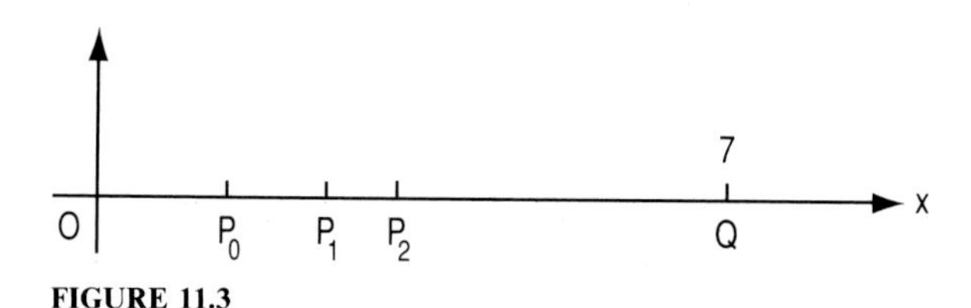

FIGURE 11.3

11.2.5 Disturbances and Stability

In our investigation of a determinate machine, we have assumed that only one transformation T is possible and that this transformation occurs many times. We now broaden our viewpoint slightly to admit the possible interjection of another transformation D. Specifically, assume that a machine has the following history:

1. Transformation T is applied repeatedly, bringing the machine to a state of equilibrium A.
2. Transformation D is then applied once, changing the state of the machine from A to B. This transformation is termed a *disturbance*.
3. Transformation T is again applied repeatedly until the machine is at a steady state.

When this series of events has been completed, three possibilities exist, and we shall now explore them. For this purpose, assume that the machine is simple, and let T be as follows:

$$T: \quad \downarrow \begin{array}{ccccccccc} a & b & c & d & e & f & g & h & i \\ g & f & c & e & e & h & e & b & c \end{array}$$

The kinematic graph appears in Fig. 11.4. Assume that the initial state is a, d, e, or g. Transformation T will eventually bring the machine to the state of equilibrium e. We recognize three types of disturbances, in this manner:

1. A disturbance that transforms e to a, d, or g. The machine will eventually revert to state e when T is again applied, and the machine is said to be *stable* under this displacement.
2. A disturbance that transforms e to b, f, h, or i. The machine will never revert to state e, and the machine is *unstable* under this displacement.
3. A disturbance that transforms e to c. The machine will remain at state c when T is again applied, and c becomes the new state of equilibrium. The machine is said to be *neutrally stable* under this displacement.

If A denotes the original state of equilibrium, the requirement for stability may be expressed symbolically in this manner:

$$\lim_{n \to \infty} T^n[D(A)] = A$$

Expressed verbally, the repeated application of T following D must ultimately return the machine to state A.

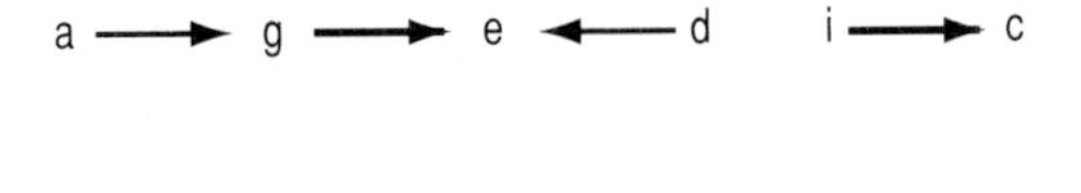

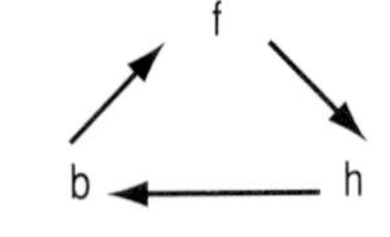

FIGURE 11.4

11.2.6 Transducers

We shall now consider a machine for which multiple transformations are available. A machine that has this characteristic is termed a *transducer*.

To illustrate this condition, assume that a simple machine has a variable x and that the *general* algebraic transformation is as follows:

$$x' = ax + b$$

where a and b can be assigned arbitrary values. Each set of values of a and b yields a unique transformation.

In this illustrative case, the quantities a and b that establish the transformation are called *parameters*. The values assigned to the parameters constitute the *input* to the machine. The state of the machine that results from the assigned values of the parameters is the *output* of the machine.

Figure 11.5 exhibits the relationships among a parameter, a transformation, and a state. The value assigned to the parameter determines which transformation will occur, and the transformation in turn determines which state the machine will assume.

Assume that the parameters have a restricted number of possible values and that we wish to determine the number of transformations that are available. If no duplications arise, the number of available transformations equals the number of ways in which the values of the parameters can be combined.

EXAMPLE 11.4 A simple machine has a variable x, and the general transformation is

$$x' = abcx$$

The parameters can assume the following values: a, 1 or 2; b, 1 or 3; c, 1, 2, or 8. How many transformations are available?

SOLUTION The number of possible combinations of a, b, and c values is $2 \times 2 \times 3 = 12$. However, several combinations are indistinguishable and therefore yield the same product. Specifically, the following pairs of values are indistinguishable: $a = 1$, $c = 2$ and $a = 2$, $c = 1$. These pairs can be combined with $b = 1$ and $b = 3$. Therefore, the number of distinguishable combinations is $12 - 2 = 10$, and that is the number of possible transformations. This conclusion can be confirmed by recording all 12 combinations of a, b, and c values and finding the corresponding products. It is found that the products are as follows: 1, 2, 3, 4, 6, 8, 12, 16, 24, and 48.

Where a set of values of the parameters yields a tranformation previously obtained, this set of values is said to be *degenerate*.

Assume that a simple machine has four possible states, a, b, c, and d, and three possible transformations, T_1, T_2, and T_3. The transformations can be exhibited

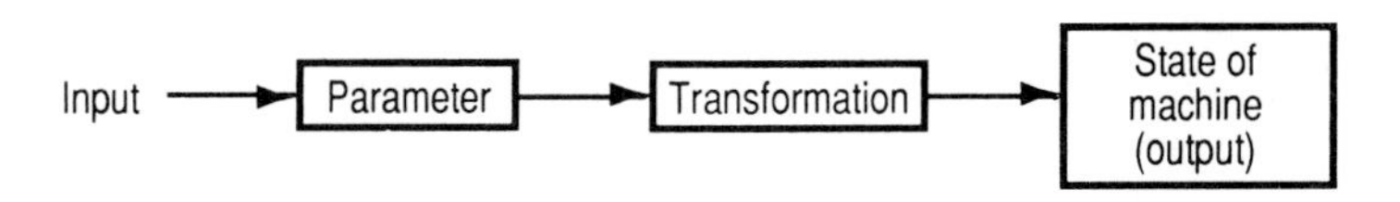

FIGURE 11.5

collectively by means of the following *transformation table*:

	$\downarrow a$	b	c	d
T_1:	b	c	a	b
T_2:	a	a	d	b
T_3:	c	b	a	c

Thus, under T_1, a becomes b; under T_2, c becomes d; under T_3, c becomes a, etc. The parameter of this machine is the subscript of T, and the parameter has three possible values.

EXAMPLE 11.5 With reference to the foregoing set of transformations, assume the following: The machine is initially at state b. It is subjected to T_1 twice, then T_3 once, then T_2 once. Establish the trajectory of the machine.

SOLUTION Under T_1, b changes to c. Under T_1, c changes to a. Under T_3, a changes to c. Under T_2, c changes to d. Thus, the trajectory is *bcacd*.

11.3 COUPLING OF MACHINES

11.3.1 Definition of Coupling

Two machines are said to be *coupled* if the machines undergo simultaneous transformations and one machine controls the behavior of the other. When machines Q and R are coupled, they form a compound machine S, and the state of S is a vector having the states of Q and R as its components. Thus, assume the following: Q and R are both simple machines; at a given instant, Q is at state e and R is at state v. The state of S is (e, v).

There are two basic methods of coupling two machines: simple coupling, and coupling with feedback. We shall define and investigate both methods.

11.3.2 Simple Coupling

The coupling of two machines is *simple* if the control is unidirectional; i.e., the first machine controls the behavior of the second, but the second machine exerts no control over the behavior of the first. The first machine is *independent* in relation to the second machine, and the first machine is said to *dominate* the second.

Let Q and R denote the coupled machines, with Q dominating R. Control is exercised in this manner: Machine R is a transducer. The present state of Q acts as input to R, determining which transformation R will next undergo. Thus, the nth state of Q establishes the $(n+1)$th transformation of R and therefore the $(n+1)$th state of R.

To establish the trajectory of R, it is necessary to have the following information: the present states of Q and R, the transformations that Q will undergo, the possible transformations of R, and the formula that links the present state of Q to the following transformation of R.

EXAMPLE 11.6 Two machines, Q and R, are simply coupled, with Q dominating R. Machine Q is subjected to transformation T repeatedly. Machine R has three possible transformations, U_1, U_2, and U_3. The machines are both simple, and their transformations are as follows: For machine Q,

$$T: \quad \downarrow \begin{matrix} i & j & k \\ k & i & j \end{matrix}$$

For machine R,

$$\begin{array}{llll} & \downarrow a & b & c \\ U_1: & b & a & b \\ U_2: & c & a & b \\ U_3: & b & b & a \end{array}$$

The transformation that R will undergo is determined in this manner:

State of Q:	i	j	k
Transformation of R:	U_2	U_1	U_3

Initially, Q and R are at states i and a, respectively. Establish the trajectory of R.

SOLUTION Refer to Table 11.1. The successive transformations are as follows:

First: i changes to k. Since Q was at state i, the transformation U_2 occurs, and a changes to c.

Second: k changes to j. Since Q was at state k, U_3 occurs, and c changes to a.

Third: j changes to i. Since Q was at state j, U_1 occurs, and a changes to b.

Fourth: i changes to k. Since Q was at state i, U_2 occurs, and b changes to a.

Fifth: k changes to j. Since Q was at state k, U_3 occurs, and a changes to b.

Sixth: j changes to i. Since Q was at state j, U_1 occurs, and b changes to a.

The composite machine has now reverted to its original state, and this cycle of events will recur indefinitely.

Example 11.7 illustrates how a set of coupled machines can be manipulated to produce a desired state.

EXAMPLE 11.7 Machine Q has the variables v and w; machine R has the variables x and y. All four variables have two possible values: 0 and 1. The machines are

TABLE 11.1 Trajectory of Coupled Machines

State of Q	Transformation of R	State of R
i		a
k	U_2	c
j	U_3	a
i	U_1	b
k	U_2	a
j	U_3	b
i	U_1	a

coupled, with Q dominating R, and machine Q is controlled by an operator. Machine R undergoes a transformation at regular intervals. Simultaneously, the operator of Q can vary either v or w, but not both, or he can leave v and w unchanged. The changes in x and y are as follows:

$$x' = y \text{ if } v = 0 \text{ and } w = 1; \text{ otherwise, } x' \neq y$$

$$y' = y \text{ if } x \neq v; \text{ otherwise, } y' \neq y$$

The present state is $v = 1$, $w = 0$, $x = 0$, $y = 1$. We wish to arrive at the state $v = 1$, $w = 1$, $x = 0$, $y = 0$. Determine whether the desired state is attainable. If it is, specify what action the operator of Q must undertake.

SOLUTION Table 11.2 is the transformation matrix of the compound machine. Each possible state of Q is assigned a row, and each possible state of R is assigned a column. The row labels give the values of v and w, in that order, and the column labels give the values of x and y, in that order. For example, the second row corresponds to $v = 0$, $w = 1$, and the fourth column corresponds to $x = 1$, $y = 1$. Each cell of the matrix is identified by a letter in superscript position, and the values of x' and y' are recorded within the matrix. As an illustration, assume that the present values are $v = 0$, $w = 1$, $x = 1$, $y = 0$. This state corresponds to cell g. Applying the transformation formula, we obtain $x' = 0$, $y' = 0$, and this set of values is recorded in the cell. Each transformation of the compound machine corresponds to a movement from one cell to another.

Consider that the transformation matrix is wrapped about a cylinder having a horizontal axis, thus making the first and fourth rows adjacent to each other. Because only v or w can be varied in one operation, we have arranged the rows of the matrix in such manner that only v or w changes as we move from one row to an adjacent row. Since it is not mandatory that the operator change v or w when x and y change, there are two possible movements corresponding to each transformation: within the present row, and from the present row to an adjacent row.

We shall assign a number subscript to each variable to indicate the sequence in which each value appears, starting with 0. Thus, x_0 denotes the initial value of x, and x_3 denotes the value of x following the third transformation. In Table 11.3, we have recorded the successive states of the compound machine, the corresponding cell in the transformation matrix, and the candidates for successor to each state. The present state is cell n in Table 11.2, and the desired state is cell i. We must discover a path leading from n to i under the imposed constraints, if one exists.

Cell n discloses that $x_1 = 0$, $y_1 = 1$. Therefore, in our first transformation, we are constrained to remain within the second column of the transformation matrix, and we now have a choice of two cells: b and j. We shall try b, making $v_1 = 0$, $w_1 = 0$.

TABLE 11.2 Transformation Matrix

	Values of x and y			
Values of v and w	00	01	10	11
00	11^a	00^b	10^c	01^d
01	01^e	10^f	00^g	11^h
11	10^i	01^j	11^k	00^l
10	10^m	01^n	11^o	00^p

TABLE 11.3 Successive States of Compound Machine

State of Q		State of R			
v	w	x	y	Cell	Possible successor cells
1	0	0	1	n	b, j
0	0	0	1	b	a, e, m
0	0	0	0	a	d, h, p
1	0	1	1	p	m, a, i
1	1	0	0	i	

Cell b discloses that $x_2 = 0$, $y_2 = 0$. Therefore, in our second transformation, we move to the first column, and we now have a choice of three cells: a, e, and m. We shall try a. Continuing in this manner, we obtain the results recorded in Table 11.3, where we proceed from n to i by the path *nbapi*. (Alternative paths may also exist.)

We have thus established that the desired state is attainable. By referring to the successive states of Q in Table 11.3, we find that the operator of Q must follow these instructions: At the first transformation of R, change v from 1 to 0. At the second transformation, refrain from action. At the third transformation, restore v to the value 1. At the fourth transformation, change w from 0 to 1.

Where simple machines are coupled and the transformations are algebraic, we shall henceforth apply the following designations: The machine will be assigned an uppercase letter, and its variable will be assigned the corresponding lowercase letter. For example, machine X has the variable x.

EXAMPLE 11.8 Machines X, Y, and Z are coupled, and the following relationships apply:

$$x' = 2x + 5$$

$$y' = xy^2 - 7$$

$$z = x(4 - 3y) \qquad (c)$$

Express the transform of z in terms of the present values of x and y.

SOLUTION Since Eq. (c) applies at every instant, it follows that

$$\begin{aligned} z' &= x'(4 - 3y') \\ &= (2x + 5)[4 - 3(xy^2 - 7)] \\ &= 50x - 15xy^2 - 6x^2y^2 + 125 \end{aligned}$$

In this compound machine, X dominates Y, and X and Y jointly dominate Z.

11.3.3 Coupling with Feedback

Two machines are said to be coupled *with feedback* if each machine influences the behavior of the other. Specifically, both machines are transducers, and one machine provides input to the other.

EXAMPLE 11.9 Two simple machines, Q and R, are coupled with feedback. The possible states are i, j, and k for Q, and a, b, c, and d for R. The possible transformations are T_1 and T_2 for Q and U_1, U_2, and U_3 for R. The transformations are as follows:

	$\downarrow$ i	j	k
T_1:	j	i	i
T_2:	i	i	j

	$\downarrow$ a	b	c	d
U_1:	c	b	a	a
U_2:	b	a	d	c
U_3:	a	c	d	a

The inputs are as follows:

State of Q:	i	j	k
Transformation of R:	U_2	U_1	U_3

State of R:	a	b	c	d
Transformation of Q:	T_1	T_2	T_2	T_1

Initially, Q is at state j and R is at state d. Establish the trajectory of the compound machine.

SOLUTION Refer to Table 11.4. The following transformations occur:

First: Since Q is at state j, machine R undergoes U_1. Since R is at state d, machine Q undergoes T_1. Under these transformations, j changes to i and d changes to a. Thus, state 1 is (i, a).

Second: Since Q is at state i, machine R undergoes U_2. Since R is at state a, machine Q undergoes T_1. Under these transformations, i changes to j and a changes to b. Thus, state 2 is (j, b).

The subsequent states are shown in Table 11.4. Since state 4 coincides with state 1, it follows that states 1, 2, and 3 will recur in an endless cycle. Therefore, the transient part of the trajectory is as follows:

$$(j, d)(i, a)(j, b)(i, b)(i, a)$$

TABLE 11.4 Successive States of Compound Machine

State number	State of Q	State of R	Subsequent transformation of Q	Subsequent transformation of R
0	j	d	T_1	U_1
1	i	a	T_1	U_2
2	j	b	T_2	U_1
3	i	b	T_2	U_2
4	i	a		

11.3.4 Coupling Diagrams

Where multiple machines are coupled, it is helpful to construct a diagram that displays the mode of coupling. This is known as a *coupling diagram* or *diagram of immediate effects*. In this diagram, two machines that are coupled are connected by an arrow that is directed toward the controlled machine.

To illustrate the construction, consider that machines W, X, Y, and Z undergo the following set of transformations:

$$\begin{aligned} w' &= 4w + 3xy^2 \\ x' &= 2w - x + 5y - 9z \\ y' &= 3xy \\ z' &= 4z - 8 \end{aligned} \tag{d}$$

Since w' is a function of x and y (as well as w), it follows that X and Y control W. Since x' is a function of w, y, and z, it follows that W, Y, and Z control X. Since y' is a function of x, it follows that X controls Y. Finally, since z' is a function of z alone, Z functions independently. The coupling diagram of this set of machines appears in Fig. 11.6. The following sets of machines are coupled with feedback: W and X, and X and Y.

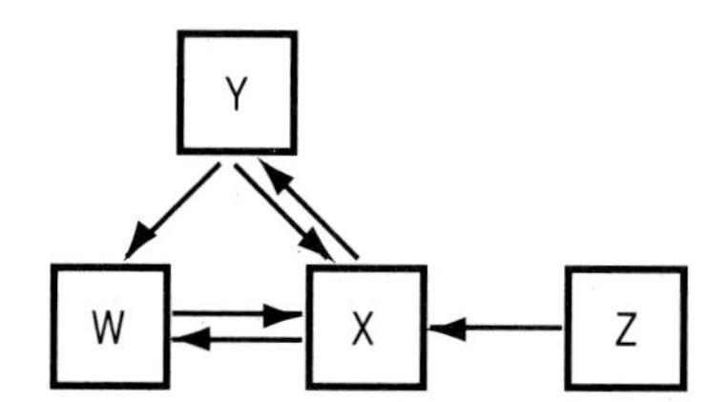

FIGURE 11.6 Coupling diagram.

11.3.5 Propagation of Effects through System of Machines

When two machines are coupled, one machine exerts an *immediate effect* on the other. For example, if X and Y are coupled and X dominates Y, the present state of X influences the succeeding state of Y. However, where a set of machines is coupled, effects are *propagated* through the entire set as a given transformation occurs repeatedly. Thus, with reference to Fig. 11.6, Z has an immediate effect on X, but not on W or Y. However, since X has an effect on both W and Y, Z has an *ultimate effect* on both W and Y. Specifically, we may deduce the following:

State 0 of Z affects state 1 of X.

State 1 of X affects state 2 of W and of Y.

Therefore, state 0 of Z affects state 2 of W and of Y.

The validity of this conclusion can be demonstrated most effectively by a simple example. Consider that machines W, X, Y, and Z are coupled in the manner shown in Fig. 11.7, where the arrows all point in one direction. These machines are said to be coupled *in series*. We shall establish the rate at which effects are propagated through the system.

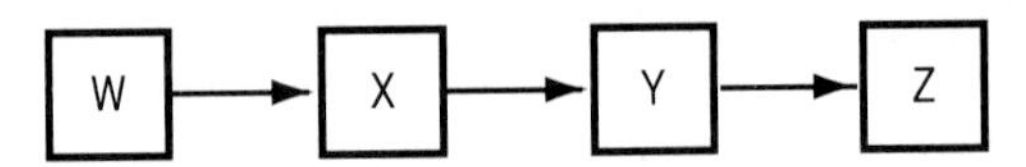

FIGURE 11.7 Coupling diagram.

EXAMPLE 11.10 The machines in Fig. 11.7 undergo the following transformations:

$$w' = 2w$$

$$x' = w + x$$

$$y' = 3x + 2y$$

$$z' = y - z$$

Applying the subscript 0 to denote the initial value of each variable, determine the values of the variables following each of the first three transformations.

SOLUTION The transformation equations yield the values recorded in Table 11.5. We find that state 0 of W affects state 1 of X, state 2 of Y, and state 3 of Z. Thus, the effect of a given state of a machine advances one step to the right in Fig. 11.7 whenever a transformation occurs.

The conclusion we reached in Example 11.10 is general, and we may state it in this form: Consider that several machines are coupled in series and that they are numbered consecutively in the direction of the arrows. The effect of the initial state of the mth machine reaches the $(m + q)$th machine on the qth transformation.

TABLE 11.5

	Variable			
State number	w	x	y	z
0	w_0	x_0	y_0	z_0
1	$2w_0$	$w_0 + x_0$	$3x_0 + 2y_0$	$y_0 - z_0$
2	$4w_0$	$3w_0 + x_0$	$3w_0 + 9x_0 + 4y_0$	$3x_0 + y_0 + z_0$
3	$8w_0$	$7w_0 + x_0$	$15w_0 + 21x_0 + 8y_0$	$3w_0 + 6x_0 + 3y_0 - z_0$

11.3.6 Establishing Mode of Coupling by Observation

Assume that several machines are known to be coupled but the precise mode of coupling is unknown. To establish the functional relationships among the machines, it is necessary to subject the compound machine to successive transformations, to observe the changes in the states of the machines, and to discern a relationship between the present state of one machine and the subsequent state of another machine. However, it is possible that a limited number of observations will induce spurious conclusions.

To illustrate this principle, consider that we have three variables, x, y and z, and that x changes in this manner:

$$x' = 2x + 5y^2z$$

If we observe changes in x while either y or z has the value 0, we would erroneously conclude that both y and z have no immediate effect on x. That x' is really a function of both y and z will emerge only when both y and z have nonzero values. Thus, in observing the behavior of an unknown compound machine, it is usually necessary to make numerous observations to establish the true relationships among its parts.

11.4 RELATED MACHINES

Many dynamic systems in the real world are related to one another, and a clear perception of their relatedness is either imperative or highly advantageous.

11.4.1 Isomorphic Machines

Two machines are *isomorphic* with respect to each other if the behavior of one is parallel to the behavior of the other. Thus, the two machines are essentially alike.

As an illustration, let Q and R be two machines that undergo transformations T and U, respectively. The kinematic graphs appear in Fig. 11.8, and they are structurally identical. Therefore, machines Q and R are isomorphic. If the states of R are assigned new designations to conform to those of Q, and if U is assigned the new designation T, machine R becomes identical with machine Q.

Isomorphism abounds in engineering and science, for many seemingly unrelated problems lead to identical mathematical formulations. As a result, the solution of one problem yields the solution to an entire class of problems. For example, consider the differential equation

$$a\frac{d^2x}{dt^2} + b\frac{dx}{dt} + cx + d = 0$$

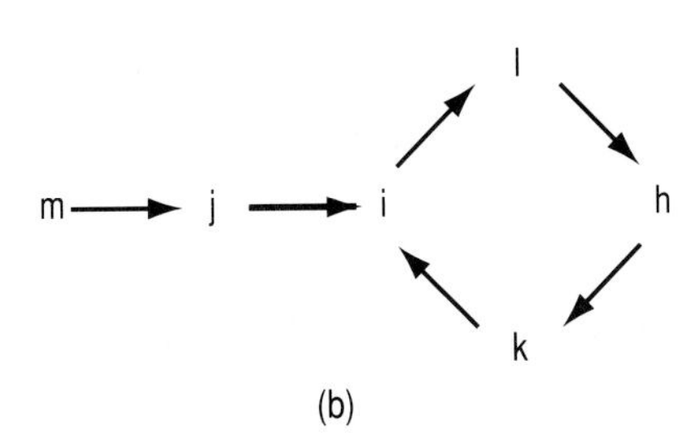

FIGURE 11.8 Kinematic graph of (*a*) machine Q and (*b*) machine R.

where t denotes elapsed time, x is a function of t, and a, b, c, and d are constants. If suitable values are assigned to the constants, this equation becomes applicable to a system in which a body vibrates vertically on a massless, linear spring and the damping force is directly proportional to the velocity of the body. Similarly, if the notation is changed in the required manner, the equation becomes applicable to an electric circuit containing resistance, inductance, and capacitance. Thus, the vibrating system and the electric circuit are isomorphic with respect to each other.

It is interesting to observe that two compound machines may be isomorphic and yet have dissimilar relationships among their parts. To illustrate this principle, let Q denote a machine that has the variables x and y, which change in this manner:

$$x' = 2(x + y)^3 + y$$

$$y' = 2(x + y)^3 + x$$

We observe the following:

$$x' + y' = 4(x + y)^3 + (x + y)$$

$$x' - y' = y - x = -(x - y)$$

We now establish these two variables:

$$u = x + y \qquad v = x - y \tag{e}$$

From the foregoing equations, we have

$$u' = x' + y' = 4u^3 + u$$

$$v' = x' - y' = -v$$

We may thus create a machine R consisting of the variables u and v. Once we assign initial values to u and v as based on the current values of x and y, machine R acquires an independent existence. However, since Eq. (*e*) always holds, the behavior of R parallels the behavior of Q, and the machines are isomorphic. Nevertheless, the machines differ in this important respect: In Q, the variables x and y are interdependent; in R, the variables u and v are independent of each other.

11.4.2 Homomorphic Machines

A machine R is *homomorphic* with respect to a machine Q if R is a simplified version of Q. To be more precise, certain details that pertain to Q are lacking in R, but the behavior of R is like that of Q in all other respects. Thus, R may be called a *model* or *abstraction* of Q.

As a very simple illustration of homomorphism, assume the following: Machine Q consists of three luminous spheres that differ in color and in diameter. When Q is transformed, each sphere undergoes a change both in diameter and in color. Machine R also consists of three luminous spheres, and each sphere always has a diameter equal to that of a corresponding sphere in Q. However, the spheres in R always retain their original color. Thus, the transformation of size in Q coincides with that in R, but the transformation of color in Q has no counterpart in R. Machine R is homomorphic with respect to Q.

The construction of a homomorphic machine from a given machine may result inadvertently from an inability to discern subtle differences, as we shall now demonstrate. Assume that a simple machine Q has seven possible states and that its behavior is described by the kinematic graph in Fig. 11.9*a*. Now assume that an

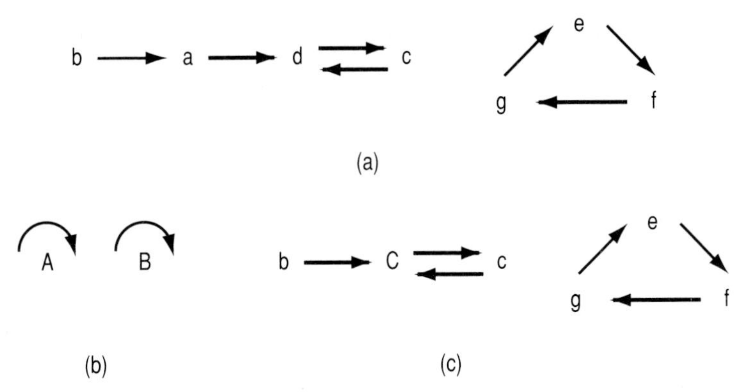

FIGURE 11.9 Kinematic graph of (*a*) machine Q, (*b*) machine R_1, and (*c*) machine R_2.

observer cannot discern the differences among states *a*, *b*, *c*, and *d*, and among states *e*, *f*, and *g*. Therefore, this observer merges states *a*, *b*, *c*, and *d* into a single state *A*, and he merges states *e*, *f*, and *g* into a single state *B*. This observer thus perceives machine Q in the form of a machine R_1 that has the kinematic graph in Fig. 11.9*b*. Now assume that another observer cannot distinguish between states *a* and *d*, but she recognizes all other states. She merges states *a* and *d* into a single state *C*, and her perception of Q is a machine R_2 that has the kinematic graph in Fig. 11.9*c*. Machines R_1 and R_2 are both homomorphic with respect to machine Q, and it is apparent that several other homomorphic machines can be constructed.

It frequently occurs that an inability to discern differences causes us to conclude that a given machine is indeterminate when in reality it is determinate. For example, with reference to the machine Q that has the kinematic graph in Fig. 11.9*a*, assume that an observer cannot distinguish state *a* from state *e* and therefore merges these states into a single state *D*. This observer would erroneously conclude that the behavior of the machine is unpredictable because on one occasion *D* is followed by *d* and on another occasion it is followed by *f*. Thus, in constructing a machine that is homomorphic with respect to Q, this restriction applies: A state in the group *a*, *b*, *c*, and *d* cannot be merged with a state in the group *e*, *f*, and *g*.

In the real world, homomorphic systems are constantly constructed that preserve only those details that are relevant to a specific task. For example, a road map is homomorphic with respect to the terrain it represents, and it exhibits only the interconnection of roads. Similarly, a contour map exhibits only the relative steepness of the terrain. In many types of investigations, it is imperative that superfluous detail be eliminated to achieve economy of time and effort. For example, if a piping system contains a complicated network of pipes in series and in parallel, we may replace the network with an equivalent single pipe if our sole objective is to study the flow between the end points under varying conditions. This single pipe is homomorphic with respect to the network.

11.5 VARIETY AND ITS TRANSMISSION

11.5.1 Meaning and Significance of Variety

Consider that a machine is subjected to a set of transformations. If we know the set of transformations but not the initial state of the machine, we cannot identify the current state of the machine. However, we can identify the *possible* current states of the machine. The number of possible current states is the *variety* of the machine at the present time.

EXAMPLE 11.11 A machine is subjected repeatedly to the following transformation:

$$\downarrow \begin{matrix} a & b & c & d & e & f & g & h & i \\ f & a & h & f & g & f & b & h & g \end{matrix}$$

Determine the variety of the machine following the first, second, and third transformations.

SOLUTION Initially, the variety is 9. The variety beyond that point can be found by two alternative methods. Under the first method, we apply the kinematic graph, which appears in Fig. 11.10. In the first transformation, e, i, d, and c vanish as possible states because they have no predecessors, and the variety following that transformation is $9 - 4 = 5$. In the second transformation, g vanishes as a possible state because it can have only one predecessor (e or i), and the variety following that transformation is $5 - 1 = 4$. In the third transformation, b vanishes as a possible state, and the variety following that transformation is $4 - 1 = 3$.

Under the second method, we count transforms at each stage. The given transformation has five distinct transforms: a, b, f, g, and h. Therefore, the variety following the first transformation is 5, and the possible transitions in the second transformation are as follows:

$$\downarrow \begin{matrix} a & b & f & g & h \\ f & a & f & b & h \end{matrix}$$

There are now four distinct transforms: a, b, f, and h. Therefore, the variety following the second transformation is 4, and the possible transitions in the third transformation are as follows:

$$\downarrow \begin{matrix} a & b & f & h \\ f & a & f & h \end{matrix}$$

There are now three distinct transforms. Therefore, the variety following the third transformation is 3.

If a given transformation is applied repeatedly to a machine, we shall say that the machine is in *phase m* immediately following the *m*th transformation. Thus, in Example 11.11, the variety of the machine is 9 in phase 0, 5 in phase 1, 4 in phase 2, and 3 in phase 3.

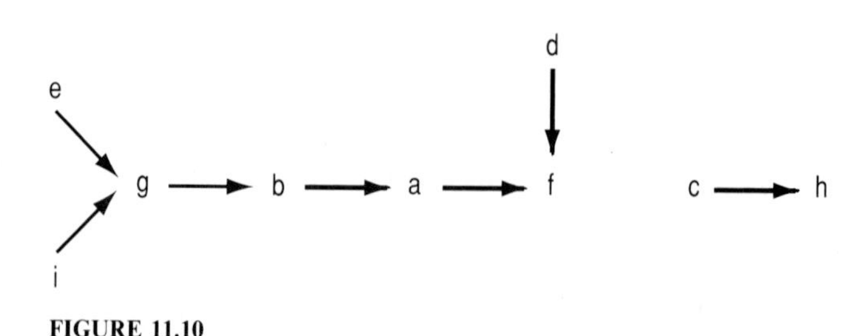

FIGURE 11.10

Example 11.11 illustrates the following principle: If an independent machine is subjected to a given transformation repeatedly, its variety remains constant if the transformation is one-one, and its variety diminishes from the initial value if the transformation is many-one.

It is to be emphasized that the definition of variety applies solely to a *group* of possible states. If only one state is possible, the variety is zero.

Consider that machine A with variety V_A and machine B with variety V_B are combined to form the compound machine C. Let V_C denote the variety of C. Since each possible state of A can be paired with each possible state of B, it follows that $V_C = V_A V_B$. However, it is advantageous to express variety in a manner that permits us to add the individual varieties rather than multiply them. This effect can be achieved by expressing the logarithm of the actual variety; by convention, the logarithms are taken to the base 2.

Let V denote variety as expressed in standard form and $V_{\log}$ denote variety as expressed in logarithmic form. Then

$$V_{\log} = \log_2 V$$

Where machines A and B are combined to form C, we have

$$V_{\log,C} = V_{\log,A} + V_{\log,B}$$

By applying Eq. (1.12), we obtain the following:

$$V_{\log} = \log_2 V = \frac{\log_{10} V}{\log_{10} 2} = 3.3219 \log_{10} V \tag{11.1}$$

Let $V = 2^n$, where n is a positive integer. Then $\log_2 V = n$, and n is the number of zeros in V when it is expressed as a binary number. For example, since $16 = 2^4$, we have $\log_2 16 = 4$, and the expression for this number in binary form is 10000. Therefore, $V_{\log}$ is referred to as the *number of bits*, the word *bit* being a contraction for *bi*nary digi*t*.

EXAMPLE 11.12 Machines A, B, and C are combined to form machine D. At a given instant, the varieties are as follows: $V_A = 5$, $V_B = 12$, $V_C = 7$. Express the variety of all four machines in logarithmic form.

SOLUTION Applying Eq. (11.1), we obtain the following:

$$V_{\log,A} = 2.322 \text{ bits} \qquad V_{\log,B} = 3.585 \text{ bits} \qquad V_{\log,C} = 2.807 \text{ bits}$$

$$V_{\log,D} = 2.322 + 3.585 + 2.807 = 8.714 \text{ bits}$$

As a check, we have the following:

$$V_D = 5 \times 12 \times 7 = 420$$

$$\log_2 420 = 8.714$$

The significance of the variety of a machine is as follows: Assume that information is to be transmitted through the machine, in coded form. In this situation, each state of the machine becomes a distinct message, such as "Lever in upright position," expressed according to some code. Thus, the variety of a machine in a given phase equals the number of alternative messages that it can transmit in that phase.

11.5.2 Transmission of Variety

Assume that a compound machine is formed by coupling a set of machines and that the compound machine is then subjected repeatedly to a given transformation. Let X and Y denote two machines in the set of coupled machines. Also assume that the initial variety of X eventually affects the variety of Y; that is, the number of distinct states that X can have at the time of coupling determines the number of distinct states that Y can have in some later phase. When the initial variety of X exerts its effect on the variety of Y, X is said to *transmit its variety* to Y.

For discussion purposes, we shall assume that the number of possible states of a machine always exceeds 1, and we shall accordingly refer to this number as the variety of the machine. We shall let V_{Xm} and V_{Ym} denote the variety of X and Y, respectively, in phase m, expressed in standard form. Thus, V_{X0} denotes the variety of X before the first transformation occurs.

11.5.3 Direct Transmission

Consider that machines X and Y are coupled, with X dominating Y. If the transformation of X is one-one, its variety remains constant at V_{X0} as this transformation is applied repeatedly. We shall demonstrate two facts: X transmits its variety to Y on the first transformation, and the variety of Y under repeated transformations has an upper limit of $V_{X0} V_{Y0}$.

EXAMPLE 11.13 Machines X and Y are transformed in this manner:

$$x' = x + 5$$

$$y' = 2x + y - 3$$

The initial value of x can be 1, 3, or 6, and the initial value of y can be 2 or 9. Thus, $V_{X0} = 3$ and $V_{Y0} = 2$. Determine the variety of Y after the first and second transformations.

SOLUTION The state of the compound machine is a vector (x, y). Since each possible x value can be paired with each possible y value, the compound machine has a variety of $3 \times 2 = 6$. The possible initial states are recorded in the first column of Table 11.6 (with parentheses omitted for simplicity). Consider that the transformation occurs twice. Applying the transformation formula, we obtain the possible states shown in the second and third columns of the table. For example, if state 0

TABLE 11.6 Possible States of Compound Machine

State 0	State 1	State 2
1, 2	6, 1	11, 10
1, 9	6, 8	11, 17
3, 2	8, 5	13, 18
3, 9	8, 12	13, 25
6, 2	11, 11	16, 30
6, 9	11, 18	16, 37

is (3, 9), we have

$$x' = 3 + 5 = 8 \qquad y' = 2 \times 3 + 9 - 3 = 12$$

After the first transformation, there are six possible y values, and therefore the variety of Y is 6. However, a transformation beyond that point does not increase the number of y values. The reason is that, once we have actually assigned initial values to x and y, we establish the state of the compound machine, and each state has only one transform. (Alternatively, we may reason that the variety of Y cannot exceed that of the compound machine, which is 6.) Thus, as the transformation is applied repeatedly, the variety of Y remains constant at 6.

EXAMPLE 11.14 With reference to Example 11.13, if the possible initial values of y are 2 and 8 and all other conditions remain unchanged, what is the variety of Y after the first and second transformations?

SOLUTION Proceeding as before, we find that the y value of 11 occurs twice following the first transformation. Therefore, the variety of Y in phase 1 is only 5. However, all y values are unique after the second transformation, and the variety of Y in phase 2 is 6.

Examples 11.13 and 11.14 demonstrate that the variety of Y has an upper limit and that it may reach this limit at state 1. Expressed symbolically,

$$V_{Y1} \leq V_{X0} V_{Y0}$$

and this limit holds as the transformations continue.

11.5.4 Transmission through Channels

Consider that three machines, X, Y, and U, are coupled in this manner: X dominates U, and U dominates Y, as shown in Fig. 11.11. Thus, X dominates Y through the intervening machine U, which is termed a *channel*. We shall investigate the manner in which the variety of Y changes under repeated transformations.

EXAMPLE 11.15 Machines X, U, and Y are transformed in this manner:

$$x' = 2x$$

$$u' = x + u$$

$$y' = 3u + y$$

The initial value of x can be x_1, x_2, or x_3; the initial value of u can be u_1 or u_2; and the initial value of y can be y_1 or y_2. Determine the variety of Y following the second transformation.

SOLUTION The initial varieties are $V_{X0} = 3$, $V_{U0} = 2$, $V_{Y0} = 2$. Therefore, the compound machine has $3 \times 2 \times 2 = 12$ possible initial states. These states and their transforms are recorded in Table 11.7. If there are no duplicate values of a variable

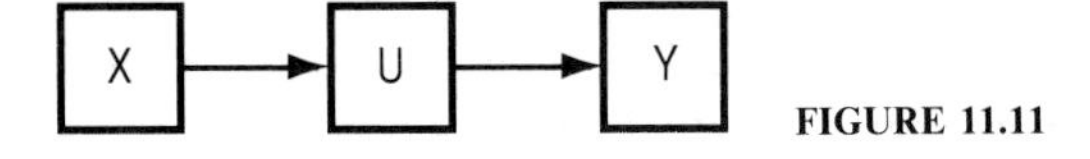

FIGURE 11.11

TABLE 11.7 Possible States of Compound Machine

State 0			State 1			State 2		
x	u	y	x	u	y	x	u	y
x_1	u_1	y_1	$2x_1$	$x_1 + u_1$	$3u_1 + y_1$	$4x_1$	$3x_1 + u_1$	$3x_1 + 6u_1 + y_1$
x_1	u_1	y_2	$2x_1$	$x_1 + u_1$	$3u_1 + y_2$	$4x_1$	$3x_1 + u_1$	$3x_1 + 6u_1 + y_2$
x_1	u_2	y_1	$2x_1$	$x_1 + u_2$	$3u_2 + y_1$	$4x_1$	$3x_1 + u_2$	$3x_1 + 6u_2 + y_1$
x_1	u_2	y_2	$2x_1$	$x_1 + u_2$	$3u_2 + y_2$	$4x_1$	$3x_1 + u_2$	$3x_1 + 6u_2 + y_2$
x_2	u_1	y_1	$2x_2$	$x_2 + u_1$	$3u_1 + y_1$	$4x_2$	$3x_2 + u_1$	$3x_2 + 6u_1 + y_1$
x_2	u_1	y_2	$2x_2$	$x_2 + u_1$	$3u_1 + y_2$	$4x_2$	$3x_2 + u_1$	$3x_2 + 6u_1 + y_2$
x_2	u_2	y_1	$2x_2$	$x_2 + u_2$	$3u_2 + y_1$	$4x_2$	$3x_2 + u_2$	$3x_2 + 6u_2 + y_1$
x_2	u_2	y_2	$2x_2$	$x_2 + u_2$	$3u_2 + y_2$	$4x_2$	$3x_2 + u_2$	$3x_2 + 6u_2 + y_2$
x_3	u_1	y_1	$2x_3$	$x_3 + u_1$	$3u_1 + y_1$	$4x_3$	$3x_3 + u_1$	$3x_3 + 6u_1 + y_1$
x_3	u_1	y_2	$2x_3$	$x_3 + u_1$	$3u_1 + y_2$	$4x_3$	$3x_3 + u_1$	$3x_3 + 6u_1 + y_2$
x_3	u_2	y_1	$2x_3$	$x_3 + u_2$	$3u_2 + y_1$	$4x_3$	$3x_3 + u_2$	$3x_3 + 6u_2 + y_1$
x_3	u_2	y_2	$2x_3$	$x_3 + u_2$	$3u_2 + y_2$	$4x_3$	$3x_3 + u_2$	$3x_3 + 6u_2 + y_2$

at any state, we have the following:

$$V_{U1} = V_{X0} V_{U0} = 3 \times 2 = 6$$

$$V_{Y1} = V_{U0} V_{Y0} = 2 \times 2 = 4$$

$$V_{Y2} = V_{U1} V_{Y0} = V_{X0} V_{U0} V_{Y0} = 3 \times 2 \times 2 = 12$$

The last result may be recast in this form:

$$V_{Y2} = V_{X0} V_{Y1} = 3 \times 4 = 12$$

In general, duplicate values may occur, and the quantities at the right represent upper limits.

Table 11.7 reveals the following: On the first transformation, X transmits its variety to U, and U transmits its variety to Y. On the second transformation, X transmits its variety to Y. As transformations continue, U and Y retain their upper limits of 4 and 12, respectively. The effect of channel U is twofold: It enhances the variety of Y, and it delays the transmission of variety from X to Y by one transformation.

We may now extend the foregoing analysis to the general case of machines that are coupled in series. With reference to Fig. 11.7, W transmits its variety to X on the first transformation, to Y on the second transformation, and to Z on the third transformation. If the machines are numbered from left to right, we may state the following: The mth machine transmits its variety to the $(m + q)$th machine on the qth transformation. This statement is wholly analogous to the statement in Art. 11.3.5 regarding the time at which the initial state of the mth machine has an impact on the $(m + q)$th machine. Thus, if A and B are two machines in the series, the impact of machine A on machine B and the transmission of variety from A to B occur simultaneously.

This simultaneity of events can be deduced by simple logic. Before machine A has an impact on machine B, the initial state of A has no bearing on the state of B. However, at the instant that A has an impact on B, the state that B assumes depends on the initial state of A. Since A initially had several possible states, the number of these states determines the number of possible states of B.

EXAMPLE 11.16 With reference to Fig. 11.7, the initial varieties are as follows: $V_{W0} = 7$, $V_{X0} = 3$, $V_{Y0} = 5$, $V_{Z0} = 2$. If the transformation of W is one-one and no duplications of state occur among X, Y, or Z, compute the subsequent varieties.

SOLUTION

$$V_{X1} = 7 \times 3 = 21 \qquad V_{Y1} = 3 \times 5 = 15 \qquad V_{Z1} = 5 \times 2 = 10$$

$$V_{X2} = 21 \qquad V_{Y2} = 7 \times 15 = 105 \qquad V_{Z2} = 3 \times 10 = 30$$

$$V_{X3} = 21 \qquad V_{Y3} = 105 \qquad V_{Z3} = 7 \times 30 = 210$$

Thus, at the third transformation, Z acquires a variety of $7 \times 3 \times 5 \times 2 = 210$, and no increase in variety occurs beyond this point.

11.5.5 Transmission between Machines with Feedback

Consider that machines X and Y are coupled with feedback. On the first transformation, each machine exerts an effect on the other, and it simultaneously transmits its variety to the other. Thus, assuming that no duplications occur, we have

$$V_{X1} = V_{Y1} = V_{X0} V_{Y0}$$

The expression at the right is the number of possible states of the compound machine, and neither X nor Y can undergo any further increase in variety.

EXAMPLE 11.17 Machines V, W, X, Y, and Z are coupled in the manner shown in Fig. 11.12. Indicate how the variety of each machine changes until it attains its upper limit.

SOLUTION The variety of V maintains its upper limit of V_{V0}.

$$V_{W1} \leq V_{V0} V_{W0} \qquad V_{X1} \leq V_{X0} V_{Y0} \qquad V_{Y1} \leq V_{W0} V_{X0} V_{Y0} \qquad V_{Z1} \leq V_{Y0} V_{Z0}$$

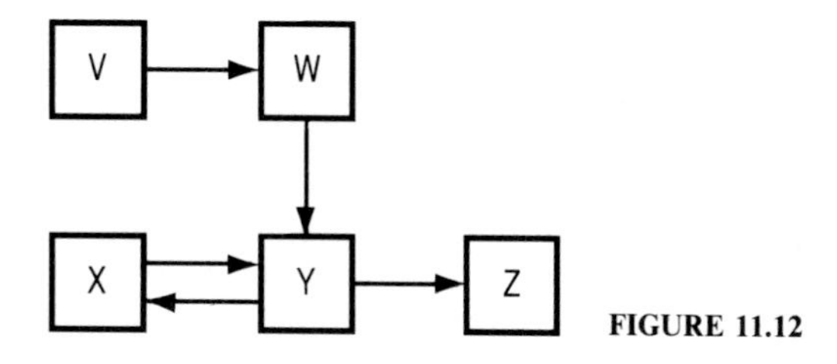

FIGURE 11.12

Machines W and X have now reached the upper limits of their variety.

$$V_{Y2} \leq V_{V0} V_{Y1} \leq V_{V0} V_{W0} V_{X0} V_{Y0}$$

$$V_{Z2} \leq V_{W0} V_{X0} V_{Z1} \leq V_{W0} V_{X0} V_{Y0} V_{Z0}$$

Machine Y has now reached the upper limit of its variety.

$$V_{Z3} \leq V_{V0} V_{Z2} \leq V_{V0} V_{W0} V_{X0} V_{Y0} V_{Z0}$$

Machine Z has now reached the upper limit of its variety.

11.6 COMMUNICATION BY CODE

One of the subjects encompassed by cybernetics is the mode of transmitting information in altered form. Thus, cybernetics is concerned with the systematic coding and decoding of messages.

11.6.1 Introduction and Definitions

There are three principal reasons why the form of a message must often be altered before it is transmitted to the receiver: To maintain secrecy, to condense a message by the use of symbols, and, if the receiver is a mechanism, to cast the message in a form that the mechanism is programmed to accept. The process of altering the form of a message is called *coding*, and the process of restoring the coded message to its original form is called *decoding*. We shall assume that decoding is a necessity. Therefore, a system of coding a message must provide a means of decoding the message.

A message is composed of *characters*, a character being a letter of the alphabet, a number, or a symbol. (We do not include blank spaces.) For example, the message THE PRICE IS $28 is composed of 13 characters; of these, 10 are letters, 2 are digits, and 1 is a symbol.

We shall discuss two coding systems where the objective of the coding is to maintain secrecy.

11.6.2 Coding by Substitution

The simplest possible way of coding a message is to replace each character in the original message with another character. For example, A may be replaced with V, B may be replaced with G, etc. Thus, each character is *translated*.

In some instances, it may be more convenient to treat an entire set of characters that compose a word, expression, or numerical quantity as a unit and to translate the unit. For example, in the message THE OBJECT IS RED, we may wish to translate the entire word RED. Similarly, in the message THE TEMPERATURE is 28°C, we may wish to translate the quantity 28°C. We shall apply the term *information unit* to denote an individual character or a set of characters that is treated as a unit.

The present coding system involves the transformation of each information unit in the original message to a new information unit. Applying the terminology of transformations, we may say that the information unit in the original message is the operand, and the corresponding information unit in the coded message is its transform. Since the coded message must have the capacity to be decoded, the original transformation must have an inverse. Therefore, the original transformation must be of the one-one type.

EXAMPLE 11.18 If x is a variable whose value is to be expressed in coded form, determine whether each of the following coding formulas is valid, and give the decoding formula.

$$x' = 3x + 5 \qquad (f)$$

$$x' = x^2 + 9 \qquad (g)$$

$$x' = \sin x \qquad \text{(where } x < 360^\circ\text{)} \qquad (h)$$

SOLUTION The decoding formula corresponding to (f) is $x = (x' - 5)/3$. There is a unique value of x corresponding to each value of x'. Therefore, decoding is possible, and the coding formula is valid.

The decoding formula corresponding to (g) is

$$x = \sqrt{x' - 9}$$

Since x can be positive or negative, ambiguity exists, and the coding formula is invalid.

The decoding formula corresponding to (h) is $x = \arcsin x'$. There are two possible values of x for each value of x'. For example, if $x' = 0.6352$, x can be 39.43° or 140.57°. Since ambiguity exists, the coding formula is invalid.

11.6.3 Coding by Transition

We shall now consider a more sophisticated coding system. Under this system, each character in the original message is represented by *a pair of adjacent characters* in the coded message. For example, assume that the coded message contains the word *CXFW*. The transitions from *C* to *X*, from *X* to *F*, and from *F* to *W* all correspond to specific characters in the original message. Thus, each character in the original message is transformed to a *transition* of characters. We shall henceforth refer to a character in the original message as a *primary character* and a character in the coded message as a *secondary character*.

This coding system operates through the use of a transducer, in this manner: With reference to Fig. 11.5, we make each primary character an input to the machine, thereby causing the machine to undergo a specific transformation. We then make the ensuing state of the machine a secondary character. To obtain a starting point, we must assign an initial state to the machine.

The present coding system has three basic requirements, which we shall now record.

1. The transducer must have a unique parameter corresponding to every primary character. We shall give the parameter the same designation as the corresponding primary character.
2. The number of possible states of the machine must be at least equal to the number of primary characters.

3. Assume the following: There are three primary characters, *Q*, *R*, and *S*, and the machine has three possible states, *A*, *B*, and *C*. The transformation table is as follows:

	↓*A*	*B*	*C*
Q:	*B*	*B*	*A*
R:	*A*	*C*	*B*
S:	*B*	*A*	*B*

Assume that the machine is initially at state *C*. The input *R* yields the transition *C* to *B*, and the input *S* also yields this transition. Thus, ambiguity exists, and decoding is precluded. Similarly, if the machine is initially at state *A*, the inputs *Q* and *S* both yield the transition *A* to *B*, and ambiguity exists. Therefore, the third requirement is as follows: A given state cannot appear more than once in any column of the transformation table.

EXAMPLE 11.19 A transducer has the following transformation table:

	↓*A*	*B*	*C*	*D*	*E*
Q:	*A*	*B*	*A*	*E*	*A*
R:	*D*	*C*	*E*	*B*	*D*
S:	*C*	*A*	*D*	*A*	*E*
T:	*B*	*E*	*B*	*C*	*B*
U:	*E*	*D*	*C*	*D*	*C*

It is the practice to start the machine at state *B*. If the message *RUST* is fed into the machine, how does the message appear in coded form? Verify the result by decoding the message.

SOLUTION Under *R*, *B* changes to *C*. Under *U*, *C* remains *C*. Under *S*, *C* changes to *D*. Under *T*, *D* changes to *C*. We omit the initial state *B*, and the coded form of the message is *CCDC*.

A diagram is helpful in performing the coding. In Fig. 11.13*a*, we record the characters in the original message across the top, and record the initial state of the transducer, which is *B*, in the position shown. By means of arrows, we indicate that *R* operates to transform *B* to *C*, *U* operates to transform *C* to *C*, *S* operates to transform *C* to *D*, and *T* operates to transform *D* to *C*.

The proof that the coded message is correct is as follows: With the initial state included, the complete coded form is *BCCDC*. The first transition is *B* to *C*. Referring to the second column of the transformation table, we find that this transition corresponds to *R*. The second transition is *C* to *C*, and this corresponds to *U*. The third transition is *C* to *D*, and this corresopnds to *S*. The fourth transition is *D* to *C*, and this corresponds to *T*. Thus, the original message is *RUST*.

EXAMPLE 11.20 With reference to Example 11.19, what is the coded message if the machine is started at state *E*?

SOLUTION Refer to Fig. 11.13*b*. The coded message is *DDAB*.

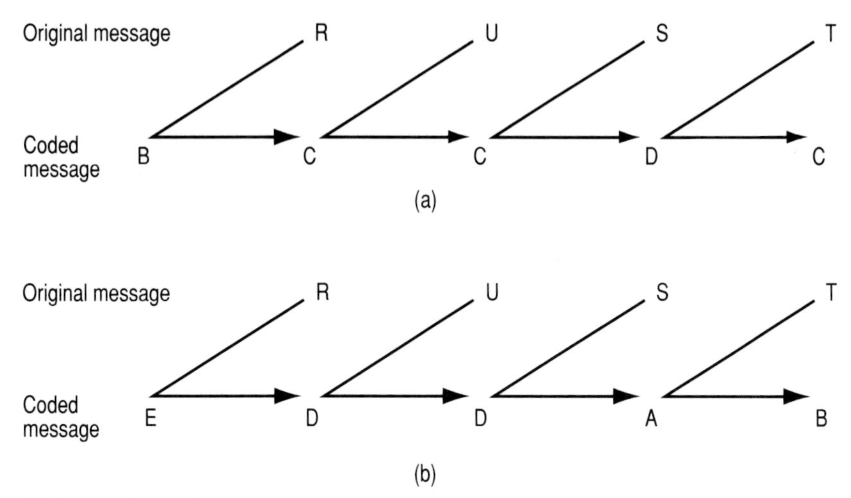

FIGURE 11.13 Coding of message with machine initially at (*a*) *B* and (*b*) *E*.

11.7 BEHAVIOR OF INDETERMINATE MACHINES

11.7.1 Multivalued Transformations

In Art. 11.1.1, we defined a multivalued transformation as one in which at least one operand has several possible transforms. Which transform will actually materialize is determined by chance, and a multivalued transformation is therefore termed *stochastic*. For generality, we shall assume that all operands have several possible transforms. Thus, each transformation is a *trial* as defined in Art. 9.1.1, and the transform is an *outcome*.

Assume that a machine will be subjected repeatedly to a multivalued transformation. Since the behavior of the machine is governed by chance, the trajectory of the machine is unpredictable, and the machine is accordingly described as *indeterminate*.

As an illustration, assume the following: A conduit contains three parallel passages, *A*, *B*, and *C*. An object moves from one end of the conduit to the other, being initially placed in one of the three passages. The object has been programmed in a manner that enables it to make a "choice" at regular intervals: It can move to a different passage, or it can continue along its present passage. The choice the object makes is governed by such environmental factors as wind velocity and humidity, and it is also influenced by its present location. For example, if it is currently in passage *A*, there is a probability of 30 percent that it will remain in *A*; if it is currently in *B*, there is a probability of 55 percent that it will remain in *B*. The operator has no control over the environmental conditions, and therefore the movement of the object is a random one.

If we call the passage in which the object is currently located its *state*, the transformation from one state to the next is stochastic. Moreover, the transformation is affected by the present state. We shall assume that the environmental conditions remain constant for the time period under consideration.

In Art. 9.4.1, we defined a Markov process as a trial in which the outcome is influenced by the preceding outcome. It is now seen that a stochastic transformation is a Markov process and that the trajectory of an indeterminate machine is a Markov chain. Associated with a Markov process is a *transition matrix*, as illustrated by Table 9.4. However, to conform with our present system for recording information, we shall transpose the rows and columns of the transition matrix, assigning a column to each operand and a row to each possible transform. Thus, the transition matrix for the object that moves through passage A, B, or C may be that shown in Table 11.8.

In Art. 9.1.6, we stated that probability may be viewed as relative frequency in the long run, and we shall view it in this manner in the present investigation. Thus, if an event E has a probability of 0.60, we consider that E will occur 60 percent of the time.

11.7.2 Constraint

Assume that a trial has n possible outcomes. Also assume that, in the absence of any restriction, all n outcomes have equal likelihood of materializing. Thus, the probability corresponding to each possible outcome is $1/n$, and the probability distribution is uniform. The extent to which the outcome is subject to restriction is termed the *constraint* of the trial. This constraint is reflected in the divergence of the true probability distribution from the uniform distribution.

As an illustration, consider the cubical die referred to in Art. 9.1.1. We define the trial as the process of tossing the die, and we define the outcome as the number of dots on the face that lands on top. There are six possible outcomes; if the die is unbiased, the probability of each possible outcome is 1/6. Now assume that the die is tossed repeatedly and that the outcome 5 is found to occur 80 percent of the time. The die is manifestly biased, and the outcome is subject to considerable constraint.

Now assume that a trial has three possible outcomes, A, B, and C. In Fig. 11.14, we exhibit the probability corresponding to each possible outcome by a vertical bar. If the probability distribution is uniform, as shown in Fig. 11.14*a*, *no constraint* is present. If the probability distribution is as shown in Fig. 11.14*b*, where only one outcome is possible, *total constraint* is present. Finally, if the probability distribu-

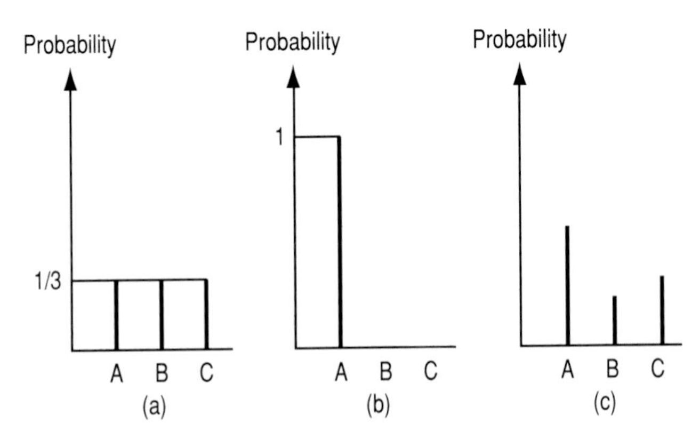

FIGURE 11.14 Probability distribution corresponding to (*a*) no constraint, (*b*) total constraint, and (*c*) partial constraint.

tion is of a type that is intermediate between these extremes, as shown in Fig. 11.14*c*, *partial constraint* is present. Where a machine is subjected to a stochastic transformation, the machine operates under constraint if the possible transforms do not have uniform probability.

11.7.3 Significance of Entropy

Assume the following: There is a set of identical indeterminate machines in operation; they are all currently at state A; this state has three possible transforms, A, B, and C. Following the next transformation, some machines will be at state A, others at state B, and others at state C. What is the variety of this set of machines following this transformation? If we say it is 3, our statement is misleading because we are assigning equal weight to all states without regard to their relative frequency.

Specifically, assume that the number of machines in the set is 100 and that the probabilities corresponding to the transforms of A are as follows: A, 0.02; B, 0.03; C, 0.95. Equating relative frequency to probability, we obtain these results: After the next transformation, 2 machines will be at A, 3 machines will be at B, and 95 machines will be at C. This set of machines is characterized by considerable uniformity, and the mere fact that it contains three distinct states is of scant significance. Manifestly, we must broaden our concept of variety as applied to this set of machines to encompass both the number of possible states it contains and its *diversity*. Our task is to devise a measure of variety in the present context.

In general, assume the following: There are N identical machines in operation, arranged in series; they are all currently at state A; this state has n possible transforms. Since the machines are arranged in series, the subsequent states form a repeated permutation, which is defined in Art. 4.1.11, and the number of possible repeated permutations is n^N.

Let P_i denote the probability corresponding to the ith transform in this set of n possible transforms. Assume initially that the n transforms have a uniform probability distribution, making $P_i = 1/n$ for all values of i. Under a uniform probability distribution, the transformation occurs without constraint, and the variety of the set of machines has its highest possible value. We may equate the variety to the number of repeated permutations. Thus, $V = n^N$. Expressed in logarithmic form,

$$V_{\log} = N \log n \tag{i}$$

We may recast Eq. (i) in this form:

$$V_{\log} = N \log \frac{1}{P_i} \tag{j}$$

We now consider a case where the transforms have a nonuniform probability distribution. Assume that state A has two possible transforms, A and B, and let P_A and P_B denote their respective probabilities. Then NP_A machines will assume state A, and NP_B machines will assume state B. Equation (j) can be adapted to this case, giving

$$V_{\log} = NP_A \log \frac{1}{P_A} + NP_B \log \frac{1}{P_B}$$

In general, if there are n possible transforms, we have

$$V_{\log} = N \sum_{i=1}^{n} P_i \log \frac{1}{P_i} \tag{11.2}$$

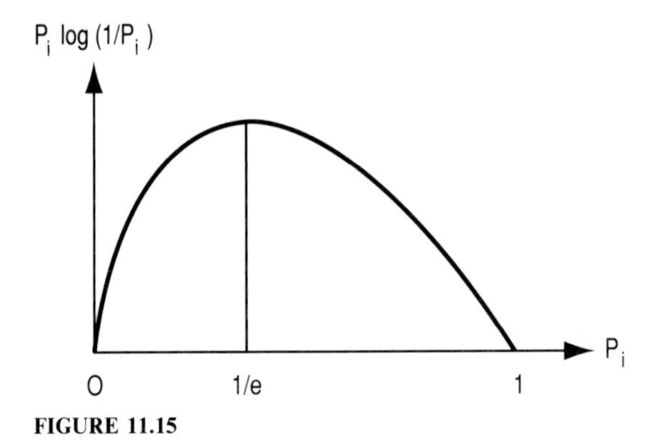

FIGURE 11.15

Figure 11.15 depicts the variation of $P_i \log(1/P_i)$. This quantity is 0 when $P_i = 0$ and when $P_i = 1$, and it attains its maximum value when $P_i = 1/e$.

That Eq. (11.2) offers an accurate measure of variety is attested by a consideration of two extreme cases. First assume that a particular transform, which we shall call M, has a probability of 1; all other transforms have probabilities of 0. Thus, P_i is either 1 or 0, and Eq. (11.2) yields $V_{\log} = 0$. Following the transformation, every machine is at state M, and the variety is nil. Thus, Eq. (11.2) yields the result that logic mandates. Now assume that all possible transforms have equal probability. It can be demonstrated that Eq. (11.2) yields the maximum value of $V_{\log}$, and we know that the variety is maximum in this case. These two extreme cases confirm the validity of Eq. (11.2).

Since $\log(1/P_i) = -\log P_i$, we may recast Eq. (11.2) in this form:

$$V_{\log} = -N\,\Sigma\,(P_i \log P_i) \tag{11.2a}$$

If we now divide the variety by N, we obtain the contribution to variety made by an individual machine. This quantity is termed the *entropy* of the machine corresponding to its present state. It is denoted by H, and its unit is bits per state. Then

$$H = -\Sigma\,(P_i \log P_i) \tag{11.3}$$

The concept of entropy was formulated by C. E. Shannon (but with a different approach). It is of fundamental importance in communication theory, where it is necessary to determine the time rate at which a given system can transmit information.

EXAMPLE 11.21 A machine has three possible states, A, B, and C, and the transition matrix appears in Table 11.8. Compute the entropy if the current state of the machine is A, B, or C.

SOLUTION We shall append a subscript to H to denote the current state. Applying Eq. (11.3) with 2 as the base of logarithms and combining this with Eq. (11.1), we obtain the following:

$$\begin{aligned} H_A &= -(0.30 \log 0.30 + 0.45 \log 0.45 + 0.25 \log 0.25) \\ &= 1.5395 \text{ bits/state} \end{aligned}$$

TABLE 11.8 Transition Matrix

Subsequent state	Present state A	B	C
A	0.30	0.18	0.33
B	0.45	0.55	0.52
C	0.25	0.27	0.15
Total	1.00	1.00	1.00

Proceeding in a similar manner, we obtain

$$H_B = 1.4297 \text{ bits/state} \qquad H_C = 1.4289 \text{ bits/state}$$

That H_A has the highest of the three values is explained by the fact that the probabilities corresponding to state A are more evenly distributed than those corresponding to states B and C.

Consider that a machine has n possible states and that their probability distribution is uniform. By Eq. (i), $H = \log n$, and this is the maximum value that H can have. Thus, in Example 11.21, the maximum possible value of H is $\log_2 3 = 1.5849$.

11.7.4 Mean Entropy at Ultimate Conditions

In Art. 9.4.1, we stated that a Markov system approaches steady-state conditions as the Markov process is performed repeatedly. Now assume that an indeterminate machine is subjected to a given transformation repeatedly. Since the machine is a Markov system, the machine approaches what we shall now term *ultimate conditions*. At ultimate conditions, the probability that the machine will be at a particular state is independent of its initial state. By the notation in Art. 9.4.3, $P^*(A)$ denotes the probability of state A at ultimate conditions.

The set of probabilities corresponding to an indeterminate machine constitute a probability distribution. In Art. 9.1.8, we defined the mean of a probability distribution, and we may apply Eq. (9.4) to compute the *mean* (or *expected*) value of entropy at ultimate conditions, which we shall denote by H^*.

EXAMPLE 11.22 With reference to the machine in Example 11.21, find the mean entropy at ultimate conditions.

SOLUTION Proceeding as in Art. 9.4.5, we obtain the following ultimate probabilities:

$$P^*(A) = 0.2449 \qquad P^*(B) = 0.5184 \qquad P^*(C) = 0.2367$$

Applying Eq. (9.4) and the results obtained in Example 11.21, we have

$$\begin{aligned} H^* &= 1.5395(0.2449) + 1.4297(0.5184) + 1.4289(0.2367) \\ &= 1.4564 \text{ bits/state} \end{aligned}$$

11.7.5 Redundancy

Again let n denote the number of possible states that a machine can assume. In accordance with the discussion in Art. 11.7.3, H^* has its maximum potential value when the probability distribution of its states at ultimate conditions is uniform, and this value is

$$H^*_{\max} = \log n \tag{11.4}$$

As an illustration, consider the stream of alphabetical letters that occurs in the English language. If all 26 letters had equal probability of following a given letter, the entropy of the language would be $\log_2 26 = 4.70$ bits/letter. An exhaustive study of the dictionary has revealed that the actual entropy of the English language is 4.15 bits/letter. This disparity stems from the fact that various sequences of letters have differing frequencies. For example e is frequently followed by d and by n, but infrequently by z.

Assume that the transformation of a machine occurs under constraint and consequently the value of H^* is below the maximum. If a similar machine can achieve the same value of entropy with a lesser number of states, the given machine is said to have *redundancy*. As we shall demonstrate, the English language is redundant in its use of letters.

EXAMPLE 11.23 Assume that we wish to design a new language, to be called *Newlang*, having this characteristic: Each word in English is to have a corresponding word in Newlang with the same number of characters. If all letters in Newlang are to have equal probability of following a given letter, what is the minimum number of letters Newlang requires?

SOLUTION Let X denote this number. The two languages must have the same entropy. Then

$$\log_2 X = 4.15 \qquad X = 2^{4.15} = 18 \text{ (to nearest integer)}$$

11.8 REFERENCES

Shannon, C. E., and W. Weaver: *The Mathematical Theory of Communication*, University of Illinois, Urbana, 1949.
Wiener, Norbert: *Cybernetics*, Wiley, New York, 1948.

SECTION 12
CALCULUS

12.1 HYPERBOLIC FUNCTIONS

12.1.1 Definitions

The trigonometric functions that are defined in Art. 2.1.2 are known as *circular functions*. Closely analogous to these functions are the *hyperbolic functions*, which we shall now define. These functions involve the quantity e, which is defined in Art. 1.28. The designation for a hyperbolic function is obtained by adding the letter h to the designation for the corresponding circular function. Thus, $\sinh u$, $\cosh u$, and $\tanh u$ denote the hyperbolic sine, hyperbolic cosine, and hyperbolic tangent, respectively. The functions are as follows:

$$\sinh u = \frac{e^{u} - e^{-u}}{2}$$

$$\cosh u = \frac{e^{u} + e^{-u}}{2}$$

$$\tanh u = \frac{\sinh u}{\cosh u} = \frac{e^{u} - e^{-u}}{e^{u} + e^{-u}}$$

$$\coth u = \frac{1}{\tanh u} = \frac{e^{u} + e^{-u}}{e^{u} - e^{-u}}$$

$$\operatorname{sech} u = \frac{1}{\cosh u} = \frac{2}{e^{u} + e^{-u}}$$

$$\operatorname{csch} u = \frac{1}{\sinh u} = \frac{2}{e^{u} - e^{-u}}$$

From these definitions, it follows that

$$\sinh(-u) = -\sinh u \qquad \cosh(-u) = \cosh u \qquad \tanh(-u) = -\tanh u$$

12.1.2 Geometric Interpretation of Hyperbolic Functions

The hyperbolic functions can be visualized readily by a simple construction. We start with the *unit hyperbola*, which has the equation $x^2 - y^2 = 1$. Figure 12.1 shows the nappe of the hyperbola corresponding to positive x values. The curve is

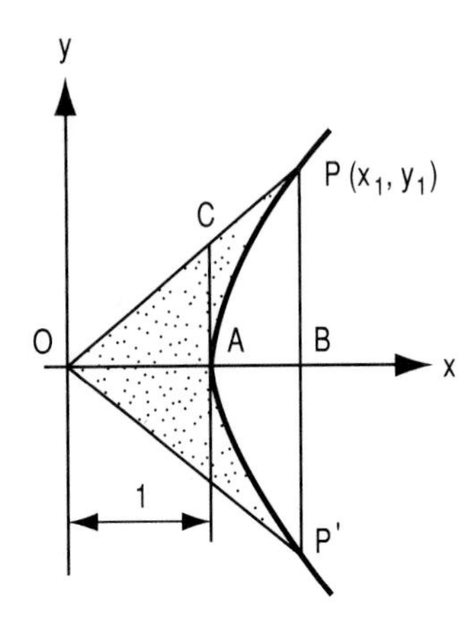

FIGURE 12.1 Geometric interpretation of the hyperbolic functions.

asymptotic to the straight lines having the equations $y = x$ and $y = -x$, and $OA = 1$. Selecting an arbitrary point $P(x_1, y_1)$ on the curve, we draw PP' and AC parallel to the y axis.

Let u denote the dotted area $OPAP'$. By setting up the expressions for the areas $OPBP'$ and APP' and subtracting, we obtain $u = \ln(x_1 + y_1)$; therefore, $e^u = x_1 + y_1$. Then $e^{-u} = 1/(x_1 + y_1) = x_1 - y_1$. It follows that $x_1 = \cosh u$ and $y_1 = \sinh u$. Also, $AC/OA = BP/OB$; therefore, $AC = \tanh u$. Thus, $\sinh u$, $\cosh u$, and $\tanh u$ are represented by BP, OB, and AC, respectively.

Figure 12.1 enables us to perceive the manner in which the hyperbolic functions vary when u is positive. Consider that P starts at A and then recedes indefinitely to the right, causing u to increase from 0 to infinity. As u changes, $\sinh u$ increases from 0 to infinity, $\cosh u$ increases from 1 to infinity, and $\tanh u$ increases from 0 to 1 as a limit. Moreover, $\cosh u$ always exceeds $\sinh u$, but their difference approaches 0 as a limit as u becomes infinite.

12.1.3 Relationships Pertaining to Hyperbolic Functions

The following relationships stem directly from the definitions of the hyperbolic functions:

$$\cosh^2 u - \sinh^2 u = 1$$

$$\tanh^2 u + \operatorname{sech}^2 u = 1$$

$$\coth^2 u - \operatorname{csch}^2 u = 1$$

The following equations are readily verifiable:

$$\sinh(u \pm v) = \sinh u \cosh v \pm \cosh u \sinh v$$

$$\cosh(u \pm v) = \cosh u \cosh v \pm \sinh u \sinh v$$

$$\tanh(u \pm v) = \frac{\tanh u \pm \tanh v}{1 \pm \tanh u \tan v}$$

The last three equations yield the following:

$$\sinh 2u = 2 \sinh u \cosh u$$

$$\cosh 2u = \cosh^2 u + \sinh^2 u$$

$$\tanh 2u = \frac{2 \tanh u}{1 + \tanh^2 u}$$

12.1.4 Equations for Hyperbolic Sine and Cosine

Equation (1.50) expresses e^x as an infinite series. By applying this series and the definitions of $\sinh u$ and $\cosh u$, we obtain the following:

$$\sinh x = x + \frac{x^3}{3!} + \frac{x^5}{5!} + \frac{x^7}{7!} + \cdots$$

$$\cosh x = 1 + \frac{x^2}{2!} + \frac{x^4}{4!} + \frac{x^6}{6!} + \cdots$$

12.1.5 Inverse Hyperbolic Functions

The definition of the inverse of a hyperbolic function is analogous to that of the inverse of a circular function. Thus, if $y = \sinh^{-1} x$, then $x = \sinh y$.

Every inverse hyperbolic function can be converted to a logarithmic function. As an illustration, let $y = \sinh^{-1} x$. Then

$$x = \sinh y = \frac{e^y - e^{-y}}{2}$$

Solving this equation for e^y and then for y, we obtain

$$y = \ln (x + \sqrt{x^2 + 1})$$

Similarly, we have the following conversions:

$$y = \cosh^{-1} x \qquad y = \ln (x + \sqrt{x^2 - 1})$$

$$y = \tanh^{-1} x \qquad y = \frac{1}{2} \ln \left(\frac{1 + x}{1 - x} \right)$$

$$y = \operatorname{csch}^{-1} x \qquad y = \ln \left(\frac{1 + \sqrt{x^2 + 1}}{x} \right)$$

$$y = \operatorname{sech}^{-1} x \qquad y = \ln \left(\frac{1 + \sqrt{1 - x^2}}{x} \right)$$

$$y = \coth^{-1} x \qquad y = \frac{1}{2} \ln \left(\frac{x + 1}{x - 1} \right)$$

12.2 DERIVATIVES

12.2.1 Definitions and Significance of the Derivative

Let y be a function of x. Expressed symbolically, $y = f(x)$. Consider that we assign a value to x, thereby establishing the value of y. Now consider that we increment x by an amount Δx, and let Δy denote the corresponding increment of y. The *derivative* of y with respect to x, which is denoted by dy/dx, is the following:

$$\frac{dy}{dx} = \lim_{\Delta x \to 0} \frac{\Delta y}{\Delta x} \tag{12.1}$$

In general, the value of the derivative is a function of the value assigned to x.

Alternatively, the derivative may be denoted by y' or $f'(x)$. Moreover, in the notation dy/dx, it may be convenient to replace y with its actual expression. Thus, if $y = x^3 + x$, dy/dx may be written in either of these forms:

$$\frac{d(x^3 + x)}{dx} \qquad \frac{d}{dx}(x^3 + x)$$

The geometric significance of the derivative is as follows: Consider that we plot the graph of the equation $y = f(x)$. We locate the point on the graph corresponding to the assigned value of x and draw the tangent to the curve at that point. The derivative dy/dx is equal to the slope of this tangent. Thus, with reference to Fig. 12.2, if $dy/dx = 0.85$ when $x = a$, the slope of the tangent to the curve at P is 0.85. In general, if θ is the angle between the tangent and the positive side of the x axis, we have $\tan \theta = dy/dx$.

Let T denote the tangent to a curve at point P. A line that contains P and is perpendicular to T is called the *normal* to the curve at P. By Eq. (3.7), the slope of the normal is $-1/(dy/dx)$.

When a point moves along a curve, the direction of its motion at any instant is along the tangent to the curve at the point it currently occupies. Therefore, dy/dx is the ratio of the instantaneous rate at which y is increasing to the instantaneous rate at which x is increasing.

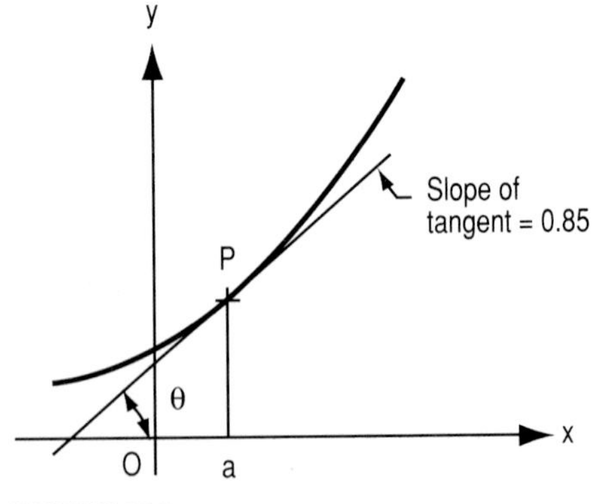

FIGURE 12.2

The process of evaluating a derivative is called *differentiation*. Specifically, when we find dy/dx, we are differentiating the equation for y *with respect to* x. Table A.1 in the appendix presents the derivatives of the basic functions, the derivatives being numbered for convenience.

Now consider that y is a function of v, where v is a function of x. Then

$$\frac{dy}{dx} = \frac{dy}{dv}\frac{dv}{dx} \tag{12.2}$$

This equation can be extended to include any number of intervening variables, and it is known as the *chain rule of derivatives*.

EXAMPLE 12.1 Find dy/dx if $y = \cos^3(3x^2)$.

SOLUTION We first rewrite the equation in this form:

$$y = [\cos(3x^2)]^3$$

We now introduce intervening functions, in this manner:

$$y = u^3 \qquad u = \cos v \qquad v = 3w \qquad w = x^2$$

Applying the chain rule, we have

$$\frac{dy}{dx} = \frac{dy}{du}\frac{du}{dv}\frac{dv}{dw}\frac{dw}{dx}$$

We now refer to Table A.1 and apply Derivatives (6), (8), (4), and again (6), in that order. For clarity, we shall enclose each derivative in parentheses. The result is

$$\frac{dy}{dx} = (3u^2)(-\sin v)(3)(2x)$$

Replacing each variable with its expression in terms of x and performing the arithmetic, we finally obtain

$$\frac{dy}{dx} = -18x\cos^2(3x^2)\sin(3x^2)$$

Where the expression for y is relatively simple, the derivative can be written directly by starting with the outermost function and proceeding to the innermost function. For example, let $y = 7x^3 + 5x^2 - 9$. Applying Derivatives (2), (4), (6), and (1), in that order, we obtain

$$\frac{dy}{dx} = 21x^2 + 10x$$

EXAMPLE 12.2 Find dy/dx if

$$y = \frac{5x^3 - x^2 - 10}{2x + 7}$$

SOLUTION Applying Derivative (5), we obtain

$$\frac{dy}{dx} = \frac{(2x+7)\dfrac{d(5x^3 - x^2 - 10)}{dx} - (5x^3 - x^2 - 10)\dfrac{d(2x+7)}{dx}}{(2x+7)^2}$$

$$= \frac{(2x+7)(15x^2 - 2x) - (5x^3 - x^2 - 10)2}{(2x+7)^2}$$

$$= \frac{20x^3 + 103x^2 - 14x + 20}{(2x+7)^2}$$

12.2.2 Differentiation of Implicit Functions

If the functional relationship between two variables is not stated explicitly, either variable may be considered to be an *implicit function* of the other. We shall illustrate the method of differentiating implicit functions.

EXAMPLE 12.3 Find dy/dx if $x^3y - 2x^2y^2 + 9y^3 = 87$.

SOLUTION For enhanced clarity, we shall enclose the derivative of each term in brackets. Differentiating both sides of the equation with respect to x and applying Derivative (3), we obtain

$$\left[x^3\frac{dy}{dx} + y(3x^2)\right] - 2\left[x^2\left(2y\frac{dy}{dx}\right) + y^2(2x)\right] + \left[27y^2\frac{dy}{dx}\right] = 0$$

Solving for dy/dx, we have

$$\frac{dy}{dx} = \frac{-3x^2y + 4xy^2}{x^3 - 4x^2y + 27y^2}$$

12.2.3 Differentiation with Parametric Equations

In Art. 3.1.8, we defined a parameter and parametric equations. Consider that the variables x and y are both functions of a parameter t. The chain rule of derivatives yields

$$\frac{dy}{dx} = \frac{dy}{dt}\frac{dt}{dx}$$

$$\therefore \frac{dy}{dx} = \frac{dy/dt}{dx/dt} \tag{12.3}$$

EXAMPLE 12.4 A point moves in a plane, and its instantaneous coordinates are $x = 9t^2$, $y = 2t^3 - 5t$, where t denotes elapsed time in seconds. Establish the direction in which the point is moving when $t = 3$ s.

SOLUTION In accordance with our previous statement, the direction of the motion is along the tangent to the curve traced by the point. Therefore, we compute dy/dx to find the slope of the tangent. We apply Eq. (12.3).

$$\frac{dx}{dt} = 18t \qquad \frac{dy}{dt} = 6t^2 - 5$$

$$\frac{dy}{dx} = \frac{6t^2 - 5}{18t}$$

When $t = 3$ s, $dy/dx = 49/54$.

Let θ denote the angle that the tangent to the curve makes with the positive side of the x axis, measured in the counterclockwise direction. Then $\tan\theta = 49/54$, and $\theta = 42.22°$.

12.2.4 Logarithmic Differentiation

In many instances, differentiation can be performed most readily by first taking the natural logarithm of both sides of the given equation and then differentiating both sides of the logarithmic equation. This process, which is known as *logarithmic differentiation*, requires an application of the laws of logarithms presented in Art. 1.6 and of Derivative (19).

EXAMPLE 12.5 Find dy/dx if $y = x^{\sin 2x}$.

SOLUTION Proceeding in the prescribed manner, we obtain the following:

$$\ln y = (\sin 2x) \ln x$$

$$\frac{1}{y}\frac{dy}{dx} = (\sin 2x)\left(\frac{1}{x}\right) + 2(\cos 2x) \ln x$$

$$\frac{dy}{dx} = y\left[\frac{\sin 2x}{x} + 2(\cos 2x) \ln x\right]$$

12.2.5 Derivatives of Higher Order

We shall now refer to dy/dx as the *first derivative* of y with respect to x. If dy/dx is itself a function of x, it also has a derivative. The derivative of dy/dx is called the *second derivative* of y with respect to x, and it is denoted by d^2y/dx^2, y'', or $f''(x)$. This process can be continued. In general, the expression obtained by applying differentiation n times is called the *nth derivative*, and it is denoted by d^ny/dx^n, $y^{(n)}$, or $f^{(n)}(x)$. Thus,

$$\frac{d^ny}{dx^n} = \frac{d}{dx}\left(\frac{d^{n-1}y}{dx}\right)$$

EXAMPLE 12.6 Find the fourth derivative of y with respect to x if $y = 3x^6 - 9x^5 + 4x^2 + 8x - 21$.

SOLUTION

$$\frac{dy}{dx} = 18x^5 - 45x^4 + 8x + 8 \qquad \frac{d^2y}{dx^2} = 90x^4 - 180x^3 + 8$$

$$\frac{d^3y}{dx^3} = 360x^3 - 540x^2 \qquad \frac{d^4y}{dx^4} = 1080x^2 - 1080x$$

The second derivative has an important geometric meaning. With reference to Fig. 12.2, the slope of the tangent to the curve at P equals the value of dy/dx corresponding to $x = a$. Consider that we move along the curve so that x increases at a uniform rate. Since d^2y/dx^2 equals the rate at which dy/dx increases with respect to x, it follows that d^2y/dx^2 is the relative rate at which the slope of the tangent to the curve increases during our left-to-right movement along the curve. In Fig. 12.2, the slope of the tangent is increasing as we pass P. Therefore, the value of d^2y/dx^2 is positive when $x = a$.

12.2.6 Effect of Rotation of Axes

Consider that the x and y axes are rotated through an angle θ to the positions x' and y' shown in Fig. 3.7*b*. This transformation of axes changes the slope of each line in the plane, and therefore it changes the derivatives of a curve.

As before, let x and y denote the original coordinates of a point, and let x' and y' denote the coordinates following the rotation of axes. Let

$$Z = 1 + (\tan\theta)\frac{dy}{dx}$$

The first and second derivatives under the original coordinates and the revised coordinates are related in this manner:

$$\frac{dy'}{dx'} = \frac{\dfrac{dy}{dx} - \tan\theta}{Z} \qquad (12.4a)$$

$$\frac{d^2y'}{dx'^2} = \frac{\dfrac{d^2y}{dx^2}}{(Z\cos\theta)^3} \qquad (12.4b)$$

Assume that dy/dx is finite. If $d^2y/dx^2 = 0$, $d^2y'/dx'^2 = 0$. If d^2y/dx^2 is infinite, d^2y'/dx'^2 is infinite. If dy/dx is infinite, the derivatives under the transformed coordinates must be evaluated by the method presented in Art. 12.3.4.

12.2.7 Differentials

In Fig. 12.3, P is a point on curve C, Q is a neighboring point on C and to the right of P, and T is the tangent to the curve at P. Lines AP and BQ are parallel to the y axis, and line PD is parallel to the x axis. Let Δx and Δy denote the increment of x and of y, respectively, as we move from P to Q.

Considering x to be the independent variable, we shall now replace Δx with dx, which is called the *differential of x*. If y were increasing along line T rather than curve C, the increment of y corresponding to dx would be DR. We label this quantity dy, and it is called the *differential of y*. Since the derivative of y with respect to x equals the slope of the tangent to the curve, the notation dy/dx can refer either to the first derivative of y with respect to x or to the ratio of the differentials, the two quantities being equal. Letting $f'(x)$ denote the first derivative, we have

$$dy = f'(x)\,dx \qquad (12.5)$$

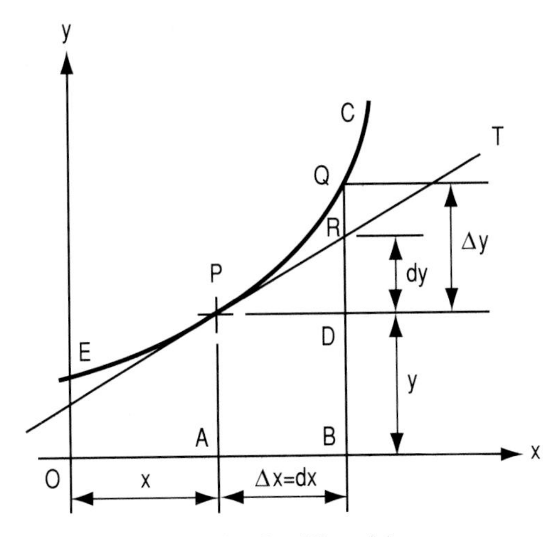

FIGURE 12.3 Notation for differentials.

EXAMPLE 12.7 If $y = 3x^2$, what is dy when $x = 14$ and $dx = 0.03$?

SOLUTION

$$f'(x) = 6x \qquad dy = 6 \times 14(0.03) = 2.52$$

In general, let u denote any variable associated with curve C in Fig. 12.3. The true increase in u when x increases by the amount dx is labeled Δu. However, the *differential* of u, which is labeled du, is the amount by which u would increase if y increased along the tangent line T rather than curve C. As dx approaches 0, du approaches Δu. Therefore, in accordance with the definition of a derivative, the ratio of du to dx becomes coincident with the derivative du/dx.

We shall now consider specific examples of variables associated with curve C in Fig. 12.3. For this purpose, consider that line AP starts at the y axis and then advances to the right while remaining parallel to the y axis and that it terminates at curve C. During this motion, point P generates the arc EP, and line AP generates the area $OAPE$ between the arc and the x axis. Let s denote the length of the arc EP. Then Δs is the length of the arc PQ, and $ds = PR$. Thus,

$$ds = \sqrt{(dx)^2 + (dy)^2} \tag{12.6}$$

Now let A denote the area $OAPE$ generated by the moving line AP. If y increased along line T while x increased by the amount dx, the increase in area would be $y\,dx + (1/2)\,dx\,dy$. Since we will divide this increase in area by dx and then allow dx to approach 0, the second term will drop out, and the differential of area is

$$dA = y\,dx \tag{12.7}$$

12.2.8 Derivative of Arc Length and Area

Equations (12.6) and (12.7) yield the following derivatives for the arc length of a curve and area under a curve as defined in Art. 12.2.7:

$$\frac{ds}{dx} = \sqrt{1 + \left(\frac{dy}{dx}\right)^2} \tag{12.8}$$

$$\frac{dA}{dx} = y \tag{12.9}$$

EXAMPLE 12.8 A point moves along the curve having the equation $y = 1/x^2$. Find the expression for ds/dx and for dA/dx.

SOLUTION The curve is plotted in Fig. 12.4. It consists of two symmetrical branches, and they are asymptotic to the x and y axes.

$$\frac{dy}{dx} = -\frac{2}{x^3} \qquad \frac{ds}{dx} = \sqrt{1 + \frac{4}{x^6}}$$

$$\frac{dA}{dx} = y = \frac{1}{x^2}$$

Consider that we start with $x = 0$ and then increase x. From the graph, we glean the following: When x is extremely small, s increases very rapidly. As x becomes infinitely large, the curve approaches parallelism with the x axis, and ds/dx approaches 1 as a limit. Similarly, when x is extremely small, A increases very rapidly. As x becomes infinitely large, A increases negligibly. The equations for ds/dx and for dA/dx are consonant with these observed facts.

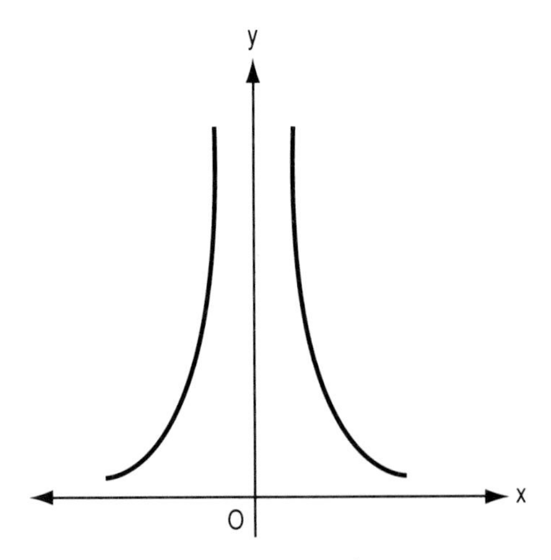

FIGURE 12.4 Graph of $y = 1/x^2$.

12.3 APPLICATIONS OF DERIVATIVES

12.3.1 Properties of a Curve

The first and second derivatives of a function enable us to identify the properties of the graph of the function, and they expedite the task of constructing the graph.

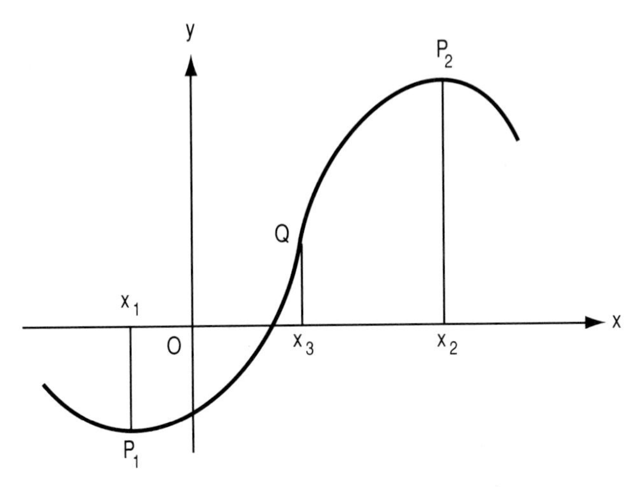

FIGURE 12.5

The properties of a curve are illustrated in Fig. 12.5. In accordance with the definition in Art. 3.1.1, this curve is concave upward to the left of Q and concave downward to the right of Q. The point Q at which the concavity changes is a *point of inflection*. Points P_1 and P_2 are the *turning points* of the curve. Point P_2, where y has a local maximum value, is a *summit*; on the other hand, point P_1, where y has a local minimum value, is a *nadir*.

To identify the properties of a curve, we apply these principles: dy/dx equals the slope of the tangent to the curve, and d^2y/dx^2 equals the rate at which the slope is increasing. To the left of Q, the slope of the tangent changes from negative to positive, and therefore the slope is *increasing*; to the right of Q, the slope of the tangent changes from positive to negative, and therefore the slope is *decreasing*. The following principles emerge:

1. In a region where the curve is concave upward, d^2y/dx^2 is positive; in a region where the curve is concave downward, d^2y/dx^2 is negative.
2. At a point of inflection, $d^2y/dx^2 = 0$.
3. At a summit, $dy/dx = 0$ and d^2y/dx^2 is negative.
4. At a nadir, $dy/dx = 0$ and d^2y/dx^2 is positive.

EXAMPLE 12.9 Determine the properties of the curve that has the equation $y = 2x^3 - 6x^2 - 90x$.

SOLUTION The curve appears in Fig. 12.6*a*.

$$\frac{dy}{dx} = 6x^2 - 12x - 90 = 6(x^2 - 2x - 15) = 6(x + 3)(x - 5)$$

$$\frac{d^2y}{dx^2} = 12x - 12$$

Setting $dy/dx = 0$, we obtain $x = -3$ and $x = 5$. When $x = -3$, $d^2y/dx^2 = -48$;

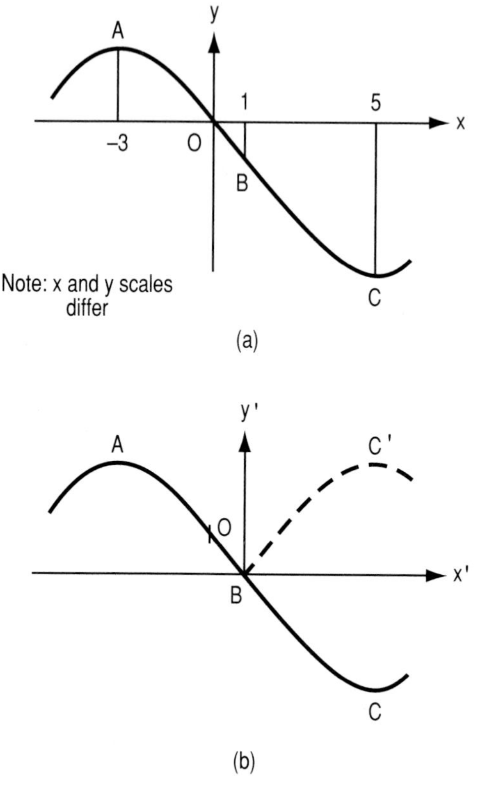

FIGURE 12.6 Graph of cubic equation (*a*) with origin in original position and (*b*) with origin displaced to point of inflection.

when $x = 5$, $d^2y/dx^2 = 48$. Finally, setting $d^2y/dx^2 = 0$, we obtain $x = 1$. We thus arrive at the following conclusions:

The curve has a summit at A, where $x = -3$, and it has a nadir at C, where $x = 5$. It has a point of inflection at B, where $x = 1$.

The curve in Fig. 12.6*a* has an interesting property, which we shall now present. In Fig. 12.6*b*, we draw x' and y' axes through B and parallel to the x and y axes, respectively. We now rotate the part of the curve that lies to the right of B about the x' axis through an angle of 180°, as shown. The entire curve is now symmetrical about the y' axis. This property is characteristic of all cubic curves.

EXAMPLE 12.10 Determine the properties of the curve that has the equation $y = x^3 - 12x^2 + 75x$.

SOLUTION The curve appears in Fig. 12.7*a*.

$$\frac{dy}{dx} = 3x^2 - 24x + 75 \qquad \frac{d^2y}{dx^2} = 6x - 24$$

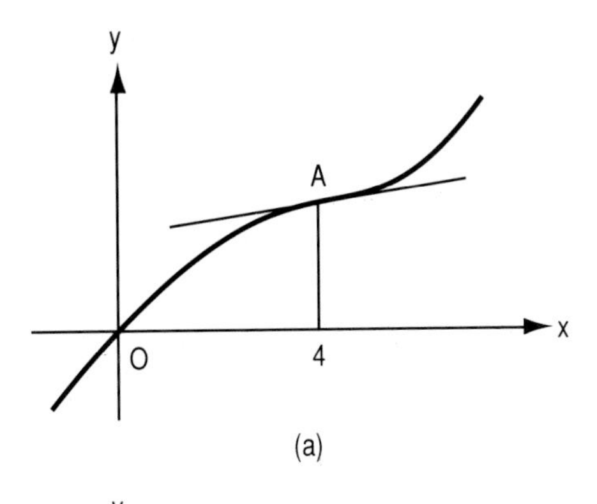

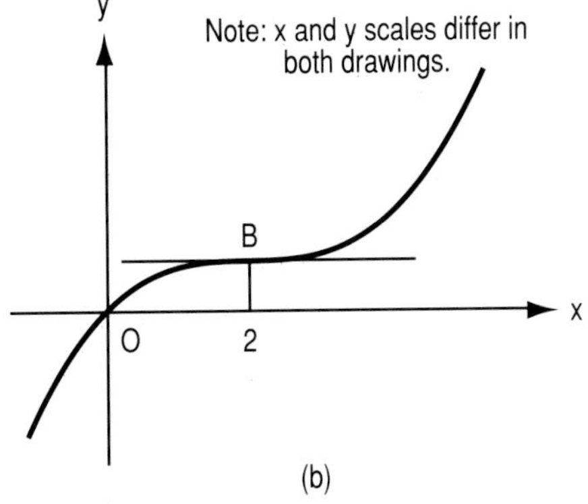

FIGURE 12.7 (*a*) Graph of $y = x^3 - 12x^2 + 75x$; (*b*) graph of $y = x^3 - 6x^2 + 12x$.

When we set $dy/dx = 0$, we obtain roots that are complex numbers. Therefore, this curve has no turning points. When we set $d^2y/dx^2 = 0$, we obtain $x = 4$. Therefore, the curve has a point of inflection at A.

EXAMPLE 12.11 Determine the properties of the curve that has the equation $y = x^3 - 6x^2 + 12x$.

SOLUTION The curve appears in Fig. 12.7*b*. When we set $dy/dx = 0$ and $d^2y/dx^2 = 0$, we obtain the same result: $x = 2$. Therefore, the curve has no turning points, and it has a point of inflection at B, where the slope of the tangent is 0.

Another special type of point that a curve may have is a *cusp point*, such as point P in Fig. 12.8. At a cusp point, the slope of the tangent changes at an infinitely rapid rate, and therefore d^2y/dx^2 is infinite in absolute value. However, both x and y have finite values.

EXAMPLE 12.12 A curve has the equation $y = (x - a)^{3/2} + b$, where a and b are constants. Test this curve for cusp points.

SOLUTION The equation may be rewritten in this form:

$$y = \sqrt{(x - a)^3} + b$$

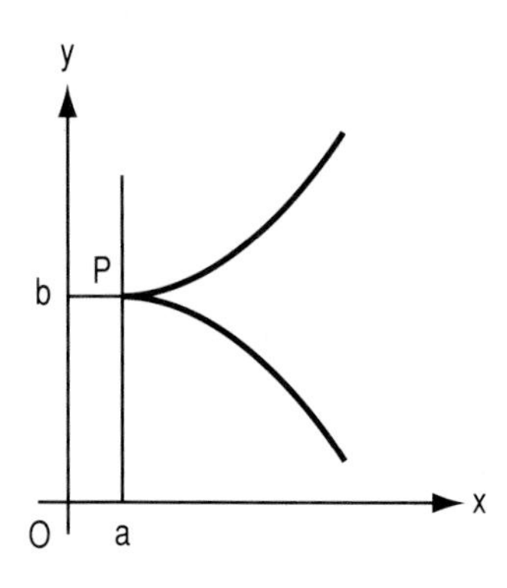

FIGURE 12.8 Graph of $y = (x - a)^{3/2} + b$.

Since y is restricted to real values, x is subject to the restriction $x \geq a$. The curve is shown in Fig. 12.8.

$$\frac{dy}{dx} = \frac{3}{2}(x - a)^{1/2} \qquad \frac{d^2y}{dx^2} = \frac{3}{4}(x - a)^{-1/2} = \frac{3}{4\sqrt{x - a}}$$

When $x = a$, $y = b$, but d^2y/dx^2 is infinite. Therefore, the curve has a cusp at P, as shown.

Assume that a curve has a cusp point at which dy/dx is infinite. The tangent to the curve at that point is parallel to the y axis, and y has a local maximum or minimum value at that point.

12.3.2 Maxima and Minima

Again let y denote a function of x. In many instances, it is necessary to establish the value of x that will cause y to assume a maximum or minimum value. The procedure consists of setting up the expression for y in terms of x, obtaining the expression for dy/dx, setting $dy/dx = 0$, and then solving for x. Whether the resulting value of x yields the maximum or minimum value of y is usually self-evident. However, if doubt exists, this doubt can be dispelled by computing d^2y/dx^2 for the computed value of x. If the second derivative is positive, the corresponding value of y is minimum. This procedure is based on the conclusions reached in Art. 12.3.1.

EXAMPLE 12.13 Let b denote the width and d denote the depth of a beam of rectangular cross section. The strength of the beam is directly proportional to bd^2. What are the dimensions of the strongest timber beam that can be cut from a cylindrical log having a diameter D?

SOLUTION Refer to Fig. 12.9.

$$b^2 + d^2 = D^2$$

Let S denote the strength of the beam and k denote the constant of proportionality. We shall express S as a function of b. Then

$$S = kbd^2 = kb(D^2 - b^2) = k(D^2b - b^3)$$

$$\frac{dS}{db} = k(D^2 - 3b^2)$$

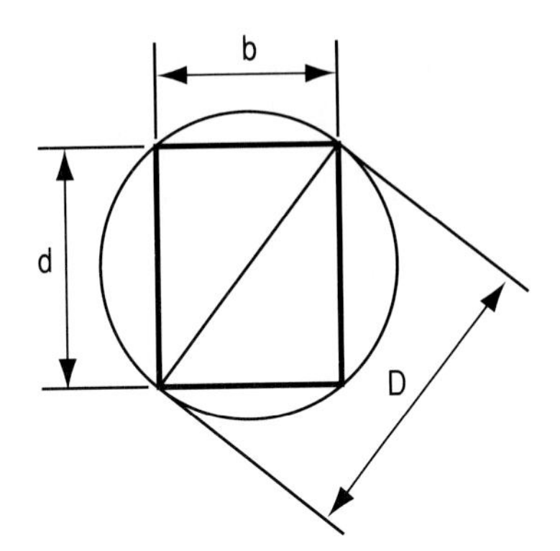

FIGURE 12.9

Setting $dS/db = 0$ and solving for b, we obtain the following as the dimensions of the strongest beam:

$$b = \sqrt{\frac{1}{3}}D \qquad d = \sqrt{\frac{2}{3}}D$$

EXAMPLE 12.14 In Fig. 12.10, points A and P lie in the same horizontal plane, the distance between them being d. Point B is vertically above A. A source of light is located at B. The illumination at P is directly proportional to the cosine of the angle of incidence θ and inversely proportional to the square of the distance from B to P. At what height should the source of light be placed if the illumination at P is to be maximum?

SOLUTION Let $AB = h$ and let $PB = s$. Let I denote the illumination at P and let k denote the constant of proportionality. We must then express I as a function of h.

$$I = k\frac{\cos\theta}{s^2} = k\frac{h}{s^3} = k\frac{h}{(h^2 + d^2)^{3/2}}$$

Applying Derivative (5) to obtain dI/dh and equating the numerator of the resulting fraction to 0, we obtain the following:

$$(h^2 + d^2)^{3/2} - 3h^2(h^2 + d^2)^{1/2} = 0$$

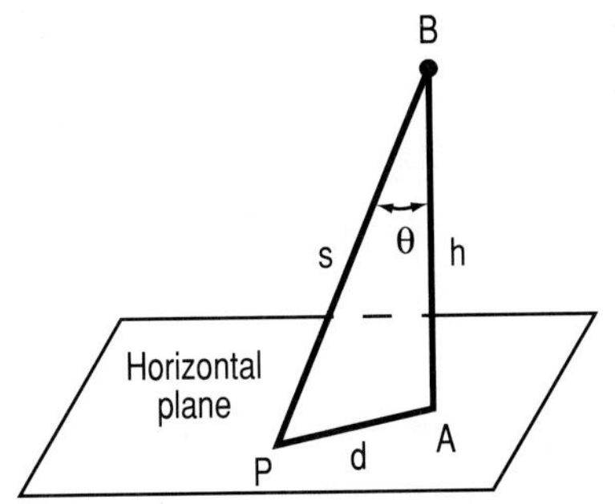

FIGURE 12.10

Dividing both sides of the equation by $(h^2+d^2)^{1/2}$, we obtain

$$h^2 + d^2 - 3h^2 = 0$$

$$h = \frac{d}{\sqrt{2}}$$

In dividing, we lost one value of h. However, this value is imaginary and therefore meaningless.

12.3.3 Related Rates of Change

Assume the following: The quantities x and y are related, they both vary with time, we are given the rate at which x is changing, and we wish to determine the rate at which y is changing. Let t denote elapsed time. Equation (12.2) becomes

$$\frac{dy}{dt} = \frac{dy}{dx}\frac{dx}{dt} \tag{12.10}$$

Thus, to find dy/dt, it is necessary to express y in terms of x and to differentiate the resulting expression.

EXAMPLE 12.15 A perfect gas undergoes a polytropic process having the equation

$$pV^{1.36} = C \tag{a}$$

where p and V denote absolute pressure and volume, respectively, and C is a constant. At a given instant, $p = 50$ lb/in², $V = 4.6$ cu ft, and the volume is increasing at the rate of 0.3 cu ft per second. Find the rate at which the pressure is changing.

SOLUTION

$$\frac{dp}{dV} = -1.36CV^{-2.36} = -1.36(pV^{1.36})V^{-2.36} = -\frac{1.36p}{V}$$

$$\frac{dp}{dt} = \frac{dp}{dV}\frac{dV}{dt} = -\frac{1.36 \times 50}{4.6}(0.3) = -4.435 \text{ lb/in}^2 \text{ per second}$$

The negative result signifies that at the given instant the pressure is decreasing, as Eq. (a) requires.

EXAMPLE 12.16 A conical storage tank is built with its vertex down, as shown in Fig. 12.11. Its depth is 12 units, and its diameter at the top is 8 units. Liquid enters the tank at the rate of 6 cubic units per min. Determine the rate at which the height of the liquid surface is rising at the instant when this height is 5 units.

SOLUTION Refer to the notation shown in Fig. 12.11, and let V denote the volume of liquid in the tank. We are given dV/dt, and we must find dh/dt. Therefore, we must establish the relationship between V and h. Applying the formula for the volume of a cone, we have

$$V = \frac{\pi r^2 h}{3} \tag{b}$$

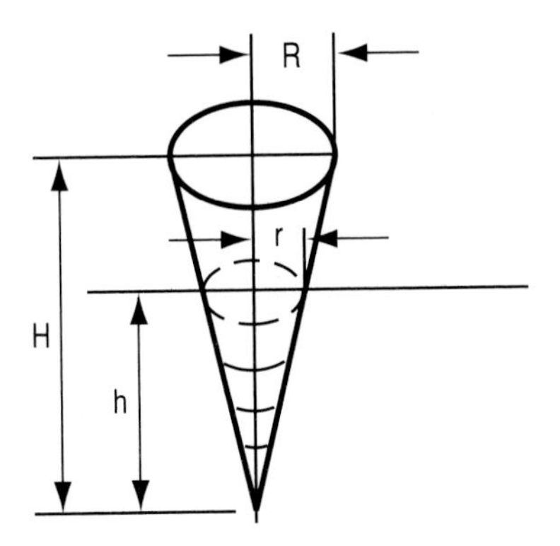

FIGURE 12.11 Conical storage tank.

The variable r can be eliminated in this manner: By similar triangles,

$$\frac{r}{R} = \frac{h}{H} \qquad \therefore r = \frac{Rh}{H}$$

Equation (b) now becomes

$$V = \frac{\pi R^2 h^3}{3H^2}$$

It will be simpler to apply dV/dh rather than dh/dV.

$$\frac{dV}{dh} = \frac{\pi R^2(3h^2)}{3H^2} = \frac{\pi R^2 h^2}{H^2}$$

$$\frac{dV}{dt} = \frac{\pi R^2 h^2}{H^2}\frac{dh}{dt} \qquad 6 = \frac{\pi \times 4^2 \times 5^2}{12^2}\frac{dh}{dt}$$

Solving, $dh/dt = 0.688$ units per min.

12.3.4 Evaluation of Indeterminate Forms

Let $f(x)$ and $g(x)$ denote functions of x. As stated in Art. 12.2.1, $f'(x)$ and $g'(x)$ denote their corresponding first derivatives. Consider that we wish to evaluate $f(x)/g(x)$ when $x = a$. If direct substitution of a for x in the two expressions yields the indeterminte form $0/0$ or ∞/∞, this obstacle can be circumvented by applying the following equation:

$$\lim_{x \to a} \frac{f(x)}{g(x)} = \lim_{x \to a} \frac{f'(x)}{g'(x)} \qquad (12.11)$$

This relationship is referred to as *l'Hôpital's rule*. If the expression at the right also yields an indeterminate form, we take the ratio of the second derivatives, continuing the process until a determinate form is reached. For brevity, we shall refer to the limiting value of y as x approaches a as the value of y when $x = a$.

EXAMPLE 12.17 Evaluate the following function when $x = 0$:

$$y = \frac{1 - \cos x}{x^2}$$

SOLUTION Direct substitution of 0 for x yields 0/0. Therefore, we apply Eq. (12.11) to obtain

$$y = \frac{\sin x}{2x}$$

This form also yields 0/0 when $x = 0$. Therefore, we apply Eq. (12.11) again to obtain

$$y = \frac{\cos x}{2}$$

When $x = 0$, $y = 1/2$. As a test, we may set $x = 0.04$, and we find that $y = 0.49993$, which is extremely close to 1/2.

Other indeterminate forms can be manipulated to make them amenable to l'Hopital's rule, as we shall now demonstrate.

EXAMPLE 12.18 Evaluate the function $y = x \ln x$ when $x = 0$.

SOLUTION Direct substitution yields the indetermintae form $0(-\infty)$. However, we may rewrite the given expression in this form:

$$y = \frac{\ln x}{1/x}$$

Applying l'Hopital's rule, we obtain

$$y = \frac{1/x}{-1/x^2} = -x$$

When $x = 0$, $y = 0$.

EXAMPLE 12.19 Evaluate the function $y = \sec x - \tan x$ when $x = \pi/2$.

SOLUTION Direct substitution yields the indeterminate form $\infty - \infty$. However, we may express y in the following alternative manner:

$$y = \frac{1}{\cos x} - \frac{\sin x}{\cos x} = \frac{1 - \sin x}{\cos x}$$

Applying l'Hopital's rule, we obtain

$$y = \frac{-\cos x}{-\sin x} = \frac{1}{\tan x}$$

When $x = \pi/2$, $y = 0$.

EXAMPLE 12.20 Evaluate the function $y = x^x$ when $x = 0$.

SOLUTION Direct substitution yields the indeterminte form 0^0. Taking the natural logarithm of both sides, we obtain

$$\ln y = x \ln x$$

In Example 12.18, we found that the expression at the right is 0 when $x = 0$.

Therefore, $\ln y = 0$, and $y = 1$. As a test, we may set $x = 0.001$, and we find that $y = 0.9931$, which is extremely close to 1.

12.3.5 Curvature

Consider that a point moves along a curve from left to right with a constant speed. At every instant, the *direction* of its motion is along the tangent to the curve at the point it currently occupies. The rate at which this direction is changing as the point moves is termed the *curvature* of the curve.

Curvature manifests itself in myriad ways. For example, when a skater rounds a curve, he must tilt his body, and the angle of tilt is related to the curvature of his path (as well as his speed). Similarly, when a motorist rounds a bend, she must hold the steering wheel in a rotated position, and the angle of rotation is related to the curvature of the road (as well as her speed).

Let K denote the curvature of a curve. In Fig. 12.12, T is the tangent to the curve S at P, θ is the angle that T makes with the positive side of the x axis, and s is the length of arc AP. The curvature is defined as $d\theta/ds$; thus, the unit of curvature is radians per unit of length. The value of curvature is

$$K = \frac{\dfrac{d^2y}{dx^2}}{\left[1 + \left(\dfrac{dy}{dx}\right)^2\right]^{3/2}} \tag{12.12}$$

The algebraic sign of K is identical with that of d^2y/dx^2, but usually it is only the absolute value that is significant.

EXAMPLE 12.21 Find the curvature of the parabola $y = x^2$.

SOLUTION

$$\frac{dy}{dx} = 2x \qquad \frac{d^2y}{dx^2} = 2$$

$$K = \frac{2}{(1 + 4x^2)^{3/2}}$$

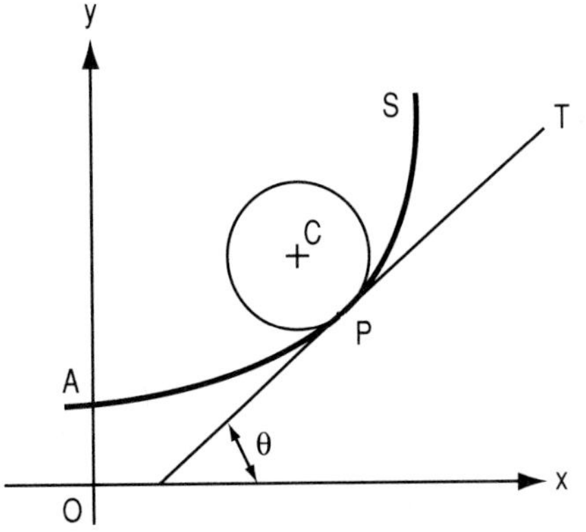

FIGURE 12.12 Curvature, circle of curvature, and center of curvature.

The curvature is maximum when $x = 0$, and it approaches 0 as x becomes infinitely large in absolute value. Thus, the parabola approaches linearity as it recedes from the y axis.

The curvature of a circle is uniform along its length and equal to the reciprocal of its radius in absolute value. Therefore, the *radius of curvature* of a curve in general, which is denoted by R, is defined in this manner:

$$R = \frac{1}{K} \tag{12.13}$$

In Fig. 12.12, we draw a circle on the concave side of the curve and tangent to T (and therefore curve S) at P, the radius of the circle being equal to the radius of curvature of S at P. This circle is the *circle of curvature* of S at P, and its center C is the *center of curvature* of S at P. In general, the centers of curvature of S form a smooth curve that is called the *evolute* of S.

If a curve approximates a horizontal straight line, $(dy/dx)^2$ is negligible, and K can be equated to d^2y/dx^2. This is the practice followed in structural engineering in analyzing the deflection of a horizontal beam.

In the design of a highway, a spiral is interposed between a straight-line segment and a circular curve to effect a gradual transition from rectilinear to circular motion, and vice versa. The type of spiral most frequently used is the *clothoid*, which has the property that K is directly proportional to s. Thus, the curvature changes at a uniform rate as we proceed along the curve.

12.3.6 Newton-Raphson Method of Finding Roots of an Equation

Consider the equation $f(x) = 0$. As defined in Art. 1.11, the root of this equation is a value of x that satisfies the equation. When an approximate value of a root is known, a more precise value can be obtained with considerable speed by an iterative process known as the *Newton-Raphson method*, which is based on the use of derivatives.

The curve in Fig. 12.13 is a plotting of $f(x)$. Our task is to find the value of x at A, where $f(x) = 0$. Let x_n denote the nth approximation (or *iterate*) of this root.

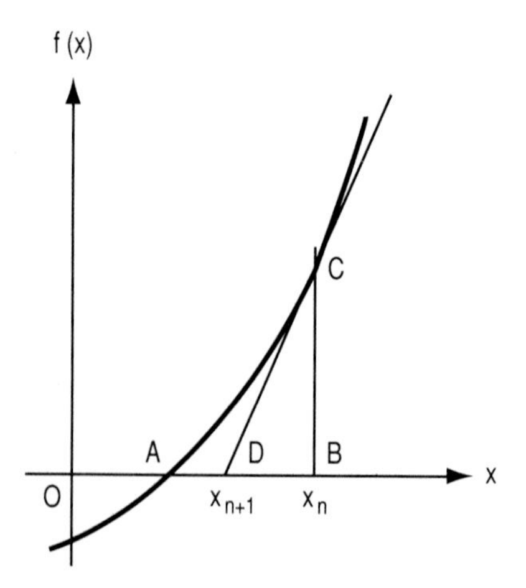

FIGURE 12.13 Newton-Raphson method.

Set $OB = x_n$. At C, draw the tangent to the curve, intersecting the x axis at D. Point D lies closer to A than does B, and we take OD as the succeeding, or $(n + 1)$th, approximation. The value of x_{n+1} is found in this manner:

$$\frac{BC}{DB} = f'(x_n)$$

$$BC = f(x_n) \qquad DB = x_n - x_{n+1}$$

Substituting and rearranging, we obtain

$$x_{n+1} = x_n - \frac{f(x_n)}{f'(x_n)} \tag{12.14}$$

We shall illustrate the process. Consider the equation $f(x) = x^4 - 2x - 15 = 0$. Then $f'(x) = 4x^3 - 2$. Assume that we start with the approximation $x_0 = 2.0$. In our first iteration, we have

$$f(2.0) = -3 \qquad f'(2.0) = 30$$

$$x_1 = 2.0 - \frac{-3}{30} = 2.1$$

In our second iteration, we have

$$f(2.1) = 0.2481 \qquad f'(2.1) = 35.044$$

$$x_2 = 2.1 - \frac{0.2481}{35.044} = 2.0929$$

This iterative process can be continued until we attain whatever degree of precision is required.

12.4 INTEGRATION

12.4.1 Meaning and Properties of Integrals

In the process of differentiation, we are given the expression for y in terms of x, and we are required to find the derivative dy/dx. The derivative yields the differential dy. The reverse process is called *integration*. In integration, we are given the differential dy expressed in terms of x, and we must construct the expression for y.

The expression for y that results from integration is called the *integral*. To signify that integration is to be performed, we place the symbol $\int$ before the differential, this symbol being a distorted S. The function that follows the integral sign is the *integrand*. For example, in

$$\int \sin^2 x \cos x \, dx$$

the integrand is $\sin^2 x \cos x$.

As an illustration, let $y = x^3$. We have the following:

$$\frac{dy}{dx} = 3x^2 \qquad dy = 3x^2 \, dx$$

Therefore, we may tentatively write

$$\int 3x^2\,dx = x^3$$

Differentiation is a many-one process. For example, consider the function

$$y = x^2 + C$$

where C is a constant. Then $dy/dx = 2x$ and $dy = 2x\,dx$ regardless of the value of C. It follows that there is an infinite number of functions having $2x\,dx$ as their differentials. Recognizing this fact, we must append a *constant of integration* C to our integral. Thus, returning to our illustrative case, we have

$$\int 3x^2\,dx = x^3 + C$$

Where an integral contains a constant of integration that can be assigned an arbitrary value, the integral is said to be *indefinite*. If we have a set of simultaneous values of x and y, we can evaluate C.

Integrals have two fundamental properties. The first property is

$$\int a\,du = a \int du \tag{12.15}$$

where a is a constant and u is a function of x. For example,

$$\int 8 \sin x\,dx = 8 \int \sin x\,dx$$

The second property is

$$\int (u + v)\,dx = \int u\,dx + \int v\,dx \tag{12.16}$$

where u and v are functions of x. For example,

$$\int (3x^4 - 9x)\,dx = \int 3x^4\,dx - \int 9x\,dx$$

Table A.2 in the appendix presents a set of integrals (with the constant of integration omitted). These are numbered for reference purposes.

An integral can always be tested for accuracy by reversing the process; that is, by equating y to the integral, and then finding the differential of y. As an illustration, consider Integral (196), and set

$$y = 2x \sin x - (x^2 - 2) \cos x$$

Differentiating, we obtain

$$\begin{aligned}\frac{dy}{dx} &= 2x \cos x + 2 \sin x - [(x^2 - 2)(-\sin x) + 2x \cos x] \\ &= x^2 \sin x\end{aligned}$$

Therefore, the integral that is presented is correct.

EXAMPLE 12.22 Evaluate the following:

$$\int \frac{dx}{9x^2 - 12x + 29}$$

SOLUTION Refer to Integrals (111) and (112). In the present case,

$$a = 9 \qquad b = -12 \qquad c = 29 \qquad 4ac - b^2 = 900 = 30^2$$

Therefore, Integral (111) applies, and the present integral is

$$\frac{2}{30} \tan^{-1} \frac{18x - 12}{30} + C = \frac{1}{15} \tan^{-1} (0.6x - 0.4) + C$$

The result can be tested by setting y equal to this expression and applying Derivative (15).

EXAMPLE 12.23 Evaluate the following:

$$\int \frac{(5x + 3)\, dx}{\sqrt{x^2 - 7}}$$

SOLUTION By resolving the integrand into two fractions and then applying Eqs. (12.15) and (12.16), we transform the given expression to the following:

$$5 \int \frac{x\, dx}{\sqrt{x^2 - 7}} + 3 \int \frac{dx}{\sqrt{x^2 - 7}}$$

Applying Integrals (79) and (81), we obtain the following as the present integral:

$$5\sqrt{x^2 - 7} + 3 \ln (x + \sqrt{x^2 - 7}) + C$$

12.4.2 Law of Natural Growth or Decay

A vast number of variables that arise in engineering, science, medicine, and economics are characterized by the following relationship:

$$\frac{dy}{dx} = ky \qquad (c)$$

where k is a constant that can be positive or negative. Expressed verbally, the variable y changes at a rate directly proportional to its own magnitude. If k is positive, y is said to obey the *law of natural* (or *organic*) *growth*; if k is negative, y is said to obey the *law of natural* (or *organic*) *decay*.

We shall now investigate such variables. Let y_0 denote the value of y when $x = 0$. In Fig. 12.14, the variation of y is represented by curve a when k is positive and by curve b when k is negative. At every point, the slope of the tangent is directly proportional to the ordinate. By transforming Eq. (c) and applying Integral (2), we obtain the following:

$$\frac{dy}{y} = k\, dx \qquad \ln y = kx + C$$

$$y = e^{kx + C} = e^{kx} e^C$$

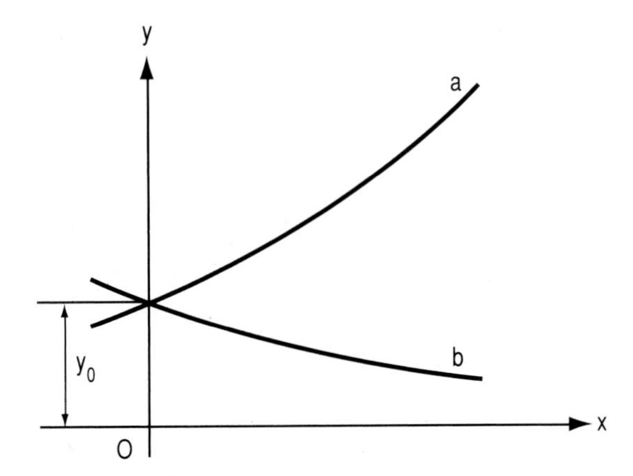

FIGURE 12.14 Graph of variable subject to natural growth or decay.

Since $y = y_0$ when $x = 0$, it follows that $e^C = y_0$, and we have

$$y = y_0 e^{kx} \tag{d}$$

Thus, y varies exponentially with x.

The following quantities increase according to the law of natural growth: the number of bacteria in a culture, the tension in a belt that is wrapped about a cylinder and subjected to unequal forces at the two ends, the principal in a savings account, the volume of a body as the temperature increases, the mass of a substance that is formed by a chemical reaction, and the amount of skin tissue that has healed following a wound.

The following quantities decrease according to the law of natural decay: the current in a circuit following the instantaneous termination of a voltage, the difference between the temperature of a given body and the temperature of its surroundings, the difference in temperature of two substances that flow through a heat exchanger, the electric charge on a capacitor when leakage occurs, the atmospheric pressure as we ascend above the earth's surface, and the mass of a radioactive substance.

EXAMPLE 12.24 A radioactive substance has a half-life period of 1200 years. How long does it take for the mass of the substance to diminish to 80 percent of its original value?

SOLUTION Scientists express the rate of radioactive disintegration of a substance by means of its *half-life period.* This is the amount of time required for the mass of the substance to diminish to one-half of the original mass.

We shall solve Eq. (d) for t, in this manner:

$$e^{kt} = \frac{y}{y_0} \qquad kt = \ln \frac{y}{y_0} \qquad t = \frac{\ln (y/y_0)}{k}$$

Let T denote the required time. Applying the last result, we have

$$\frac{T}{1200} = \frac{\ln 0.8}{\ln 0.5} \qquad T = 386.3 \text{ years}$$

12.4.3 Calculation of a Plane Area

In Art. 12.2.7, we considered that line AP in Fig. 12.3 starts at the y axis and then moves to the right while remaining parallel to the y axis. As the line moves, it generates an area bounded by the curve, the x axis, the y axis, and its present position. This area is denoted by A. Applying Eq. (12.9), we obtain

$$A = \int y\,dx \tag{12.17}$$

EXAMPLE 12.25 A curve has the equation $y = 3x^2$. Compute the generated area when $x = 4$.

SOLUTION

$$A = \int 3x^2\,dx = x^3 + C$$

Since $A = 0$ when $x = 0$, it follows that $C = 0$. Replacing x with 4, we obtain $A = 64$. The unit of A is the product of the units of x and y.

12.4.4 The Definite Integral

With reference to Fig. 12.15, assume that we wish to find the area bounded by the curve, the x axis, and the straight lines having the equations $x = a$ and $x = b$. Let A denote this area. The procedure is as follows: Applying Eq. (12.17), compute the area A_a from the y axis to $x = a$, compute the area A_b from the y axis to $x = b$, and then set $A = A_b - A_a$. The values a and b are termed the *limits of integration*, a being the lower limit and b the upper limit. Symbolically, the area is represented by placing the limits of integration after the integral sign, with the upper limit above the line and the lower limit below, in this manner:

$$A = \int_a^b y\,dx \tag{e}$$

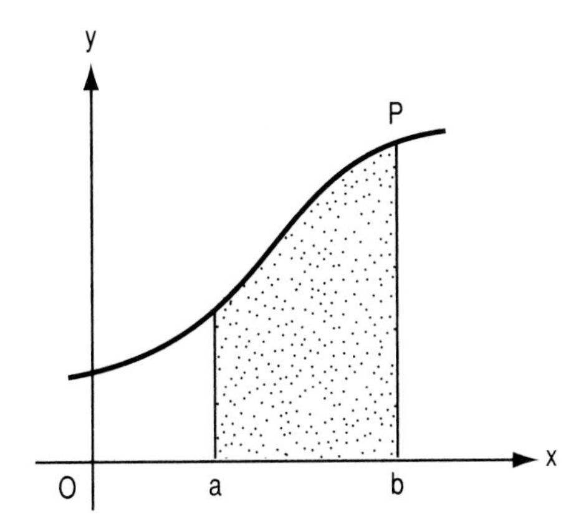

FIGURE 12.15

If the expression that results from integration is a single term, it is followed by a closing bracket, and the limits of integration are placed after the bracket. If the expression contains several terms, the entire expression is enclosed in brackets.

When A_a is subtracted from A_b, the constant of integration vanishes. Consequently, an integral that contains limits of integration is termed a *definite integral.*

If we consider that the area is generated by a line that starts at the y axis and moves to the right, Eq. (e) yields the *increase* in the generated area. Consequently, a definite integral can always be regarded as the increase in the value of the integral across the specified interval.

EXAMPLE 12.26 A curve has the equation $y = x^2 + x$. Compute the area bounded by the curve, the x axis, and the lines $x = 2$ and $x = 5$.

SOLUTION

$$A = \int_2^5 (x^2 + x)\, dx = \left[\frac{x^3}{3} + \frac{x^2}{2}\right]_2^5$$

The subtraction can be performed within each term, giving

$$A = \frac{5^3 - 2^3}{3} + \frac{5^2 - 2^2}{2} = 49.5$$

EXAMPLE 12.27 A curve has the equation $y = 3x^2 - 75$. Find the total amount of area (in absolute value) bounded by the curve, the x axis, and the lines $x = -2$ and $x = 6$.

SOLUTION Care must be exercised where the curve lies partly above and partly below the x axis, as straight integration between the given limits will yield simply the *net* area. Setting $y = 0$, we find that the curve lies below the x axis in the interval $-5 < x < 5$ and above the x axis elsewhere. Therefore, we must divide the given interval into two subintervals: from $x = -2$ to $x = 5$, and from $x = 5$ to $x = 6$.

$$\int (3x^2 - 75)\, dx = x^3 - 75x$$

$$\begin{aligned} A &= -[x^3 - 75x]_{-2}^{5} + [x^3 - 75x]_5^6 \\ &= -[5^3 - (-2)^3] + 75[5 - (-2)] + (6^3 - 5^3) - 75(6 - 5) \\ &= 408 \end{aligned}$$

From $x = -2$ to $x = 5$, the area is -392; from $x = 5$ to $x = 6$, the area is 16. Therefore, straight integration between $x = -2$ and $x = 6$ would have yielded the misleading value -376.

Now assume that we wish to find the area bounded by a curve, the y axis, and the straight lines having the equations $y = c$ and $y = d$, where $d > c$. By analogy with Eq. (e), we have

$$A = \int_c^d x\, dy$$

Where it is necessary to compute the area encompassed by intersecting curves, the value can usually be obtained by simple subtraction. For example, assume that we wish to find the area bounded by the two curves in Fig. 12.16. It can be found

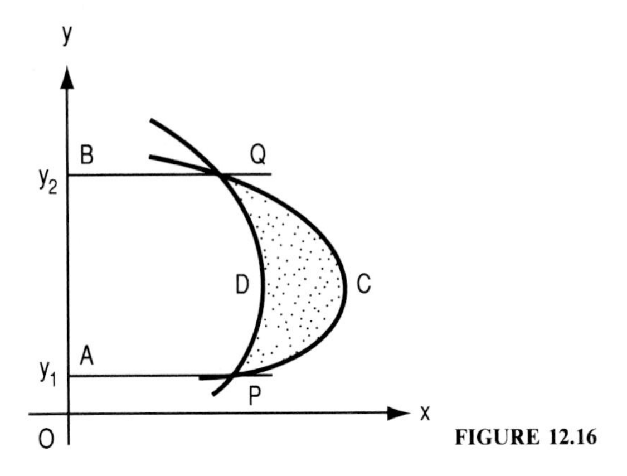

FIGURE 12.16

by this procedure: Determine the ordinates y_1 and y_2 of the intersection points P and Q, respectively. Then compute the areas $APCQB$ and $APDQB$, and subtract the second from the first. The values of y_1 and y_2 are found by solving the equations of the curves for y, equating the two expressions for y, and solving the resulting equation.

12.4.5 Interpretation of Integral as an Area

Since the amount of area between a curve and the x axis is an integral, we can invert our point of view to conceive that a given integral is equal to an area.

To illustrate this concept, assume that a point moves in a straight line. Let s, v, and t denote distance traversed, velocity, and elapsed time, respectively. Then

$$v = \frac{ds}{dt} \qquad \therefore\ ds = v\,dt \qquad s = \int v\,dt$$

If we draw a diagram in which t is plotted on the horizontal axis and v is plotted on the vertical axis, the area between the resulting curve and the horizontal axis equals the distance traversed.

In many instances, it is very helpful to conceive of an integral as an area. By drawing the appropriate diagram, we can instantly visualize how the integral varies in numerical value. Moreover, we can sometimes analyze the integral more readily by applying the properties of an area that we shall develop subsequently.

12.4.6 Improper Integrals

A definite integral having the limits a and b is said to be *improper* if either a or b is infinite or if both are infinite, or if the integrand becomes infinite within the interval a to b.

EXAMPLE 12.28 Compute the area that is bounded by the curve $y = 1/x^2$, the x axis, and the line $x = a$, where a is positive, and that extends indefinitely to the right.

SOLUTION The curve is plotted in Fig. 12.4.

$$A = \int_a^\infty x^{-2}\,dx = -\frac{1}{x}\bigg]_a^\infty = \frac{1}{a}$$

Thus, although the area is of infinite extent, it has a finite value, and the integral is said to *converge* to this value.

12.4.7 The Fundamental Theorem

We have found that an area can be viewed as an integral, and we shall now view an area as a sum. These alternative ways of viewing an area lead to a principle that is of vast significance.

Referring to Fig. 12.17, consider again that we wish to evaluate the area bounded by the curve, the x axis, and the lines $x = a$ and $x = b$. We proceed in this manner: Divide the interval $x = a$ to $x = b$ into n parts, and number the parts from left to right. (The parts need not be of uniform width.) Let Δx_i denote the width of the ith part. Draw lines parallel to the y axis at the boundaries. Now select an arbitrary point on the curve within each part, and draw a line through this point parallel to the x axis. Let y_i denote the ordinate to the arbitrary point in the ith part. We have thus constructed n rectangles. The total area of these rectangles is

$$\sum_{i=1}^{n} y_i\,\Delta x_i$$

Now consider that n is increased by making the widths of the parts progressively smaller. As the widths approach 0 (and n becomes infinitely large), the total area of the rectangles approaches the area between the curve and the x axis as a limit. Therefore, applying our previous expression for the area, we obtain

$$\lim_{\Delta x_i \to 0} \Sigma\, y_i\,\Delta x_i = \int_a^b y\,dx \qquad (12.18)$$

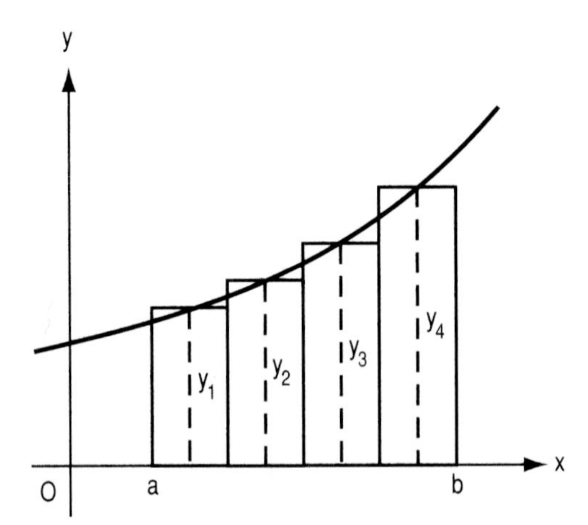

FIGURE 12.17 Development of the fundamental theorem.

This relationship, which is known as the *fundamental theorem* of integral calculus, enables us to view the definite integral as a sum. Although we have developed this theorem by formulating alternative expressions for an area, it can be proved rigorously without recourse to the concept of an area. We now present two simple illustrations of the application of the theorem.

EXAMPLE 12.29 A force is applied to a body, causing the body to move in the direction of the force. The magnitude of the force varies with the displacement of the body, in this manner:

$$F = 1.7s + 0.13s^2$$

where F is the force and s is the displacement of the body. Compute the work performed in displacing the body 12 units.

SOLUTION If the force acting on a body remains constant, the work performed by the force is the product of the force and the displacement of the body in the direction of the force. In the present case, the force varies.

Referring to Fig. 12.18, consider that the distance through which the body moves is divided into small parts. Let Δs denote the length of one part, and let s denote the displacement to the center of this part. Assume that the force remains constant across this part, and let ΔW denote the work performed in moving the body through this part. Then

$$\Delta W = (1.7s + 0.13s^2)\Delta s$$

To find the total work W performed, we sum the work performed across all the parts, allow the parts to become progressively smaller, and apply Eq. (12.18). Then

$$W = \int_0^{12} (1.7s + 0.13s^2)\, ds = \left[\frac{1.7s^2}{2} + \frac{0.13s^3}{3}\right]_0^{12} = 197.3$$

The unit of work is the product of the units of force and distance.

EXAMPLE 12.30 The resistance of a plane wall of unit length to heat transmission is $R = t/kA$, where R is the resistance, t is the thickness, k is the thermal conductivity, and A is the area of the wall normal to the direction of heat flow. Determine the resistance to heat transmission of a cylindrical pipe of unit length having an internal radius r and external radius R.

SOLUTION The pipe can be divided into concentric rings of minute thickness. Let x denote the interior radius of a ring, Δx denote its thickness, and ΔR denote the resistance of the ring. Since Δx is infinitesimal, an approximation is obtained by

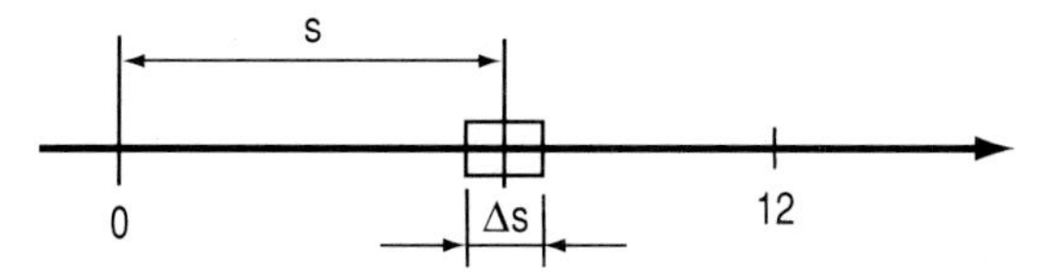

FIGURE 12.18

setting A equal to the area at the interior surface of the ring, which is $2\pi x$. Then

$$\Delta R = \frac{1}{k}\frac{\Delta x}{2\pi x}$$

To find the total resistance of the pipe, we sum the resistances of the rings, allow the rings to become progressively thinner, and apply Eq. (12.18). Then

$$R = \frac{1}{2\pi k}\int_r^R \frac{dx}{x} = \frac{1}{2\pi k}\ln x\Big]_r^R = \frac{\ln R - \ln r}{2\pi k} = \frac{\ln (R/r)}{2\pi k}$$

12.4.8 Calculation of Area with Polar Coordinates

Many curves are described more effectively by the use of polar coordinates, which are defined in Art. 3.1.9. We must therefore formulate an expression for area in terms of polar coordinates. We shall apply the following relationship: The area of a circular sector of radius r and central angle θ is $(1/2)r^2\theta$.

With reference to Fig. 12.19, assume that we wish to evaluate the dotted area A bounded by the curve, the positive side of the x axis, and the radius vector OP, where P has the coordinates (r, θ). Let θ assume an increment $d\theta$ and let Q denote the corresponding point on the curve. Draw the circular arc PR with O as center, and let dA denote the area of the sector OPR. Then

$$dA = \frac{1}{2}r^2\, d\theta \qquad \frac{dA}{d\theta} = \frac{1}{2}r^2$$

As $d\theta$ approaches 0, area OPR approaches area OPQ. Therefore,

$$A = \frac{1}{2}\int r^2\, d\theta \tag{12.19}$$

If we wish to find the area bounded by a curve and two radius vectors, we simply apply Eq. (12.19) with the corresponding vectorial angles as the limits of integration.

As previously stated, the limits of integration must be carefully established in calculating an area to avoid erroneous conclusions. To illustrate this principle in the present context, assume that we wish to find the area enclosed by the curve $r = b \sin n\theta$, where b is a constant and n is a positive integer. The curve comprises several loops. Consider the points $P_1(r_1, \theta_1)$ and $P_2(r_2, \theta_1 + \pi)$. Then $r_1 = b \sin n\theta_1$

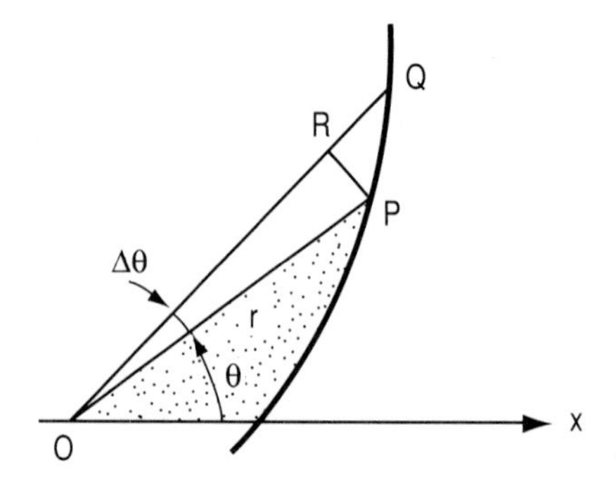

FIGURE 12.19

and $r_2 = b \sin(n\theta_1 + n\pi)$. If n is even, $r_2 = r_1$, and P_1 and P_2 are distinct points that are symmetric about the origin. The curve contains $2n$ loops, and it is completed when θ ranges from 0 to 2π. On the other hand, if n is odd, $r_2 = -r_1$, and P_1 and P_2 coincide. The curve contains only n loops, and it is completed when θ ranges from 0 to π.

EXAMPLE 12.31 Find the area enclosed by the curve $r = a + b \sin 3\theta$, where a and b are positive constants and $a > b$.

SOLUTION The curve is shown in Fig. 12.20 for the case where $a = 2b$. Again consider the points $P_1(r_1, \theta_1)$ and $P_2(r_2, \theta_1 + \pi)$. Since $\sin(3\theta_1 + 3\pi) = \sin(3\theta_1 + \pi) = -\sin 3\theta_1$, we have the following:

$$r_1 = a + b \sin 3\theta_1 \qquad r_2 = a - b \sin 3\theta_1$$

These distances are both positive, and P_1 and P_2 are distinct points that lie on opposite sides of the origin. Therefore, the curve is completed when θ ranges from 0 to 2π.

Set $u = 3\theta$. Then $d\theta = (du)/3$, and the range of u is from 0 to 6π. We now have

$$r = a + b \sin u \qquad r^2 = a^2 + 2ab \sin u + b^2 \sin^2 u$$

$$\text{Area} = \frac{1}{2}\int_0^{2\pi} r^2 \, d\theta = \frac{1}{6}\int_0^{6\pi} r^2 \, du$$

By substituting the expression for r^2 and applying Integrals (150) and (151), we find that

$$\text{Area} = \frac{\pi}{2}(2a^2 + b^2) = \pi a^2 + \frac{\pi b^2}{2}$$

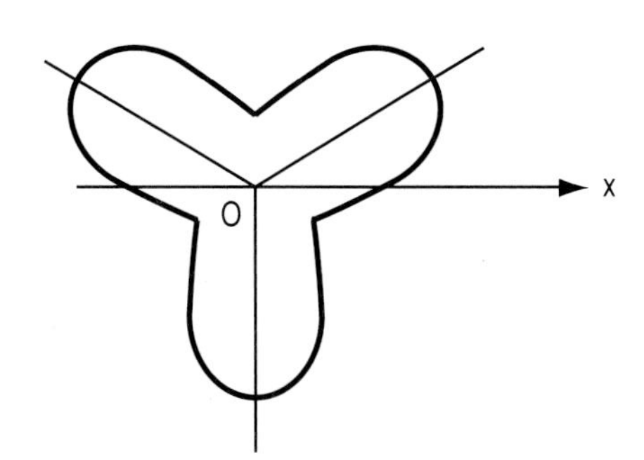

FIGURE 12.20 Graph of $r = a + b \sin 3\theta$ with $a = 2b$.

12.4.9 Approximate Integration

Many expressions are not susceptible of exact integration. However, the concept of an integral as the area under a curve provides a means of approximating the value of an integral by finding an approximate value of the area.

Assume that we wish to evaluate $\int y \, dx$ between the limits a and b, and assume that the curve in Fig. 12.21 is the graph of $y = f(x)$. The integral we require is the area $ABQP$, which we denote by A. Consider that we divide this area into n strips

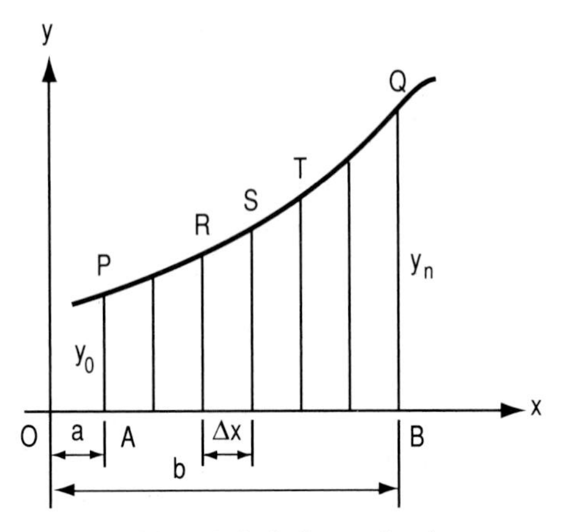

FIGURE 12.21 Method of approximating an area.

of uniform width Δx by drawing lines parallel to the y axis. Let y_r denote the ordinate at the right boundary of the rth strip. The area A can be approximated by making a simplifying assumption.

Under the *trapezoidal rule*, it is assumed that each strip is a trapezoid, and the area becomes

$$A = \frac{\Delta x}{2}[y_0 + 2(y_1 + y_2 + \cdots + y_{n-1}) + y_n] \tag{12.20}$$

Under *Simpson's rule*, it is assumed that three successive boundary points on the curve, such as R, S, and T, lie on a parabolic arc. The following relationship is applied: In Fig. 12.22, DEF is a parabolic arc, and $AB = BC = d$. The ordinates are numbered as shown. It can be readily demonstrated that the area $ACFD$ is expressible in this form:

$$A = \frac{d}{3}(y_0 + 4y_1 + y_2)$$

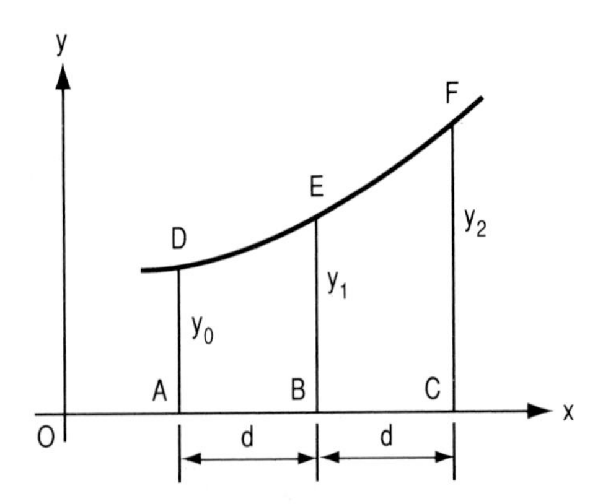

FIGURE 12.22 Area under a parabolic arc.

Returning to Fig. 12.21, if the number of strips is even, the area $ABQP$ becomes

$$A = \frac{\Delta x}{3}(y_0 + 4\,\Sigma\, y_{\text{odd}} + 2\,\Sigma\, y_{\text{even}} + y_n) \tag{12.21}$$

where "odd" and "even" refer to the numbers in the subscripts and $\Sigma\, y_{\text{even}}$ excludes y_n.

For a given value of n, Simpson's rule generally yields more accurate results than the trapezoidal rule. We shall illustrate the application of Simpson's rule.

EXAMPLE 12.32 Approximate

$$\int_2^5 \sqrt{3+2x}\,dx$$

by applying Simpson's rule and using $n = 6$. Then compare the result with the true value. Compute values of the function to three decimal places.

SOLUTION With $n = 6$, $\Delta x = (5-2)/6 = 0.5$. Set $y = \sqrt{3+2x}$. We number the ordinates from left to right, starting with 0. Let m denote the ordinate number. The corresponding x and y values are as follows:

m	0	1	2	3	4	5	6
x	2.0	2.5	3.0	3.5	4.0	4.5	5.0
y	2.646	2.828	3.000	3.162	3.317	3.464	3.606

$$\Sigma\, y_{\text{odd}} = 2.828 + 3.162 + 3.464 = 9.454$$

$$\Sigma\, y_{\text{even}} = 3.000 + 3.317 = 6.317$$

Equation (12.21) yields

$$A = \frac{0.5}{3}(2.646 + 4 \times 9.454 + 2 \times 6.317 + 3.606) = 9.450$$

This is the required integral.

If we now apply Integral (32), we obtain the following:

$$\int_2^5 \sqrt{3+2x}\,dx = \frac{2}{3 \times 2}(\sqrt{13^3} - \sqrt{7^3}) = 9.451$$

and this is the exact value of the integral.

There are instances where y values change rapidly in one region and slowly in another region. In these situations, the precision can be improved if we divide the interval between the limits of integration into two or more regions, divide each region into strips of uniform width, and then calculate the area of each region by means of Eq. (12.21).

12.5 MOTION IN A PLANE

As our first application of the principles thus far developed, we shall investigate the motion of a point in a plane.

12.5.1 Definitions, Notation, and Units

The notation is as follows:

Let t = elapsed time
s = distance traversed
v = velocity
a = acceleration (time rate of change of velocity)

If the unit of time is the second (s) and the unit of distance is the centimeter (cm), the unit of velocity is centimeters per second (cm/s), and the unit of acceleration is centimeters per second per second [(cm/s)/s], or cm/s^2.

A quantity that has magnitude only is termed a *scalar*. Thus, the temperature of a body and the number of electrons in an atom are scalars. A quantity that has both magnitude and direction is termed a *vector*. Thus, velocity and acceleration are both vectors. If a moving point undergoes a change in the magnitude or direction of its velocity, or both, it has acceleration. The direction of a vector has two characteristics: *inclination* (e.g., 40° with the horizontal) and *sense* (e.g., southwestward).

12.5.2 Rectilinear Motion

The motion of a point in a straight line is called *rectilinear*. Since the direction of motion remains constant, we may view velocity and acceleration as scalars in the present discussion.

The motion of a point can be depicted by means of three diagrams, in each of which time is plotted on the horizontal axis. In the *s-t diagram*, distance traversed is plotted on the vertical axis; in the *v-t diagram*, velocity is plotted on the vertical axis; in the *a-t diagram*, acceleration is plotted on the vertical axis.

Let Δs and Δv denote the change in distance and velocity, respectively, during a given time interval. In the following material, we shall omit the limits of integration, and we shall refer to the slope of the tangent to a curve as simply the slope of the curve. The basic relationships are as follows:

Let $v = \dfrac{ds}{dt}$ = slope of s-t diagram

$a = \dfrac{dv}{dt} = \dfrac{d^2s}{dt^2}$ = slope of v-t diagram

$\Delta s = \displaystyle\int v\,dt$ = area between v-t diagram and t axis

$\Delta v = \displaystyle\int a\,dt$ = area between a-t diagram and t axis

Dividing v by a and then cross-multiplying, we obtain the following:

$$\int a\,ds = \int v\,dv \tag{12.22}$$

Equation (12.22) is applicable to a situation where the acceleration is directly related to the distance traversed.

EXAMPLE 12.33 A test vehicle moves on a straight track, and its acceleration is $a = 0.8s^3$, the unit of time being the second. What distance does the vehicle traverse as its speed increases from 60 to 95 units/s?

SOLUTION Let S denote the distance. Applying Eq. (12.22) and replacing a with the given expression, we obtain

$$0.8 \int_0^S s^3 \, ds = \int_{60}^{95} v \, dv$$

$$\frac{0.8S^4}{4} = \frac{v^2}{2}\bigg]_{60}^{95} = \frac{95^2 - 60^2}{2} = 2712.5 \qquad S = 10.79 \text{ units}$$

In many instances, the acceleration of a point is uniform. Let v_i and v_f denote the initial and final velocity, respectively, for a time interval t. Then

$$a = \frac{v_f - v_i}{t} \tag{12.23}$$

$$s = v_i t + \frac{at^2}{2} = v_f t - \frac{at^2}{2} = \frac{v_f^2 - v_i^2}{2a} \tag{12.24}$$

EXAMPLE 12.34 The upward movement of a cam follower occurs in three phases. There is a constant acceleration in phase 1, a constant velocity of 56 units/s in phase 2, and a constant deceleration in phase 3. The distances traversed are 0.7 units in phase 1, 2.8 units in phase 2, and 1.75 units in phase 3. Compute the following: the acceleration of the follower in phases 1 and 3, and the total time T that the follower is ascending.

SOLUTION The cam follower executes an oscillatory motion in a vertical line. The upward stroke begins and ends with zero velocity. Refer to the v-t diagram in Fig. 12.23. We shall apply subscripts corresponding to the phase numbers. Applying

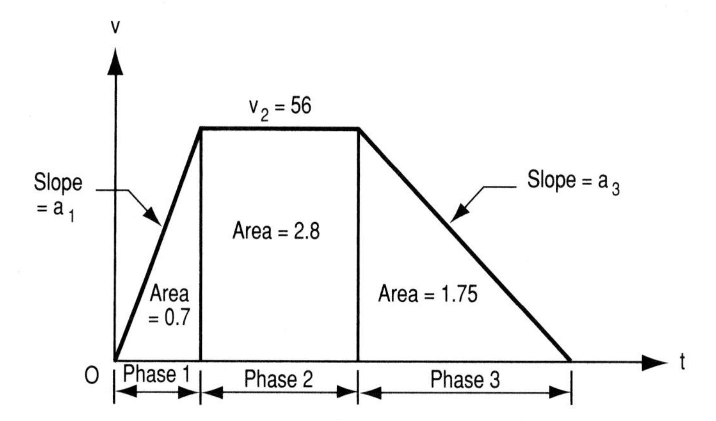

FIGURE 12.23 Velocity-time diagram of cam follower on upward stroke.

the foregoing equations, we obtain the following:

$$s_1 = \frac{56^2 - 0}{2a_1} = 0.7 \qquad a_1 = 2240 \text{ units/s}^2$$

$$s_3 = \frac{0 - 56^2}{2a_3} = 1.75 \qquad a_3 = -896 \text{ units/s}^2$$

$$t_1 = \frac{v_2}{a_1} = \frac{56}{2240} = 0.0250 \text{ s} \qquad t_3 = -\frac{v_2}{a_3} = \frac{56}{896} = 0.0625 \text{ s}$$

$$t_2 = \frac{s_2}{v_2} = \frac{2.8}{56} = 0.0500 \text{ s}$$

$$T = 0.0250 + 0.0500 + 0.0625 = 0.1375 \text{ s}$$

When a body falls vertically downward, its acceleration due to the earth's gravitational pull is labeled g. For a body near the earth's surface, the mean value of g is 32.16 ft/s^2 (9.802 m/s^2).

EXAMPLE 12.35 A projectile is fired vertically upward with a velocity of 230 ft/s (70.10 m/s) at a height of 65 ft (19.81 m) above ground. Neglecting air resistance, compute the following: the time required for the projectile to attain a height of 800 ft (243.8 m) above ground; the maximum height above ground to which the projectile rises; the time required for the projectile to reach its summit; the velocity of the projectile when it strikes the ground; the interval of time T between the instant the projectile is fired and the instant it strikes the ground.

SOLUTION Take the upward direction as positive, and let y and s denote the elevation of the projectile above the ground and above the point of firing, respectively. Then $s = y - 65$ ft.

Refer to the diagrams in Fig. 12.24. At A, $y = 800$ ft; at B, $y = y_{max}$; at C, the projectile strikes the ground.

When $y = 800$ ft, $s = 735$ ft. Applying Eqs. (12.23) and (12.24) with subscripts to identify the points, we obtain the following:

$$v_A = \sqrt{v_i^2 + 2as} = \sqrt{230^2 + 2(-32.16)735} = 75.00 \text{ ft/s}$$

$$t_A = \frac{75 - 230}{-32.16} = 4.82 \text{ s}$$

Alternatively, t_A can be found directly by applying the first form of Eq. (12.24) and solving for t. We take the smaller of the two values that result.

At B, $v = 0$. Then

$$s_{max} = \frac{0 - 230^2}{2(-32.16)} = 822.5 \text{ ft } (250.7 \text{ m})$$

$$y_{max} = 822.5 + 65 = 887.5 \text{ ft } (270.5 \text{ m})$$

$$t_B = \frac{0 - 230}{-32.16} = 7.15 \text{ s}$$

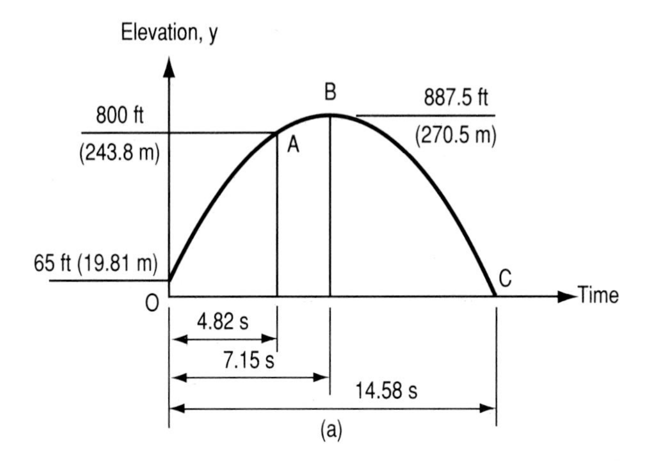

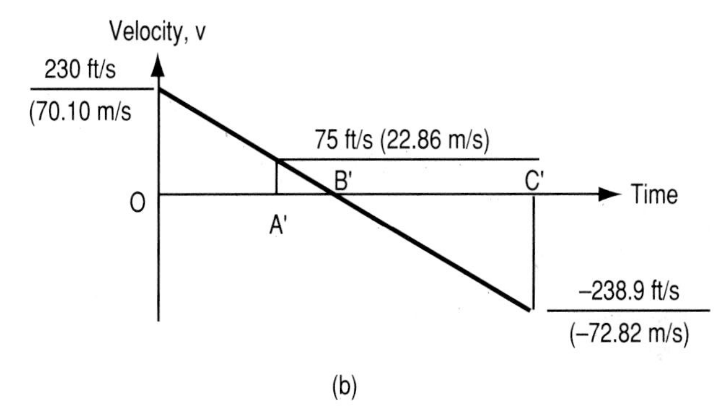

FIGURE 12.24 Diagrams for projectile fired vertically upward. (*a*) Elevation-time diagram; (*b*) velocity-time diagram.

At C, $s = -65$ ft. Then

$$v_C = -\sqrt{230^2 + 2(-32.16)(-65)} = -238.9 \text{ ft/s } (-72.82 \text{ m/s})$$

$$T = \frac{-238.9 - 230}{-32.16} = 14.58 \text{ s}$$

12.5.3 Curvilinear Motion

Consider that a particle moves along the curve in Fig. 12.25*a*. Let P be its present position, and let s denote the distance along the curve from an arbitrary reference point A to P. The magnitude of the velocity and acceleration of the particle are as follows:

$$v = \frac{ds}{dt} \qquad a = \frac{d^2s}{dt^2}$$

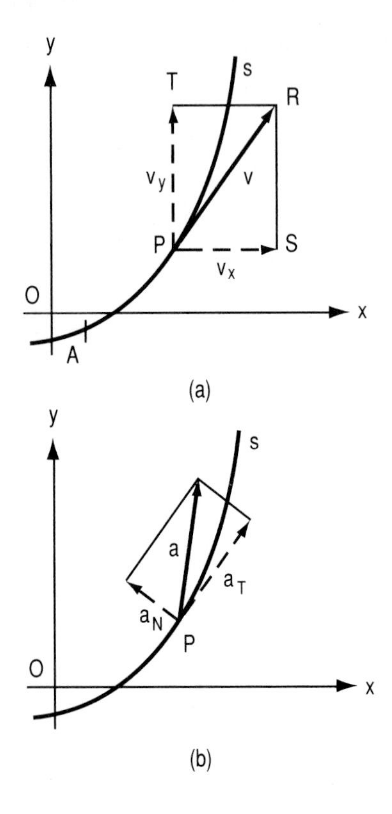

FIGURE 12.25 Motion along a curve. (*a*) Components of velocity; (*b*) components of acceleration.

As previously stated, the direction of the velocity is along the tangent at *P*. If the force that constrains the particle to traverse the curve were withdrawn at the instant the particle reaches *P*, its motion would be along the tangent.

It is helpful to resolve the velocity into components v_x and v_y parallel to the x and y axes, respectively. In Fig. 12.25*a*, the velocity is represented by the vector *PR*, and its components are represented by the dashed vectors *PS* and *PT*. Then

$$\frac{v_y}{v_x} = \frac{dy/dt}{dx/dt} = \frac{dy}{dx} \qquad v^2 = v_x^2 + v_y^2$$

Similarly, it is helpful to resolve the acceleration into components a_T and a_N in the tangential and normal directions, respectively, as shown in Fig. 12.25*b*. Then

$$a^2 = a_T^2 + a_N^2$$

The acceleration components are as follows:

$$a_T = \frac{dv}{dt} \tag{12.25}$$

$$a_N = v^2 K = \frac{v^2}{R} \tag{12.26}$$

where K and R are the curvature and radius of curvature of the curve at the given point. (Refer to Art. 12.3.5.) The algebraic sign of a_N coincides with that of K (and therefore of d^2y/dx^2), and the sense of a_N is toward the center of curvature. An alternative expression for a_N is available. Let

$$m = \frac{v_y}{v_x} = \frac{dy}{dx}$$

$$a_N = \frac{v\dfrac{dm}{dt}}{1+m^2} \tag{12.27}$$

Where a particle moves along a curve with constant velocity, $a_T = 0$, and $a = a_N$.

EXAMPLE 12.36 The motion of a particle in a plane is described by the parametric equations $x = t^2$, $y = t^3$. Develop expressions for the following: the velocity; the x and y components of acceleration; the tangential and normal components of acceleration. Test the results for consistency.

SOLUTION We shall apply Eq. (12.27) to find a_N. Differentiating the given equations twice, we obtain the following:

$$v_x = 2t \qquad a_x = 2$$

$$v_y = 3t^2 \qquad a_y = 6t$$

$$v = \sqrt{4t^2 + 9t^4} = t\sqrt{4 + 9t^2}$$

$$a_T = \frac{dv}{dt} = \frac{2(2+9t^2)}{\sqrt{4+9t^2}}$$

$$m = \frac{v_y}{v_x} = \frac{3t}{2} \qquad \frac{dm}{dt} = \frac{3}{2}$$

$$a_N = \frac{t\sqrt{4+9t^2}\,(3/2)}{1 + \dfrac{9t^2}{4}} = \frac{6t}{\sqrt{4+9t^2}}$$

The results can be tested in this manner:

$$a^2 = a_x^2 + a_y^2 = 4 + 36t^2$$

$$a^2 = a_T^2 + a_N^2 = 4 + 36t^2$$

This consistency in the values of a^2 proves that our results are accurate.

12.5.4 Rotation

Consider that line L in Fig. 12.26 rotates about the origin in the counterclockwise direction, and let θ denote the angle between L and the positive side of the x axis. Let ω and α denote the angular velocity and angular acceleration, respectively, and

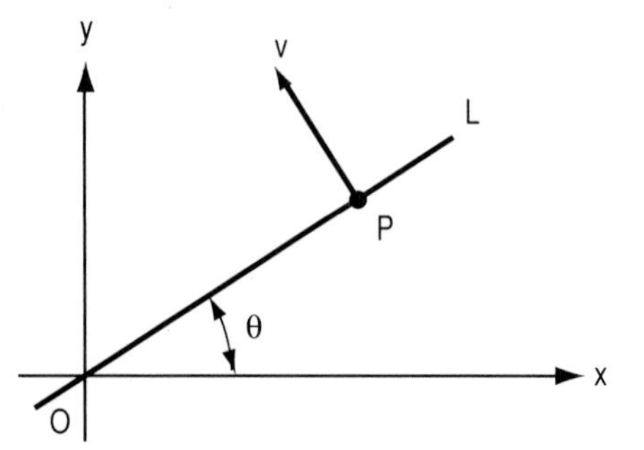

FIGURE 12.26

again let t denote elapsed time. Then

$$\omega = \frac{d\theta}{dt} \qquad \alpha = \frac{d\omega}{dt} = \frac{d^2\theta}{dt^2}$$

The increases in θ and in ω during a given time interval are as follows:

$$\Delta\theta = \int \omega \, dt \qquad \Delta\omega = \int \alpha \, dt$$

If time is measured in seconds, the unit of ω is rad/s, and the unit of α is rad/s^2.

Mathematically, the rotation of a line is exactly analogous to the rectilinear motion of a point, and the equations of rectilinear motion can be adapted to rotation simply by replacing s with θ, v with ω, and a with α.

The linear velocity of point P on L is

$$v = (OP)\omega$$

12.5.5 Simple Harmonic Motion

Let Q denote a point that repeatedly traverses the circumference of a circle with constant speed, and let P denote the projection of Q onto a straight line. The cyclic movement of P is referred to as *simple harmonic motion.*

The maximum displacement of P from its central position is its *amplitude*, the time required for P to complete one cycle is the *period*, and the number of cycles completed in a unit time is the *frequency*.

In Fig. 12.27, we place the origin at the center of the circle that Q traverses and take P as the projection of Q onto the x axis. Let r denote the radius and x denote OP. Consider that Q moves in the counterclockwise direction;

let θ = angle between OQ and the positive side of the x axis
ω = angular velocity of OQ
v = velocity of P
a = acceleration of P
T = period

f = frequency
t = time elapsed since Q was at A, s
A = amplitude

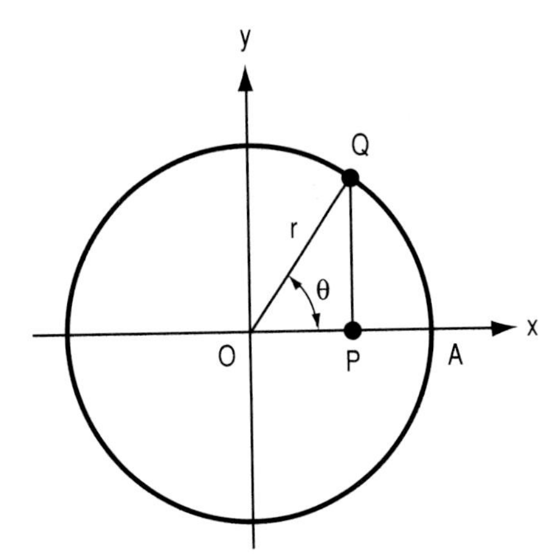

FIGURE 12.27 Generating simple harmonic motion.

The unit of f is cycles/s, or hertzes (Hz).

Since there are 2π rad in a circle, we have

$$T = \frac{2\pi}{\omega} \qquad f = \frac{1}{T} = \frac{\omega}{2\pi} \tag{12.28}$$

We also have the following:

Let $x = r \cos \theta = A \cos \omega t$

$$v = \frac{dx}{dt} = -A\omega \sin \omega t$$

$$a = \frac{dv}{dt} = -A\omega^2 \cos \omega t$$

The first and third equations yield

$$a = -\omega^2 x$$

Thus, the acceleration is directly proportional to the displacement in absolute value, but the direction of acceleration is opposite to that of displacement.

The first and second equations yield

$$v^2 = \omega^2(A^2 - x^2)$$

This relationship expresses velocity as a function of displacement, and it discloses that velocity is maximum in absolute value when P is at the origin.

EXAMPLE 12.37 A particle executes simple harmonic motion with an amplitude of six units, and the duration of a cycle is 1.2 s. Find the velocity and acceleration of the particle when it is located four units from the origin in the positive direction. Verify the results.

SOLUTION

$$A = 6 \text{ units} \qquad T = 1.2 \text{ s} \qquad x = 4 \text{ units}$$

Applying the foregoing equations, we obtain the following:

$$\omega = \frac{2\pi}{T} = \frac{2\pi}{1.2} = 5.236 \text{ rad/s}$$

$$a = -\omega^2 x = -(5.236)^2 \times 4 = -109.7 \text{ units/s}^2$$

$$v^2 = \omega^2(A^2 - x^2) = (5.236)^2(6^2 - 4^2)$$

$$v = \pm 23.42 \text{ units/s}$$

The velocity is positive or negative according to whether the particle is moving away from or toward the origin, respectively.

The calculated values can be verified by applying the equations for x, v, and a to obtain the following:

$$\cos \omega t = \frac{x}{A} = \frac{4}{6} \qquad \sin \omega t = \sqrt{1 - \left(\frac{4}{6}\right)^2} = \pm 0.74536$$

$$v = -6(5.236) \sin \omega t = \pm 23.42 \text{ units/s}$$

$$a = -6(5.236)^2 \cos \omega t = -109.7 \text{ units/s}^2$$

12.5.6 Analysis of Absolute Motion

Assume that a point P moves in some prescribed manner relative to another point Q, where Q itself is in motion. For example, P and Q may be the centers of the moon and earth, respectively. The *absolute* motion of P is a composite of the motion of Q and the motion of P relative to Q. The path traced by a moving point is termed its *trajectory*.

EXAMPLE 12.38 Point Q moves along the x axis in the positive direction with a constant velocity u. Simultaneously, P revolves about Q in a circle of radius a with a constant angular velocity ω. Initially, Q is at the origin and P is at $(a, 0)$. Construct a specimen trajectory of P and discuss its properties.

SOLUTION Let x and y denote the instantaneous coordinates of P, let t denote elapsed time, and let v_x and v_y denote the velocity of P in the x and y directions, respectively. Refer to Fig. 12.28*a*.

$$x = OQ + QA = ut + a \cos \omega t \qquad y = AP = a \sin \omega t$$

$$v_x = \frac{dx}{dt} = u - a\omega \sin \omega t \qquad v_y = \frac{dy}{dt} = a\omega \cos \omega t$$

$$\frac{dy}{dx} = \frac{v_y}{v_x} = \frac{a\omega \cos \omega t}{u - a\omega \sin \omega t}$$

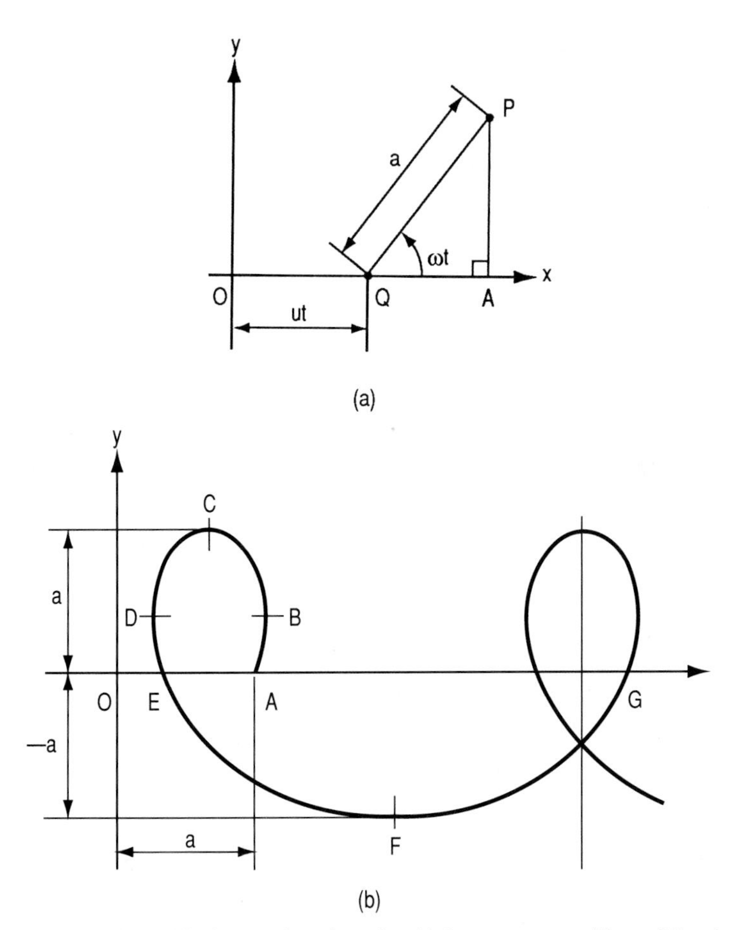

FIGURE 12.28 Absolute motion of a point. (*a*) Instantaneous positions of P and Q; (*b*) trajectory of P with $u < a\omega$.

The motion of P is cyclic, and we may divide each cycle into two phases. In phase 1, $0 < \omega t < \pi$, and P is moving to the left relative to Q. In phase 2, $\pi < \omega t < 2\pi$, and P is moving to the right relative to Q. Thus, the absolute velocity of P in the x direction is greater in phase 2 than in phase 1.

There are three possible conditions, as follows:

Case 1. $u < a\omega$. A specimen trajectory is shown in Fig. 12.28*b*, where A is the initial position of P. The expression for v_x reveals the following: P starts with a velocity u in the x direction. However, as time elapses, this velocity decreases and eventually becomes negative. Thus, there is an interval during which P moves to the left, and consequently the trajectory contains a loop. The quantity ωt has the following values at the indicated points: at A, 0; at C, $\pi/2$; at E, π; at F, $3\pi/2$; at G, 2π. At B and D, $v_x = 0$, and $\sin \omega t = u/a\omega$. If u increases while ω remains constant, the width BD of the loop diminishes and B is displaced to the right, as can readily be seen by drawing the v_x-t diagram.

Case 2. $u > a\omega$. The expression for v_x reveals that this quantity is always positive, and consequently P moves to the right during the entire cycle. The trajectory does not contain a loop.

Case 3. $u = a\omega$. In this intermediate case,

$$\frac{dy}{dx} = \frac{\cos \omega t}{1 - \sin \omega t}$$

Applying the result of Example 12.19, we find that dy/dx is infinite when $\omega t = \pi/2$. Therefore, the curve contains a cusp when $y = a$. We may view this cusp as a degenerate loop of zero width.

12.6 VOLUMES, ARC LENGTHS, AND SURFACE AREAS

In the subsequent material, we shall apply the following relationships: Let r, C, and A denote, respectively, the radius, circumference, and area of a circle. Then $C = 2\pi r$ and $A = \pi r^2$.

12.6.1 Volume of a Solid of Revolution

Many solid objects may be conceived to be generated by revolving a portion of a plane about an axis within the plane. Such objects are referred to as *solids of revolution*. The plane surface that is revolved is called the *section* of the solid.

In Fig. 12.29, PQ is a continuous curve, and lines AP and BQ are parallel to the y axis. Consider that the region $ABQP$ is revolved about the x axis through an angle of 2π. We wish to evaluate the volume V of the solid thus formed.

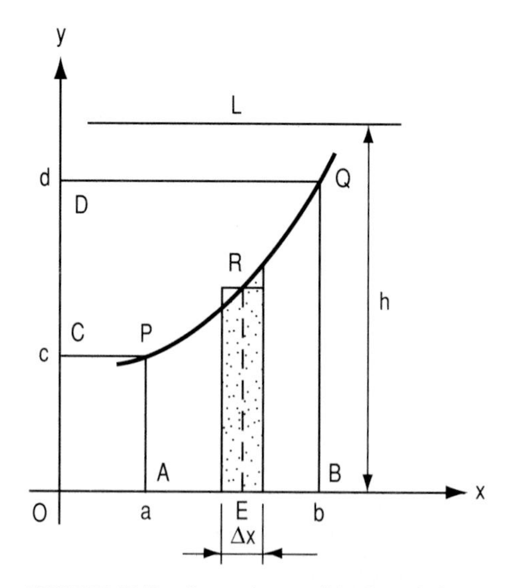

FIGURE 12.29 Generating a solid of revolution.

For this purpose, we divide the region into n parts by means of lines parallel to the y axis. One part, of width Δx, is shown dotted. We select an arbitrary point E within this part, draw ER parallel to the y axis, where R lies on the curve, and draw a line through R parallel to the x axis, thereby forming a rectangle. Let $y = ER$. When this rectangle revolves about the x axis, line ER generates a circle of area πy^2, and the rectangle generates a right circular cylinder, or *disk*, of volume $\pi y^2 \Delta x$. By summing the volumes of these disks, allowing n to become infinite, and applying Eq. (12.18), we obtain

$$V = \pi \int_a^b y^2\, dx \tag{12.29a}$$

Similarly, if the region $CPQD$ is revolved about the y axis, the volume of the solid of revolution thus generated is

$$V = \pi \int_c^d x^2\, dy \tag{12.29b}$$

EXAMPLE 12.39 Prove that the volume of a sphere of radius r is $(4/3)\pi r^3$.

SOLUTION Figure 12.30 shows a semicircle of radius r having its center at the origin. We may consider that the sphere is generated by revolving the semicircular region about the x axis. The equation of the semicircle is

$$x^2 + y^2 = r^2$$

Equation (12.29a) yields

$$V = \pi \int y^2\, dx = \pi \int (r^2 - x^2)\, dx$$

$$V = \pi \left[r^2 x - \frac{x^3}{3} \right]_{-r}^{r} = \pi \left[2r^3 - \frac{2r^3}{3} \right] = \frac{4}{3} \pi r^3$$

Now consider that the region $ABQP$ in Fig. 12.29 is revolved about line L, which has the equation $y = h$. In this case, the "solid" of revolution is actually a hollow body. Point E describes a circular arc of radius h, and R describes a circular arc of radius $h - y$. Therefore, the volume ΔV generated by the dotted rectangle is

$$\Delta V = \pi[h^2 - (h - y)^2]\Delta x = \pi(2hy - y^2)\, \Delta x$$

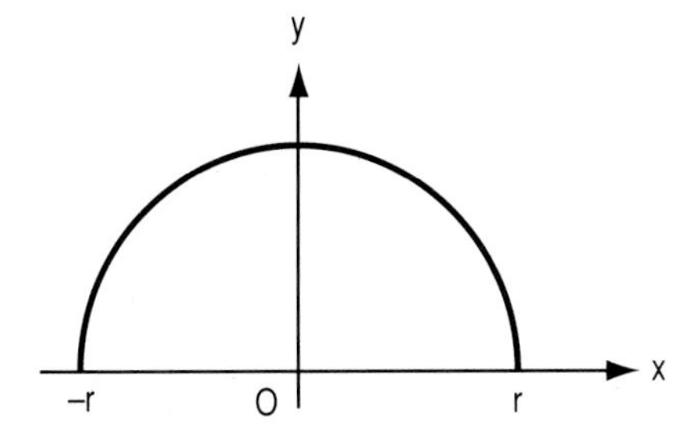

FIGURE 12.30

and the volume generated by the region $ABQP$ is

$$V = \pi \int_a^b (2hy - y^2)\, dx$$

Similar expressions can be obtained for other axes of revolution.

EXAMPLE 12.40 If a circle is revolved about an axis in the same plane but outside the circle, the solid thus generated is called a *torus*. Prove that the volume of a torus is $2\pi^2 Rr^2$, where r is the radius of the circle and R is the distance from the center of the circle to the axis of revolution.

SOLUTION Place the center of the circle at the origin, and place the axis of revolution at a distance R above the x axis. Taking only the positive value of y, we have

$$V = \pi \int [(R + y)^2 - (R - y)^2]\, dx = 4\pi R \int y\, dx$$

$$= 4\pi R \int \sqrt{r^2 - x^2}\, dx$$

Applying Integral (58) with the limits of integration r and $-r$, and setting $\sin^{-1} 1 = \pi/2$ and $\sin^{-1}(-1) = -\pi/2$, we obtain the given equation for the volume.

Now consider that the region $ABQP$ in Fig. 12.29 is revolved about the y axis through an angle of 2π. Let $OE = x$. Point E describes a circle of circumference $2\pi x$, and the dotted rectangle forms a *cylindrical shell* of thickness Δx. If this thickness is infinitesimal, we may set the volume of this shell equal to $2\pi xy\, \Delta x$. By summing the volumes of these shells, allowing n to become infinite, and applying Eq. (12.18), we obtain

$$V = 2\pi \int_a^b xy\, dx \tag{12.30}$$

EXAMPLE 12.41 The curve in Fig. 12.31 has the equation $y = x^{1/2}$, AB is parallel to the y axis, and $OA = X$. If the region OAB is revolved about the y axis, what volume is generated? Verify the result.

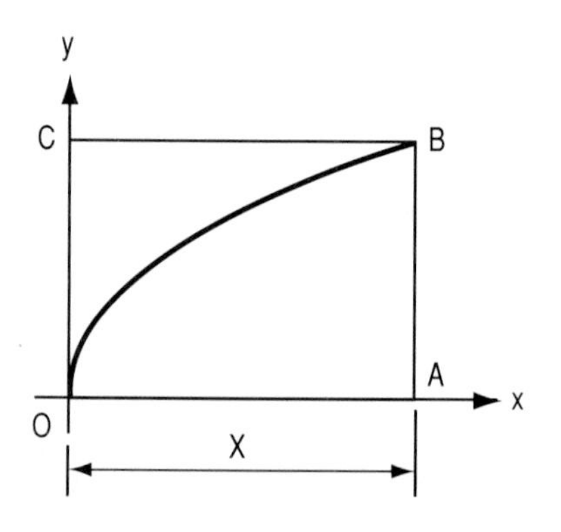

FIGURE 12.31

SOLUTION Applying Eq. (12.30), we obtain

$$V = 2\pi \int_0^X x(x^{1/2})\,dx = 2\pi \int_0^X x^{3/2}\,dx = 2\pi\,\frac{X^{5/2}}{5/2} = \frac{4\pi X^{5/2}}{5}$$

We shall now verify this result. In Fig. 12.31, draw CB parallel to the x axis, and consider that the entire rectangle $OABC$ is revolved about the y axis. A right circular cylinder is generated, and its volume V_T is

$$V_T = \pi(OA)^2(AB) = \pi X^2(X^{1/2}) = \pi X^{5/2}$$

Now consider that the region OBC alone is revolved about the y axis. Applying Eq. (12.29*b*), we find that its volume V' is

$$V' = \pi \int x^2\,dy = \pi \int_0^X x^2\left(\frac{1}{2}x^{-1/2}\,dx\right) = \frac{\pi}{2}\int_0^X x^{3/2}\,dx = \frac{\pi X^{5/2}}{5}$$

Therefore, the volume V generated by the region OAB alone is $V = V_T - V'$, and the result is identical with that previously obtained.

12.6.2 Volume of a Solid of Translation

A line or plane is said to be *translated* when it moves while remaining parallel to its original position. Many solid objects may be conceived to be generated by translating a planar surface along a straight line while this surface changes its size in a prescribed manner. Such objects are termed *solids of translation*, and the planar surface that generates the solid is called its *section*.

As an illustration, consider that plane P in Fig. 12.32 starts at the yz plane and is then translated to the right, thus remaining parallel to the yz plane. This plane contains a circle having its center on the x axis. If the diameter of the circle is directly proportional to the displacement of P from the yz plane, the circle generates a right circular cone.

Assume that the translating surface remains parallel to the yz plane in Fig. 12.32. Let V denote the volume of the solid generated in this manner, and let

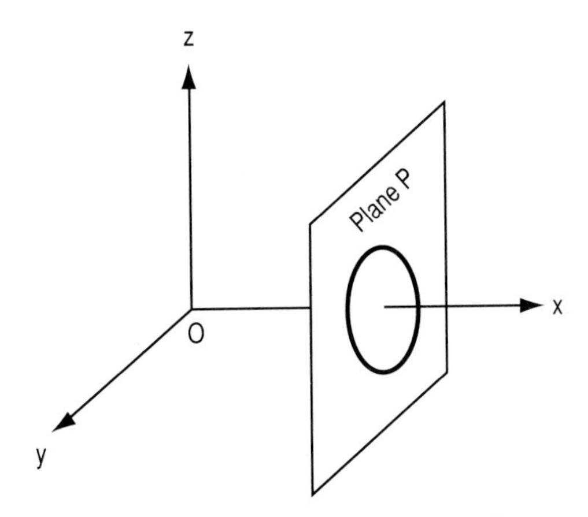

FIGURE 12.32 Generating a solid of translation.

A denote the area of the section corresponding to a given value of x. We divide the solid into n parts having sides parallel to the yz plane. In Fig. 12.32, assume that A remained constant across the distance Δx. The volume of this part would be $A\,\Delta x$. By summing the volumes of all parts, allowing n to become infinite, and then applying Eq. (12.18), we obtain

$$V = \int A\,dx \tag{12.31}$$

EXAMPLE 12.42 The base of a solid is a circle of radius r. A section perpendicular to a reference diameter is an isosceles right triangle having its hypotenuse in the base. Find the volume of the solid.

SOLUTION We may conceive that the solid is generated by the triangle ABC in Fig. 12.33, in this manner: AB is the hypotenuse, the triangle moves while remaining parallel to the yz plane, and vertices A and B lie on a circle of radius r having its center at the origin.

Draw the altitude CD. Since the triangle is isosceles, $AD = DB$, and angle BAC is 45°. Then $DC = AD$. The area of the triangle is $(1/2)(AB)(DC) = (AD)^2 = y^2 = r^2 - x^2$. Then

$$V = \int_{-r}^{r} (r^2 - x^2)\,dx = \left[r^2 x - \frac{x^3}{3} \right]_{-r}^{r} = \frac{4}{3} r^3$$

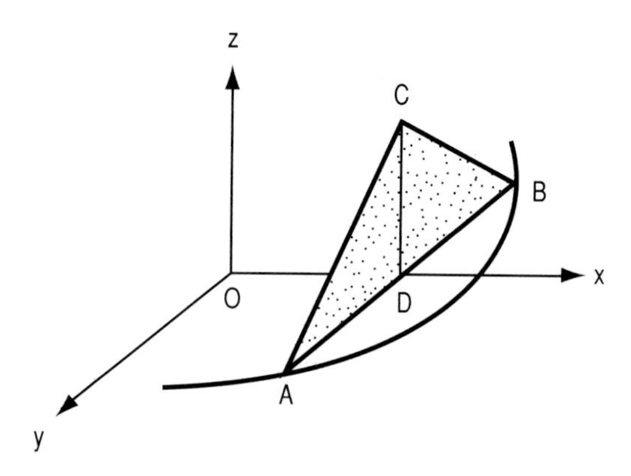

FIGURE 12.33

12.6.3 Length of Arc

Let s denote the length of an arc in a plane extending from the point where $x = a$ to the point where $x = b$. By Eq. (12.6), we have

$$s = \int_a^b \sqrt{1 + \left(\frac{dy}{dx}\right)^2}\,dx \tag{12.32}$$

EXAMPLE 12.43 Determine the length of the parabolic arc $y = x^2/2p$ from $x = 0$ to $x = X$.

SOLUTION Equation (12.32) yields

$$s = \int \sqrt{1 + \frac{x^2}{p^2}}\, dx = \frac{1}{p} \int \sqrt{p^2 + x^2}\, dx$$

Applying Integral (42) with the specified limits of integration and applying the principle that $\ln a - \ln b = \ln (a/b)$, we obtain the following:

$$s = \frac{1}{2p} \left[X\sqrt{p^2 + X^2} + p^2 \ln \left(\frac{X + \sqrt{p^2 + X^2}}{p} \right) \right]$$

In polar coordinates, the length of arc from the point where $\theta = \alpha$ to the point where $\theta = \beta$ is

$$s = \int_{\alpha}^{\beta} \sqrt{r^2 + \left(\frac{dr}{d\theta} \right)^2}\, d\theta \qquad (12.33)$$

12.6.4 Area of a Surface of Revolution

We shall require the equation for the surface area of a frustum of a right circular cone. Let r denote the radius at the midsection and l denote the slant height. Then

$$A = 2\pi r l \qquad (12.34)$$

This equation is derived by considering the frustum of the cone to be the limiting position of a frustum of an inscribed pyramid as the number of sides of the pyramid increases without limit. (It may also be considered to be intuitively self-evident from the equation for the surface area of a right circular cylinder.)

Consider that the continuous arc CD in Fig. 12.34 is revolved about the x axis through an angle of 2π, thereby generating a *surface of revolution*. We wish to evaluate the area S of this surface. For this purpose, consider that the region $ABDC$ is divided into n parts by lines parallel to the y axis. One part, of width Δx, is shown dotted. Draw the chord PR, and let y denote the ordinate at the midpoint

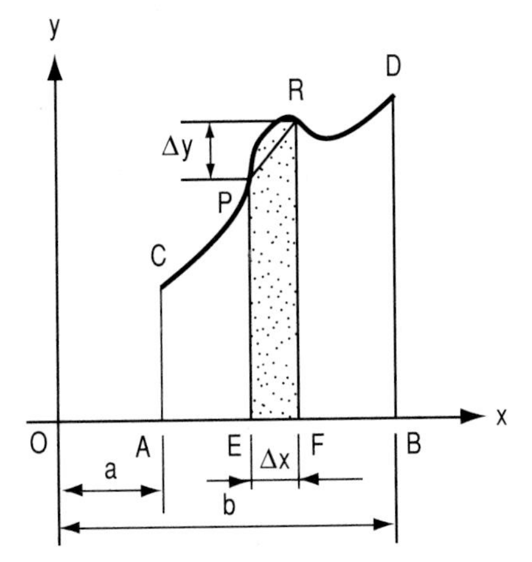

FIGURE 12.34 Generating a surface of revolution.

of PR. When the trapezoid $EFRP$ revolves about the x axis, it generates a frustum of a cone having a surface area of

$$2\pi y\sqrt{(\Delta x)^2+(\Delta y)^2}=2\pi y\sqrt{1+\left(\frac{\Delta y}{\Delta x}\right)^2}\;\Delta x$$

By summing the surface areas of all the frustums of cones thus formed, allowing n to become infinite, and applying Eq. (12.18), we obtain

$$S=2\pi\int_a^b y\sqrt{1+\left(\frac{dy}{dx}\right)^2}\,dx \tag{12.35}$$

EXAMPLE 12.44 Find the surface area of a sphere of radius r.

SOLUTION As in Example 12.39, we may consider that the sphere is generated by revolving the semicircular region in Fig. 12.30 about the x axis. Then

$$y=\sqrt{r^2-x^2}\qquad \frac{dy}{dx}=-\frac{x}{\sqrt{r^2-x^2}}$$

Substituting in Eq. (12.35) and simplifying, we obtain

$$S=2\pi\int_{-r}^{r} r\,dx=2\pi rx\Big]_{-r}^{r}=4\pi r^2$$

12.7 PARTIAL DIFFERENTIATION

12.7.1 Introduction and Basic Notation

Thus far, we have investigated situations where a given variable is a function of a single variable. We now extend the scope of our investigation to situations where a given variable is a function of multiple variables. Although for discussion purposes we shall take the number of independent variables as two, the results we obtain are entirely general and apply where this number has some value greater than two.

Let x and y denote quantities that vary independently of each other, and let z denote a function of x and y. Expressed symbolically, $z=f(x, y)$. The graph of this equation is a surface, such as the surfaces discussed in Arts. 3.4.1 through 3.4.10. Each point on this surface has coordinates that satisfy the given equation.

Although in general x and y vary simultaneously, it is helpful to isolate the effects of their variations. Let Δx and Δy denote increments of x and y, respectively. We apply the following notation:

Let $\Delta_x z=$ increment of z due solely to Δx
$\Delta_y z=$ increment of z due solely to Δy

We may consider that $\Delta_x z$ occurs while x varies and y remains constant, and $\Delta_y z$ occurs while y varies and x remains constant.

12.7.2 Partial Derivatives

The *partial derivative* of z with respect to x, which is denoted by $\partial z/\partial x$, is defined in this manner:

$$\lim_{\Delta x \to 0} \frac{\Delta_x z}{\Delta x} = \frac{\partial z}{\partial x} \tag{12.36a}$$

Similarly, the partial derivative of z with respect to y is

$$\lim_{\Delta y \to 0} \frac{\Delta_y z}{\Delta y} = \frac{\partial z}{\partial y} \tag{12.36b}$$

The geometric interpretation of partial derivatives is as follows: Let S denote the surface having the equation $z = f(x, y)$. Figure 12.35 shows the part of S that lies in the first octant. Let $P(x_1, y_1, z_1)$ denote a point on this surface. To hold y constant and allow x to vary, we pass the plane $y = y_1$ through S, intersecting S along the curve AB. We now draw line PQ through P tangent to the curve AB. Then $\partial z/\partial x$ equals the slope of PQ. Similarly, to hold x constant and allow y to vary, we pass the plane $x = x_1$ through S, intersecting S along the curve CD. We now draw line PR through P tangent to the curve CD. Then $\partial z/\partial y$ equals the slope of PR.

EXAMPLE 12.45 The *equation of state* (or *characteristic equation*) of a perfect gas of constant mass is $pV = kT$, where p, V, and T denote pressure, volume, and temperature, respectively, and k is a constant. If V and T can be varied at will, find the partial derivatives of p.

SOLUTION Solving the given equation for p, we have

$$p = \frac{kT}{V} = kTV^{-1}$$

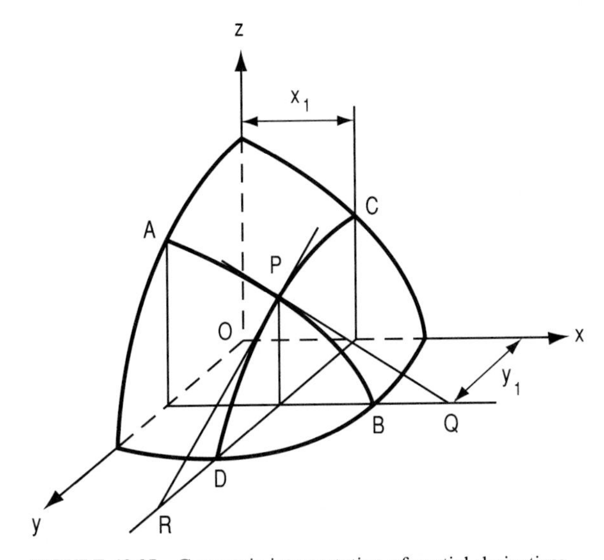

FIGURE 12.35 Geometric interpretation of partial derivatives.

Holding T constant while V varies, we have

$$\frac{\partial p}{\partial V} = -\frac{kT}{V^2}$$

Holding V constant while T varies, we have

$$\frac{\partial p}{\partial T} = \frac{k}{V}$$

12.7.3 Related Rates of Change

Now consider that x and y vary with time but are independent of each other, again let z denote a function of x and y, and let t denote elapsed time. By an extension of Eq. (12.10), we obtain the following as the rate of change of z:

$$\frac{dz}{dt} = \frac{\partial z}{\partial x}\frac{dx}{dt} + \frac{\partial z}{\partial y}\frac{dy}{dt} \tag{12.37}$$

The quantity dz/dt is the *total derivative* of z with respect to t.

EXAMPLE 12.46 A right circular cone undergoes a continuous change in size. At a given instant, the following data apply: The altitude is 40 units, and it is increasing at the rate of 3 units/s; the radius of the base is 6 units, and it is decreasing at the rate of 0.8 units/s. Find the instantaneous rate of change of the volume. Apply this formula: The volume of a right circular cone is

$$V = \frac{\pi r^2 h}{3}$$

where h, r, and V denote the altitude, the radius of the base, and the volume, respectively.

SOLUTION

$$\frac{\partial V}{\partial h} = \frac{\pi r^2}{3} = \frac{\pi \times 6^2}{3} = 12\pi$$

$$\frac{\partial V}{\partial r} = \frac{2\pi rh}{3} = \frac{2\pi \times 6 \times 40}{3} = 160\pi$$

$$\frac{dV}{dt} = \frac{\partial V}{\partial h}\frac{dh}{dt} + \frac{\partial V}{\partial r}\frac{dr}{dt} = 12\pi \times 3 + 160\pi(-0.8)$$

$$= -92\pi \text{ cu units/s}$$

The negative result signifies that the volume is decreasing at the given instant.

This result can be tested approximately by assuming that these rates of change remain constant during the first second. The original volume is 480π cu units, the volume at the end of 1 s would be 387.57π cu units, and the difference is 92.43π cu units. The slight discrepancy arises from the fact that the second result inlcudes products of differentials and dV/dt does not.

12.7.4 Partial Derivatives of Higher Order

The definition of a derivative of higher order in Art. 12.2.5 can be extended to partial derivatives as well. The notation is illustrated by the following forms:

$$\frac{\partial^2 z}{\partial x^2} = \frac{\partial}{\partial x}\left(\frac{\partial z}{\partial x}\right) \qquad \frac{\partial^2 z}{\partial y\, \partial x} = \frac{\partial}{\partial y}\left(\frac{\partial z}{\partial x}\right)$$

As an illustration, let

$$z = \sin x - \cos xy$$

Treating y as a constant and x as a variable, we obtain

$$\frac{\partial z}{\partial x} = \cos x + y \sin xy \tag{f}$$

Performing this operation again, we obtain

$$\frac{\partial^2 z}{\partial x^2} = -\sin x + y^2 \cos xy$$

Now returning to Eq. (f) and treating x as a constant and y as a variable, we obtain

$$\frac{\partial^2 z}{\partial y\, \partial x} = y \frac{\partial}{\partial y}(\sin xy) + \sin xy \frac{\partial y}{\partial y}$$
$$= xy \cos xy + \sin xy$$

The geometric significance of second-order partial derivatives is as follows: In Fig. 12.35, let m denote the slope of the tangent PQ that is parallel to the xz plane. Then $\partial z/\partial x = m$, and $\partial^2 z/\partial x^2$ is the rate at which m changes as we move along curve AB. On the other hand, $\partial^2 z/(\partial y\, \partial x)$ is the rate at which m changes as we move along curve CD.

12.7.5 Partial and Total Differentials

We now set $dx = \Delta x$ and $dy = \Delta y$, where dx and dy are the differentials of x and of y, respectively. By analogy with Eq. (12.5), we define the following quantities:

$$d_x z = \frac{\partial z}{\partial x} dx \qquad d_y z = \frac{\partial z}{\partial y} dy \tag{12.38}$$

The quantities $d_x z$ and $d_y z$ are termed the *partial differentials* of z with respect to x and y, respectively.

These partial differentials can be interpreted geometrically in this manner: Starting at point P in Fig. 12.35, assume that x increases by an amount dx while y remains constant. If z increased along the tangent PQ rather than curve AB, the increase in z would be $d_x z$. An analogous statement applies with respect to $d_y z$. (In this diagram, of course, z decreases as x and y increase.) The *total differential* dz of z is simply

$$dz = d_x z + d_y z \tag{12.39}$$

EXAMPLE 12.47 A rectangular parallelepiped having the following dimensions is to be constructed: width, 7 units; height, 4 units; length, 25 units. The true dimensions are allowed to deviate from the required dimensions, in either direction, by these amounts: width, 0.02 units; height, 0.01 units; length, 0.04 units. Assuming that the true dimension always exceeds the required dimension by the allowable amount, compute the following, to four significant figures: the total differential of the volume; the exact difference between the true volume and the theoretical volume.

SOLUTION Let w, h, l, and V denote the width, height, length, and volume, respectively. Then $V = whl$. Each deviation may be regarded as a differential, and we have the following:

$$\begin{aligned} dV &= d_w V + d_h V + d_l V = \frac{\partial V}{\partial w} dw + \frac{\partial V}{\partial h} dh + \frac{\partial V}{\partial l} dl \\ &= hl\, dw + wl\, dh + wh\, dl \\ &= 4 \times 25(0.02) + 7 \times 25(0.01) + 7 \times 4(0.04) \\ &= 4.870 \text{ cu units} \end{aligned}$$

The exact difference in volume is

$$\Delta V = 7.02 \times 4.01 \times 25.04 - 7 \times 4 \times 25 = 4.881 \text{ cu units}$$

12.7.6 Maxima and Minima

Again let S denote the surface having the equation $z = f(x, y)$. If S contains a summit or nadir, the plane that is tangent to S at this point is parallel to the xy plane. It follows that $\partial z/\partial x = \partial z/\partial y = 0$.

In general, if both partial derivatives are 0 at a given point, that point is a summit, a nadir, or a saddle point, such as that of the hyperbolic paraboloid discussed in Art. 3.4.10. If the characteristic of the point is not self-evident, it is possible to apply a test that involves second-order differentials. However, the characteristic can be found quickly by this practical test: Let $P(x_1, y_1, z_1)$ be the point where both partial derivatives are 0, and let h denote some minute quantity. Compute z_1, and then compute $z_2 = f(x_1 + h, y_1)$ and $z_3 = f(x_1, y_1 + h)$. If $z_2 > z_1$ and $z_3 > z_1$, the point is a nadir. If $z_2 < z_1$ and $z_3 < z_1$, the point is a summit. If $z_2 > z_1$ but $z_3 < z_1$, or vice versa, the point is a saddle point.

This geometric principle enables us to find maximum and minimum values of a function by equating all partial derivatives to 0. This technique also applies where a variable is a function of three or more independent variables.

EXAMPLE 12.48 A plane has the equation $3x - 2y - z = 5$. Identify the point M on this plane that lies closest to the point $P(3, 7, -2)$, and verify the solution.

SOLUTION We first rewrite the given equation in the form $z = 3x - 2y - 5$. Let $R(x, y, z)$ be an arbitrary point on the plane, and let $w = (RP)^2$. Applying Eq. (3.38) for the distance between two points, we obtain

$$\begin{aligned} w &= (x-3)^2 + (y-7)^2 + (3x - 2y - 3)^2 \\ &= 10x^2 + 5y^2 - 12xy - 24x - 2y + 67 \end{aligned}$$

To minimize w, we set

$$\frac{\partial w}{\partial x} = 20x - 12y - 24 = 0$$

$$\frac{\partial w}{\partial y} = -12x + 10y - 2 = 0$$

The solution of this system of simultaneous equations is $x = 33/7$ and $y = 41/7$. Then $z = -18/7$. These are the coordinates of point M.

The verification is as follows: Applying Eq. (3.38), we obtain $MP = 4\sqrt{14}/7$. Applying Eq. (3.51) for the distance from a point to a plane, we obtain the same result. It follows that the coordinates of M are correct.

12.7.7 Properties of a Space Curve

We shall divide space curves into two types. A type 1 curve is formed when x, y, and z are functions of a parameter t and the points representing simultaneous values of these functions are plotted. A type 2 curve is formed by the intersection of two surfaces.

Let $P(x_1, y_1, z_1)$ denote a point on a space curve and let T denote the tangent to the curve at P. The directions of T are expressed by means of its *direction components*, which are defined in Art. 3.3.5. For a type 1 space curve, the direction components in the x, y, and z directions may be taken as dx/dt, dy/dt, and dz/dt, respectively. By Eq. (3.59), the equation of T becomes

$$\frac{x - x_1}{dx/dt} = \frac{y - y_1}{dy/dt} = \frac{z - z_1}{dz/dt} \tag{12.40}$$

EXAMPLE 12.49 A space curve has the parametric equations $x = 2t^2$, $y = t^3$, $z = -5t$. Find the equation of the tangent to this curve at the point corresponding to $t = 4$.

SOLUTION The point is $P(32, 64, -20)$.

$$\frac{dx}{dt} = 4t = 16 \qquad \frac{dy}{dt} = 3t^2 = 48 \qquad \frac{dz}{dt} = -5$$

By Eq. (12.40), the equation of the tangent is

$$\frac{x - 32}{16} = \frac{y - 64}{48} = \frac{z + 20}{-5}$$

For a type 2 space curve, the direction components in the x, y, and z directions may be taken as the differentials dx, dy, and dz, respectively.

Let s denote the length of arc of a space curve between two points. For a type 1 curve, we have

$$s = \int \left[\left(\frac{dx}{dt} \right)^2 + \left(\frac{dy}{dt} \right)^2 + \left(\frac{dz}{dt} \right)^2 \right]^{1/2} dt \tag{12.41a}$$

where the limits of integration are the values of t at the end points. For a type 2 curve, we have

$$s = \int \left[1 + \left(\frac{dy}{dx} \right)^2 + \left(\frac{dz}{dx} \right)^2 \right]^{1/2} dx \qquad (12.41b)$$

where the limits of integration are the values of x at the end points.

Now let P and Q denote two points on the space curve at a distance Δs apart. We draw tangents to the curve at these points, and let $\Delta\theta$ denote the angle between these tangents. The *curvature* K of the curve at P is defined in this manner:

$$K = \lim_{\Delta s \to 0} \frac{\Delta\theta}{\Delta s}$$

In Art. 3.3.5, we defined the *direction cosines* of a line in space. Let a, b, and c denote the direction cosines of the tangent to the curve at P. Then

$$K = \sqrt{\left(\frac{da}{ds} \right)^2 + \left(\frac{db}{ds} \right)^2 + \left(\frac{dc}{ds} \right)^2} \qquad (12.42)$$

12.7.8 Properties of a Tangent Plane

We return to Fig. 12.35, where $P(x_1, y_1, z_1)$ is a point on the surface S having the equation $z = f(x, y)$, and PQ and PR are the tangents at P parallel to the xz and yz planes, respectively. Consider that we pass a plane T through these tangent lines. This plane itself is tangent to the surface S at P (i.e., all lines in this plane that pass through P are tangent to S). Let θ_{xy} denote the angle that plane T makes with the xy coordinate plane. From the trigonometry of the drawing, we obtain

$$\tan \theta_{xy} = \sqrt{\left(\frac{\partial z}{\partial x} \right)^2 + \left(\frac{\partial z}{\partial y} \right)^2} \qquad (12.43a)$$

Analogous equations apply to the angles that plane T makes with the xz and yz planes. In Fig. 12.2, we have $\tan \theta = dy/dx$, and Eq. (12.43a) is merely an extension of that relationship.

Applying the relationship $\sec^2 \theta = 1 + \tan^2 \theta$, we obtain

$$\sec \theta_{xy} = \sqrt{1 + \left(\frac{\partial z}{\partial x} \right)^2 + \left(\frac{\partial z}{\partial y} \right)^2} \qquad (12.43b)$$

We shall now develop the equation of plane T, and we apply the following principle: All lines in T that are parallel to a given coordinate plane are parallel to one another. Consider that we start at P and then move along T on a line parallel to the xz plane. The slope of this line is $\partial z/\partial x_1$, and the increase in z equals the increase in x multiplied by this derivative. Consider that we next move along T on a line parallel to the yz plane. The slope of this line is $\partial z/\partial y_1$, and the increase in z equals the increase in y multiplied by this derivative. Combining these effects, we have

$$z - z_1 = (x - x_1) \frac{\partial z}{\partial x_1} + (y - y_1) \frac{\partial z}{\partial y_1} \qquad (12.44)$$

This is the equation of plane T.

EXAMPLE 12.50 Find the equation of the plane that is tangent to the surface $z = 3x^2 + 2xy + y^2 - 6$ at $P(2, 3, 27)$.

SOLUTION

$$\frac{\partial z}{\partial x} = 6x + 2y = 18 \text{ at } P$$

$$\frac{\partial z}{\partial y} = 2x + 2y = 10 \text{ at } P$$

Applying Eq. (12.44) and then simplifying, we obtain

$$z - 27 = 18(x - 2) + 10(y - 3)$$

$$z = 18x + 10y - 39$$

12.8 MULTIPLE INTEGRATION

12.8.1 Meaning of Multiple Integration

Consider again that z is a function of x and y and that x and y can vary independently of one another. Multiple (or repeated) integration is the reverse of partial differentiation. In multiple integration, we integrate in steps, allowing only one independent variable to vary in each step.

The integral that results is termed a *double integral* if the number of independent variables is two, and a *triple integral* if this number is three. The general notation for double integration is as follows:

$$\int_c^d \int_a^b f(x, y)\, dy\, dx$$

In step 1, we hold x constant, let y vary, and integrate the expression $f(x, y)\, dy$ between the limits a and b. Call I the resulting integral. In step 2, we let x vary, and integrate the expression $I\, dx$ between the limits c and d. Thus, the sequence in which the quantities vary corresponds to the order in which the differentials are recorded. A limit of integration of a variable may be a function of the value of the succeeding variable.

EXAMPLE 12.51 Evaluate $\int_2^4 \int_x^{2x} \int_0^6 (xy + z^2)\, dz\, dy\, dx$.

SOLUTION

$$\int_0^6 (xy + z^2)\, dz = \left[xyz + \frac{z^3}{3} \right]_0^6 = 6xy + 72$$

$$\int_x^{2x} (6xy + 72)\, dy = [3xy^2 + 72y]_x^{2x} = 9x^3 + 72x$$

$$\int_2^4 (9x^3 + 72x)\, dx = \left[\frac{9x^4}{4} + 36x^2 \right]_2^4 = 972$$

12.8.2 Volume of a Solid

As our first application of multiple integration, we undertake the task of calculating the volume of a general type of solid with sides parallel to the z axis. We shall assume initially that the base of the solid lies in the xy plane; the top face of the solid can be a plane or a curved surface.

In Fig. 12.36, consider that we pass closely spaced planes through the solid parallel to the xz and yz planes, thereby dividing the solid into strips parallel to the z axis. One such strip is shown dotted. Let Δx and Δy denote the dimensions of the section of this strip, as shown, and let z be the height of the strip at its centerline. The volume ΔV of this strip is approximately $z\,\Delta x\,\Delta y$. If we sum the volumes of these strips, allow the number of strips to become infinite, and apply Eq. (12.18), we obtain the following as the total volume V of the solid:

$$V \int\int z\,dy\,dx \tag{12.45}$$

The limits of integration are obtained from the properties of the solid.

EXAMPLE 12.52 A solid has its base in the region of the xy plane bounded by the parabola $y = x^2 - 8$ and the straight line $y = 2x$. The top face is the plane $z = x + 5$. Compute the volume of the solid and its mean height z_m (i.e., the height the solid would have if its top face were made parallel to the xy plane while its volume remained unchanged). Verify the results.

SOLUTION The base of the solid is shown in Fig. 12.37. The range of y is from $x^2 - 8$ to $2x$. To find the range of x, we must locate the intersection points P and Q. We therefore set $x^2 - 8 = 2x$. The solution of this equation is $x = -2$, $x = 4$.

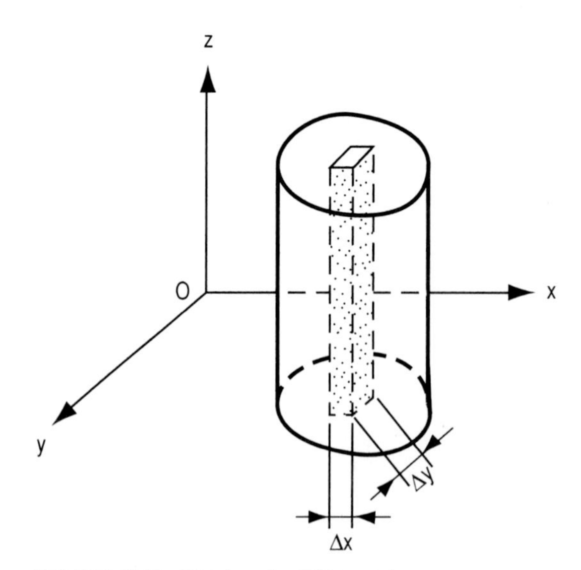

FIGURE 12.36 Division of solid into strips.

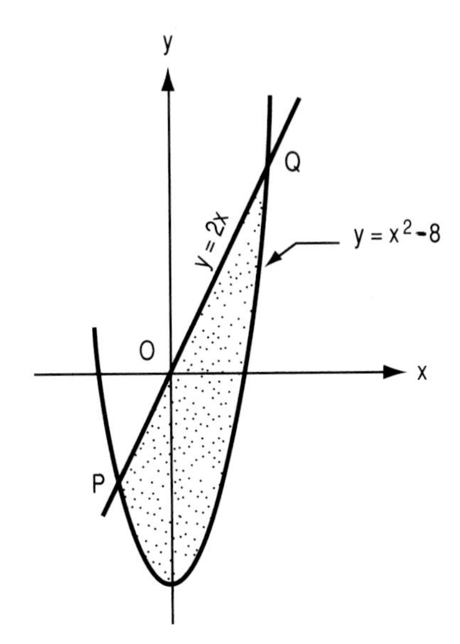

FIGURE 12.37 Base of solid.

Then

$$V = \int_{-2}^{4} \int_{x^2-8}^{2x} (x+5)\,dy\,dx$$

The first integration yields

$$[(x+5)y]_{x^2-8}^{2x} = -x^3 - 3x^2 + 18x + 40$$

The second integration yields

$$\left[-\frac{x^4}{4} - x^3 + 9x^2 + 40x \right]_{-2}^{4} = 216$$

Thus, $V = 216$ cu units.

Let A denote the area of the base. From Fig. 12.37, we have

$$A = \int_{-2}^{4} (2x - x^2 + 8)\,dx = 36 \text{ sq units}$$

If the solid had a uniform height z_m, its volume would be Az_m. Then

$$z_m = \frac{V}{A} = \frac{216}{36} = 6 \text{ units}$$

We shall now verify our results. Refer to Fig. 12.38. Let Q denote the plane where the height of the solid is z_m. At Q, $z = x + 5 = 6$, and $x = 1$. We label the regions to the left and to the right of Q between the top face and the plane $z = 6$ as shown. We must demonstrate that these two regions have equal volumes.

For region 1, the height of a strip parallel to the z axis is $6 - z = 6 - (x+5) = 1 - x$. Then

$$V_1 = \int_{-2}^{1} \int_{x^2-8}^{2x} (1-x)\,dy\,dx = \frac{81}{4} \text{ cu units}$$

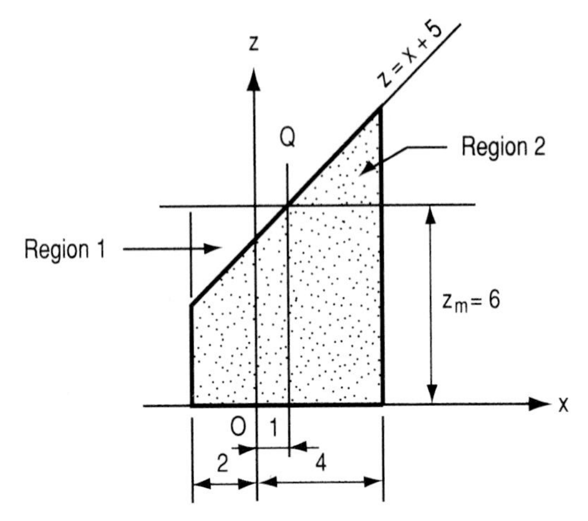

FIGURE 12.38 View normal to *xz* plane.

For region 2, the height of a strip is $z - 6 = x + 5 - 6 = x - 1$. Then

$$V_2 = \int_1^4 \int_{x^2-8}^{2x} (x - 1)\, dy\, dx = \frac{81}{4} \text{ cu units}$$

Thus, $V_1 = V_2$, and this equality proves that all our results are correct.

In the general type of solid, both the top and bottom faces are planes that are inclined to the *xy* plane, or they are curved surfaces. The volume of the solid can be found by computing the volume between each face and the *xy* plane and then combining the results. The same effect is achieved by expressing the volume as a triple integral, in this manner:

$$V = \iiint dz\, dy\, dx$$

The range of z is from the lower face to the upper face.

12.8.3 Area of a Surface

We shall develop an expression for the area S of a surface in space. In Fig. 12.36, let R denote the region that lies on the top face of the solid and is bounded by the sides of the strip that is shown dotted. Let ΔS denote the area of this region. Now consider that we draw a plane T tangent to the top face at any point within R. The area dS that lies on T and is bounded by the sides of the strip (prolonged if necessary) is a close approximation to ΔS. Let θ_{xy} denote the angle that T makes with the *xy* plane. The projection of S onto the *xy* plane is $\Delta x\, \Delta y$, and

$$dS = \Delta x\, \Delta y \sec \theta_{xy}$$

As the number of strips increases, the sum of these plane areas approaches the total area S. Applying Eq. (12.43b), we obtain

$$S = \iint \sqrt{1 + \left(\frac{\partial z}{\partial x}\right)^2 + \left(\frac{\partial z}{\partial y}\right)^2}\, dy\, dx \tag{12.46}$$

The boundaries of the surface establish the limits of integration.

12.8.4 Gravitational Force on a Particle

A problem that is of fundamental importance in engineering and science is that of calculating the gravitational force that two bodies exert on each other. We start by considering two particles (bodies of infinitesimal volume). Let m_1 and m_2 denote their masses, r the distance between them, and F the gravitational force. By Newton's law,

$$F = \frac{km_1 m_2}{r^2}$$

where k is a constant whose value depends on the units of force and distance. This force acts along the line connecting the bodies.

In Fig. 12.39, P is a particle of mass m located at the origin, and A is a body of mass M lying in the xy plane. We wish to find the gravitational force F that A exerts on P. We divide body A into infinitesimal parts, one of which is shown darkened. Let ΔM denote the mass of this part, r denote its distance from P, and x and y denote the coordinates of its center. The force ΔF that this part exerts on P is

$$\Delta F = \frac{km\, \Delta M}{r^2}$$

We shall resolve this force into its x component ΔF_x and y component ΔF_y, which are shown dashed. To find the x component, we multiply the force by $\cos\theta$, which is x/r. To find the x component F_x of the total force F, we sum the forces exerted by the parts. If body A is a solid rather than a plane surface, there is also a z component of F. The components are as follows:

$$F_x = km \int \frac{x}{r^3}\, dM \qquad F_y = km \int \frac{y}{r^3}\, dM \qquad F_z = km \int \frac{z}{r^3}\, dM \tag{12.47}$$

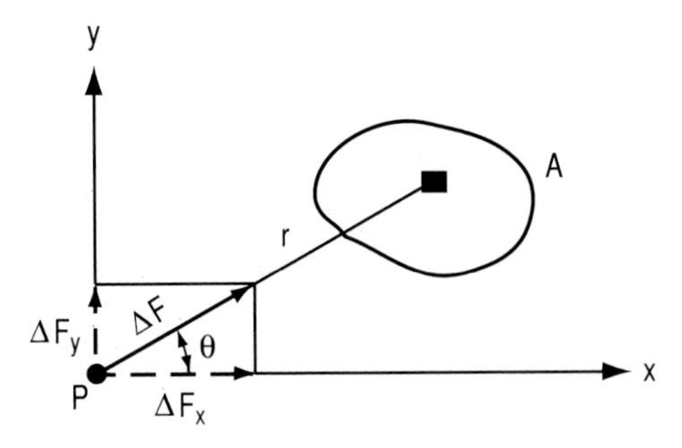

FIGURE 12.39 Components of gravitational force.

If the thickness of a body is minuscule in relation to the other dimensions, we may consider that its mass is distributed along a line or plane, as the case may be. We shall let μ denote the unit mass (mass per unit volume) of the body. If the body is homogeneous, its unit mass is constant across the body. In the general case where body A is a solid, it is necessary to divide the body into parts in the form of rectangular parallelepipeds having dimensions Δx, Δy, and Δz. If the body is homogeneous, the mass ΔM of a part is $\mu \, \Delta x \, \Delta y \, \Delta z$. Therefore, to calculate the gravitational force, it is necessary to apply triple integration.

In Examples 12.53 through 12.55, it is understood that a particle P of mass m is located at the origin.

EXAMPLE 12.53 A slender homogeneous rod of mass M has the location shown in Fig. 12.40. Compute the x component of the force the rod exerts on P.

SOLUTION We divide the rod into parts of width Δx. The mass ΔM of a part is $\mu \, \Delta x$.

$$r^2 = c^2 + x^2 \qquad r^3 = (c^2 + x^2)^{3/2}$$

$$F_x = km \int \frac{x}{r^3} \, dM = km\mu \int_a^{a+b} \frac{x \, dx}{(c^2 + x^2)^{3/2}}$$

Applying Integral (50) and replacing μ with M/b, we obtain

$$F_x = \frac{kmM}{b} \left[\frac{1}{\sqrt{c^2 + a^2}} - \frac{1}{\sqrt{c^2 + (a+b)^2}} \right]$$

Let r_A and r_B denote the distances from the origin to points A and B, respectively. Then

$$F_x = \frac{kmM}{b} \left(\frac{1}{r_A} - \frac{1}{r_B} \right)$$

EXAMPLE 12.54 A thin rectangular homogeneous plate of mass M has the location shown in Fig. 12.41. Compute the x component of the force the plate exerts on P. Express the result in terms of μ.

SOLUTION We can apply the result obtained in Example 12.53 by expressing F_x in terms of μ, replacing c with y, and allowing y to vary from 2 to 3. We must apply

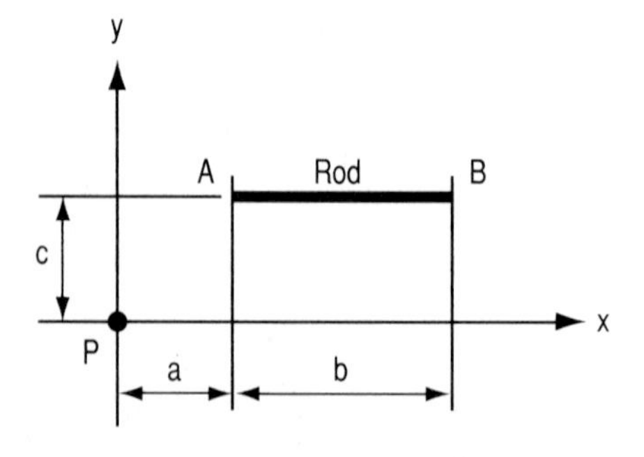

FIGURE 12.40

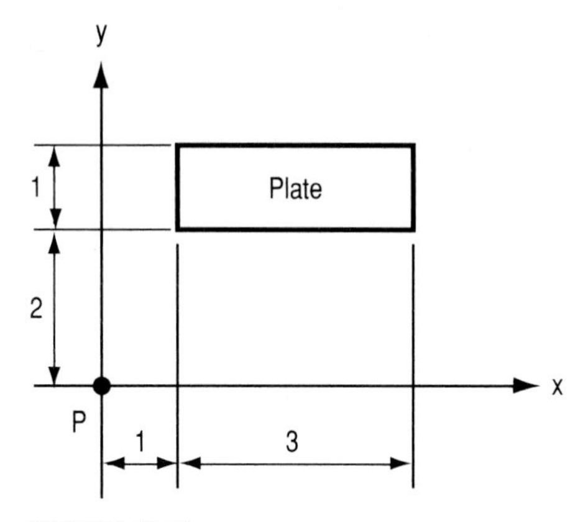

FIGURE 12.41

Integral (47). Then

$$F_x = km\mu \int_2^3 \left(\frac{1}{\sqrt{y^2+1}} - \frac{1}{\sqrt{y^2+16}} \right) dy$$

$$= km\mu \left(\ln \frac{3+\sqrt{10}}{2+\sqrt{5}} - \ln \frac{3+5}{2+\sqrt{20}} \right) = 0.1629 km\mu$$

EXAMPLE 12.55 A homogeneous wire is bent into the form of a semicircular arc of radius r and placed with its center at the origin. Compute the force this wire exerts on P.

SOLUTION Refer to Fig. 12.42. We divide the wire into parts of length Δs. Then $\Delta s = r\, \Delta\theta$, and $\Delta M = \mu r\, \Delta\theta$. By symmetry, $F_x = 0$, and we have the following:

$$F = F_y = km \int \frac{y}{r^3}\, dM = km \int \frac{r \sin\theta}{r^3}\, \mu r\, d\theta$$

$$= \frac{km\mu}{r} \int_0^{\pi} \sin\theta\, d\theta = \frac{km\mu}{r} [-\cos\theta]_0^{\pi} = \frac{2km\mu}{r}$$

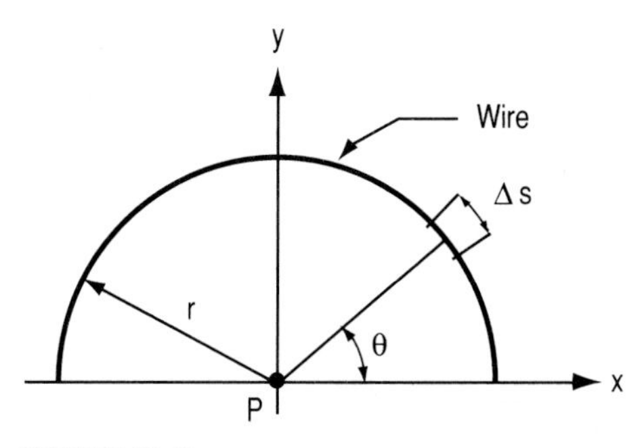

FIGURE 12.42

Replacing μ with $M/\pi r$, we obtain

$$F = \frac{2kmM}{\pi r^2}$$

12.8.5 Product of Integrals

The product of two integrals can be expressed as a double integral, and vice versa. For example, consider the following:

$$\int_a^b x^2\,dx \int_c^d \cos y\,dy \qquad \text{and} \qquad \int_a^b \int_c^d x^2 \cos y\,dy\,dx$$

In both instances, we obtain $(1/3)(b^3 - a^3)/(\sin d - \sin c)$.

In general, we have

$$\int u\,du \int v\,dv = \int\int uv\,dv\,du$$

This law can be extended to include the product of three or more integrals.

12.8.6 Change of Coordinate System

When the magnitude of a plane area is expressed in the form of a double integral, the expression is $\int\int dx\,dy$ if rectangular coordinates are used, and $\int\int r\,dr\,d\theta$ if polar coordinates are used. Therefore, applying the proper limits of integration, we have

$$\int\int dx\,dy = \int\int r\,dr\,d\theta$$

In performing a double integration with rectangular coordinates, it is sometimes desirable to transfer to the use of polar coordinates. This transfer can be effected by making the following substitutions: $r\cos\theta$ for x, $r\sin\theta$ for y, and $r\,dr\,d\theta$ for $dx\,dy$. The limits of integration must of course be changed to those of r and θ.

12.9 PROPERTIES OF AREAS

In engineering analysis, certain expressions pertaining to areas and masses recur so frequently that it becomes advantageous to assign names to these expressions and to develop their characteristics. We shall first present the properties of areas, and then we shall extend the concepts to arcs, volumes, and masses. In the subsequent material, we shall express values of an element of an area in terms of differentials.

12.9.1 Statical Moment

Consider that the area in Fig. 12.43 is divided into n elements, one of which is shown darkened. Let A denote the total area and dA the area of an element. Also let y denote the ordinate of the center of the element. We form the product $y\,dA$ of each element and sum these products. We now make n progressively larger by

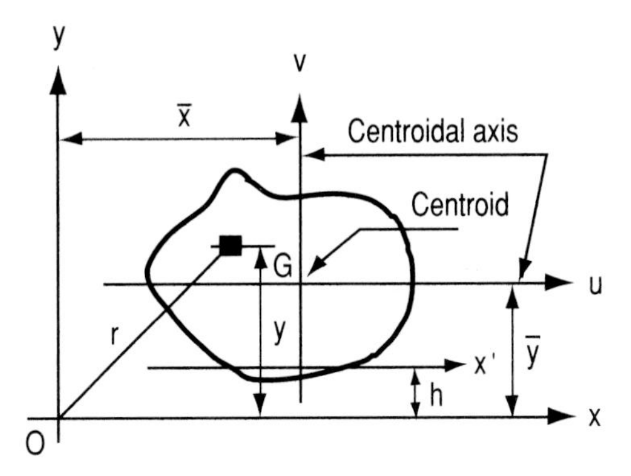

FIGURE 12.43 Centroidal axes and centroid.

making the elements progressively smaller. By the fundamental theorem, we have

$$\lim_{n \to \infty} \Sigma\, y\, dA = \int y\, dA$$

where integration is performed across the entire area. This integral is called the *statical moment* (or *first moment*) of the area with respect to the x axis, and it is denoted by Q_x. Then

$$Q_x = \int y\, dA \tag{12.48a}$$

Analogously, the statical moment of the area with respect to the y axis is

$$Q_y = \int x\, dA \tag{12.48b}$$

An area is always considered positive. Therefore, the algebraic sign of $y\, dA$ is identical with the algebraic sign of y, and a statical moment can be positive, negative, or zero. The unit of statical moment is the cube of the unit of length.

12.9.2 Centroidal Axes and Centroid

In Fig. 12.43, let x' denote an axis that is parallel to the x axis and at a distance h above it, as shown. The statical moment of the area with respect to the x' axis is

$$Q_{x'} = \int (y - h)\, dA = \int y\, dA - h \int dA = Q_x - Ah$$

Now let u denote an axis parallel to the x axis that has the property $Q_u = 0$, and let $\bar{y}$ denote the distance between the x and u axes. Then

$$Q_u = Q_x - A\bar{y} = 0$$

$$\therefore\ \bar{y} = \frac{Q_x}{A} \tag{12.49a}$$

The u axis is called the *centroidal axis* of the area in the x direction.

Now let v denote an axis parallel to the y axis that has the property $Q_v = 0$, and let $\bar{x}$ denote the distance between the y and v axes. Then

$$\bar{x} = \frac{Q_y}{A} \tag{12.49b}$$

The v axis is the centroidal axis of the area in the y direction.

The point G at which the two centroidal axes intersect is called the *centroid* of the area. If the x and y axes are rotated about O, a new set of centroidal axes is created, but these axes also intersect at G. Thus, the position of the centroid is invariant; it is an inherent characteristic of the area. Therefore, in locating the centroid, we may place the coordinate axes in any suitable position.

It is helpful to recast Eqs. (12.49) in this form:

$$Q_x = A\bar{y} \qquad Q_y = A\bar{x} \tag{12.50}$$

Assume that the area has an axis of symmetry s. To each element that lies at a distance w from s, there is a corresponding element that lies at a distance $-w$ from s, and it follows that $Q_s = 0$. We thus arrive at the following principle:

Theorem 12.1. If an area has an axis of symmetry, this axis is a centroidal axis.

EXAMPLE 12.56 Curve OB in Fig. 12.44 is a half parabola having its vertex at O. Locate the centroid of the area OBC.

SOLUTION We must first develop the equation of the parabola in terms of a and b. Set $y = mx^2$. Then $b = ma^2$. Solving the second equation for m and substituting the result in the first equation, we obtain

$$y = \frac{bx^2}{a^2} \qquad x = \frac{a}{b^{1/2}}y^{1/2}$$

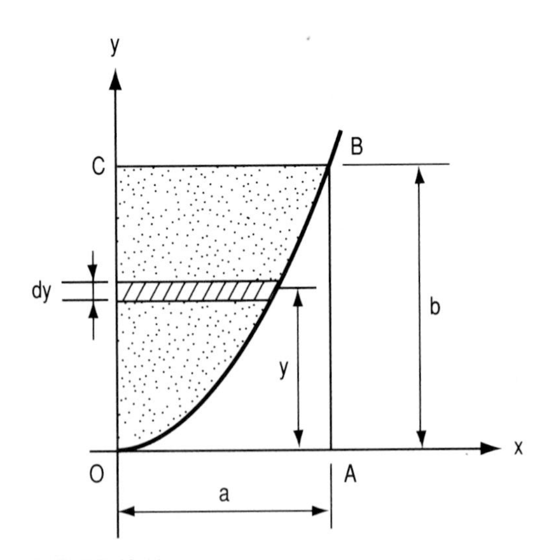

FIGURE 12.44

The problem can be solved by setting $dA = dx\,dy$ and applying double integration, but a more direct approach is possible. We divide the area into strips parallel to the x axis, of width dy. With reference to a strip, we have the following:

$$dA = x\,dy$$

$$dQ_x = y\,dA = xy\,dy = \frac{a}{b^{1/2}}y^{3/2}\,dy$$

$$dQ_y = \frac{x}{2}\,dA = \frac{x^2}{2}\,dy = \frac{a^2 y}{2b}\,dy$$

With reference to the total area, we have the following:

$$Q_x = \frac{a}{b^{1/2}}\int_0^b y^{3/2}\,dy = \frac{a}{b^{1/2}}\left.\frac{y^{5/2}}{5/2}\right]_0^b = \frac{2}{5}ab^2$$

$$Q_y = \frac{a^2}{2b}\int_0^b y\,dy = \frac{a^2}{2b}\left.\frac{y^2}{2}\right]_0^b = \frac{1}{4}a^2 b$$

$$A = \int_0^b x\,dy = \frac{a}{b^{1/2}}\int_0^b y^{1/2}\,dy = \frac{a}{b^{1/2}}\,\frac{b^{3/2}}{3/2} = \frac{2ab}{3}$$

$$\bar{y} = \frac{Q_x}{A} = \frac{(2/5)ab^2}{(2/3)ab} = \frac{3}{5}b$$

$$\bar{x} = \frac{Q_y}{A} = \frac{(1/4)a^2 b}{(2/3)ab} = \frac{3}{8}a$$

EXAMPLE 12.57 Locate the centroid of the area $OADCFE$ in Fig. 12.45.

SOLUTION

Method 1. We divide the area into the segments $OADB$ and $BCFE$, which we label 1 and 2, respectively. Since statical moment is a sum, it follows that the statical moment of a built-up area equals the sum of the statical moments of the segments that compose it. We shall apply subscripts to identify the segments, and we shall omit a subscript when referring to the total area. Because an axis of symmetry is a centroidal axis, the centroid of a rectangle coincides with its center.

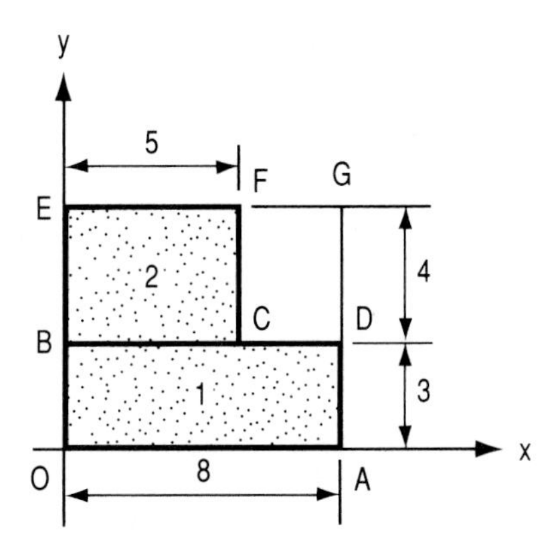

FIGURE 12.45

$$A_1 = 8 \times 3 = 24 \qquad A_2 = 5 \times 4 = 20$$

$$A = 24 + 20 = 44$$

Applying Eq. (12.50), we have

$$Q_{x,1} = 24 \times 1.5 = 36 \qquad Q_{x,2} = 20(3 + 2) = 100$$

$$Q_x = 36 + 100 = 136$$

$$Q_{y,1} = 24 \times 4 = 96 \qquad Q_{y,2} = 20 \times 2.5 = 50$$

$$Q_y = 96 + 50 = 146$$

$$\bar{y} = \frac{136}{44} = 3.091 \qquad \bar{x} = \frac{146}{44} = 3.318$$

Method 2. We may consider that the area is the difference between *OAGE* and *CDGF*. Let the subscripts 3 and 4, respectively, refer to these areas.

$$A_3 = 8(3 + 4) = 56 \qquad A_4 = (8 - 5)4 = 12$$

$$A = 56 - 12 = 44$$

$$Q_{x,3} = 56 \times 3.5 = 196 \qquad Q_{x,4} = 12(3 + 2) = 60$$

$$Q_x = 196 - 60 = 136$$

$$Q_{y,3} = 56 \times 4 = 224 \qquad Q_{y,4} = 12(5 + 1.5) = 78$$

$$Q_y = 224 - 78 = 146$$

The remaining calculations are identical with those under Method 1.

12.9.3 Moment of Inertia and Radius of Gyration

In Fig. 12.43, we now form the product $y^2\,dA$ of each element and sum these products. We again make the number of elements n progressively larger by making the elements progressively smaller. The limit of the sum is termed the *moment of inertia* (or *second moment*) of the area with respect to the x axis, and it is denoted by I_x. Then

$$I_x = \int y^2\,dA \qquad (12.51a)$$

Analogously, the moment of inertia of the area with respect to the y axis is

$$I_y = \int x^2\,dA \qquad (12.51b)$$

A moment of inertia is restricted to positive values. Its unit is the fourth power of the unit of length.

The moment of inertia of the area in Fig. 12.43 with respect to the x' axis is

$$I_{x'} = \int (y - h)^2\,dA = \int (y^2 - 2hy + h^2)\,dA$$

$$= I_x - 2hQ_x + Ah^2$$

If we replace x' with u and transpose terms, we obtain

$$I_x = I_u + A\bar{y}^2 \tag{12.52a}$$

This relationship is known as the *transfer formula*. Similarly,

$$I_y = I_v + A\bar{x}^2 \tag{12.52b}$$

It is advantageous to express moment of inertia in terms of the total area. By analogy with Eq. (12.50), we may write

$$I_x = Ak_x^2 \qquad I_y = Ak_y^2 \tag{12.53}$$

The distances k_x and k_y that we have introduced for convenience are termed the *radii of gyration* of the area with respect to the x and y axes, respectively. In general, the radius of gyration of the area with respect to an arbitrary axis m is

$$k_m = \sqrt{\frac{I_m}{A}} \tag{12.53a}$$

EXAMPLE 12.58 With reference to Fig. 12.44, compute the radius of gryation of the area OBC with respect to its centroidal axis u that is parallel to the x axis.

SOLUTION To simplify the calculations, we shall first compute I_x and then find I_u by applying Eq. (12.52*a*). We shall apply the results obtained in Example 12.56. With reference to a strip parallel to the x axis, we have

$$dI_x = y^2\,dA = y^2(x\,dy) = \frac{ay^{5/2}}{b^{1/2}}$$

$$I_x = \frac{a}{b^{1/2}}\int_0^b y^{5/2}\,dy = \frac{2ab^3}{7}$$

$$I_u = I_x - A\bar{y}^2 = \frac{2ab^3}{7} - \frac{2ab}{3}\left(\frac{3b}{5}\right)^2 = \frac{8}{175}ab^3$$

$$k_u^2 = \frac{I_u}{A} = \frac{(8/175)ab^3}{(2/3)ab} = \frac{12}{175}b^2 \qquad k_u = 0.2619b$$

This radius of gyration is an index of the extent to which the area is dispersed about the u axis.

12.9.4 Product of Inertia

Returning to Fig. 12.43, we now form the product $xy\,dA$ for each element, sum these products, and make the number of elements progressively larger. The limit of this sum is called the *product of inertia* of the area with respect to the x and y axes, and it is denoted by P_{xy}. Then

$$P_{xy} = \int xy\,dA \tag{12.54}$$

Theorem 12.2. If an area has an axis of symmetry, the product of inertia of the area with respect to this axis and an axis perpendicular thereto is zero.

The proof is similar to that of Theorem 12.1.

EXAMPLE 12.59 Find the product of inertia of the right triangle in Fig. 12.46 with respect to the x and y axes.

SOLUTION We shall first sum the products for a strip parallel to the y axis. For a point on OB, $y{:}b = x{:}a$, and $y = bx/a$. Then

$$P_{xy} = \int xy\, dA = \int_0^a \int_0^{bx/a} xy\, dy\, dx$$

The first integration yields $b^2x^3/(2a^2)$. Then

$$P_{xy} = \frac{b^2}{2a^2}\int_0^a x^3\, dx = \frac{a^2b^2}{8}$$

12.9.5 Effect of Rotation of Axes

Assume the following: We are given an area and the positions of the x and y axes; we know I_x, I_y, and P_{xy}; we also require the values of moment of inertia and product of inertia with respect to other axes through the origin. To obtain these values, consider that the x and y axes are rotated about the origin in a counterclockwise direction through an angle θ, as shown in Fig. 3.7b. Again let x' and y' denote the new positions of the axes. By applying Eq. (3.14), we obtain the following:

$$I_{x'} = \frac{I_x + I_y}{2} + \left(\frac{I_x - I_y}{2}\right)\cos 2\theta - P_{xy}\sin 2\theta \qquad (12.55a)$$

$$P_{x'y'} = \left(\frac{I_x - I_y}{2}\right)\sin 2\theta + P_{xy}\cos 2\theta \qquad (12.55b)$$

Figure 12.47 exhibits visually the manner in which the moment of inertia and product of inertia vary with θ. This diagram, which is known as *Mohr's circle of moment of inertia*, is similar to Fig. 3.8a. In Fig. 12.47, values of I appear on the

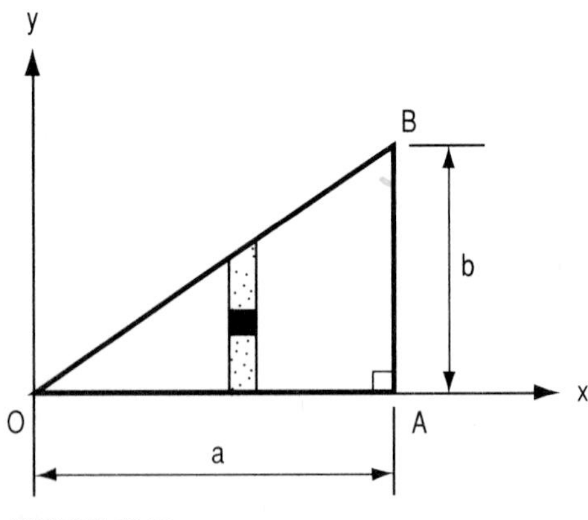

FIGURE 12.46

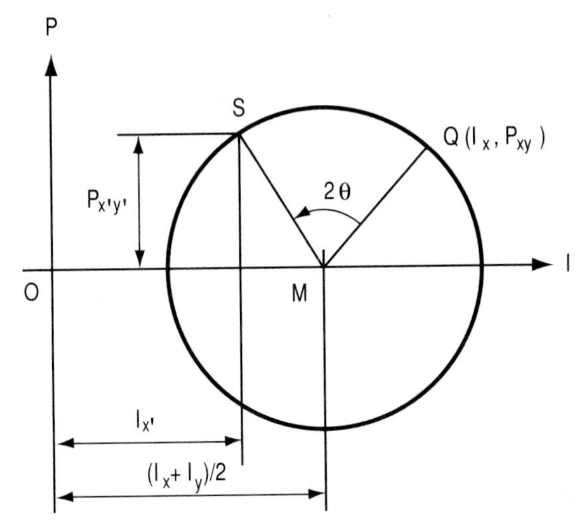

FIGURE 12.47 Mohr's circle of moment of inertia.

horizontal axis, and values of P appear on the vertical axis. The steps in the construction are as follows: Plot point Q having the coordinates I_x and P_{xy}. Locate point M on the horizontal axis, with $OM = (I_x + I_y)/2$. Draw a circle having M as center and MQ as radius. Now draw the radius MS making an angle of 2θ with MQ as measured in the counterclockwise direction. The coordinates of S are $I_{x'}$ and $P_{x'y'}$.

Now consider that we are given an area and a point O in the plane of the area. We may select an arbitrary pair of mutually perpendicular axes through O and assign positive directions to these axes. Call the axes a and b. Associated with each pair of axes is a specific value of I_a, I_b, and P_{ab}. A *principal axis* is an axis for which I is maximum or minimum. Figure 12.47 reveals the following facts:

1. Through a given point, there are two principal axes, and they are mutually perpendicular.
2. The product of inertia with respect to the principal axes is zero.
3. The perpendicular axes for which the product of inertia is maximum or minimum make 45° angles with the principal axes.

Let R denote the radius of the circle in Fig. 12.47. Then

$$R = \sqrt{\left(\frac{I_x - I_y}{2}\right)^2 + P_{xy}^2} \tag{12.56}$$

$$P_{\max} = R \tag{12.57a}$$

$$I_{\max} = \frac{I_x + I_y}{2} + R \qquad I_{\min} = \frac{I_x + I_y}{2} - R \tag{12.57b}$$

12.9.6 Polar Moment of Inertia

Returning to Fig. 12.43, let r denote the distance from the origin to the center of the element. We now form the product $r^2\,dA$ for each element, sum these products, and make the number of elements progressively larger. The limit of this sum is called the *polar moment of inertia* of the area with respect to the pole O, and it is denoted by J_O. Then

$$J_O = \int r^2\,dA \tag{12.58}$$

Since $r^2 = y^2 + x^2$, it follows that

$$J_O = I_x + I_y \tag{12.59}$$

By analogy with Eq. (12.53), we may write

$$J_O = Ak_O^2 \tag{12.60}$$

and the distance k_O is termed the radius of gyration of the area with respect to O.

12.10 PROPERTIES OF ARCS, VOLUMES, AND MASSES

12.10.1 Properties of an Arc

We have defined the properties of an area, and an arc of a curve has analogous properties. Let s denote the length of the arc. Consider that the arc is divided into elements of length ds. The statical moments of the arc are as follows:

$$Q_x = \int y\,ds \qquad Q_y = \int x\,ds$$

By applying Eq. (12.6) for ds, we obtain the following:

$$Q_x = \int y\sqrt{1 + \left(\frac{dx}{dy}\right)^2}\,dy \tag{12.61a}$$

$$Q_y = \int x\sqrt{1 + \left(\frac{dy}{dx}\right)^2}\,dx \tag{12.61b}$$

Again let u and v denote axes that are parallel to the x and y axes, respectively, and have the property $Q_u = Q_v = 0$. Axes u and v are centroidal axes of the arc, and their point of intersection is the centroid of the arc. With due regard for algebraic signs, let $\bar{y}$ denote the distance between the u and x axes; let $\bar{x}$ denote the distance between the v and y axes. Then

$$\bar{x} = \frac{Q_y}{s} \qquad \bar{y} = \frac{Q_x}{s} \tag{12.62}$$

EXAMPLE 12.60 Locate the centroid of an arc that is a quarter circle of radius r.

SOLUTION Place the origin at the center of the circle, with the arc falling in the first quadrant. By symmetry, $\bar{x} = \bar{y}$.

$$x^2 + y^2 = r^2 \qquad \frac{dy}{dx} = -\frac{x}{\sqrt{r^2 + x^2}}$$

$$Q_y = \int_0^r x \sqrt{1 + \frac{x^2}{r^2 - x^2}}\, dx = r \int_0^r \frac{x}{\sqrt{r^2 - x^2}}\, dx = r^2$$

$$\bar{x} = \frac{Q_y}{s} = \frac{r^2}{\pi r/2} = \frac{2r}{\pi}$$

The centroid lies on a line that makes 45° angles with the x and y axes.

12.10.2 Properties of a Volume

The volume of a solid body has properties analogous to those of an area. Consider that the body is divided into elements of volume dV. The statical moment and moment of inertia of the entire volume with respect to the xy plane are as follows:

$$Q_{xy} = \int z\, dV \qquad I_{xy} = \int z^2\, dV$$

Let v denote a plane that is parallel to the xy plane and has the property $I_v = 0$. Then v is a *centroidal plane*, and its distance $\bar{z}$ from the xy plane is

$$\bar{z} = \frac{Q_{xy}}{V}$$

The point of intersection of the three centroidal planes parallel to the coordinate planes is the centroid of the volume.

Let r denote the distance from the origin to the center of an element. The polar moment of inertia of the volume with respect to the origin is

$$J_O = \int r^2\, dV = I_{xy} + I_{xz} + I_{yz}$$

EXAMPLE 12.61 Locate the centroid of a solid in the form of a hemisphere of radius r.

SOLUTION Place the coordinate axes in the position shown in Fig. 12.48, where the origin is at the center of the hemisphere. Because they are planes of symmetry, the xy and xz planes are centroidal planes, and therefore $\bar{y} = \bar{z} = 0$. Thus, the centroid lies on the x axis, and it is simply necessary to find $\bar{x}$. To do this, we must find Q_{yz}. In Example 12.39, we proved that the volume of a sphere is $(4/3)\pi r^3$.

The problem can be solved by setting $dQ_{yz} = x\, dx\, dy\, dz$ and then applying triple integration. However, a more direct approach consists of evaluating the derivative dQ_{yz}/dx. We divide the body into strips parallel to the yz plane, of width dx, and consider the radius of a strip to be constant, as shown in

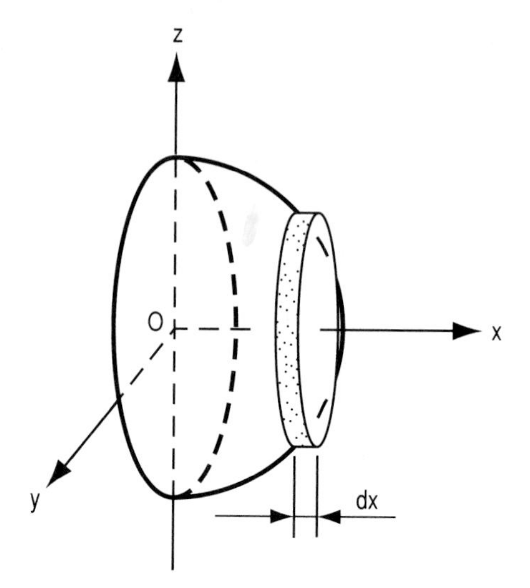

FIGURE 12.48

Fig. 12.48. Let A denote the area of the face of a strip. For a particular strip, we have the following:

$$dQ_{yz} = x\, dV = xA\, dx = x(\pi z^2)\, dx = \pi x(r^2 - x^2)\, dx$$

$$Q_{yz} = \pi \int_0^r (r^2 x - x^3)\, dx = \frac{\pi r^4}{4}$$

$$\bar{x} = \frac{Q_{yz}}{V} = \frac{(1/4)\pi r^4}{(2/3)\pi r^3} = \frac{3}{8} r$$

The manner in which the volume of the hemisphere is distributed makes it apparent that the value of $\bar{x}$ is less than $(1/2)r$. Therefore, our calculated result seems reasonable.

12.10.3 Properties of a Mass

The properties of a mass are analogous to those of a volume. Again let μ denote the unit mass of the material, m denote the mass of the body, and dm denote the mass of an element of the body. The statical moment of the mass with respect to the xy plane is

$$Q_{xy} = \int z\, dm = \int z\mu\, dV$$

If the body is homogeneous, the centroid of its mass coincides with the centroid of its volume.

EXAMPLE 12.62 A sphere of radius r was made by joining two hemispheres. The unit mass of one hemisphere is 1.5 times the unit mass of the other. Compute the eccentricity of the sphere (i.e., the distance from the centroid of the mass to the center of the volume).

SOLUTION We shall present two methods of solution. The first method is more time-consuming, but it illustrates the procedure for obtaining the properties of a mass.

Method 1. Place the sphere in such position that the plane separating the two parts coincides with the yz plane, the heavier part being to the right of this plane. Let μ denote the unit mass of the lighter part. Refer to Fig. 12.48. We apply the result obtained in Example 12.61 and the following equation, which is the analog of Eq. (12.50):

$$Q_{yz} = m\bar{x}$$

where m is the mass of the given body. Let V denote the volume of each hemisphere. Then Q_{yz} of the sphere is

$$Q_{yz} = 1.5\mu V\left(\frac{3}{8}r\right) - \mu V\left(\frac{3}{8}r\right) = \frac{3}{16}\mu Vr$$

$$\bar{x} = \frac{Q_{yz}}{m} = \frac{(3/16)\mu Vr}{V(1.5\mu + \mu)} = \frac{3}{40}r$$

Method 2. Let m_1 and m_2 denote the masses of the heavier and lighter hemispheres, respectively, and let x_h denote the $\bar{x}$ value of the hemisphere, in absolute value. It can be demonstrated that the $\bar{x}$ value of the sphere is

$$\bar{x} = x_h\left(\frac{m_1 - m_2}{m_1 + m_2}\right)$$

Setting $x_h = (3/8)r$ and $m_1 = 1.5m_2$, we obtain $\bar{x} = (3/40)r$.

12.10.4 Theorems of Pappus

The following principles are helpful in calculating surface areas and volumes that result from revolution:

Theorem 12.3. If the arc of a plane curve is revolved about an axis that lies in the same plane but does not intersect the arc, the area of the surface generated is equal to the product of the length of the arc and the distance traversed by the centroid of the arc.

Theorem 12.4. If a plane region is revolved about an axis that lies in that plane but outside this region, the volume generated is equal to the product of the area of the region and the distance traversed by the centroid of the region.

EXAMPLE 12.63 With reference to the parabolic arc in Fig. 12.44, $a = 4$ and $b = 8$. The region OBC is revolved about the straight line $y = x - 5$. Applying the results obtained in Example 12.56, compute the volume of the solid that is formed.

SOLUTION The equation of the parabola is $y = 0.5x^2$. It is apparent that the straight line does not intersect the parabola, but this property can be proved rigorously by setting $y = 0.5x^2 = x - 5$ and solving this equation for x. The results are complex numbers; therefore, no points of intersection exist.

From Example 12.56, we have $\bar{x} = (3/8)a = (3/8)4 = 1.5$, $\bar{y} = (3/5)b = (3/5)8 = 4.8$, and the area is $(2/3)4 \times 8 = 64/3$. We must find the distance d from the centroid to the straight line. First, rewriting the equation of the straight line in the form $Ax + By + C = 0$, we have $x - y - 5 = 0$. Thus, $A = 1$, $B = -1$, $C = -5$. By Eq. (3.10), we have $c = 1.5 - 4.8 - 5 = -8.3$. By Eq. (3.11), $d = 8.3/\sqrt{2}$. The centroid describes a circle having this distance as its radius. Therefore, the distance traversed by the centroid is $2\pi d = 36.876$. By Theorem 12.4, the volume of the solid of revolution is $36.876(64/3) = 786.7$ cu units.

12.11 INFINITE SERIES

12.11.1 Definitions and Notation

In Art. 1.17, we defined the following terms: series, infinite series, convergent and divergent series, power series, alternating series, and various types of series. A *positive series* is one that contains plus signs exclusively and in which each term is positive.

The general notation for an infinite series is

$$u_1 + u_2 + u_3 + \cdots + u_n + \cdots$$

Let S_n denote the sum of the first n terms of the series, which is termed a *partial sum*. If the series is convergent, S_n approaches a finite limit as n becomes infinite, and we shall denote this limit by S. Hereafter, the term *series* will always refer to an infinite series.

In dealing with an infinite series, the fundamental problem that arises is to determine whether the series is convergent or divergent. Various tests are available that enable us to classify the series, and we shall now discuss these tests. We shall first consider positive series and then alternating series.

12.11.2 Tests for Positive Series

The principles that follow provide a means of testing a positive series for convergence.

Theorem 12.5. If a series is convergent, u_n approaches 0 as a limit as n becomes infinite.

As an illustration, consider the series

$$\frac{1}{5} + \frac{2}{7} + \frac{3}{9} + \cdots + \frac{n}{2n+3} + \cdots$$

We may rewrite u_n as

$$u_n = \frac{1}{2 + 3/n}$$

As n becomes infinite, u_n approaches the limit 1/2. Therefore, the series is divergent.

It is to be emphasized that the requirement in Theorem 12.5 screens out *some* divergent series, but not all. In other words, this theorem establishes a *necessary* condition for convergence, but not a *sufficient* one. Thus, the harmonic series recorded in Art. 1.17 satisfies the requirement of Theorem 12.5, but nevertheless the series is divergent.

Theorem 12.6. A positive series is convergent if S_n always remains less than some finite number as n increases.

Theorem 12.7. A positive series is convergent if each term is less than or at most equal to the corresponding term of a positive series that is known to be convergent. A positive series is divergent if each term is greater than or at least equal to the corresponding term of a positive series that is known to be divergent.

Theorem 12.7 provides a *comparison test* for convergence or divergence. There are two types of series that are particularly useful for comparison purposes. One is the geometric series

$$a + ar + ar^2 + \cdots + ar^{n-1} + \cdots$$

which is convergent if $|r| < 1$ and divergent otherwise. The other is the p *series*

$$1 + \frac{1}{2^p} + \frac{1}{3^p} + \cdots + \frac{1}{n^p} + \cdots$$

which is convergent if $p > 1$ and divergent otherwise. In applying a comparison test, we may discard a finite number of terms if it is convenient to do so.

As an illustration, consider the series

$$3^3 + \frac{3^2}{2} + \frac{3}{3} + \frac{1}{4} + \frac{1}{5 \cdot 3} + \frac{1}{6 \cdot 3^2} + \cdots + \frac{3^{4-n}}{n} + \cdots$$

As our comparison series, we take

$$3^3 + 3^2 + 3 + 1 + \frac{1}{3} + \frac{1}{3^2} + \cdots$$

This is a geometric series in which $r = 1/3$; therefore, it is convergent. Beyond the first term, each term of the given series is less than the corresponding term of the comparison series. It follows that the given series is convergent.

In the expression for u_n, let P_n and P_d denote the highest power of n in the numerator and denominator, respectively, and assume that these values remain constant from term to term. By applying the p series, we arrive at the following principle:

Theorem 12.8. A positive series is convergent if $P_d > P_n + 1$; otherwise, it is divergent.

For example, in the series

$$\frac{2}{1 \cdot 4} + \frac{3}{2 \cdot 5} + \frac{4}{3 \cdot 6} + \cdots + \frac{n+1}{n(n+3)} + \cdots$$

the denominator of u_n may be rewritten as $n^2 + 3n$, and we have $P_n = 1$ and $P_d = 2$.

Since $P_d = P_n + 1$, the series is divergent. On the other hand, in the series

$$\frac{5}{1\cdot 2\cdot 4}+\frac{8}{2\cdot 3\cdot 5}+\frac{11}{3\cdot 4\cdot 6}+\cdots+\frac{3n+2}{n(n+1)(n+3)}+\cdots$$

we have $P_n = 1$ and $P_d = 3$. Since $P_d > P_n + 1$, the series is convergent. Similarly, in the series

$$\frac{1^{3/2}}{5\cdot 2^2}+\frac{2^{3/2}}{6\cdot 3^2}+\frac{3^{3/2}}{7\cdot 4^2}+\cdots+\frac{n^{3/2}}{(n+4)(n+1)^2}+\cdots$$

we have $P_n = 3/2$ and $P_d = 3$. Therefore, the series is convergent.

A series may also be tested for convergence by taking the ratio of one term to the preceding term. Let $R = u_{n+1}/u_n$. We have the following:

Theorem 12.9. A positive series is convergent if R is always less than some number r where $0 < r < 1$.

Theorem 12.10. Let L denote the limit of R as n becomes infinite. A positive series is convergent if $L < 1$ and divergent if $L > 1$. If $R = 1$, no conclusion can be drawn.

Theorem 12.10 provides the *Cauchy ratio test*.

In the subsequent material, we shall apply the following relationship: In an arithmetic progression, let a denote the first term and d denote the common difference. The nth term is $a + (n - 1)d$.

EXAMPLE 12.64 Test the following series for convergence:

$$\frac{1}{5}+\frac{1\cdot 3}{5\cdot 8}+\frac{1\cdot 3\cdot 5}{5\cdot 8\cdot 11}+\frac{1\cdot 3\cdot 5\cdot 7}{5\cdot 8\cdot 11\cdot 14}+\cdots$$

SOLUTION The factors in both the numerator and denominator form arithmetic progressions, and we have the following:

$$u_n = \frac{1\cdot 3\cdot 5\cdots(2n-1)}{5\cdot 8\cdot 11\cdots(3n+2)}$$

$$R = \frac{u_{n+1}}{u_n} = \frac{2(n+1)-1}{3(n+1)+2} = \frac{2n+1}{3n+5} = \frac{2+1/n}{3+5/n} \qquad L = \frac{2}{3}$$

Thus, the series is convergent.

For the case where $L = 1$ and the Cauchy ratio test is inconclusive, the following principle can be applied:

Theorem 12.11. If a positive series has $L = 1$, assume that R can be expressed in this manner:

$$R = \frac{n^p + an^{p-1} + \cdots}{n^p + a'n^{p-1} + \cdots}$$

The series is convergent if $a' > a + 1$; otherwise, it is divergent.

EXAMPLE 12.65 Test the following series for convergence:

$$\frac{1}{6}+\frac{1\cdot 2}{6\cdot 7}+\frac{1\cdot 2\cdot 3}{6\cdot 7\cdot 8}+\cdots$$

SOLUTION

$$u_n=\frac{1\cdot 2\cdot 3\cdots n}{6\cdot 7\cdot 8\cdots(n+5)}$$

$$R=\frac{u_{n+1}}{u_n}=\frac{n+1}{(n+1)+5}=\frac{n+1}{n+6}$$

Then $a=1$, $a'=6$, and the series is convergent.

12.11.3 Alternating Series

The general notation for an alternating series is

$$u_1-u_2+u_3-u_4+\cdots$$

where all the u's are positive. The principles that follow provide a means of testing an alternating series for convergence.

Theorem 12.12. An alternating series is convergent if the corresponding positive series is convergent.

For example, since the series

$$1+\frac{1}{2^2}+\frac{1}{3^2}+\frac{1}{4^2}+\cdots+\frac{1}{n^2}+\cdots$$

is convergent, it follows that

$$1-\frac{1}{2^2}+\frac{1}{3^2}-\frac{1}{4^2}+\cdots+(-1)^{n-1}\frac{1}{n^2}+\cdots$$

is also convergent.

Theorem 12.13. An alternating series is convergent if each term is numerically less than the preceding term and u_n approaches 0 as n becomes infinite.

Theorem 12.14. If an alternating series is convergent, S is intermediate between S_n and S_{n+1}, and consequently $|S-S_n|<u_{n+1}$.

Figure 12.49 demonstrates the validity of Theorems 12.13 and 12.14. In this diagram, $OA=u_1$, $|AB|=u_2$, $BC=u_3$, etc. Thus, $OA=S_1$, $OB=S_2$, $OC=S_3$,

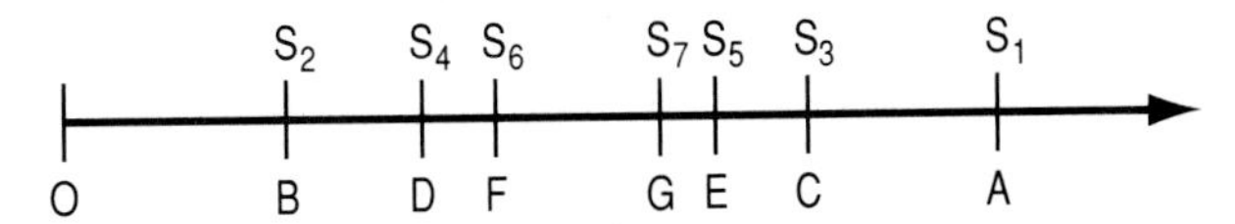

FIGURE 12.49 Geometric proof of properties of an alternating series.

etc. If n is even, $S_n < S$; if n is odd, $S < S_n$. Assume that n is even. Then

$$S_n < S < S_{n+1} \qquad \text{or} \qquad S_n < S < S_n + u_{n+1}$$

$$\therefore\ S - S_n < u_{n+1}$$

If n is odd, we simply take the absolute value of the expression at the left. This relationship discloses that if we approximate S by summing only the first n terms, the resulting error is less than the first term that is ignored.

A convergent alternating series is *absolutely convergent* if the corresponding positive series is also convergent, and *conditionally convergent* if such is not the case.

EXAMPLE 12.66 Test the following series for convergence. If it is convergent, determine whether its sum is greater than or less than 0.40.

$$\frac{2}{1 \cdot 4} - \frac{3}{2 \cdot 5} + \frac{4}{3 \cdot 6} - \cdots + (-1)^{n-1} \frac{n+1}{n(n+3)}$$

SOLUTION Since this series satisfies the requirements of Theorem 12.13, it is convergent. However, we previously found that the corresponding positive series is divergent. Consequently, the present series is conditionally convergent.

Taking partial sums, we obtain these results:

$$S_1 = 0.5 \qquad S_2 = 0.2 \qquad S_3 = 0.4222$$

$$S_4 = 0.2437 \qquad S_5 = 0.3937$$

Since $S < S_5$, it follows that $S < 0.40$.

12.11.4 Power Series

Let y be a function of x. The power series in y is

$$a_0 + a_1 y + a_2 y^2 + \cdots + a_n y^n + \cdots$$

where the a's are all constant. Whether this series converges or diverges depends on the value of y, and thus on the value assigned to x. Assume that the ratio $|a_n / a_{n+1}|$ approaches a finite limit as n becomes infinite, and let L' denote this limit.

Theorem 12.15. A power series in y is absolutely convergent if $|y| < L'$ and divergent if $|y| > L'$. If $|y| = L'$, no conclusion can be drawn.

This principle stems from the Cauchy ratio test. The range of x values that makes the series convergent is termed the *interval of convergence.*

EXAMPLE 12.67 Without testing for boundary values, establish the interval of convergence of the following series:

$$\frac{x+5}{1 \cdot 2} + \frac{(x+5)^2}{3 \cdot 4} + \frac{(x+5)^3}{5 \cdot 6} + \cdots$$

SOLUTION

$$a_n = \frac{1}{(2n-1)2n} \qquad a_{n+1} = \frac{1}{(2n+1)(2n+2)}$$

$$\frac{a_n}{a_{n+1}} = \frac{2n^2+3n+1}{2n^2-n} = \frac{2+3/n+1/n^2}{2-1/n} \qquad L' = \frac{2}{2} = 1$$

Thus, $x+5$ must lie between -1 and 1. Expressing the inequality and then subtracting 5 from each member, we obtain

$$-1 < x+5 < 1 \qquad -6 < x < -4$$

The interval of convergence is from -6 to -4.

12.11.5 Taylor and Maclaurin Series

Assume that a function of x can be expressed as the sum of a power series in $x-a$, where a is a constant. Then

$$f(x) = a_0 + a_1(x-a) + a_2(x-a)^2 + a_3(x-a)^3 + \cdots$$

By differentiating this equation successively and setting $x = a$, we obtain the values of the coefficients. Substituting these values, we obtain the following:

$$f(x) = f(a) + f'(a)(x-a) + \frac{f''(a)}{2!}(x-a)^2 + \frac{f'''(a)}{3!}(x-a)^3 + \cdots + \frac{f^n(a)}{n!}(x-a)^n + \cdots \qquad (12.63)$$

The series at the right is termed a *Taylor series*, and $f(x)$ is said to be *expanded* in a Taylor series.

If we now set $a = 0$, Eq. (12.63) reduces to

$$f(x) = f(0) + f'(0)x + \frac{f''(0)}{2!}x^2 + \frac{f'''(0)}{3!}x^3 + \cdots + \frac{f^n(0)}{n!}x^n + \cdots \qquad (12.64)$$

The series at the right is termed a *Maclaurin series*, and it is a special form of the Taylor series. The Taylor and Maclaurin series are of extreme importance, for they provide a means of evaluating many functions.

EXAMPLE 12.68 Expand $\ln(1+x)$ in powers of x.

SOLUTION

$$f(x) = \ln(1+x) \qquad f(0) = \ln 1 = 0$$

$$f'(x) = (1+x)^{-1} \qquad f'(0) = 1$$

$$f''(x) = -(1+x)^{-2} \qquad f''(0) = -1$$

$$f'''(x) = 2!(1+x)^{-3} \qquad f'''(0) = 2!$$

$$f^{\text{iv}}(x) = -3!(1+x)^{-4} \qquad f^{\text{iv}}(0) = -3!$$

The pattern that emerges reveals that $f^n(0) = (-1)^{n-1}[(n-1)!]$. Substituting in Eq. (12.64), we obtain

$$\ln(1+x) = x - \frac{x^2}{2} + \frac{x^3}{3} - \frac{x^4}{4} + \cdots$$

12.11.6 Even and Odd Functions

A function is described as *even* if $f(-x) = f(x)$ and *odd* if $f(-x) = -f(x)$. Thus, $\cos x$ is even because $\cos(-x) = \cos x$, and $\sin x$ is odd because $\sin(-x) = -\sin x$. If a function is neither even nor odd, $f(x)$ and $f(-x)$ differ in absolute value.

The function x^n is even or odd according to whether n is even or odd, respectively. We therefore deduce the following: Consider that a function is expanded in a Maclaurin series. If the function is even, the series contains solely even powers of x. If the function is odd, the series contains solely odd powers of x. If the function is neither even nor odd, the series contains a mix of even and odd powers of x.

As an illustration, consider the functions $x/(1+x^2)$ and $x/(1+x^3)$. The first is odd and the second is neither even nor odd. If $-1 < x < 1$, the expansions of these functions are the following infinite geometric series, respectively:

$$x - x^3 + x^5 - x^7 + \cdots$$

$$x - x^4 + x^7 - x^{10} + \cdots$$

In the first series, all exponents are odd; in the second series, the exponents are both even and odd.

In some instances, the task of expanding a function in a Maclaurin series can be simplified if we know in advance that the series will contain solely even or odd powers of x.

12.11.7 Euler's Formula

When $\sin x$ and $\cos x$ are expanded in accordance with Eq. (12.64), the results are as follows:

$$\sin x = x - \frac{x^3}{3!} + \frac{x^5}{5!} - \frac{x^7}{7!} + \cdots$$

$$\cos x = 1 - \frac{x^2}{2!} + \frac{x^4}{4!} - \frac{x^6}{6!} + \cdots$$

Equation (1.50) is the power series for e^x. Again let i denote the unit of imaginary numbers, which are discussed in Art. 1.8. If we replace x with ix in this equation, we find that

$$e^{ix} = \cos x + i \sin x \tag{12.65}$$

This relationship, which is known as *Euler's formula*, has numerous applications. For example, in electrical engineering, where quantities vary sinusoidally with time, it enables us to express these quantities in terms of $e^{i\theta}$ rather than in terms of $\sin\theta$ or $\cos\theta$. Thus, it provides an alternative mode of expression.

12.11.8 Evaluation of Power Series

Assume that we wish to find the sum of a power series for numerous values of the independent variable. It is first necessary to determine how many terms are to be included in the summation. The computer time required to obtain the sum of the truncated series can be reduced substantially by applying either of two devices. The first device consists of placing the series in nested form, in the manner described in Art. 1.18. The second device consists of transforming the truncated series to a polynomial fraction whose value closely approximates the sum of the series and then transforming this fraction to a continued fraction, in the manner described in Art. 1.21.

We shall illustrate the second device, and we shall use the following series for this purpose:

$$S = \frac{x}{1 \cdot 2} - \frac{x^3}{3 \cdot 4} + \frac{x^5}{5 \cdot 6} - \cdots + (-1)^{n-1} \frac{x^{2n-1}}{(2n-1)2n}$$

For brevity, we shall include only the first three terms in our summation.

The polynomial fraction that replaces the truncated series must satisfy three requirements. First, since this series contains three terms, the fraction must contain three parameters. Second, since $S = 0$ when $x = 0$, the numerator of the fraction must be devoid of any constant terms. Third, since S is an odd function of x, the fraction must also be an odd function of x. Let S_3 denote the sum of the truncated series and Y denote the value of the fraction. We contrive the following:

$$Y = \frac{Ax^3 + Bx}{Cx^2 + 1}$$

where A, B, and C are constants that must be evaluated. We now cross-multiply and replace Y with S_3. Then

$$Ax^3 + Bx = (Cx^2 + 1)\left(\frac{x}{2} - \frac{x^3}{12} + \frac{x^5}{30}\right)$$

Performing the multiplication and equating terms that contain identical powers of x from 1 through 5, we obtain these results:

$$Bx = \frac{x}{2} \qquad Ax^3 = \frac{Cx^3}{2} - \frac{x^3}{12}$$

$$0 = -\frac{Cx^5}{12} + \frac{x^5}{30}$$

The first equation yields $B = 1/2$, the third equation yields $C = 2/5$, and the second equation now yields $A = 7/60$. Our fraction now becomes

$$Y = \frac{(7/60)x^3 + (1/2)x}{(2/5)x^2 + 1} = \frac{(7/60)[x^3 + (30/7)x]}{(2/5)(x^2 + 5/2)}$$

$$= \frac{7}{24}\left[\frac{x^3 + (30/7)x}{x^2 + 5/2}\right]$$

At this point, we shall test the precision of Y. Arbitrarily setting $x = 0.3$, we have $S_3 = 0.147831$ and $Y = 0.147828$. The difference is negligible.

The contrived fraction is now expanded to a continued fraction by following the instructions in Art. 1.21. The result is

$$Y = \frac{7}{24}\left(x + \cfrac{25/14}{x + \cfrac{5/2}{x}}\right)$$

The computer time required to find Y is less than that required to find S_3 in a straightforward manner.

12.11.9 Fourier Series for Periodic Functions

A *periodic function* is one whose values recur in cycles. The *interval* of the function is the amount by which the independent variable increases in value in one cycle. For example, $\sin x$ and $\cos x$ are periodic functions having an interval of 2π. Many periodic functions that arise in engineering and science are characterized by discrete changes in value at specific points.

This question suggests itself: Is it possible to express a general type of periodic function as the sum of an infinite series consisting of trigonometric functions? We shall find that this possibility often exists.

For convenience, we record here the following integrals:

$$\int \sin^2 nx\, dx = \frac{x}{2} - \frac{\sin 2nx}{4n} \qquad \int \cos^2 nx\, dx = \frac{x}{2} + \frac{\sin 2nx}{4n}$$

where n is any positive integer (or zero). If we integrate between the limits c and $c + 2\pi$, where c is a constant, we obtain π in both cases.

Let $f(x)$ be a periodic function having an interval of 2π, and assume that it can be expressed in this manner:

$$f(x) = \frac{a_0}{2} + a_1 \cos x + a_2 \cos 2x + \cdots + a_n \cos nx + \cdots$$
$$+ b_1 \sin x + b_2 \sin 2x + \cdots + b_n \sin nx + \cdots \qquad (12.66)$$

The series at the right is termed a *Fourier series*, and the function is said to be *expanded* in its Fourier series. Our task is to evaluate the coefficients associated with a given function. If we multiply both sides of Eq. (12.66) by $\cos nx\, dx$ and integrate between the limits c and $c + 2\pi$, all terms at the right vanish except that which was originally $a_n \cos nx$. If we now multiply both sides of Eq. (12.66) by $\sin nx\, dx$ and integrate between the limits c and $c + 2\pi$, all terms at the right vanish except that which was originally $b_n \sin nx$. By solving the equations that result from integration for a_n and b_n, we obtain the following results:

$$a_n = \frac{1}{\pi} \int_c^{c+2\pi} f(x) \cos nx\, dx \qquad (12.67a)$$

$$b_n = \frac{1}{\pi} \int_c^{c+2\pi} f(x) \sin nx\, dx \qquad (12.67b)$$

These equations are also valid when $n = 0$. In applying these equations, we equate c to the value of x at the start of a cycle.

EXAMPLE 12.69 A periodic function has the following values:

$$f(x) = 9 \qquad 0 < x < \pi$$

$$f(x) = 3 \qquad \pi < x < 2\pi$$

Expand the function as a Fourier series.

SOLUTION The function is plotted in Fig. 12.50. We shall apply the following integrals:

$$\int \sin nx\, dx = -\frac{\cos nx}{n} \qquad \int \cos nx\, dx = \frac{\sin nx}{n}$$

Let m denote an integer. Then $\sin m\pi = 0$, $\cos m\pi = 1$ when m is even, and $\cos m\pi = -1$ when m is odd.

Since the cycle starts with $x = 0$, $c = 0$. Applying Eq. (12.67a) with $n = 0$, we obtain

$$a_0 = \frac{1}{\pi}\left[\int_0^{\pi} 9\, dx + \int_{\pi}^{2\pi} 3\, dx\right] = \frac{9(\pi - 0) + 3(2\pi - \pi)}{\pi} = 12$$

Applying Eq. (12.67a) with $n > 0$, we obtain $a_n = 0$ because $\sin m\pi = 0$. Applying Eq. (12.67b), we obtain

$$b_n = \frac{1}{\pi}\left[\int_0^{\pi} 9 \sin nx\, dx + \int_{\pi}^{2\pi} 3 \sin nx\, dx\right]$$

$$= -\frac{1}{n\pi}[9(\cos n\pi - \cos 0) + 3(\cos 2n\pi - \cos n\pi)]$$

$$= -\frac{1}{n\pi}[9(\cos n\pi - 1) + 3(1 - \cos n\pi)]$$

$$= -\frac{1}{n\pi}(6 \cos n\pi - 6)$$

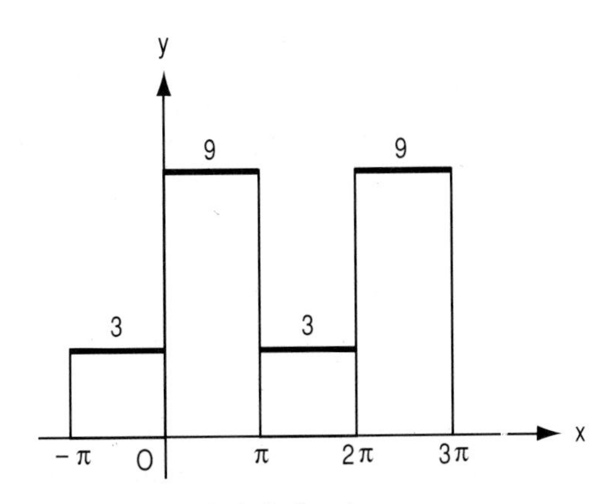

FIGURE 12.50 Periodic function.

If n is even, $\cos n\pi = 1$, and $b_n = 0$. If n is odd, $\cos n\pi = -1$, and $b_n = 12/n\pi$. Substituting in Eq. (12.66), we have

$$f(x) = 6 + \frac{12}{\pi}\left(\sin x + \frac{\sin 3x}{3} + \frac{\sin 5x}{5} + \cdots\right)$$

Let K denote the sum of the series in parentheses. By substituting the values of the function, we find that $K = \pi/4$ when $0 < x < \pi$ and $K = -\pi/4$ when $\pi < x < 2\pi$. When $x = \pi$, this equation yields $f(x) = 6$, which is the arithmetic mean of 9 and 3.

Generally, the interval of a periodic function has some value other than 2π. Let $2i$ denote the interval. The foregoing results can be generalized by replacing x in the Fourier series with a new function z, where $z{:}2\pi = x{:}2i$. Thus, when x increases by $2i$, z increases by 2π. Then

$$z = \frac{\pi x}{i} \qquad dz = \frac{\pi}{i}\,dx$$

We now replace x with z in the foregoing equations, but we express z and dz in terms of x and dx, respectively. Equation (12.66) becomes

$$f(x) = \frac{a_0}{2} + a_1 \cos\frac{\pi x}{i} + a_2 \cos\frac{2\pi x}{i} + \cdots + a_n \cos\frac{n\pi x}{i} + \cdots$$
$$+ b_1 \sin\frac{\pi x}{i} + b_2 \sin\frac{2\pi x}{i} + \cdots + b_n \sin\frac{n\pi x}{i} + \cdots \tag{12.68}$$

Equation (12.67) becomes

$$a_n = \frac{1}{i}\int_c^{c+2i} f(x)\cos\frac{n\pi x}{i}\,dx \tag{12.69a}$$

$$b_n = \frac{1}{i}\int_c^{c+2i} f(x)\sin\frac{n\pi x}{i}\,dx \tag{12.69b}$$

Again, we equate c to the value of x at the start of a cycle.

12.12 ORDINARY DIFFERENTIAL EQUATIONS

12.12.1 Definitions and Classification

A *differential equation* is one that contains derivatives or differentials. An *ordinary* differential equation is one that is devoid of partial derivatives. The *order* of a differential equation is the highest order of a derivative in the equation. In the subsequent material, we shall consider x and y to be the independent and dependent variables, respectively.

The *degree* of a term in a differential equation is found by adding the exponents of y and its derivatives. For example, the term $x(d^3y/dx^3)$ is of degree 1, and the term $y(dy/dx)^2$ is of degree 3. A *linear* differential equation is one in which each term that contains y or its derivatives is of degree 1.

The general form of a linear equation of order n is

$$a_0\frac{d^ny}{dx^n} + a_1\frac{d^{n-1}y}{dx^{n-1}} + a_2\frac{d^{n-2}y}{dx^{n-2}} + \cdots + a_{n-1}\frac{dy}{dx} + a_ny = f(x)$$

where the coefficients are functions of x or constants.

A functional relationship that contains no derivatives and that satisfies the given differential equation is termed a *solution* of that equation. In the absence of a set of simultaneous values of x and y, a differential equation generally has an infinite number of solutions. A functional relationship that encompasses the entire set of solutions is termed the *general solution*; any other solution is a *particular solution.* The general solution contains arbitrary constants that result from integration, and the number of these constants equals the order of the differential equation.

As an illustration, consider the second-order differential equation

$$\frac{d^2y}{dx^2} + 2\frac{dy}{dx} - 15y = 0$$

A particular solution of this equation is $y = e^{3x}$, as we can verify in this manner:

$$\frac{dy}{dx} = 3e^{3x} \qquad \frac{d^2y}{dx^2} = 9e^{3x}$$

Substituting in the given equation, we obtain

$$9e^{3x} + 6e^{3x} - 15e^{3x} = 0$$

and the solution is thus confirmed. However, another particular solution is $y = e^{-5x}$, and the general solution is $y = c_1e^{3x} + c_2e^{-5x}$, where c_1 and c_2 are the arbitrary constants.

The subject of differential equations is very broad, and the coverage in this text must perforce be restricted. We divide the equations we shall investigate into the following categories, with the indicated subdivisions:

Type 1: First-order equations: variables separable; integrable combinations; homogeneous; linear

Type 2: Special second-order equations

Type 3: Linear equations with constant coefficients: homogeneous; nonhomogeneous

12.12.2 Type 1 Equations with Variables Separable

If a first-order equation can be restructured so that all x terms are associated with dx and all y terms are associated with dy, we obtain an equation of the form

$$F(x)\,dx + G(y)\,dy = 0$$

and the equation can be solved by integration.

EXAMPLE 12.70 Solve the equation

$$8xe^{2y}\,dy + \frac{2x^2+5}{y}\,dx = 0$$

SOLUTION To separate the variables, we multiply both sides of the equation by y/x, giving

$$8ye^{2y}\,dy + \left(2x + \frac{5}{x}\right)dx = 0$$

We now integrate both sides of the equation. Integral (136) applies to the first term, and we obtain

$$4ye^{2y} - 2e^{2y} + x^2 + 5 \ln x = c$$

where c is an arbitrary constant.

12.12.3 Type 1 Equations with Integrable Combinations

Differential equations that contain certain combinations of terms associated with dx and dy can be solved readily by integration. For example, $x\,dy + y\,dx$ is the differential of xy, and $x\,dy - y\,dx$ is the numerator of the differential of y/x and of $-x/y$. Numerous other integrable combinations are possible, but they may not be readily discernible.

We shall now consider integrable combinations of powers of x and y. Let

$$z = [f(x)]^m [g(y)]^n$$

Differentiating, we obtain

$$dz = mf'(x)[f(x)]^{m-1}[g(y)]^n\,dx + ng'(y)[f(x)]^m[g(y)]^{n-1}\,dy$$

As an illustration, consider the equation

$$4x^3(y^2 - 3y + 9)^2\,dx + (2x^4 - 5)(2y^3 - 9y^2 + 27y - 27)\,dy = 0$$

A comparison of this equation with the standard one suggests that we try $f(x) = 2x^4 - 5$, $g(y) = y^2 - 3y + 9$, $z = c$, where c is a constant, $m = 1$, and $n = 2$. The tentative solution, therefore, is

$$(2x^4 - 5)(y^2 - 3y + 9)^2 = c$$

Upon differentiating and performing the resulting multiplication, we obtain the given equation, and our tentative solution is correct.

12.12.4 Type 1 Homogeneous Equations

Let $z = f(x, y)$, and let r denote an arbitrary constant. If the replacement of x and y with rx and ry, respectively, transforms z to $r^m z$, the function z is said to be *homogeneous of degree m*. If this replacement leaves z unchanged, $m = 0$, and z is homogeneous of degree zero.

Now consider that we are given the expression for dy/dx. If this expression is homogeneous of degree zero, the differential equation is described as homogeneous. Equations of this type can be solved by replacing x with uy or replacing y with vx. In this manner, the given equation is transformed to one that can be solved by separating the variables. We shall illustrate the procedure.

EXAMPLE 12.71 Solve the following equation and verify the result.

$$\frac{dy}{dx} = \frac{-5x^2 + 4xy + 9y^2}{2x^2 + 6xy} \qquad (g)$$

SOLUTION The given equation is homogeneous. Set $y = vx$. Making this substitution in Eq. (g) and then dividing by x^2, we obtain the following:

$$v + x\frac{dv}{dx} = \frac{-5x^2 + 4vx^2 + 9v^2x^2}{2x^2 + 6vx^2} = \frac{-5 + 4v + 9v^2}{2 + 6v}$$

$$x\frac{dv}{dx} = \frac{-5 + 4v + 9v^2}{2 + 6v} - v = \frac{-5 + 2v + 3v^2}{2 + 6v}$$

We now rearrange the terms in this manner:

$$\frac{(2 + 6v)\,dv}{-5 + 2v + 3v^2} = \frac{dx}{x}$$

In the fraction at the left, the numerator is the differential of the denominator. Integrating and introducing the constant of integration, we obtain

$$\ln(3v^2 + 2v - 5) = \ln x + \ln c = \ln cx$$

$$3v^2 + 2v - 5 = cx$$

By replacing v with y/x and then multiplying by x^2, we obtain as our solution

$$3y^2 + 2xy - 5x^2 = cx^3 \qquad (h)$$

The verification is as follows: We differentiate Eq. (h) and solve for dy/dx to obtain

$$\frac{dy}{dx} = \frac{3cx^2 + 10x - 2y}{6y + 2x} \qquad (i)$$

We now solve Eq. (h) for c and substitute the resulting expression in Eq. (i). This process culminates in Eq. (g), and the accuracy of Eq. (h) is thus confirmed.

12.12.5 Type 1 Linear Equations

An equation of this type can be expressed in the general form

$$\frac{dy}{dx} + Py = Q$$

where P and Q are either functions of x alone or constants. The left side of this equation can be transformed to the derivative of a product if both sides of the equation are multiplied by an *integrating factor* R that has the value

$$R = e^{\int P\,dx}$$

The differential equation is then solved by integrating both sides.

In solving a linear differential equation, it is often necessary to apply the identity $e^{\ln A} = A$. This relationship stems from the definition of $\ln A$: the power to which e must be raised to obtain A.

EXAMPLE 12.72 Solve the equation

$$\frac{dy}{dx} = \frac{2x^5 + 3y}{x}$$

SOLUTION To cast the equation in standard form, we replace the fraction at the right with $2x^4 + 3y/x$ and then transpose. The result is

$$\frac{dy}{dx} - \frac{3}{x}y = 2x^4 \qquad (j)$$

Then $P = -3/x$, and we have

$$\int P\,dx = -3 \ln x = \ln x^{-3}$$

$$R = e^{\int P\,dx} = x^{-3}$$

We now multiply both sides of Eq. (j) by R to obtain

$$x^{-3}\frac{dy}{dx} - 3x^{-4}y = 2x$$

The expression at the left is an exact integral, and integration yields

$$x^{-3}y = \int 2x\,dx = x^2 + c$$

We now multiply by x^3, and the solution of the differential equation becomes

$$y = x^5 + cx^3$$

The solution can be verified by a procedure similar to that followed in Example 12.71.

If a differential equation is linear with respect to x rather than y, the same method of solution can be followed by simply reversing the roles of the two variables.

12.12.6 Type 2 Equations

If a second-order differential equation is devoid of any term containing y, it becomes amenable to solution by reducing the order of the equation to one. The procedure consists of replacing dy/dx with p, solving the transformed equation for p, and then finding y by integration.

EXAMPLE 12.73 Solve the following equation:

$$5x^3\frac{d^2y}{dx^2} + 4x^3\left(\frac{dy}{dx}\right)^2 - 15x^2\frac{dy}{dx} = 0$$

SOLUTION When we set $p = dy/dx$, the given equation is transformed to

$$5x^3\frac{dp}{dx} + 4x^3p^2 - 15x^2p = 0$$

Multiplying by dx, rearranging terms, and dividing by p^2, we obtain

$$\frac{5x^3\,dp - 15px^2\,dx}{p^2} + 4x^3\,dx = 0$$

The fraction at the left is the differential of $-5x^3/p$. Integrating and then solving for p, we obtain the following:

$$-\frac{5x^3}{p} + x^4 + c_1 = 0$$

$$p = \frac{dy}{dx} = \frac{5x^3}{x^4 + c_1} = \frac{5}{4}\frac{4x^3}{x^4 + c_1}$$

A second integration yields

$$y = \frac{5}{4}\ln(x^4 + c_1) + c_2$$

If a differential equation is devoid of any term containing x, the same procedure can be followed. However, the derivative of p must be expressed in terms of y rather than x. Therefore, we set

$$\frac{d^2y}{dx^2} = \frac{dp}{dx} = \frac{dp}{dy}\frac{dy}{dx} = p\frac{dp}{dy}$$

12.12.7 Second-Order Homogeneous Type 3 Equations

With reference to the general form of a linear equation recorded in Art. 12.12.1, the equation is described as *homogeneous* if the coefficients are constants and $f(x) = 0$. A second-order linear homogeneous equation with constant coefficients has the form

$$\frac{d^2y}{dx^2} + a_1\frac{dy}{dx} + a_2 y = 0$$

Associated with this differential equation is an *auxiliary* (or *characteristic*) *equation* that is formed in this manner: Replace y with 1, replace dy/dx with m, and replace d^2y/dx^2 with m^2. Thus, the auxiliary equation is

$$m^2 + a_1 m + a_2 = 0$$

The solution of the differential equation inheres in the solution of the auxiliary equation, and three possibilities exist.

Case 1: The roots of the auxiliary equation are real and distinct. Let r and s denote the roots. The solution of the differential equation is

$$y = c_1 e^{rx} + c_2 e^{sx} \tag{12.70}$$

where c_1 and c_2 are the arbitrary constants.

Case 2: The roots of the auxiliary equation are real and identical. Let r denote the roots. The solution of the differential equation is

$$y = (c_1 x + c_2)e^{rx} \tag{12.71}$$

Case 3: The roots of the auxiliary equation are conjugate complex numbers. Let the roots be $\alpha + i\beta$ and $\alpha - i\beta$. The solution of the differential equation is

$$y = e^{\alpha x}(c_1 \cos \beta x + c_2 \sin \beta x) \tag{12.72}$$

We shall illustrate each case.

The differential equation

$$\frac{d^2y}{dx^2} + 2\frac{dy}{dx} - 15y = 0$$

has the auxiliary equation $m^2 + 2m - 15 = 0$, and the roots are 3 and -5. By Eq. (12.70),

$$y = c_1e^{3x} + c_2e^{-5x}$$

The differential equation

$$\frac{d^2y}{dx^2} - 8\frac{dy}{dx} + 16y = 0$$

has the auxiliary equation $m^2 - 8m + 16 = 0$, and the roots are both 4. By Eq. (12.71),

$$y = (c_1x + c_2)e^{4x}$$

The differential equation

$$\frac{d^2y}{dx^2} + 6\frac{dy}{dx} + 73y = 0$$

has the auxiliary equation $m^2 + 6m + 73 = 0$, and the roots are $-3 + 8i$ and $-3 - 8i$. By Eq. (12.72),

$$y = e^{-3x}(c_1 \cos 8x + c_2 \sin 8x)$$

12.12.8 Second-Order Nonhomogeneous Type 3 Equations

A linear differential equation with constant coefficients that is of the second order and nonhomogeneous has the form

$$\frac{d^2y}{dx^2} + a_1\frac{dy}{dx} + a_2y = f(x)$$

where a_1 and a_2 are constants. We shall present a method of solution that is known as the *variation of parameters.*

By replacing $f(x)$ with 0, we obtain the homogeneous equation that corresponds to the given one. This homogeneous equation has a solution of the form $y = c_1u + c_2v$, where c_1 and c_2 are arbitrary constants and u and v are functions of x. We now assume that the given equation has a solution of the form

$$y = Pu + Qv \qquad (k)$$

where P and Q are also functions of x. Our task is to evaluate P and Q.

Without loss of generality, we may set

$$P'u + Q'v = 0 \qquad (l)$$

where the primes denote first derivatives. By formulating the expressions for y' and y'' and substituting in the given differential equation, we obtain

$$P'u' + Q'v' = f(x) \qquad (m)$$

Equations (l) and (m) constitute a system of simultaneous equations. The solution of this system is as follows:

$$P' = \frac{v\,f(x)}{-uv' + u'v} \qquad Q' = \frac{u\,f(x)}{uv' - u'v} \tag{12.73}$$

EXAMPLE 12.74 Solve the following equation:

$$\frac{d^2y}{dx^2} - 3\frac{dy}{dx} - 10y = 14x - 7$$

SOLUTION The corresponding homogeneous equation has the auxiliary equation $m^2 - 3m - 10 = 0$. The roots of this equation are 5 and -2. Therefore, by Eq. (12.70),

$$u = e^{5x} \qquad v = e^{-2x}$$

$$u' = 5e^{5x} \qquad v' = -2e^{-2x} \qquad uv' = -2e^{3x} \qquad u'v = 5e^{3x}$$

Equation (12.73) yields

$$P' = (2x - 1)e^{-5x} \qquad Q' = -(2x - 1)e^{2x}$$

Integrating these expressions with Integrals (136) and (131), we obtain

$$P = \frac{e^{-5x}(-10x + 3)}{25} + c_1 \qquad Q = e^{2x}(-x + 1) + c_2$$

Equation (k) yields

$$y = -\frac{7}{5}x + c_1e^{5x} + c_2e^{-2x} + \frac{28}{25}$$

12.12.9 Type 3 Equations of Higher Order

The methods of solving second-order linear equations with constant coefficients can be readily extended to equations of the third or higher order. For example, consider the equation

$$\frac{d^3y}{dx^3} - 6\frac{d^2y}{dx^2} + 5\frac{dy}{dx} + 12y = 0$$

The auxiliary equation is $m^3 - 6m^2 + 5m + 12 = 0$, and the roots are 3, 4, and -1. An extension of Eq. (12.70) yields

$$y = c_1e^{3x} + c_2e^{4x} + c_3e^{-x}$$

Now consider the equation

$$\frac{d^4y}{dx^4} - 11\frac{d^3y}{dx^3} + 42\frac{d^2y}{dx^2} - 68\frac{dy}{dx} + 40y = 0$$

The auxiliary equation is $m^4 - 11m^3 + 42m^2 - 68m + 40 = 0$. Three roots are 2, and one root is 5. When Eq. (12.71) is extended and combined with Eq. (12.70), the result is

$$y = (c_1 x^2 + c_2 x + c_3)e^{2x} + c_4 e^{5x}$$

12.12.10 Solution of Differential Equation as an Infinite Series

Many differential equations that arise in practice do not fall into the categories we enumerated in Art. 12.12.1. Moreover, we often have a set of simultaneous values of x and y, and our interest is restricted to the particular solution that satisfies these values. In these situations, it is often possible to solve the differential equation by a method of successive approximations. The method is known as *Picard's method*, and it generates an expression for y in the form of an infinite series. The value of y corresponding to a given value of x can then be obtained to the required degree of precision by applying the methods discussed in Art. 12.11.8. If the series is positive, it is necessary to determine how many terms must be included in the evaluation.

Let y_r denote the rth approximation of y. This approximate value is substituted in the given equation, and the equation is then solved to obtain y_{r+1}, the process being continued until the pattern becomes discernible. We shall illustrate the procedure.

EXAMPLE 12.75 Solve the following equation:

$$\frac{d^2y}{dx^2} = -xy \qquad (n)$$

It is known that $y = 1$ when $x = 0$.

SOLUTION We set $y_0 = 1$, and Eq. (n) becomes $d^2y/dx^2 = -x$. Integrating this equation twice and applying the given set of values, we obtain

$$y_1 = 1 - \frac{x^3}{3!}$$

Equation (n) now becomes $d^2y/dx^2 = -x + x^4/3!$. Integrating this equation twice and applying the given set of values, we obtain

$$y_2 = 1 - \frac{x^3}{3!} + \frac{4x^6}{6!}$$

Continuing this process, we obtain the following as our solution:

$$y = 1 - \frac{x^3}{3!} + \frac{4x^6}{6!} - \frac{4 \cdot 7x^9}{9!} + \frac{4 \cdot 7 \cdot 10x^{12}}{12!} - \cdots$$
$$+ (-1)^{n-1} \frac{4 \cdot 7 \cdot 10 \cdots (3n-5)x^{3n-3}}{(3n-3)!} + \cdots$$

The validity of this result is readily tested. By obtaining the second derivative, we find that Eq. (n) is satisfied.

EXAMPLE 12.76 Solve the following equation:

$$\frac{dy}{dx} = x + x^2 y \tag{o}$$

It is known that $y = 2$ when $x = 0$.

SOLUTION We set $y_0 = 2$, and Eq. (o) becomes $dy/dx = x + 2x^2$. Then

$$y_1 = 2 + \frac{x^2}{2} + \frac{2x^3}{3}$$

Continuing this process, we obtain the solution

$$y = 2\left(1 + \frac{x^3}{3} + \frac{x^6}{3 \cdot 6} + \frac{x^9}{3 \cdot 6 \cdot 9} + \frac{x^{12}}{3 \cdot 6 \cdot 9 \cdot 12} + \cdots\right)$$

$$+ \frac{x^2}{2} + \frac{x^5}{2 \cdot 5} + \frac{x^8}{2 \cdot 5 \cdot 8} + \frac{x^{11}}{2 \cdot 5 \cdot 8 \cdot 11} + \cdots$$

The denominator of the nth term of the expression in parentheses can be written in the more compact form $3^{n-1}[(n-1)!]$.

12.12.11 Solution of Initial-Value Problems by Runge-Kutta Methods

In the general type of first-order differential equation, the expression for dy/dx contains both x and y. Therefore, replacing dy/dx with k, we have $k = f(x, y)$.

Assume that we have the expression for k and a set of simultaneous values of x and y, which we denote by x_n and y_n, respectively. These values are termed the *initial values*. Now assume that x increases to x_{n+1} and we wish to find the corresponding value of y, which we denote by y_{n+1}. The *Runge-Kutta methods* offer an approximate solution. Under these methods, y is expanded in a Taylor series and all terms beyond the mth are discarded. The result obtained is termed an mth-order solution. However, for $m > 1$, multiple solutions exist. We shall present the fourth-order solution as based on Runge's coefficients.

The procedure is as follows: Let $h = x_{n+1} - x_n$. Applying the functional relationship for k, let

$$k_1 = f(x_n, y_n)$$

$$x_1 = x_n + \frac{h}{2} \qquad y_1 = y_n + \frac{h}{2} k_1 \qquad k_2 = f(x_1, y_1)$$

$$y_2 = y_n + \frac{h}{2} k_2 \qquad k_3 = f(x_1, y_2)$$

$$y_3 = y_n + hk_3 \qquad k_4 = f(x_{n+1}, y_3)$$

We now set

$$y_{n+1} = y_n + \frac{h}{6}(k_1 + 2k_2 + 2k_3 + k_4)$$

EXAMPLE 12.77 It is known that $y = -3.72160$ when $\theta = 0.20$ rad, and $dy/d\theta = 3 \sin \theta + 5 \cos \theta + y$. Find the value of y corresponding to $\theta = 0.50$ rad by applying the Runge-Kutta method. Then compute the precise value of y by applying the solution to the differential equation, which is $y = \sin \theta - 4 \cos \theta$.

SOLUTION

$$\theta_n = 0.20 \text{ rad} \qquad y_n = -3.72160 \qquad \theta_{n+1} = 0.50 \text{ rad}$$

$$h = 0.50 - 0.20 = 0.30 \text{ rad}$$

$$k_1 = 3 \sin 0.20 + 5 \cos 0.20 - 3.72160 = 1.77474$$

$$\theta_1 = 0.35 \text{ rad} \qquad y_1 = -3.72160 + 0.15(1.77474) = -3.45539$$

$$k_2 = 3 \sin 0.35 + 5 \cos 0.35 - 3.45539 = 2.27017$$

$$y_2 = -3.72160 + 0.15(2.27017) = -3.38107$$

$$k_3 = 3 \sin 0.35 + 5 \cos 0.35 - 3.38107 = 2.34449$$

$$y_3 = -3.72160 + 0.30(2.34449) = -3.01825$$

$$k_4 = 3 \sin 0.50 + 5 \cos 0.50 - 3.01825 = 2.80794$$

$$k_1 + 2k_2 + 2k_3 + k_4 = 13.81200$$

$$y_{n+1} = -3.72160 + \frac{0.30}{6}(13.81200) = -3.03100$$

The precise value is $y_{n+1} = \sin 0.50 - 4 \cos 0.50 = -3.03090$.

In the special case where dy/dx is a function of x alone, the fourth-order Runge-Kutta solution coincides with Simpson's rule for evaluating an integral, which is presented in Art. 12.4.9.

The Runge-Kutta methods are particularly useful in obtaining numerical results by computer. Their range can be extended to second-order differential equations if such equations can be transformed to a series of simultaneous first-order equations. For example, an equation that expresses the acceleration of a point can be transformed to one that expresses the rate of change of its velocity.

12.13 SPECIAL FUNCTIONS

12.13.1 The Gamma Function

Consider the function

$$y = x^{n-1} e^{-x} \tag{12.74}$$

where n is a constant and x is restricted to nonnegative values. Then

$$\frac{dy}{dx} = x^{n-2} e^{-x} (n - 1 - x)$$

and it follows that y is maximum when $x = n - 1$. By successive application of l'Hôpital's rule in Art. 12.3.4, we find that y approaches 0 as x becomes infinite. Figure 12.51 is the graph of Eq. (12.74) for a positive value of n.

Now assume that we wish to evaluate the area A under this graph and the x axis. Since there is a unique curve for each value of n, this area is a function of n. Applying Integral (137), we find that the area has the following value:

$$\int_0^\infty x^{n-1}e^{-x}\,dx = (n-1)\int_0^\infty x^{n-2}e^{-x}\,dx$$

The integral at the left is termed the *gamma function* of n, and it is denoted by $\Gamma(n)$. In the foregoing equation, the integral at the right is seen to be $\Gamma(n-1)$, and we thus have the recursive relationship

$$\Gamma(n) = (n-1)\Gamma(n-1) \tag{12.75}$$

By applying this relationship repeatedly, we obtain the following:

$$\Gamma(n) = (n-1)(n-2)(n-3)\cdots(n-k+1)(n-k)\Gamma(n-k) \tag{12.76}$$

Expressed more compactly,

$$\Gamma(n) = \frac{(n-1)!}{(n-k-1)!}\Gamma(n-k) \tag{12.76a}$$

By direct substitution, we find that $\Gamma(1) = 1$. Therefore, if we now assign a positive integral value to n, set $k = n - 1$, and apply Eq. (12.76*a*), we obtain

$$\Gamma(n) = (n-1)! \tag{12.77}$$

Consider that we start with $n = 1$ and then reduce n successively by 1. Equation (12.75) reveals that $\Gamma(n)$ is infinite in absolute value if n is 0 or a negative integer.

If we consider that Eq. (12.77) is valid for all values of n except 0 and negative integers, it follows that the gamma function is a generalization of factorial numbers. It transforms factorial numbers from a set of discrete values to a continuum. Values of $\Gamma(n)$ for $1 < n \leq 2$ are available in a table of gamma functions. A particularly important value is $\Gamma(1/2) = \sqrt{\pi}$, which is obtained by performing the integration.

EXAMPLE 12.78 Evaluate $\Gamma(7/2)$.

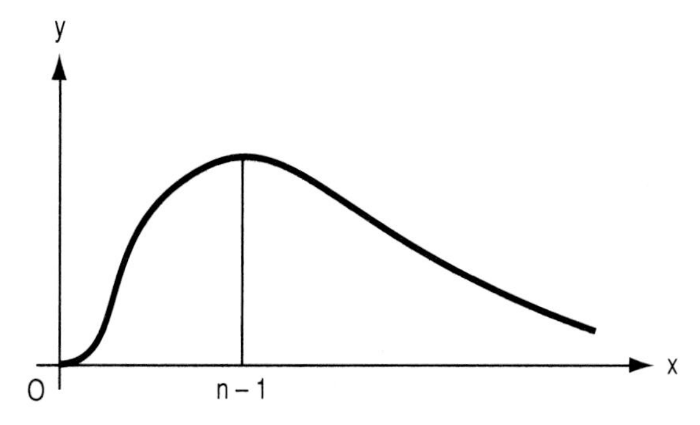

FIGURE 12.51 Graph of $y = x^{n-1}e^{-x}$.

SOLUTION We apply Eq. (12.76) with $n = 7/2$ and $k = 3$. Then

$$\Gamma\left(\frac{7}{2}\right) = \frac{5}{2}\frac{3}{2}\frac{1}{2}\Gamma\left(\frac{1}{2}\right) = \frac{15}{8}\sqrt{\pi}$$

EXAMPLE 12.79 Evaluate $\Gamma(-5/2)$.

SOLUTION Equation (12.76) yields

$$\Gamma\left(\frac{1}{2}\right) = \left(-\frac{1}{2}\right)\left(-\frac{3}{2}\right)\left(-\frac{5}{2}\right)\Gamma\left(-\frac{5}{2}\right) = \sqrt{\pi} \qquad \Gamma\left(-\frac{5}{2}\right) = -\frac{8}{15}\sqrt{\pi}$$

If we replace x with some other variable in the expression that defines $\Gamma(n)$, many useful results ensue. As an illustration, set $x = u^m$, where m is a constant. The result is

$$\int_0^\infty u^{mn-1} e^{-(u^m)}\, du = \frac{1}{m}\Gamma(n)$$

Moreover, if we now set $n = 1/m$, this equation reduces to

$$\int_0^\infty e^{-(u^m)}\, du = \Gamma\left(1 + \frac{1}{m}\right)$$

The gamma function enables us to evaluate certain integrals, and it is a tool in the formulation of other functions. The probability-density function defined by Eq. (9.35) resembles the integrand in the definition of the gamma function.

12.13.2 The Beta Function

Let m and n denote two positive numbers. The *beta function* of m and n, which is denoted by $B(m, n)$, is defined in this manner:

$$B(m, n) = \int_0^1 x^{m-1}(1 - x)^{n-1}\, dx$$

It can be demonstrated that

$$B(m, n) = \frac{\Gamma(m)\Gamma(n)}{\Gamma(m + n)}$$

and the beta function is thus related to the gamma function.

The beta function is also highly useful in enabling us to evaluate various integrals. For example, if we set $x = u/(1 + u)$, we obtain

$$B(m, n) = \int_0^\infty \frac{u^{m-1}}{(1 + u)^{m+n}}\, du$$

12.13.3 Bessel Functions

Consider the differential equation

$$x^2\frac{d^2y}{dx^2} + x\frac{dy}{dx} + (x^2 - n^2)y = 0 \qquad (12.78)$$

where n is restricted to nonnegative values. Although the differential equation is of order 2, it is also referred to as a Bessel equation of order n. The *Bessel function* of order n is a particular solution of this equation. Since it is a function of both x and n, y is replaced by the notation $J_n(x)$.

By expanding y in the form of a power series, we obtain the following as the simplest particular solution:

$$J_n(x) = \frac{(x/2)^n}{0!\Gamma(n+1)} - \frac{(x/2)^{n+2}}{1!\Gamma(n+2)} + \frac{(x/2)^{n+4}}{2!\Gamma(n+3)} - \frac{(x/2)^{n+6}}{3!\Gamma(n+4)} + \cdots \qquad (12.79)$$

In many applications, n is a positive integer or 0. Equation (12.79) then reduces to the following:

$$J_n(x) = \frac{(x/2)^n}{0!n!} - \frac{(x/2)^{n+2}}{1!(n+1)!} + \frac{(x/2)^{n+4}}{2!(n+2)!} - \frac{(x/2)^{n+6}}{3!(n+3)!} + \cdots \qquad (12.79a)$$

In Art. 12.11.6, we defined even and odd functions. Equation (12.79*a*) reveals that $J_n(x)$ is even or odd according to whether n is even or odd, respectively. For $n = 0$, $J_n(0) = 1$; for all other values of n, $J_n(0) = 0$. If we replace $x/2$ with u, the series for $J_0(x)$ and $J_1(x)$ substantially resemble the series for $\cos u$ and $\sin u$, respectively.

Figure 12.52 is the graph of the indicated Bessel functions. Values of $J_0(x)$ and $J_1(x)$ are available in tables, but other values can be calculated readily when n is an integer.

EXAMPLE 12.80 Evaluate $J_2(3.5)$ to four decimal places.

SOLUTION Taking the first seven terms of the series, we have

$$J_2(3.5) = \frac{1.75^2}{0!2!} - \frac{1.75^4}{1!3!} + \frac{1.75^6}{2!4!} - \frac{1.75^8}{3!5!} + \frac{1.75^{10}}{4!6!} - \frac{1.75^{12}}{5!7!} + \frac{1.75^{14}}{6!8!} = 0.4586$$

There are numerous relationships that link Bessel functions of consecutive order. Among these are the following:

$$\frac{d}{dx}[x^n J_n(x)] = x^n J_{n-1}(x) \qquad \frac{d}{dx}[x^{-n} J_n(x)] = -x^{-n} J_{n+1}(x)$$

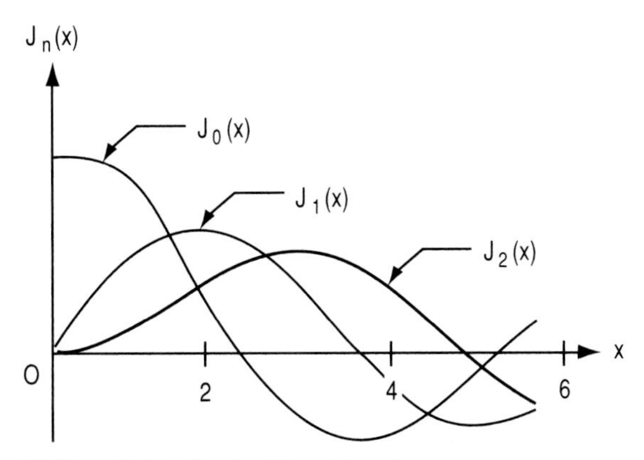

FIGURE 12.52 Graph of Bessel functions.

For example, if we set $n = 1$ in the first equation and $n = 0$ in the second equation, we obtain

$$\frac{d}{dx}[xJ_1(x)] = xJ_0(x) \qquad \frac{d}{dx}[J_0(x)] = -J_1(x)$$

The *three-term recurrence formula* is

$$J_{n-1}(x) - \frac{2n}{x}J_n(x) + J_{n+1}(x) = 0$$

If n has a nonintegral value, we may replace n with $-n$ in Eq. (12.79) to obtain a definition of $J_{-n}(x)$. Since the replacement of n with $-n$ in Eq. (12.78) does not affect the equation, it follows that $J_{-n}(x)$ is also a solution of the Bessel equation. Thus, if n is nonintegral, we may consider the general solution of the Bessel equation to be $AJ_n(x) + BJ_{-n}(x)$, where A and B are arbitrary constants.

The functions $J_{1/2}(x)$ and $J_{-1/2}(x)$ are highly interesting because they reduce to the following finite forms:

$$J_{1/2}(x) = \sqrt{\frac{2}{\pi x}}\sin x \qquad J_{-1/2}(x) = \sqrt{\frac{2}{\pi x}}\cos x$$

It follows that

$$[J_{1/2}(x)]^2 + [J_{-1/2}(x)]^2 = \frac{2}{\pi x}$$

12.13.4 Elliptic Integrals

Let k denote a constant such that $0 < k < 1$. The elliptic integrals of the first and second kind and their designations are as follows, respectively:

$$F(k, \phi) = \int_0^\phi \frac{d\phi}{\sqrt{1 - k^2 \sin^2 \phi}} \tag{12.80a}$$

$$E(k, \phi) = \int_0^\phi \sqrt{1 - k^2 \sin^2 \phi}\, d\phi \tag{12.80b}$$

The quantities k and ϕ are called the *modulus* and *amplitude*, respectively, of the integral. Now let n denote a constant that can assume any value except k. The elliptic integral of the third kind is as follows:

$$\Pi(n, k, \phi) = \int_0^\phi \frac{d\phi}{(1 + n^2 \sin^2 \phi)\sqrt{1 - k^2 \sin^2 \phi}} \tag{12.80c}$$

Let

$$y = \sqrt{1 - k^2 \sin^2 \phi}$$

Then y is a periodic function of ϕ with a period of π. Moreover, the graph of y versus ϕ is symmetric about a line parallel to the y axis at $\phi = \pi/2$. Therefore, if the upper limit of integration in Eqs. (12.80) is $\pi/2$, the integral is termed the *complete* elliptic integral. The designations for complete elliptic integrals are as follows: $K(k)$ for the first kind, $E(k)$ for the second kind, and $\Pi(n, k)$ for the third kind.

The form of an elliptic integral can be modified by changing the variable to $x = \sin \phi$. Then $\phi = \sin^{-1} x$, and

$$d\phi = \frac{dx}{\sqrt{1-x^2}}$$

Making the substitutions in Eq. (12.80) and adding the subscript 1 in the designation, we obtain the following:

$$F_1(k, x) = \int_0^x \frac{dx}{\sqrt{(1-x^2)(1-k^2x^2)}} \tag{12.81a}$$

$$E_1(k, x) = \int_0^x \sqrt{\frac{1-k^2x^2}{1-x^2}}\, dx \tag{12.81b}$$

$$\Pi_1(n, k, x) = \int_0^x \frac{dx}{(1+n^2x^2)\sqrt{(1-x^2)(1-k^2x^2)}} \tag{12.81c}$$

The elliptic integrals cannot be expressed in finite form; they can be expressed only as infinite series. We shall illustrate the procedure for the integral of the first kind. With reference to Eq. (12.80*a*), we first write the following:

$$\frac{1}{\sqrt{1-k^2 \sin^2 \phi}} = (1-k^2 \sin^2 \phi)^{-1/2}$$

We now apply Eq. (1.35) with $x = -k^2 \sin^2 \phi$. Since $|x| < 1$, this equation is valid, and the binomial expansion is the following infinite series:

$$F(k, \phi) = \left[1 + \frac{1}{2}k^2 \sin^2 \phi + \frac{1 \cdot 3}{2^2(2!)} k^4 \sin^4 \phi \right.$$
$$\left. + \frac{1 \cdot 3 \cdot 5}{2^3(3!)} k^6 \sin^6 \phi + \frac{1 \cdot 3 \cdot 5 \cdot 7}{2^4(4!)} k^8 \sin^8 \phi + \cdots \right] d\phi$$

We now integrate each term by applying Integral (152), which is recursive. Tables that give the numerical values of elliptic integrals of the first and second kind are available.

It is a simple matter to evaluate the complete elliptic integral. When the limits of integration are 0 and $\pi/2$, the first term in the expression for Integral (152) vanishes because $\sin 0 = 0$ and $\cos(\pi/2) = 0$. Applying this integral repeatedly with values of n that increase successively by 2, we obtain the following:

$$\int_0^{\pi/2} \sin^n \phi \, d\phi = \frac{\pi}{2}\left[\frac{1 \cdot 3 \cdot 5 \cdots (n-1)}{2 \cdot 4 \cdot 6 \cdots n}\right]$$

The result is

$$K(k) = \frac{\pi}{2}\left[1 + \left(\frac{1}{2}\right)^2 k^2 + \left(\frac{1 \cdot 3}{2 \cdot 4}\right)^2 k^4 + \left(\frac{1 \cdot 3 \cdot 5}{2 \cdot 4 \cdot 6}\right)^2 k^6 + \cdots\right]$$

We shall now cite two investigations that give rise to elliptic integrals. The first is an analysis of the oscillatory motion of a simple pendulum. In a simple pendulum, the rod is assumed to be weightless and the bob is considered to be a particle (i.e., a body of infinitesimally small volume). Thus, the mass of the pendulum is concentrated at a point. In Fig. 12.53, we place the origin at the

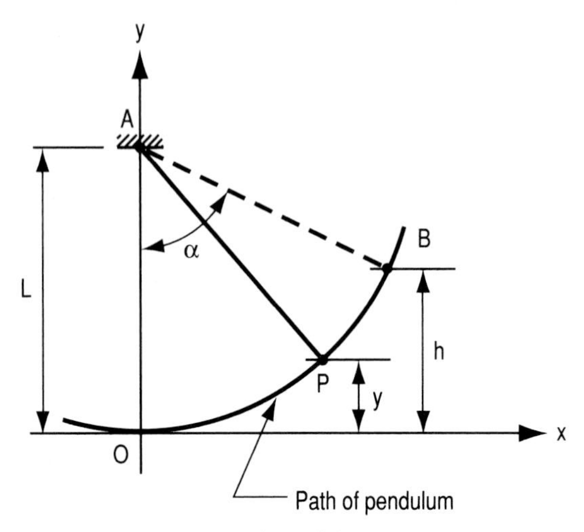

FIGURE 12.53 The simple pendulum.

lowest point of the bob. The point at which the pendulum is supported is A, the highest point to which the pendulum rises is B, and the instantaneous position of the bob is P. The length of the pendulum is L. Let

$$k = \sin\frac{\alpha}{2} \qquad \sin\phi = \sqrt{\frac{y}{h}}$$

The time t required for the bob to traverse the arc OP (in either direction) is

$$t = \sqrt{\frac{L}{g}} \int_0^{\phi} \frac{d\phi}{\sqrt{1 - k^2 \sin^2\phi}}$$

where g is the gravitational acceleration. This expression for t contains the elliptic integral of the first kind.

The second investigation pertains to the length of an elliptic arc. (The ellipse is analyzed in Art. 3.2.4.) In Fig. 3.18, let ϕ denote the angle between OP' and the positive side of the y axis; as before, let e denote the eccentricity of the ellipse. The length of the arc BP is

$$\widehat{BP} = a \int_0^{\phi} \sqrt{1 - e^2 \sin^2\phi}\, d\phi$$

This expression for the arc length contains the elliptic integral of the second kind, with k replaced by e.

Where an integrand can be reduced to an expression of the types appearing in Eq. (12.81), the corresponding integral can be transformed to an elliptic integral or combination of elliptic integrals. We shall present several such cases, but we shall apply the variable ϕ rather than x.

First, by subtracting the elliptic integral of the second kind from that of the first kind and simplifying the resulting expression, we find that

$$\int_0^{\phi} \frac{\sin^2\phi\, d\phi}{\sqrt{1 - k^2 \sin^2\phi}} = \frac{F(k, \phi) - E(k, \phi)}{k^2} \tag{12.82}$$

Now consider that an integral has the same form as that in Eq. (12.80*a*) but $k > 1$. We introduce two new variables, θ and m, where $\sin\theta = k\sin\phi$ and $m = 1/k$. Then

$$d\phi = m\sqrt{\frac{1-\sin^2\theta}{1-m^2\sin^2\theta}}\,d\theta$$

Substitution in the original integral yields

$$\int_0^{\phi}\frac{d\phi}{\sqrt{1-k^2\sin^2\phi}} = m\int_0^{\theta}\frac{d\theta}{\sqrt{1-m^2\sin^2\theta}}$$

Since $m < 1$, the integral at the right is elliptic, and the original integral is thus transformed to $mF(m, \theta)$. However, this transformation to an elliptic integral is possible only if $\sin\phi \le 1/k$. The upper limit of ϕ must be converted to the corresponding value of θ.

Now consider that an integral has the same form as that in Eq. (12.80*b*) but $k > 1$. We again set $\sin\theta = k\sin\phi$ and $m = 1/k$. Making the substitutions and applying Eq. (12.82), we arrive at the following transformation:

$$\int_0^{\phi}\sqrt{1-k^2\sin^2\phi}\,d\phi = \left(m - \frac{1}{m}\right)F(m, \theta) + \frac{E(m, \theta)}{m}$$

Again, the transformation is possible only if $\sin\phi \le 1/k$.

Finally, consider the integral

$$\int_0^{\phi}\frac{d\phi}{\sqrt{a-\cos\phi}}$$

where $a > 1$. By Eq. (2.15),

$$\cos\phi = 2\cos^2\left(\frac{\phi}{2}\right) - 1$$

Now let

$$\theta = \frac{\pi}{2} - \frac{\phi}{2}$$

Then $$d\phi = -2\,d\theta \qquad \cos\phi = 2\sin^2\theta - 1$$

With the limits of integration omitted, the orginal integral is transformed to the following:

$$\int\frac{-2\,d\theta}{\sqrt{a+1-2\sin^2\theta}} = -\frac{2}{\sqrt{a+1}}\int\frac{d\theta}{\sqrt{1-\left(\dfrac{2}{a+1}\right)\sin^2\theta}}$$

Let $k = \sqrt{2/(a+1)}$. Since $k < 1$, the last integral is elliptic if the lower limit of integration is 0. Now assume that the limits of integration are $\phi = 0$ and $\phi = \pi/2$. The corresponding values of θ are $\pi/2$ and $\pi/4$, respectively. Let D denote the differential in the last integral. The original integral is thus transformed to the following:

$$-k\sqrt{2}\int_{\pi/2}^{\pi/4} D = k\sqrt{2}\int_{\pi/4}^{\pi/2} D = k\sqrt{2}\left(\int_0^{\pi/2} D - \int_0^{\pi/4} D\right)$$

$$= k\sqrt{2}\left[K(k) - F\left(k, \frac{\pi}{4}\right)\right]$$

12.14 VECTOR ANALYSIS

12.14.1 Definitions and Notation

As stated in Art. 12.5.1, a quantity that has magnitude only is termed a *scalar*, and one that has both magnitude and direction is termed a *vector*. Thus, the temperature of a body is a scalar; a force, a velocity, and an acceleration are vectors. The direction of a vector has two characteristics: *inclination* and *sense* (e.g., southwestward). Vectors that are parallel to each other have the same inclination, but their senses may be alike or opposite to each other.

A vector is represented by an arrow. The length of the arrow equals the magnitude of the vector as based on a convenient scale. The arrow is given the same inclination as the vector, and its arrowhead shows the sense of the vector. The term *vector* as used hereafter will refer to the arrow that represents the vector. Thus, we shall speak of the length of a vector rather than its magnitude.

If we are concerned solely with the length and direction of a vector and not its precise location in space, the vector is described as *free*, and we shall deal exclusively with free vectors. Thus, a vector can be translated (i.e., displaced while remaining parallel to its original position). It follows that two vectors are equal to each other if their lengths and directions are identical, even though they are at different locations. In our discussion, we shall assume for convenience that the vectors under consideration have a common initial point. If such is not the case, we can translate vectors to achieve this assumed condition. This translation will not affect the relationships we shall develop.

When a letter denotes a vector rather than a scalar, the letter is shown in roman boldface type. Therefore, in the subsequent material, a letter that appears in italics denotes a scalar, and a letter that appears in roman boldface denotes a vector. If A is the initial point and B the terminal point of a vector, the vector is often denoted by $\overrightarrow{AB}$, but we shall use the notation $(\mathbf{AB})$. The length of a vector $\mathbf{v}$ is denoted by $|\mathbf{v}|$ or v; we shall have occasion to use both designations. A vector having zero length is termed a *null* or *zero vector*, and it is denoted by $\mathbf{0}$.

The *negative* of vector $\mathbf{v}$ is a vector that has the same length and inclination as $\mathbf{v}$ but is oppositely sensed; it is denoted by $-\mathbf{v}$. Thus, if $\mathbf{v}$ is northwestward, $-\mathbf{v}$ is southeastward. It follows that $-(\mathbf{AB}) = (\mathbf{BA})$. Vectors that are perpendicular to one another are called *orthogonal vectors*.

12.14.2 Product of Scalar and Vector

Consider that a vector $\mathbf{v}$ is multiplied by a scalar s. The product is denoted by $s\mathbf{v}$. Let $\mathbf{w} = s\mathbf{v}$. Vector $\mathbf{w}$ has these characteristics: It is parallel to $\mathbf{v}$; its length is s times that of $\mathbf{v}$; its sense is the same as that of $\mathbf{v}$ if s is positive and opposite to that of $\mathbf{v}$ if s is negative.

In some instances, we shall denote the product of a scalar s and a vector $\mathbf{v}$ as $\mathbf{v}s$ for the sake of clarity. For example, if $s = \cos\theta$, it is preferable to write $\mathbf{v}\cos\theta$.

12.14.3 Unit Vectors

A *unit vector* is a vector that has a length of 1 unit. The unit vectors that have the same direction as the x, y, and z axes are denoted by **i**, **j**, and **k**, respectively, and they are referred to as the *basic unit vectors*. In vector analysis, the coordinate axes are assigned the positions shown in Fig. 12.54. This diagram also shows the basic unit vectors.

Assume that a vector $\mathbf{v}$ is parallel to the x axis and has a length a. Then $\mathbf{v} = a\mathbf{i}$. Thus, the function of a unit vector is to specify the inclination of a given vector.

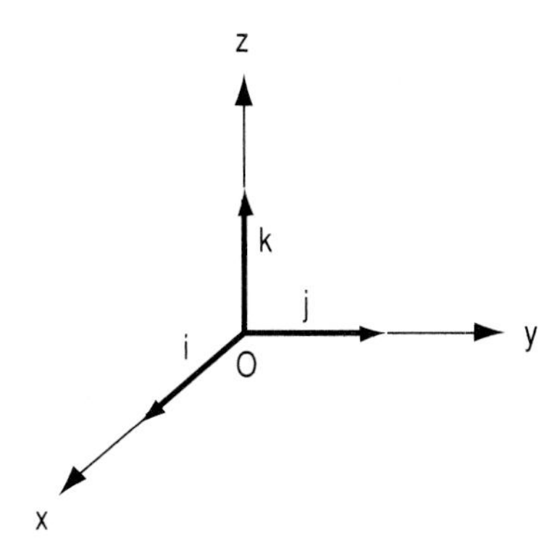

FIGURE 12.54 Position of coordinate axes and location of basic unit vectors.

12.14.4 Addition and Subtraction of Vectors

Some texts represent the addition and subtraction of vectors by the symbols $\dotplus$ and $\rightarrow$, respectively. However, we shall use simply plus and minus signs for this purpose. Let $\mathbf{v}_3 = \mathbf{v}_1 + \mathbf{v}_2$. Vector $\mathbf{v}_3$ is found graphically by the following procedure: In Fig. 12.55*a*, place $\mathbf{v}_1$ and $\mathbf{v}_2$ in the positions shown, with the terminal point of $\mathbf{v}_1$ becoming the initial point of $\mathbf{v}_2$. Now draw a vector to close the chain, the initial point of the closing vector being the initial point of $\mathbf{v}_1$ and the terminal point of the closing vector being the terminal point of $\mathbf{v}_2$. The closing vector is $\mathbf{v}_3$.

The foregoing procedure can be extended to the addition of numerous vectors. Thus, let $\mathbf{v}_7 = \mathbf{v}_4 + \mathbf{v}_5 + \mathbf{v}_6$. The addition is performed graphically in Fig. 12.55*b*.

Now let $\mathbf{v}_8 = \mathbf{v}_1 - \mathbf{v}_2$, where $\mathbf{v}_1$ and $\mathbf{v}_2$ are the vectors shown in Fig. 12.55*a*. Then $\mathbf{v}_8 = \mathbf{v}_1 + (-\mathbf{v}_2)$, and $\mathbf{v}_8$ is obtained in the manner shown in Fig. 12.55*c*.

The following laws pertaining to vector addition are immediately evident:

$$\mathbf{v}_1 + \mathbf{v}_2 = \mathbf{v}_2 + \mathbf{v}_1$$

$$\mathbf{v}_1 + (\mathbf{v}_2 + \mathbf{v}_3) = (\mathbf{v}_1 + \mathbf{v}_2) + \mathbf{v}_3$$

$$a(\mathbf{v}_1 + \mathbf{v}_2) = a\mathbf{v}_1 + a\mathbf{v}_2$$

$$(a + b)\mathbf{v} = a\mathbf{v} + b\mathbf{v}$$

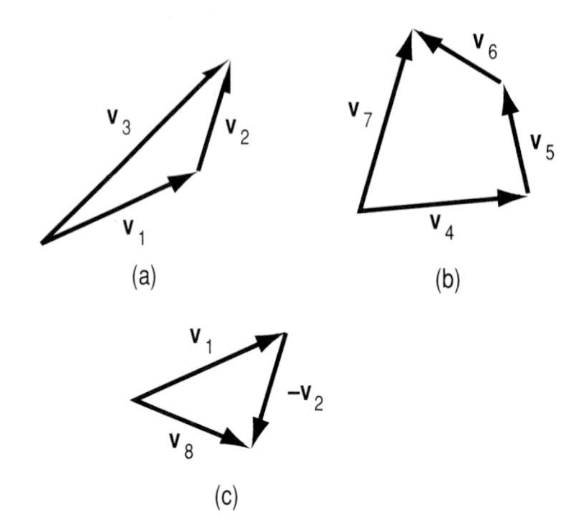

FIGURE 12.55 (*a*) Addition of two vectors; (*b*) addition of three vectors; (*c*) subtraction of vectors.

The first law states that vector addition is *commutative*, the second law states that it is *associative*, and the third and fourth laws state that the multiplication of scalars and vectors is *distributive*. By combining the third and fourth laws, we obtain

$$(a+b)(\mathbf{v}_1+\mathbf{v}_2) = a\mathbf{v}_1 + a\mathbf{v}_2 + b\mathbf{v}_1 + b\mathbf{v}_2$$

We also have the following: If $\mathbf{v}_1 + \mathbf{v}_2 = \mathbf{v}_3$, then $\mathbf{v}_1 = \mathbf{v}_3 - \mathbf{v}_2$, as can readily be seen in Fig. 12.55*a*. Thus, in a vector equation, quantities can be transposed in the same manner as in an algebraic equation.

With reference to Fig. 12.55*a*, $\mathbf{v}_1$ and $\mathbf{v}_2$ are the *components* of $\mathbf{v}_3$ in the indicated directions, and the process of adding vectors is called the *composition* of vectors. Conversely, it is possible to replace a given vector $\mathbf{v}_3$ with two vectors $\mathbf{v}_1$ and $\mathbf{v}_2$ such that $\mathbf{v}_1 + \mathbf{v}_2 = \mathbf{v}_3$ and $\mathbf{v}_1$ and $\mathbf{v}_2$ have specified inclinations. The latter process is called the *resolution* of the given vector.

12.14.5 Basic Components of Vectors

Let v_1, v_2, and v_3 denote the lengths of vector $\mathbf{v}$ as projected onto the x, y, and z axes, respectively, with proper regard for algebraic signs. Then

$$\mathbf{v} = v_1\mathbf{i} + v_2\mathbf{j} + v_3\mathbf{k}$$

Vectors $v_1\mathbf{i}$, $v_2\mathbf{j}$, and $v_3\mathbf{k}$ are the *basic components of* $\mathbf{v}$. The length of $\mathbf{v}$ is

$$v = \sqrt{v_1^2 + v_2^2 + v_3^2} \tag{12.83}$$

When a vector is expressed as the sum of its basic components, the vector is completely defined because the expression allows us to calculate the length of the vector and it discloses the direction of the vector. For example, assume that $\mathbf{v} = 5\mathbf{i} - 3\mathbf{j} + 2\mathbf{k}$. If we move along the vector so that x increases by 5 units, y decreases by 3 units and z increases by 2 units.

In accordance with the definitions in Art. 3.3.5, v_1, v_2, and v_3 are direction components of **v**, and v_1/v, v_2/v, and v_3/v are the direction cosines of **v** with reference to the x, y, and z axes, respectively. Two vectors are equal to each other if and only if the basic components of one are equal to the corresponding components of the other.

Two vectors can be added by adding their corresponding basic components. Thus, let

$$\mathbf{A} = a_1\mathbf{i} + a_2\mathbf{j} + a_3\mathbf{k} \qquad \mathbf{B} = b_1\mathbf{i} + b_2\mathbf{j} + b_3\mathbf{k}$$

Then

$$\mathbf{A} + \mathbf{B} = (a_1 + b_1)\mathbf{i} + (a_2 + b_2)\mathbf{j} + (a_3 + b_3)\mathbf{k}$$

Similarly,

$$\mathbf{A} - \mathbf{B} = (a_1 - b_1)\mathbf{i} + (a_2 - b_2)\mathbf{j} + (a_3 - b_3)\mathbf{k}$$

EXAMPLE 12.81 Vector **v** starts at $P_1(2, -5, 9)$ and terminates at $P_2(13, 7, 3)$. Express **v** in terms of its basic components.

SOLUTION Let $a_1\mathbf{i}$, $a_2\mathbf{j}$, and $a_3\mathbf{k}$ denote the components.

$$a_1 = 13 - 2 = 11 \qquad a_2 = 7 - (-5) = 12 \qquad a_3 = 3 - 9 = -6$$

$$\mathbf{v} = 11\mathbf{i} + 12\mathbf{j} - 6\mathbf{k}$$

EXAMPLE 12.82 Vector **v** is tangent to the parabola $y = -(3/32)x^2 + 7$ at the point where $x = 4$. If the length of **v** is 20 and its x component is positive, express **v** in terms of its basic components.

SOLUTION The slope of **v** is $dy/dx = -2(3/32)x = -(3/16)4 = -3/4$. The slope of **v** is also the ratio of its y component to its x component. Then $a_2 = -(3/4)a_1$. We thus have a 3-4-5 right triangle, and $a_1 = (4/5)20 = 16$, $a_2 = -(3/5)20 = -12$. Then $\mathbf{v} = 16\mathbf{i} - 12\mathbf{j}$.

12.14.6 Scalar Product of Vectors

Let **A** and **B** denote two vectors, and let θ denote the angle between them, as shown in Fig. 12.56. The *scalar* (or *dot*) *product* of **A** and **B** is denoted by $\mathbf{A} \cdot \mathbf{B}$, and it is defined in this manner:

$$\mathbf{A} \cdot \mathbf{B} = AB \cos\theta \qquad (12.84)$$

As its name indicates, this product is a scalar.

In Fig. 12.57, vectors **A** and **B** have the positions Oa and Ob, respectively. Line ad is perpendicular to Ob, and be is perpendicular to Oa. Then $Od = A\cos\theta$ and

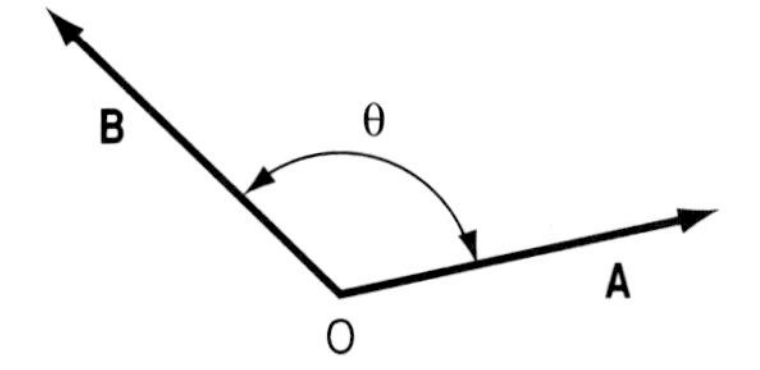

FIGURE 12.56

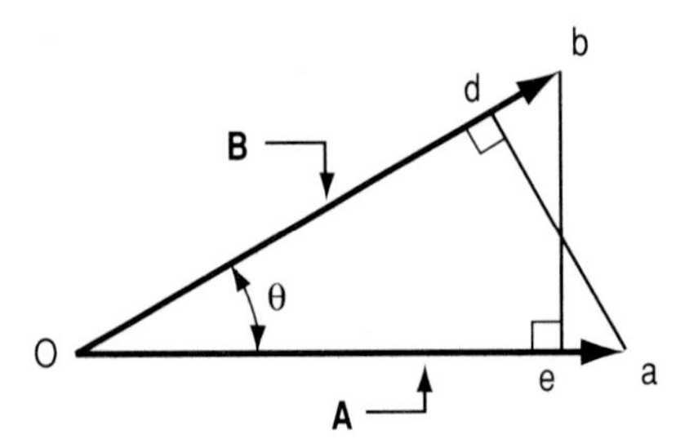

FIGURE 12.57 Interpretation of scalar product of vectors.

$Oe = B \cos \theta$, and it follows that $\mathbf{A} \cdot \mathbf{B} = (Oa)(Oe) = (Ob)(Od)$. Thus, the scalar product can be interpreted as the product of either of the following: the length of **A** and the length of **B** as projected onto **A**; the length of **B** and the length of **A** as projected onto **B**. Manifestly, $\mathbf{A} \cdot \mathbf{B} = \mathbf{B} \cdot \mathbf{A}$, and scalar multiplication is commutative.

If **A** and **B** are perpendicular to each other, $\cos \theta = 0$, and $\mathbf{A} \cdot \mathbf{B} = 0$. Conversely, if $\mathbf{A} \cdot \mathbf{B} = 0$ and neither vector is a null vector, the vectors are perpendicular to each other. It follows that

$$\mathbf{i} \cdot \mathbf{j} = \mathbf{j} \cdot \mathbf{k} = \mathbf{k} \cdot \mathbf{i} = 0 \tag{12.85}$$

If **A** and **B** are parallel to each other, $\theta = 0$ and $\cos \theta = 1$. Then $\mathbf{A} \cdot \mathbf{B} = AB$. It follows that

$$\mathbf{i} \cdot \mathbf{i} = \mathbf{j} \cdot \mathbf{j} = \mathbf{k} \cdot \mathbf{k} = 1 \tag{12.86}$$

In Fig. 12.58, vector **A** has the position Oa, **B** has the position Ob, and **C** has the position bc. Then $\mathbf{D} = \mathbf{B} + \mathbf{C}$ has the position Oc. Lines $b'b$ and $c'c$ are perpendicular to Oa. We have the following:

$$\mathbf{A} \cdot \mathbf{B} = (Oa)(Ob') \qquad \mathbf{A} \cdot \mathbf{C} = (Oa)(b'c') \qquad \mathbf{A} \cdot \mathbf{D} = (Oa)(Oc')$$

Therefore,

$$\mathbf{A} \cdot (\mathbf{B} + \mathbf{C}) = \mathbf{A} \cdot \mathbf{B} + \mathbf{A} \cdot \mathbf{C} \tag{12.87}$$

Thus, scalar multiplication is distributive. Extending this law, we have

$$(\mathbf{A} + \mathbf{B}) \cdot (\mathbf{C} + \mathbf{D}) = \mathbf{A} \cdot \mathbf{C} + \mathbf{A} \cdot \mathbf{D} + \mathbf{B} \cdot \mathbf{C} + \mathbf{B} \cdot \mathbf{D} \tag{12.87a}$$

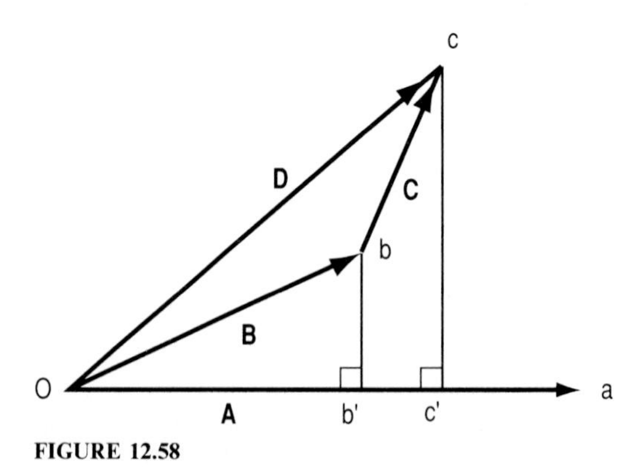

FIGURE 12.58

Now let

$$\mathbf{A} = a_1\mathbf{i} + a_2\mathbf{j} + a_3\mathbf{k} \qquad \mathbf{B} = b_1\mathbf{i} + b_2\mathbf{j} + b_3\mathbf{k}$$

$$\mathbf{A} \cdot \mathbf{B} = (a_1\mathbf{i} + a_2\mathbf{j} + a_3\mathbf{k}) \cdot (b_1\mathbf{i} + b_2\mathbf{j} + b_3\mathbf{k})$$

Applying the distributive law and Eqs. (12.85) and (12.86), we obtain

$$\mathbf{A} \cdot \mathbf{B} = a_1 b_1 + a_2 b_2 + a_3 b_3 \tag{12.88}$$

But

$$\mathbf{A} \cdot \mathbf{B} = AB \cos\theta$$

and it follows that

$$\cos\theta = \frac{a_1}{A}\frac{b_1}{B} + \frac{a_2}{A}\frac{b_2}{B} + \frac{a_3}{A}\frac{b_3}{B} \tag{12.89}$$

Since each fraction is a direction cosine, this equation is merely a restatement of Eq. (3.44*a*) in modified form. If $\mathbf{A} \cdot \mathbf{B}$ is negative, $\cos\theta$ is negative, and $90° < \theta < 180°$.

Consider that vectors **A** and **B** are known and that we wish to resolve **B** into a component $\mathbf{B}_1$ parallel to **A** and a component $\mathbf{B}_2$ perpendicular to **A**, as shown in Fig. 12.59. Since B_1 is the length of **B** as projected onto **A**, it follows that $B_1 = (\mathbf{A} \cdot \mathbf{B})/A$, and

$$\frac{B_1}{A} = \frac{\mathbf{A} \cdot \mathbf{B}}{A^2} \tag{12.90}$$

EXAMPLE 12.83 If $\mathbf{A} = 2\mathbf{i} - 5\mathbf{j} + 6\mathbf{k}$ and $\mathbf{B} = -6\mathbf{i} + 3\mathbf{j} - 4\mathbf{k}$, identify the components $\mathbf{B}_1$ and $\mathbf{B}_2$ of **B** that are parallel and perpendicular to **A**, respectively. Verify the solution.

SOLUTION Applying Eqs. (12.83) and (12.88), we obtain

$$A^2 = 2^2 + (-5)^2 + 6^2 = 65$$

$$\mathbf{A} \cdot \mathbf{B} = 2(-6) + (-5)3 + 6(-4) = -51$$

By Eq. (12.90), $B_1/A = -51/65$. Multiplying each component of **A** by this factor, we obtain

$$\mathbf{B}_1 = -\left(\frac{102}{65}\right)\mathbf{i} + \left(\frac{51}{13}\right)\mathbf{j} - \left(\frac{306}{65}\right)\mathbf{k}$$

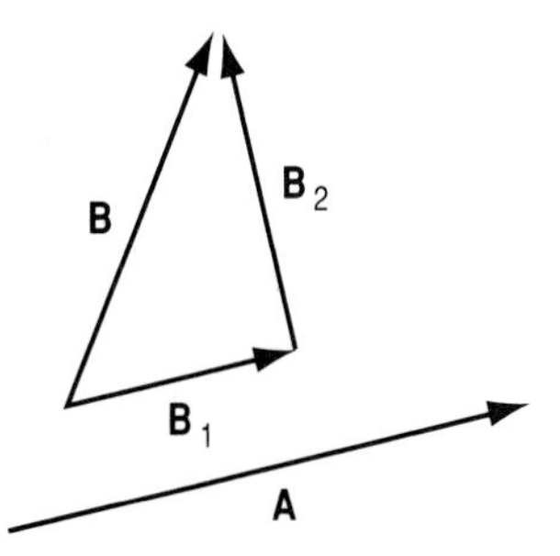

FIGURE 12.59 Resolution of vector into components of specified inclinations.

Setting $\mathbf{B}_2 = \mathbf{B} - \mathbf{B}_1$ and subtracting corresponding components, we obtain

$$\mathbf{B}_2 = -\left(\frac{288}{65}\right)\mathbf{i} - \left(\frac{12}{13}\right)\mathbf{j} + \left(\frac{46}{65}\right)\mathbf{k}$$

The sense of $\mathbf{B}_1$ is opposite to that of **A**.

Let θ_1 and θ_2 denote the angle between **A** and $\mathbf{B}_1$ and between **A** and $\mathbf{B}_2$, respectively. Applying Eq. (12.89), we find that $\cos\theta_1 = -1$ and $\cos\theta_2 = 0$. The results are thus confirmed.

12.14.7 Vector Product of Vectors

In Fig. 12.60, vectors **A** and **B** lie in plane Q, and θ is the angle between them. We impose the restriction $0 < \theta < 180°$. Let **n** denote a unit vector that is normal to Q. The sense of **n** is established in this manner: Consider that **A** is rotated through angle θ to bring it into alignment with **B**. If this rotation is counterclockwise as perceived by an observer, as is true in Fig. 12.60, **n** is directed *toward* the observer; if this rotation is clockwise, **n** is directed *away from* the observer. The *vector* (or *cross*) *product* of **A** and **B** is denoted by $\mathbf{A} \times \mathbf{B}$ and it is defined in this manner:

$$\mathbf{A} \times \mathbf{B} = \mathbf{n}AB \sin\theta \tag{12.91}$$

As its name indicates, this product is a vector, and it is shown in Fig. 12.60. It follows at once that

$$\mathbf{B} \times \mathbf{A} = -\mathbf{n}AB \sin\theta = -(\mathbf{A} \times \mathbf{B})$$

Thus, this form of multiplication is not commutative, and it is imperative that the order of multiplication be specified.

If **A** and **B** are perpendicular to each other, $\sin\theta = 1$, and $\mathbf{A} \times \mathbf{B} = \mathbf{n}AB$. It follows that

$$\mathbf{i} \times \mathbf{j} = \mathbf{k} \qquad \mathbf{j} \times \mathbf{k} = \mathbf{i} \qquad \mathbf{k} \times \mathbf{i} = \mathbf{j} \tag{12.92}$$

If **A** and **B** are parallel to each other, $\sin\theta = 0$, and $\mathbf{A} \times \mathbf{B} = \mathbf{0}$. Conversely, if $\mathbf{A} \times \mathbf{B} = \mathbf{0}$ and neither is a null vector, **A** and **B** are parallel to each other. It

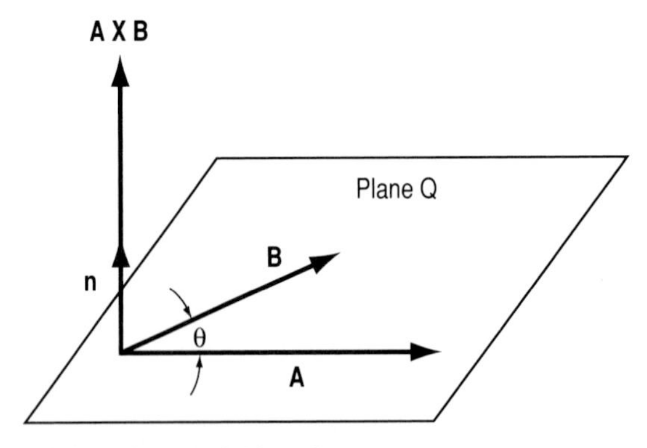

FIGURE 12.60 Definition of vector **n**.

follows that

$$\mathbf{i} \times \mathbf{i} = \mathbf{j} \times \mathbf{j} = \mathbf{k} \times \mathbf{k} = \mathbf{0} \tag{12.93}$$

In Fig. 12.61, we form a parallelogram having **A** and **B** as sides. Let S denote the area of this parallelogram. Then

$$S = hA = (B \sin \theta)A = AB \sin \theta = |\mathbf{A} \times \mathbf{B}|$$

Thus, $|\mathbf{A} \times \mathbf{B}|$ equals the area of the parallelogram having **A** and **B** as sides.

It is advantageous to consider that the vector $\mathbf{A} \times \mathbf{B}$ is constructed in this manner: In Fig. 12.62, pass plane P through O perpendicular to **A**. The angle between **B** and P is $90° - \theta$. Now project **B** onto P, and call the projection $\mathbf{B}'$. Then $B' = B \sin \theta$. Now rotate $\mathbf{B}'$ about O through 90°, in the manner shown, to form $\mathbf{B}''$. Vector $\mathbf{B}''$ is perpendicular to the plane that contains **A** and **B**. Now multiply $\mathbf{B}''$ by A to form **T**. Then $T = AB'' = AB \sin \theta$. Since **T** has the same length and direction as $\mathbf{A} \times \mathbf{B}$, it follows that $\mathbf{T} = \mathbf{A} \times \mathbf{B}$.

By applying the construction in Fig. 12.62, we can demonstrate that

$$\mathbf{A} \times (\mathbf{B} + \mathbf{C}) = \mathbf{A} \times \mathbf{B} + \mathbf{A} \times \mathbf{C} \tag{12.94a}$$

If we now multiply both sides of the equation by -1, the order of the factors is reversed, and we obtain

$$(\mathbf{B} + \mathbf{C}) \times \mathbf{A} = \mathbf{B} \times \mathbf{A} + \mathbf{C} \times \mathbf{A} \tag{12.94b}$$

An extension of these equations yields

$$(\mathbf{A} + \mathbf{B}) \times (\mathbf{C} + \mathbf{D}) = \mathbf{A} \times \mathbf{C} + \mathbf{A} \times \mathbf{D} + \mathbf{B} \times \mathbf{C} + \mathbf{B} \times \mathbf{D} \tag{12.94c}$$

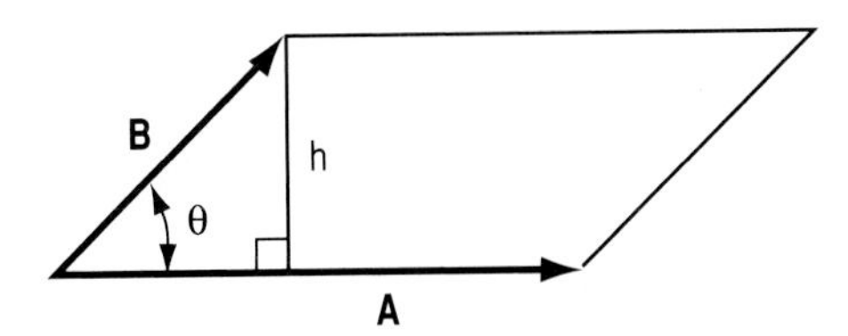

FIGURE 12.61

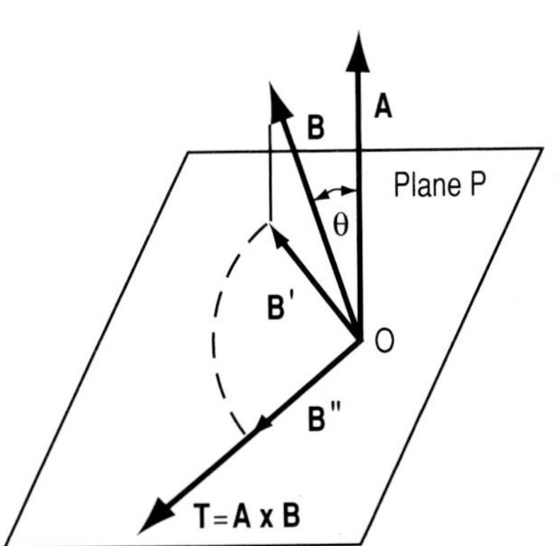

FIGURE 12.62 Construction for forming the vector product.

As before, let

$$\mathbf{A} = a_1\mathbf{i} + a_2\mathbf{j} + a_3\mathbf{k} \qquad \mathbf{B} = b_1\mathbf{i} + b_2\mathbf{j} + b_3\mathbf{k}$$

By performing the multiplication and applying the foregoing relationships, we obtain the following:

$$\mathbf{A} \times \mathbf{B} = (a_2 b_3 - a_3 b_2)\mathbf{i} + (a_3 b_1 - a_1 b_3)\mathbf{j} + (a_1 b_2 - a_2 b_1)\mathbf{k} \tag{12.95}$$

In Art. 5.5.1, we demonstrated how the determinant of a third-order matrix can be formed. The expression in Eq. (12.95) is a determinant, and we have the following:

$$\mathbf{A} \times \mathbf{B} = \begin{bmatrix} \mathbf{i} & \mathbf{j} & \mathbf{k} \\ a_1 & a_2 & a_3 \\ b_1 & b_2 & b_3 \end{bmatrix} \tag{12.95a}$$

EXAMPLE 12.84 If $\mathbf{A} = 3\mathbf{i} - 6\mathbf{j} + 2\mathbf{k}$, $\mathbf{B} = -4\mathbf{i} + 5\mathbf{j} - 9\mathbf{k}$, and $\mathbf{C} = \mathbf{A} \times \mathbf{B}$, identify $\mathbf{C}$.

SOLUTION

$$a_1 = 3 \qquad a_2 = -6 \qquad a_3 = 2 \qquad b_1 = -4 \qquad b_2 = 5 \qquad b_3 = -9$$

Substituting in Eq. (12.95), we obtain

$$\mathbf{C} = 44\mathbf{i} + 19\mathbf{j} - 9\mathbf{k}$$

EXAMPLE 12.85 With reference to Example 12.84, vector $\mathbf{D}$ is perpendicular to both $\mathbf{A}$ and $\mathbf{B}$, and its length is 100. Identify $\mathbf{D}$.

SOLUTION Vector $\mathbf{C}$ in Example 12.84 is perpendicular to both $\mathbf{A}$ and $\mathbf{B}$. By Eq. (12.83),

$$C = \sqrt{44^2 + 19^2 + (-9)^2} = 48.76$$

Then $D/C = 100/48.76 = 2.051$. Since $\mathbf{D}$ is parallel to $\mathbf{C}$, it is simply necessary to multiply the components of $\mathbf{C}$ by the factor 2.051. However, because the sense of $\mathbf{D}$ is not specified, two solutions are available. Then

$$\mathbf{D} = \pm(90.24\mathbf{i} + 38.97\mathbf{j} - 18.46\mathbf{k})$$

12.14.8 Product of Three Vectors

The definitions of the vector products can be extended to encompass the product of three vectors as well as products of a mixed type. Such products arise frequently in engineering applications. We consider three cases.

1. Let $\mathbf{D} = (\mathbf{A} \cdot \mathbf{B})\mathbf{C}$. Since $\mathbf{A} \cdot \mathbf{B}$ is a scalar s, $\mathbf{D}$ is obtained by multiplying $\mathbf{C}$ by s; that is, $\mathbf{D} = s\mathbf{C}$.
2. Let $D = (\mathbf{A} \times \mathbf{B}) \cdot \mathbf{C}$. This product is termed the *triple scalar product*. Its geometric significance is as follows: In Fig. 12.63, we form a parallelepiped having $\mathbf{A}$, $\mathbf{B}$, and $\mathbf{C}$ as edges. Let $\mathbf{E} = \mathbf{A} \times \mathbf{B}$. Then $\mathbf{E}$ is perpendicular to the plane of $\mathbf{A}$ and $\mathbf{B}$, as shown, and E is the area of the base $Oabc$. Let θ denote

the angle between **E** and **C**. Then, in absolute value, $D = EC \cos\theta =$ (area $Oabc$)(Od) = volume of parallelepiped. In Fig. 12.63, D is positive. If **C** were located below $Oabc$, D would be negative.

Since any face of the parallelepiped in Fig. 12.63 can be regarded as its base, it follows that

$$(\mathbf{A} \times \mathbf{B}) \cdot \mathbf{C} = (\mathbf{B} \times \mathbf{C}) \cdot \mathbf{A} = (\mathbf{C} \times \mathbf{A}) \cdot \mathbf{B} \qquad (12.96)$$

Moreover, since the scalar product is commutative, the second expression in Eq. (12.96) can be reversed, and the result is

$$(\mathbf{A} \times \mathbf{B}) \cdot \mathbf{C} = \mathbf{A} \cdot (\mathbf{B} \times \mathbf{C}) \qquad (12.97)$$

Now let

$$\mathbf{A} = a_1\mathbf{i} + a_2\mathbf{j} + a_3\mathbf{k} \qquad \mathbf{B} = b_1\mathbf{i} + b_2\mathbf{j} + b_3\mathbf{k}$$

$$\mathbf{C} = c_1\mathbf{i} + c_2\mathbf{j} + c_3\mathbf{k}$$

By performing the multiplication and applying the foregoing relationships, we obtain the following determinant:

$$(\mathbf{A} \times \mathbf{B}) \cdot \mathbf{C} = \begin{bmatrix} a_1 & a_2 & a_3 \\ b_1 & b_2 & b_3 \\ c_1 & c_2 & c_3 \end{bmatrix} \qquad (12.98)$$

3. Let $\mathbf{D} = (\mathbf{A} \times \mathbf{B}) \times \mathbf{C}$. This product is termed the *triple vector product*. The location of this vector is found in this manner: Let $\mathbf{E} = \mathbf{A} \times \mathbf{B}$; then $\mathbf{D} = \mathbf{E} \times \mathbf{C}$. Now let Q denote the plane that contains **A** and **B**. Then **E** is perpendicular to Q. Since **D** is perpendicular to the plane that contains **E** and **C**, it follows that **D** lies in Q.

It can be demonstrated that

$$(\mathbf{A} \times \mathbf{B}) \times \mathbf{C} = (\mathbf{A} \cdot \mathbf{C})\mathbf{B} - (\mathbf{B} \cdot \mathbf{C})\mathbf{A} \qquad (12.99a)$$

If we multiply both sides of this equation by -1, the factors at the left change position. If we now interchange letters, we obtain

$$\mathbf{A} \times (\mathbf{B} \times \mathbf{C}) = (\mathbf{A} \cdot \mathbf{C})\mathbf{B} - (\mathbf{A} \cdot \mathbf{B})\mathbf{C} \qquad (12.99b)$$

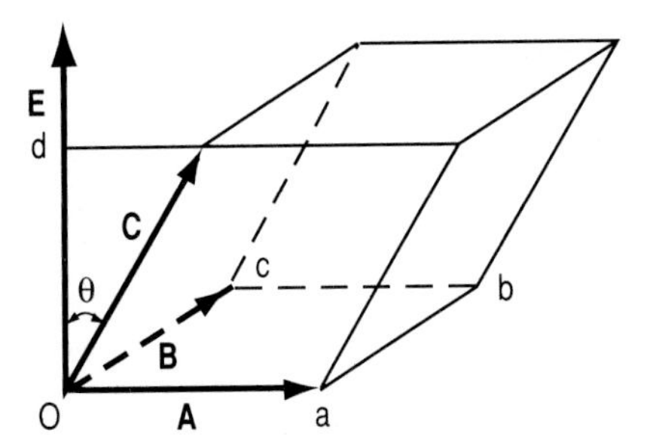

FIGURE 12.63 Geometric interpretation of triple scalar product.

EXAMPLE 12.86 Given the following vectors:

$$\mathbf{A} = -2\mathbf{i} + 7\mathbf{j} - 4\mathbf{k}$$

$$\mathbf{B} = 6\mathbf{i} + 3\mathbf{j} - 5\mathbf{k}$$

$$\mathbf{C} = 9\mathbf{i} - 8\mathbf{j} + \mathbf{k}$$

Find $\mathbf{A} \times (\mathbf{B} \times \mathbf{C})$ (*a*) by applying Eq. (12.99*b*); (*b*) without recourse to this equation.

SOLUTION

Part a. Equation (12.88) yields the following:

$$\mathbf{A} \cdot \mathbf{C} = (-2)9 + 7(-8) + (-4)1 = -78$$

$$\mathbf{A} \cdot \mathbf{B} = (-2)6 + 7 \times 3 + (-4)(-5) = 29$$

By Eq. (12.99*b*),

$$\mathbf{A} \times (\mathbf{B} \times \mathbf{C}) = -78(6\mathbf{i} + 3\mathbf{j} - 5\mathbf{k}) - 29(9\mathbf{i} - 8\mathbf{j} + \mathbf{k})$$

$$= -729\mathbf{i} - 2\mathbf{j} + 361\mathbf{k}$$

Part b. By Eq. (12.95*a*),

$$\mathbf{B} \times \mathbf{C} = \begin{bmatrix} \mathbf{i} & \mathbf{j} & \mathbf{k} \\ 6 & 3 & -5 \\ 9 & -8 & 1 \end{bmatrix} = -37\mathbf{i} - 51\mathbf{j} - 75\mathbf{k}$$

By Eq. (12.95*a*),

$$\mathbf{A} \times (\mathbf{B} \times \mathbf{C}) = \begin{bmatrix} \mathbf{i} & \mathbf{j} & \mathbf{k} \\ -2 & 7 & -4 \\ -37 & -51 & -75 \end{bmatrix} = -729\mathbf{i} - 2\mathbf{j} + 361\mathbf{k}$$

12.14.9 Product of Four Vectors

The relationships involving products of three vectors can readily be extended to yield products of four vectors. As an illustration, let $\mathbf{E} = (\mathbf{A} \times \mathbf{B}) \times (\mathbf{C} \times \mathbf{D})$. To establish the direction of **E**, let $\mathbf{F} = \mathbf{A} \times \mathbf{B}$ and $\mathbf{G} = \mathbf{C} \times \mathbf{D}$. Let P denote the plane containing **A** and **B**, Q denote the plane containing **C** and **D**, and R denote the plane containing **F** and **G**. As shown in Fig. 12.64, **F** is perpendicular to P and **G** is perpendicular to Q. Therefore, **R** is perpendicular to both P and Q. Since **E** is perpendicular to R, **E** is parallel to the line of intersection of P and Q.

Applying Eq. (12.99*a*), we obtain the following:

$$(\mathbf{A} \times \mathbf{B}) \times (\mathbf{C} \times \mathbf{D}) = [\mathbf{A} \cdot (\mathbf{C} \times \mathbf{D})]\mathbf{B} - [\mathbf{B} \cdot (\mathbf{C} \times \mathbf{D})]\mathbf{A} \qquad (12.100a)$$

An alternative expression is

$$(\mathbf{A} \times \mathbf{B}) \times (\mathbf{C} \times \mathbf{D}) = [(\mathbf{A} \times \mathbf{B}) \cdot \mathbf{D}]\mathbf{C} - [(\mathbf{A} \times \mathbf{B}) \cdot \mathbf{C}]\mathbf{D} \qquad (12.100b)$$

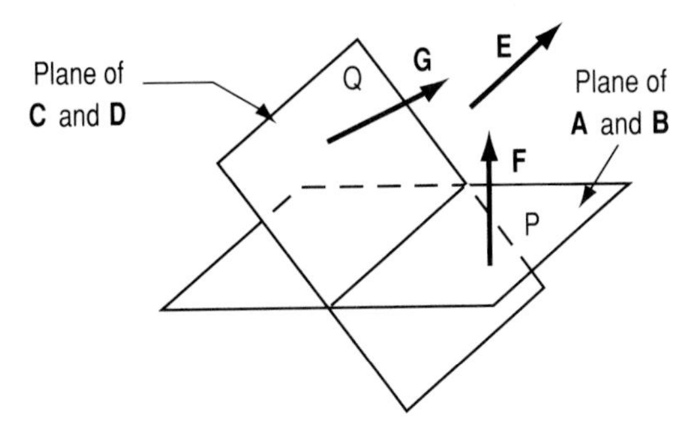

FIGURE 12.64 Method of determining direction of vector product of four vectors.

12.14.10 Derivatives and Integrals of Vectors

The definition of the derivative of a vector is analogous to that of a scalar. Consider that $\mathbf{v}$ is a function of an independent variable t. Now consider that t increases by an amount Δt, causing $\mathbf{v}$ to increase by an amount $\Delta\mathbf{v}$, which may or may not be collinear with $\mathbf{v}$. The derivative of $\mathbf{v}$ with respect to t is denoted by $d\mathbf{v}/dt$, and its value is

$$\frac{d\mathbf{v}}{dt} = \lim_{\Delta t \to 0} \frac{\Delta\mathbf{v}}{\Delta t} \tag{12.101}$$

Now let $\mathbf{v} = s\mathbf{u}$, where both s and $\mathbf{u}$ are functions of t. Then

$$\mathbf{v} + \Delta\mathbf{v} = (s + \Delta s)(\mathbf{u} + \Delta\mathbf{u})$$

and it follows that

$$\frac{d}{dt}(s\mathbf{u}) = s\frac{d\mathbf{u}}{dt} + \frac{ds}{dt}\mathbf{u} \tag{12.102}$$

In the special case where $\mathbf{u}$ is constant and s is the independent variable, this equation reduces to

$$\frac{d\mathbf{v}}{ds} = \mathbf{u} \tag{12.102a}$$

By applying Eqs. (12.87a) and (12.94c) and proceeding as before, we obtain the following results:

$$\frac{d}{dt}(\mathbf{v}\cdot\mathbf{w}) = \mathbf{v}\cdot\frac{d\mathbf{w}}{dt} - \frac{d\mathbf{v}}{dt}\cdot\mathbf{w} \tag{12.103}$$

$$\frac{d}{dt}(\mathbf{v}\times\mathbf{w}) = \mathbf{v}\times\frac{d\mathbf{w}}{dt} + \frac{d\mathbf{v}}{dt}\times\mathbf{w} \tag{12.104}$$

In the last equation, it is mandatory that the order of the factors be maintained, since a reversal of the order causes a change in algebraic sign.

Consider the derivative

$$\frac{d}{dt}\left(\frac{d\mathbf{v}}{dt} \times \mathbf{v}\right)$$

Applying Eq. (12.104), we obtain this product:

$$\frac{d\mathbf{v}}{dt} \times \frac{d\mathbf{v}}{dt} + \frac{d^2\mathbf{v}}{dt^2} \times \mathbf{v}$$

The first term is the null vector, and it follows that

$$\frac{d}{dt}\left(\frac{d\mathbf{v}}{dt} \times \mathbf{v}\right) = \frac{d^2\mathbf{v}}{dt^2} \times \mathbf{v} \tag{12.105}$$

Vector integration is simply the inverse of vector differentiation. Thus, if $d\mathbf{w}/dt = \mathbf{v}$, then

$$\int \mathbf{v}\, dt = \mathbf{w} + \mathbf{c}$$

where $\mathbf{c}$ is an arbitrary constant vector.

Now let $\mathbf{w} = x\mathbf{i} + y\mathbf{j} + z\mathbf{k}$, where the scalars are functions of t. By Eq. (12.102*a*), the partial derivatives of $\mathbf{w}$ are as follows:

$$\frac{\partial \mathbf{w}}{\partial x} = \mathbf{i} \qquad \frac{\partial \mathbf{w}}{\partial y} = \mathbf{j} \qquad \frac{\partial \mathbf{w}}{\partial z} = \mathbf{k} \tag{12.106}$$

12.14.11 Vectors Related to Planar Curves and Motion

Consider that a point moves along a curve in the xy plane, and let $P(x, y)$ denote the present position of the point. Let $\mathbf{r}$ denote the vector from the origin to P, as shown in Fig. 12.65*a*; this is termed the *position vector* (or *radius vector*) of the moving point. Then

$$\mathbf{r} = x\mathbf{i} + y\mathbf{j}$$

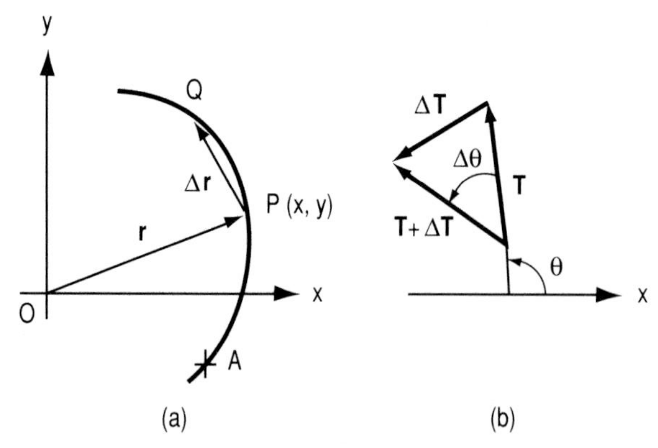

FIGURE 12.65 Development of (*a*) unit tangential vector and (*b*) unit normal vector.

At P, we may draw a *velocity vector* $\mathbf{v}$ and an *acceleration vector* $\mathbf{a}$ that give the magnitude and direction of these quantities when the moving point is at P. To obtain these vectors, consider that the point moves from P to Q during a time interval Δt, causing $\mathbf{r}$ to undergo a change $\Delta\mathbf{r}$. Then $\Delta\mathbf{r} = \mathbf{i}\,\Delta x + \mathbf{j}\,\Delta y$. Manifestly, $\mathbf{v} = d\mathbf{r}/dt$, and we have

$$\mathbf{v} = \frac{dx}{dt}\mathbf{i} + \frac{dy}{dt}\mathbf{j} \tag{12.107a}$$

The slope of $\mathbf{v}$ is dy/dx, and $\mathbf{v}$ is tangent to the curve at P. Setting $\mathbf{a} = d\mathbf{v}/dt$, we obtain

$$\mathbf{a} = \frac{d^2x}{dt^2}\mathbf{i} + \frac{d^2y}{dt^2}\mathbf{j} \tag{12.107b}$$

In Fig. 12.65a, let s denote the arc length from some arbitrary point A to P. The arc length PQ is Δs. Let

$$\mathbf{T} = \frac{d\mathbf{r}}{ds}$$

As Δs approaches 0, Q approaches P, the length Δr approaches Δs, and $\Delta\mathbf{r}$ approaches a position tangent to the curve at P. Thus, $T = 1$, and $\mathbf{T}$ is tangent to the curve at P. Accordingly, $\mathbf{T}$ is called the *unit tangential vector* at P.

Let θ denote the angle that $\mathbf{T}$ makes with the positive side of the x axis. As the moving point traverses arc PQ in Fig. 12.65a, $\mathbf{T}$ changes by an amount $\Delta\mathbf{T}$ and θ increases by an amount $\Delta\theta$, as shown in Fig. 12.65b. Now let

$$\mathbf{N} = \frac{d\mathbf{T}}{d\theta}$$

As Q approaches P, ΔT approaches $T\,\Delta\theta$ and the angle between $\mathbf{T}$ and $\Delta\mathbf{T}$ approaches 90°. Setting $T = 1$, we find that $N = 1$, $\mathbf{N}$ is normal to the curve at P, and the sense of $\mathbf{N}$ is to the concave side of the curve. Accordingly, $\mathbf{N}$ is termed the *unit normal vector* at P.

In Art. 12.3.5, we defined the curvature K of a curve as $K = d\theta/ds$. We also defined the radius of curvature R, where $R = 1/K$. Let

$$\mathbf{K} = \frac{d\mathbf{T}}{ds} = \frac{d\theta}{ds}\frac{d\mathbf{T}}{d\theta}$$

Thus,

$$\mathbf{K} = K\mathbf{N} \tag{12.108}$$

and $\mathbf{K}$ is accordingly termed the *curvature vector* at P.

Since the velocity vector $\mathbf{v}$ is in the direction of $\mathbf{T}$, we may write

$$\mathbf{v} = v\mathbf{T}$$

where v is the speed of the moving point when it is at P. Then $v = ds/dt$. Now differentiating $\mathbf{T}$ with respect to time, we obtain

$$\frac{d\mathbf{T}}{dt} = \frac{ds}{dt}\frac{d\mathbf{T}}{ds} = v\mathbf{K} = vK\mathbf{N}$$

The acceleration vector can now be expressed in this form:

$$\mathbf{a} = \frac{d\mathbf{v}}{dt} = \frac{d}{dt}(v\mathbf{T}) = v\frac{d\mathbf{T}}{dt} + \frac{dv}{dt}\mathbf{T}$$

Applying the foregoing relationship and setting $R = 1/K$, we obtain

$$\mathbf{a} = \frac{v^2}{R}\mathbf{N} + \frac{dv}{dt}\mathbf{T} \qquad (12.109)$$

This equation resolves the acceleration into its normal and tangential components, and Eqs. (12.25) and (12.26) have thus been proved.

EXAMPLE 12.87 A particle of mass m is located in a force field that remains constant in intensity but rotates with a constant angular velocity, the force being

$$\mathbf{F} = A(\mathbf{i}\cos ut + \mathbf{j}\sin ut)$$

where A is a constant, u is the angular velocity of the field, and t denotes elapsed time. Initially, the particle is at $P(0, h)$, where $h > 0$, and it is moving toward the origin with a speed w. Applying the equation $\mathbf{F} = m\mathbf{a}$, formulate the parametric equations of the trajectory of the particle.

SOLUTION This problem can be solved without recourse to vector methods by resolving F and a into their x and y components, setting $F_x = ma_x$ and $F_y = ma_y$, and then integrating each equation twice. However, vector methods enable us to evaluate the x and y components of each quantity simultaneously.

We shall apply the relationships $\int \mathbf{a}\,dt = \mathbf{v}$ and $\int \mathbf{v}\,dt = \mathbf{r}$. At the initial state, $\mathbf{v} = -\mathbf{j}w$ and $\mathbf{r} = \mathbf{j}h$. We start by replacing $\mathbf{F}$ with $m\mathbf{a}$ to obtain

$$m\mathbf{a} = \mathbf{i}A\cos ut + \mathbf{j}A\sin ut$$

Integrating,

$$m\mathbf{v} = \mathbf{i}\frac{A\sin ut}{u} - \mathbf{j}\frac{A\cos ut}{u} + \mathbf{c}_1$$

where $\mathbf{c}_1$ is a constant of integration. Setting $t = 0$ and replacing $\mathbf{v}$ with its initial value, we obtain

$$-\mathbf{j}mw = -\mathbf{j}\frac{A}{u} + \mathbf{c}_1 \qquad \mathbf{c}_1 = \mathbf{j}\left(\frac{A}{u} - mw\right)$$

Substituting this value,

$$m\mathbf{v} = \mathbf{i}\frac{A\sin ut}{u} + \mathbf{j}\left(\frac{A}{u} - \frac{A\cos ut}{u} - mw\right)$$

Integrating again,

$$m\mathbf{r} = -\mathbf{i}\frac{A\cos ut}{u^2} + \mathbf{j}\left(\frac{At}{u} - \frac{A\sin ut}{u^2} - mwt\right) + \mathbf{c}_2$$

Setting $t = 0$ and replacing $\mathbf{r}$ with its initial value, we obtain

$$\mathbf{j}mh = -\mathbf{i}\frac{A}{u^2} + \mathbf{c}_2 \qquad \mathbf{c}_2 = \mathbf{i}\frac{A}{u^2} + \mathbf{j}mh$$

Substituting this value and dividing by m,

$$\mathbf{r} = \frac{1}{m}\left[\mathbf{i}\frac{A}{u^2}(1 - \cos ut) + \mathbf{j}\left(\frac{At}{u} - \frac{A \sin ut}{u^2} + mh - mwt\right)\right]$$

Taking the standard expression $\mathbf{r} = \mathbf{i}x + \mathbf{j}y$, we obtain the following parametric equations:

$$x = \frac{A}{mu^2}(1 - \cos ut) \qquad y = \frac{At}{mu} - \frac{A \sin ut}{mu^2} + h - wt$$

Setting $t = 0$, we obtain $x = 0$ and $y = h$. These results are consistent with the given information.

EXAMPLE 12.88 A moving particle is subjected to an attractive force that is always directed toward a fixed point O. Prove that the line joining O and the particle sweeps out equal areas in equal intervals of time.

SOLUTION In Fig. 12.66, let P denote the present position of the particle. We draw the position vector $\mathbf{r}$ from O to P. The velocity vector $\mathbf{v}$ is tangent to the trajectory of the particle. Since $\mathbf{F} = m\mathbf{a}$, the acceleration vector is collinear with $\mathbf{F}$; therefore, $\mathbf{a}$ is directed from P to O. Since $\mathbf{r}$ and $\mathbf{a}$ make an angle of 180° with each other, it follows that $\mathbf{a} \times \mathbf{r} = \mathbf{0}$.

Applying Eq. (12.105), we may write the following:

$$\mathbf{a} \times \mathbf{r} = \frac{d^2\mathbf{r}}{dt^2} \times \mathbf{r} = \frac{d}{dt}\left(\frac{d\mathbf{r}}{dt} \times \mathbf{r}\right) = \frac{d}{dt}(\mathbf{v} \times \mathbf{r}) = \mathbf{0}$$

Integrating, we now obtain

$$\mathbf{v} \times \mathbf{r} = \mathbf{c}$$

where $\mathbf{c}$ is a constant vector.

Let ϕ denote the angle between $\mathbf{r}$ and $\mathbf{v}$. The foregoing result yields $vr \sin \phi = c$, or $(dr/dt)r \sin \phi = c$. Figure 12.65a reveals that the area swept out by $\mathbf{r}$ in a time interval dt is $(1/2)r\, dr \sin \phi$, and the foregoing equation shows that this area is constant. It follows that the area swept out by $\mathbf{r}$ in a specific time interval t is constant.

The rate at which the rotating vector $\mathbf{r}$ generates area is termed the *areal velocity* of the moving particle. When applied to planetary motion, the principle we have

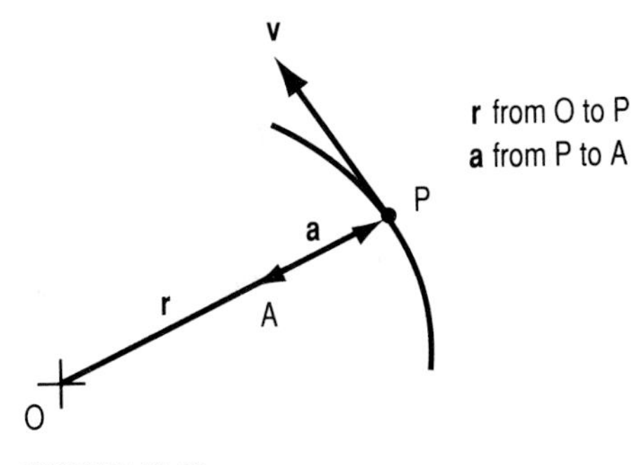

FIGURE 12.66

proved is known as Kepler's second law. We shall prove this law without recourse to vector analysis in Art. 12.15.4.

12.14.12 Radial and Transverse Components of Curvilinear Motion

In analyzing the motion of a point along a planar curve, we previously resolved the acceleration of the point into components normal and tangential to the curve. However, if the equation of the curve is expressed in polar coordinates as described in Art. 3.1.9, it is more appropriate to use a different system of components. Let $P(r, \theta)$ denote the instantaneous position of the moving point, and refer to Fig. 12.67*a*. We shall resolve both the velocity and acceleration of the point into two components: one along OP, and one perpendicular to OP. The first is the *radial component*, and the second is the *transverse component*.

Again let **v** and **a** denote, respectively, the velocity and acceleration of the point, and let the subscripts r and tr refer to the radial and transverse components, respectively. Let **u** denote a unit vector that is located on OP and sensed outward, and let **w** denote a unit vector normal to OP and sensed toward increasing values of θ. These vectors are shown in Fig. 12.67*a*.

Now consider that the point moves from P to Q during a time interval Δt, and let $r + \Delta r$ and $\theta + \Delta\theta$ be the coordinates of Q. During this displacement, **u** and **w** increase by the amount $\Delta\mathbf{u}$ and $\Delta\mathbf{w}$ shown in Fig. 12.67*b*. The lengths are as follows:

$$\Delta u = u\,\Delta\theta = \Delta\theta \qquad \Delta w = w\,\Delta\theta = \Delta\theta$$

We have the following:

$$\mathbf{v}_r = \frac{dr}{dt}\mathbf{u} \qquad \mathbf{v}_{tr} = r\frac{d\theta}{dt}\mathbf{w}$$

$$\mathbf{v} = \frac{dr}{dt}\mathbf{u} + r\frac{d\theta}{dt}\mathbf{w}$$

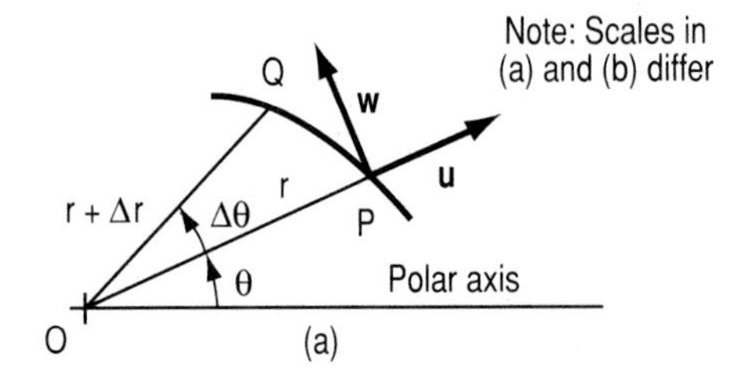

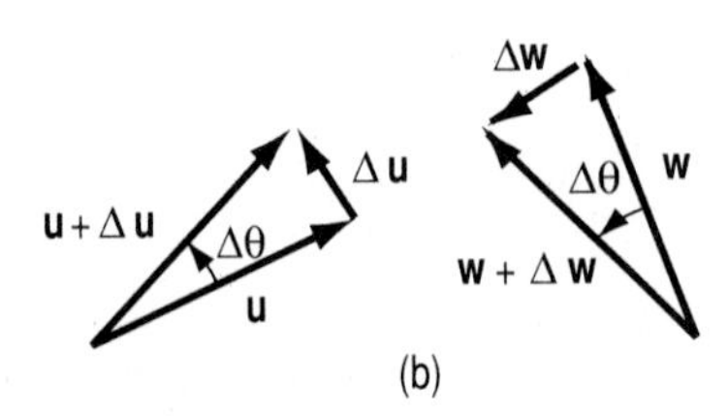

FIGURE 12.67 (*a*) Unit radial and transverse vectors; (*b*) changes in the unit vectors.

$$\mathbf{a} = \frac{d\mathbf{v}}{dt} = \frac{d^2r}{dt^2}\mathbf{u} + \frac{dr}{dt}\frac{d\mathbf{u}}{dt} + \frac{dr}{dt}\frac{d\theta}{dt}\mathbf{w} + r\frac{d^2\theta}{dt^2}\mathbf{w} + r\frac{d\theta}{dt}\frac{d\mathbf{w}}{dt}$$

From Fig. 12.67*b*,

$$\frac{d\mathbf{u}}{dt} = \frac{d\theta}{dt}\mathbf{w} \qquad \frac{d\mathbf{w}}{dt} = -\frac{d\theta}{dt}\mathbf{u}$$

Substituting these expressions in the foregoing equation and combining terms, we obtain the following radial and transverse components of acceleration:

$$\mathbf{a}_r = \left[\frac{d^2r}{dt^2} - r\left(\frac{d\theta}{dt}\right)^2\right]\mathbf{u} \tag{12.110a}$$

$$\mathbf{a}_{tr} = \left(2\frac{dr}{dt}\frac{d\theta}{dt} + r\frac{d^2\theta}{dt^2}\right)\mathbf{w} = \frac{1}{r}\frac{d}{dt}\left(r^2\frac{d\theta}{dt}\right)\mathbf{w} \tag{12.110b}$$

EXAMPLE 12.89 A point moves along the cardioid having the equation $r = a(1 - \cos\theta)$, where a is a constant. Its motion is such that the radius vector (the line joining the origin and the moving point) rotates in a counterclockwise direction with a constant angular velocity ω. Formulate expressions for the radial and transverse components of velocity and acceleration in terms of θ.

SOLUTION

$$\frac{d\theta}{dt} = \omega \qquad \frac{d^2\theta}{dt^2} = 0$$

$$\frac{dr}{dt} = \frac{dr}{d\theta}\frac{d\theta}{dt} = a\omega\sin\theta$$

$$\frac{d^2r}{dt^2} = \frac{d(dr/dt)}{dt} = \frac{d(dr/dt)}{d\theta}\frac{d\theta}{dt} = a\omega^2\cos\theta$$

When these expressions are substituted in the equations of velocity and acceleration, the results are as follows:

$$v_r = a\omega\sin\theta \qquad v_{tr} = r\omega = a\omega(1 - \cos\theta)$$

$$a_r = a\omega^2\cos\theta - a\omega^2(1 - \cos\theta) = a\omega^2(2\cos\theta - 1)$$

$$a_{tr} = 2a\omega^2\sin\theta$$

12.14.13 Point Functions and Fields

Let u and $\mathbf{v}$ denote a scalar and a vector, respectively, whose values vary across space. These quantities are known as *point functions* (or *functions of position*), u being a scalar point function and $\mathbf{v}$ being a vector point function. For example, u may be the temperature, atmospheric pressure, or luminosity at a given point, and $\mathbf{v}$ may be the velocity of a fluid or the force acting on a unit electric charge at a given point.

We assume that the variation of u and $\mathbf{v}$ across space is determinate rather than random. Therefore, u and $\mathbf{v}$ are functions of the coordinates of the point at which they are evaluated. Let $P(x, y, z)$ denote this point. Then u is a function of x, y, and z, and

$$\mathbf{v} = v_1\mathbf{i} + v_2\mathbf{j} + v_3\mathbf{k} \tag{p}$$

where v_1, v_2, and v_3 are collectively functions of x, y, and z.

The region of space across which u or $\mathbf{v}$ varies is termed a *field.* Moreover, the field is described as a scalar field if u is the variable and a vector field if $\mathbf{v}$ is the variable. For example, if u is the luminosity within a room, that room is the scalar field.

Consider that we connect all points in the field at which u or $\mathbf{v}$ has some assigned value. The surface thus formed is called a *level surface.* Thus, if u is the pressure within a fluid at rest, a level surface is a horizontal plane; if u is the speed of a fluid particle in a cylindrical pipe in which there is uniform laminar flow, a level surface is a cylindrical surface that is concentric with the pipe.

12.14.14 Gradient, Divergence, and Curl

The symbol ∇ (read *del*) is a vector operator that causes the quantity by which it is multiplied to be differentiated. Specifically,

$$\nabla = \frac{\partial}{\partial x}\mathbf{i} + \frac{\partial}{\partial y}\mathbf{j} + \frac{\partial}{\partial z}\mathbf{k} \tag{q}$$

Thus,

$$\nabla u = \frac{\partial u}{\partial x}\mathbf{i} + \frac{\partial u}{\partial y}\mathbf{j} + \frac{\partial u}{\partial z}\mathbf{k} \tag{12.111}$$

Similarly, by applying the expressions in Eqs. (*p*) and (*q*) and performing the multiplication, we obtain the following:

$$\nabla \cdot \mathbf{v} = \frac{\partial v_1}{\partial x} + \frac{\partial v_2}{\partial y} + \frac{\partial v_3}{\partial z} \tag{12.112}$$

$$\nabla \times \mathbf{v} = \left(\frac{\partial v_3}{\partial y} - \frac{\partial v_2}{\partial z}\right)\mathbf{i} + \left(\frac{\partial v_1}{\partial z} - \frac{\partial v_3}{\partial x}\right)\mathbf{j} + \left(\frac{\partial v_2}{\partial x} - \frac{\partial v_1}{\partial y}\right)\mathbf{k} \tag{12.113}$$

The vector ∇u is called the *gradient* of u, and it is written grad u. The scalar $\nabla \cdot \mathbf{v}$ is called the *divergence* of $\mathbf{v}$, and it is written div $\mathbf{v}$. The vector $\nabla \times \mathbf{v}$ is called the *curl* of $\mathbf{v}$, and it is written curl $\mathbf{v}$.

We shall now form the divergence of the gradient of u. Symbolically, this is $\nabla \cdot \nabla u$. By performing the multiplication, we find that

$$\nabla \cdot \nabla u = \frac{\partial^2 u}{\partial x^2} + \frac{\partial^2 u}{\partial y^2} + \frac{\partial^2 u}{\partial z^2} \tag{12.114}$$

Thus, the symbol $\nabla \cdot \nabla$, which is usually written ∇^2, is an operator that causes the quantity by which it is multiplied to be subjected to double differentiation. It is referred to as the Laplacian operator. The equation $\nabla^2 u = \mathbf{0}$, which arises in many engineering problems, is known as the Laplace equation.

We now consider the significance of ∇u. First, we may regard this as a vector whose basic components equal the corresponding partial derivatives of u. Now let P denote the point at which u is evaluated. Extending the definition of the position vector $\mathbf{r}$ in Fig. 12.65 to three-dimensional space, we have

$$\mathbf{r} = x\mathbf{i} + y\mathbf{j} + z\mathbf{k}$$

When we move from P to a neighboring point Q, u increases by an amount du and $\mathbf{r}$ increases by an amount $d\mathbf{r}$. Then

$$du = \frac{\partial u}{\partial x} dx + \frac{\partial u}{\partial y} dy + \frac{\partial u}{\partial z} dz$$

$$d\mathbf{r} = dx\ \mathbf{i} + dy\ \mathbf{j} + dz\ \mathbf{k}$$

Applying Eq. (12.111) and performing the multiplication, we find that

$$du = \nabla u \cdot d\mathbf{r} \qquad (12.115)$$

This relationship yields the true significance of the gradient. In Fig. 12.68, S is the level surface at which $u = u_1$, and S' is the level surface at which $u = u_1 + du$. Consider that we start at point P on S and move to point Q on S'. Then u increases by du and $\mathbf{r}$ increases by $d\mathbf{r}$. Let ds denote the length of $d\mathbf{r}$. The rate of increase of u is du/ds, and it is a function of the path traversed from S to S'. Manifestly, this rate of increase is maximum if we move along the line that is normal to S at P. Let R denote the point where this line intersects S', let $dN = PR$, and let $\mathbf{N}$ denote a unit vector on this line that is sensed toward increasing values of u. Then

$$dN = \mathbf{N} \cdot d\mathbf{r}$$

$$du = \frac{du}{dN} dN = \frac{du}{dN} \mathbf{N} \cdot d\mathbf{r}$$

Comparing this result with Eq. (12.115), we obtain

$$\nabla u = \frac{du}{dN} \mathbf{N} \qquad (12.116)$$

Thus, the gradient of u is a vector that has the following properties: Its inclination is normal to the level surface, its sense is toward increasing values of u, and its length equals the maximum rate of increase of u.

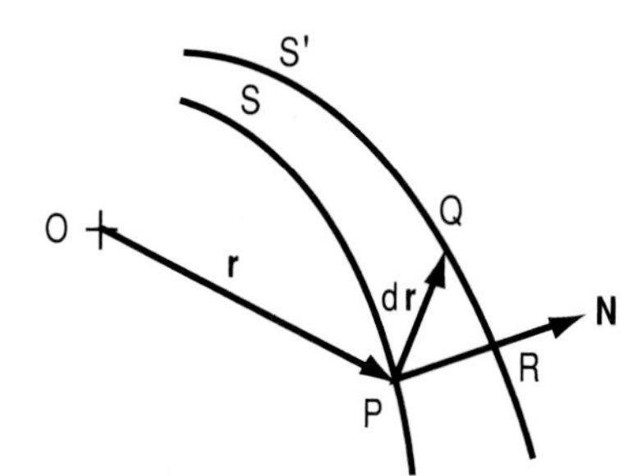

FIGURE 12.68 Significance of gradient of a scalar point function.

12.15 MOTION OF PLANETS AND ROCKETS

12.15.1 Introduction, Units, and Notation

Consider that the motion of body A is influenced by the gravitational attraction of body B. Our task is to analyze the motion of body A on the basis of observed conditions. For this purpose, we shall apply newtonian physics, and we shall make the following simplifying assumptions:

1. The motion of body A is influenced by body B exclusively. Thus, in tracing the orbit of a planet about the sun, we ignore the gravitational effects of other bodies in our solar system.
2. The mass of body A is extremely small in relation to the mass of body B. Therefore, we may ignore the effect that body A exerts on the motion of body B and, for our present purpose, we may consider that body B remains stationary.
3. Bodies A and B are both perfect spheres, and they are homogeneous. It can be demonstrated that the gravitational force of a solid spherical body of uniform composition is the same as it would be if its entire mass were concentrated at its center. Consequently, we may treat the two bodies as if their masses were indeed concentrated at their centers.
4. There is no frictional resistance to the motion of body A.

The point at which body A is at its minimum distance from B is called the *perihelion*. If the orbit of A is a closed curve, the point at which A is at its maximum distance from B is called the *aphelion*.

In the International System of Units (SI), the base units that concern us in our present investigation are the meter (m) for length, the kilogram (kg) for mass, and the second (s) for time. The unit of force is the newton (N). This is a derived unit; it is the force that is required to give a mass of 1 kg an acceleration of 1 m/s^2. Since force is the product of mass and acceleration, it follows that $1\ \text{N} = 1\ \text{kg} \cdot \text{m/s}^2$. Therefore, if the unit of a physical quantity is a compound that contains N, we can replace N with kg · m/s^2 if we wish to obtain a combination of base units exclusively.

The basic notation is as follows:

Let m and M = mass of bodies A and B, respectively
F = gravitational force that B exerts on A (and A exerts on B)
G = universal gravitational constant
r = distance between the centers of the bodies
v = velocity of body A
t = elapsed time

In SI, $G = 6.672 \times 10^{-11}\ \text{N} \cdot \text{m}^2/\text{kg}^2$. If we replace N with kg · m/s^2, the unit of G becomes m^3/(kg · s^2). We record the following for future reference: The unit of GM is m^3/s^2; the unit of rv^2 is m^3/s^2; the unit of r^2v^2 is m^4/s^2.

12.15.2 Mathematical Principles

In our present investigation, we shall use mainly the polar coordinate system defined in Art. 3.1.9. For convenience, we shall now state several mathematical relationships that will be applied in this investigation.

In Art. 12.4.8, we calculated the area bounded by the polar axis, a curve, and a radius vector, as shown in Fig. 12.19. We found that

$$dA = \frac{1}{2} r^2\, d\theta$$

Therefore, a rotating radius vector generates area at this rate:

$$\frac{dA}{dt} = \frac{1}{2} r^2 \frac{d\theta}{dt} \tag{12.117}$$

In Art. 3.2.7, we presented a general definition of a conic section. Associated with each conic is a point called a *focus* and a line called a *directrix*. The line that contains the focus and is perpendicular to the directrix is the *principal axis*. Let F denote the focus, L denote the directrix, and P denote a point on the conic. The ratio of the distance from F to P to the perpendicular distance from P to L is the *eccentricity* e. Corresponding to each type of conic is a specific value or range of values of e. The ellipse and hyperbola contain a second focus F' and a second directrix L', where F' and L' are mirror images of F and L, respectively, with reference to a particular line that is perpendicular to the principal axis.

To formulate a general equation of a conic section, we shall first position the conic in the following manner, which is shown in Fig. 12.69*a*: Place the focus F at the pole O and make the principal axis coincident with the polar axis. Let p denote the distance from F to L. Then

$$e = \frac{OP}{PR} = \frac{r}{p - r\cos\theta}$$

Solving for r,

$$r = \frac{ep}{1 + e\cos\theta}$$

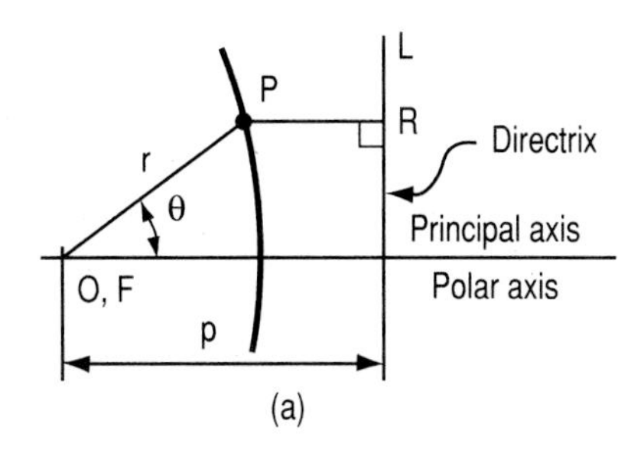

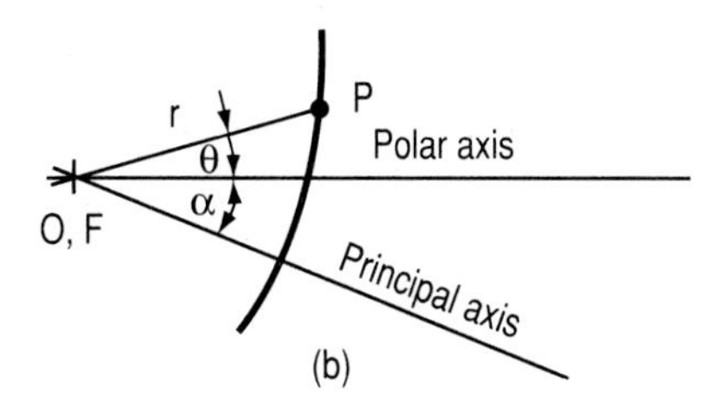

FIGURE 12.69 (*a*) Original position of conic; (*b*) position of conic after rotation.

The denominator of this fraction discloses the following:

1. If $e < 1$, r remains finite as θ ranges from 0 to 360°, and the curve is closed. (Specifically, the curve is an ellipse.)
2. If $e > 1$, r tends to infinity, and the curve is open. (Specifically, the curve is a hyperbola.) Angle θ is subject to the restriction $\cos\theta > -1/e$. This restriction corresponds with the fact that the hyperbola is asymptotic to two straight lines, as shown in Fig. 3.21.

For a reason that will soon become apparent, we shall now rotate the conic in Fig. 12.69*a* about O through a clockwise angle α, as shown in Fig. 12.69*b*. This rotation decreases the value of θ by α, and the equation of the conic now becomes

$$r = \frac{ep}{1 + e\cos(\theta + \alpha)} \tag{12.118}$$

In Art. 3.2.4, we analyzed the ellipse in terms of rectangular coordinates, and Figs. 3.17 and 3.19 display the dimensions of the ellipse. For convenience, the relevant dimensions are repeated in Fig. 12.70, where the center of the ellipse lies at the origin. We have $b^2 = a^2 - c^2$, and the eccentricity is $e = c/a$. Line $A'A$ is the *major axis* of the ellipse, and OA is the *semimajor axis*. Again let p denote the distance from F to L. Then

$$p = \frac{a}{e} - c = \frac{a^2}{c} - c = \frac{a^2 - c^2}{c} = \frac{b^2}{c} \qquad ep = \frac{b^2}{a}$$

The area of the ellipse is πab.

We shall resolve velocities and accelerations into their radial and transverse components, and we shall apply the notation in Art. 12.14.12.

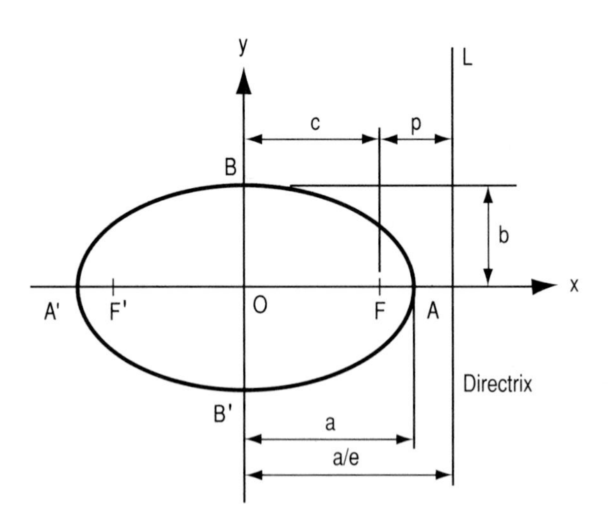

FIGURE 12.70 Dimensions of the ellipse.

12.15.3 Kepler's Laws of Planetary Motion

On the basis of available astronomical data, Kepler formulated the following laws governing the motion of each planet relative to the sun:

1. The orbit of the planet is an ellipse with the sun at one focus.
2. The line joining the sun and the planet sweeps out equal areas in equal intervals of time.
3. The square of the period of revolution of the planet is proportional to the cube of the length of the semimajor axis of the ellipse.

These laws were formulated empirically, but the development of calculus and newtonian physics enabled mathematicians to prove them rigorously. In our present generalized study, we view the first and third laws as pertaining to the special case where the orbit of body A relative to B is a closed curve.

12.15.4 Properties of the Orbit in the General Case

To identify the orbit of body A relative to body B, we shall place the pole at the position of body B, select an arbitrary polar axis, and call P the present position of body A, as shown in Fig. 12.71*a*. The gravitational force of B on A acts from P toward O; thus, it creates radial acceleration but not transverse acceleration. Newton's basic laws yield

$$F = -\frac{GMm}{r^2} = ma_r$$

Then $a_r = -GM/r^2$. Equating this expression to that in Eq. (12.110*a*), we obtain

$$\frac{d^2r}{dt^2} - r\left(\frac{d\theta}{dt}\right)^2 = -\frac{GM}{r^2} \qquad (r)$$

Now applying Eq. (12.110*b*) and setting the transverse acceleration equal to zero, we obtain

$$\frac{1}{r}\frac{d}{dt}\left(r^2\frac{d\theta}{dt}\right) = 0 \qquad (s)$$

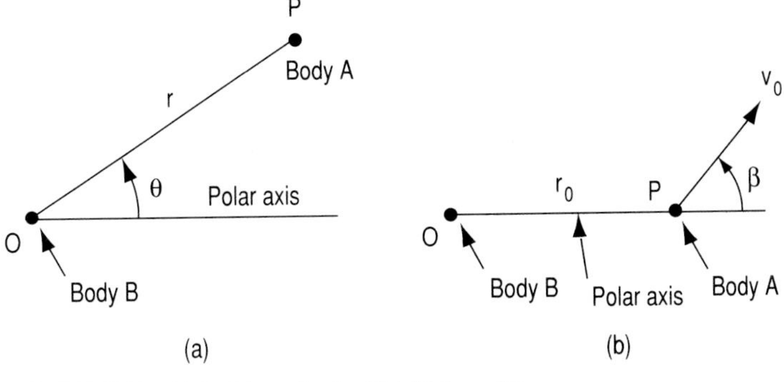

FIGURE 12.71 (*a*) Position of pole; (*b*) initial conditions.

Multiplying Eq. (s) by r and integrating, we obtain

$$r^2 \frac{d\theta}{dt} = h \tag{12.119}$$

where h is a constant of integration having the unit m^2/s. Equation (12.117) gives the rate at which a rotating radius vector generates area. Comparing this equation with Eq. (12.119), we see that $dA/dt = h/2$, and the area generated in a time interval t is

$$A = \frac{ht}{2} \tag{12.120}$$

Thus, the area generated in a given time interval is constant, and Kepler's second law is thus confirmed. (This law was also proved in Example 12.88 by vector analysis.)

By Eq. (12.119), $d\theta/dt = h/r^2$. Substituting this expression in Eq. (s) and rearranging, we obtain

$$\frac{d^2r}{dt^2} = \frac{h^2}{r^3} - \frac{GM}{r^2}$$

By setting $dr/dt = (dr/d\theta)(d\theta/dt) = (dr/d\theta)h/r^2$, we can transform the foregoing differential equation to one that contains $d^2r/d\theta^2$. The solution of the latter equation is

$$r = \frac{h^2/(GM)}{1 + e \cos(\theta + \alpha)} \tag{12.121}$$

where e and α are arbitrary constants.

The three constants in this equation are evaluated by introducing the initial conditions. Let r_0 and v_0 denote the value of r and v, respectively, when $\theta = 0$. Let β denote the angle that v_0 makes with the polar axis, as shown in Fig. 12.71b. The transverse component of v_0 is $v_0 \sin \beta = r_0(d\theta/dt)$. Equation (12.119) yields

$$h = r_0 v_0 \sin \beta \tag{12.122}$$

Equation (12.121) now yields

$$e \cos \alpha = \frac{r_0 v_0^2 \sin^2 \beta}{GM} - 1 \tag{12.123a}$$

It can also be demonstrated that

$$e \sin \alpha = \frac{r_0 v_0^2 \sin 2\beta}{2GM} \tag{12.123b}$$

Equations (12.123a) and (12.123b) enable us to isolate e and α to obtain the following:

$$e^2 = 1 + \frac{r_0^2 v_0^2 \sin^2 \beta (v_0^2 - 2GM/r_0)}{G^2 M^2} \tag{12.124}$$

$$\tan \alpha = \frac{r_0 v_0^2 \sin 2\beta}{2(r_0 v_0^2 \sin^2 \beta - GM)} \tag{12.125}$$

A comparison of Eq. (12.121) with Eq. (12.118) discloses that the orbit of body A relative to B is a conic that has these properties: The focus F is at body B; the eccentricity equals the quantity e as given by Eq. (12.124). We shall now classify the conic, applying the criteria given in Art. 3.2.7.

The definition of the eccentricity e restricts it to nonnegative values, and the square of any real number is also nonnegative. Thus, it follows that the expression in parentheses in Eq. (12.124) governs the algebraic sign of the second term and thereby establishes the relationship of e to 1. Let

$$D = \sqrt{\frac{2GM}{r_0}}$$

We arrive at these conclusions:

If $v_0 < D$, $e < 1$, and the orbit is an ellipse.
If $v_0 = D$, $e = 1$, and the orbit is a parabola.
If $v_0 > D$, $e > 1$, and the orbit is a hyperbola.

It is interesting to observe that the type of conic is independent of angle β.

Applying the units of quantities as recorded in Art. 12.15.1, we find that each of the foregoing equations is dimensionally homogeneous; i.e., the two sides of the equation have identical units.

12.15.5 Properties of the Orbit when $\beta = 90°$

We shall now investigate the special case where the velocity of body A is perpendicular to the polar axis when $\theta = 0$. This condition is achieved simply by measuring θ from the instant that the velocity of body A is perpendicular to the line joining bodies A and B. Equation (12.122) becomes

$$h = r_0 v_0 \tag{12.122a}$$

and Eq. (12.125) becomes $\tan \alpha = 0$. Two possibilities exist: $\alpha = 0$ and $\alpha = 180°$. We shall apply the designations Case 1 and Case 2 in referring to the conditions where $\alpha = 0$ and where $\alpha = 180°$, respectively. In Case 1, $\cos \alpha = 1$; in Case 2, $\cos \alpha = -1$.

We start with Case 1. Figure 12.69*b* reveals that the principal axis of the conic is now coincident with the polar axis. Equation (12.123*a*) becomes

$$e = \frac{r_0 v_0^2}{GM} - 1 \tag{12.123c}$$

Equation (12.121) now becomes

$$r = \frac{r_0^2 v_0^2}{GM + (r_0 v_0^2 - GM) \cos \theta} \tag{12.121a}$$

Since e is nonnegative, it follows from Eq. (12.123*c*) that

$$v_0 \geq \sqrt{\frac{GM}{r_0}} \tag{12.126}$$

This relationship establishes the minimum value of v_0 corresponding to Case 1.

With reference to Eq. (12.121a), the expression in parentheses is positive or zero. If it is zero, r is constant, and the orbit is a circle. If the expression has a nonzero value, r is minimum (and body A is at its perihelion) when $\theta = 0$. If the orbit is elliptic, r is maximum (and body A is at its aphelion) when $\theta = 180°$.

Figure 12.72a shows three possible orbits corresponding to Case 1. Since r is minimum when $\theta = 0$, point C lies closer to the focus F than to F'. If the orbit is elliptic, F' lies to the left of F; if the orbit is hyperbolic, the reverse is true.

We shall now investigate Case 2. The transition from Case 1 to Case 2 can be effected by rotating the principal axis of the conic about O through an angle of 180°. This rotation causes the foci F and F' to change their relative positions, as the expression for r will confirm.

Equation (12.123a) now becomes

$$e = 1 - \frac{r_0 v_0^2}{GM} \qquad (12.123d)$$

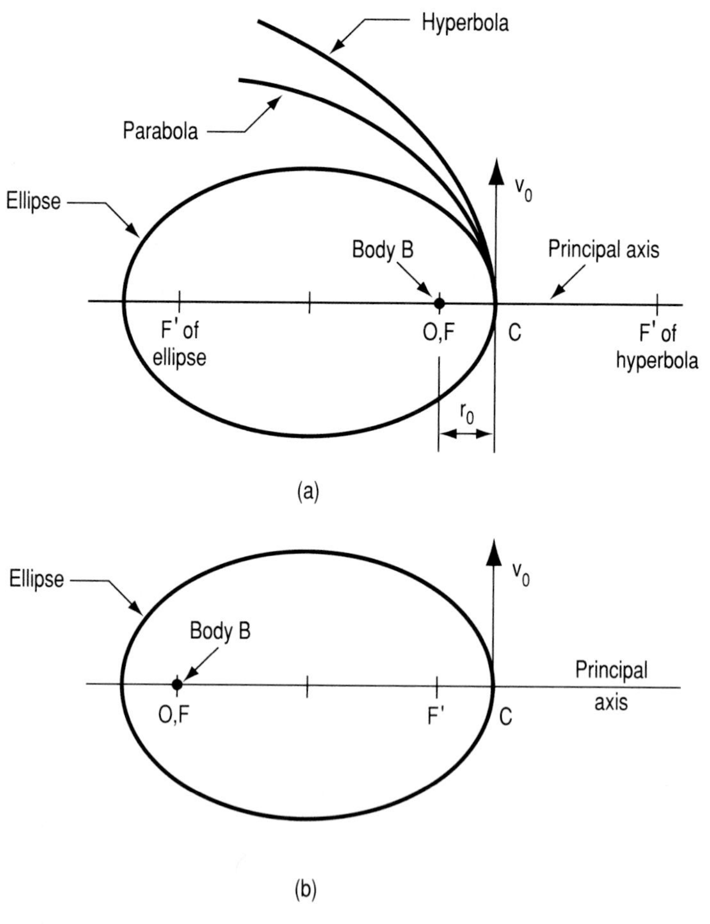

FIGURE 12.72 Possible orbits for (a) Case 1 and (b) Case 2.

Since e is nonnegative, it follows that

$$v_0 \le \sqrt{\frac{GM}{r_0}}$$

Therefore, Case 2 is restricted to an elliptic orbit.

Equation (12.121a) applies to Case 2 as well as Case 1. However, in Case 2, the expression in parentheses is either negative or zero. If it is negative, r is maximum when $\theta = 0$ and minimum when $\theta = 180°$. Therefore, in Case 2, the focus F' lies to the right of F, as shown in Fig. 12.72b.

For either Case 1 or Case 2, assume that the orbit of body A is an ellipse. Since body B is located at a focus, Fig. 12.70 discloses that the length of the major axis $A'A$ is the sum of the maximum and minimum values of r. By computing these values of r, we find that

$$\text{Length of major axis} = \frac{2r_0^2 v_0^2}{GM(1-e^2)} \tag{12.127}$$

EXAMPLE 12.90 When body A is located 4 m from body B, its velocity is 1.6 m/s and perpendicular to the line joining the two bodies. If $GM = 2.1\ \text{m}^3/\text{s}^2$, determine the equation of the orbit of body A. Classify the orbit.

SOLUTION

$$r_0 v_0^2 = 4 \times 1.6^2 = 10.24 \qquad r_0^2 v_0^2 = 4 \times 10.24 = 40.96$$

Equation (12.121a) yields

$$r = \frac{40.96}{2.1 + 8.14 \cos\theta} = \frac{19.50}{1 + 3.876 \cos\theta}$$

Since the eccentricity of this conic is 3.876, the orbit is a hyperbola.

EXAMPLE 12.91 With reference to Example 12.90, assume that the initial velocity of body A can be varied at will while other conditions remain unchanged. Determine the initial velocity that will cause the orbit to be (a) an ellipse with an eccentricity of 0.3; (b) a circle; (c) a parabola; (d) a hyperbola with an eccentricity of 1.6. Compute the length of the major axis of the ellipse and the diameter of the circle.

SOLUTION When Eqs. (12.123c) and (12.123d) are solved for v_0, the results can be combined in this manner:

$$v_0 = \sqrt{\frac{GM(1 \pm e)}{r_0}} \tag{t}$$

The plus sign applies to Case 1, the minus sign to Case 2.

Part a. Since the problem does not state whether body A is at its perihelion or aphelion at the initial state, the problem is ambiguous. We shall compute v_0 for both cases. In Case 1,

$$v_0^2 = \frac{2.1(1+0.3)}{4} = 0.6825 \qquad v_0 = 0.8261\ \text{m/s}$$

By Eq. (12.127),

$$\text{Length of major axis} = \frac{2 \times 4^2(0.6825)}{2.1(1 - 0.3^2)} = 11.43 \text{ m}$$

In Case 2,

$$v_0^2 = \frac{2.1(1 - 0.3)}{4} = 0.3675 \qquad v_0 = 0.6062 \text{ m/s}$$

From the foregoing result,

$$\text{Length of major axis} = 11.43\left(\frac{0.3675}{0.6825}\right) = 6.15 \text{ m}$$

Part b. Applying Eq. (t) with $e = 0$, we obtain

$$v_0 = \sqrt{\frac{GM}{r_0}} = \sqrt{\frac{2.1}{4}} = 0.7246 \text{ m/s}$$

Since body A remains at the fixed distance of 4 m from body B, the diameter of the circle is 8 m.

Part c. Applying Eq. (t) with the plus sign and with $e = 1$, we obtain

$$v_0 = \sqrt{\frac{2.1 \times 2}{4}} = 1.0247 \text{ m/s}$$

Part d. Applying Eq. (t) with the plus sign and with $e = 1.6$, we obtain $v_0 = 1.1683$ m/s.

Example 12.91 illuminates the manner in which the orbit of body A varies with changes in the initial velocity of this body. Assume that v_0 can be varied at will while holding r_0 constant and holding $\beta = 90°$. Again, body B is located at the focus F. When v_0 is relatively small, the orbit is an ellipse of relatively small extent, and F' lies between the two bodies. As v_0 increases, F' is displaced toward F and the ellipse elongates. When v_0 becomes equal to $\sqrt{GM/r_0}$, F' becomes coincident with F, and the orbit becomes a circle. As v_0 continues to increase, F' continues its movement, and the ellipse continues to elongate. When v_0 becomes equal to $\sqrt{2GM/r_0}$, the orbit becomes a parabola, and body A ceases to be a satellite of body B. As v_0 continues to increase, the orbit becomes a hyperbola, and the hyperbola then becomes steadily broader.

In a general sense, the foregoing conclusions are intuitively self-evident, for the extent to which body A is deflected from its original course under the influence of body B is determined by its initial velocity. The higher the initial velocity, the smaller the deflection.

The minimum value that v_0 requires if body A is to traverse an open trajectory is termed the *escape velocity* of body A. Let v_e denote this velocity. When $v_0 = v_e$, the orbit is a parabola, and therefore

$$v_e = \sqrt{\frac{2GM}{r_0}} \tag{12.128}$$

12.15.6 Period of Revolution in an Elliptic Orbit

Assume that the orbit of body A is an ellipse. The time required for body A to complete one revolution is called the *period of revolution*. We shall now compute this quantity, which we denote by T.

By Eq. (12.120), the area generated by the radius vector in time T is $hT/2$. We now equate this to the area of an ellipse; as stated in Art. 12.15.2, this is πab. Then

$$\frac{hT}{2} = \pi ab \qquad T^2 = \frac{4\pi^2 a^2 b^2}{h^2}$$

By comparing Eq. (12.121) with Eq. (12.118), we find that $h^2/(GM) = ep$. By applying the expression for ep as given in Art. 12.15.2, we obtain the following:

$$\frac{h^2}{GM} = ep = \frac{b^2}{a} \qquad h^2 = \frac{GMb^2}{a}$$

By subtracting this expression for h^2 in the equation for T^2, we obtain

$$T^2 = \frac{4\pi^2}{GM} a^3 \tag{12.129}$$

With reference to a planet in our solar system, M is the mass of the sun; therefore, T^2 is directly proportional to a^3. As stated in Art. 12.15.2, a is the length of the semimajor axis of the ellipse. Thus, Kepler's third law of planetary motion is now confirmed.

In accordance with our previous discussion, a is the arithmetic mean of the maximum and minimum values of r. For this reason, Kepler's third law is sometimes stated in this alternative manner: The square of the period of revolution is proportional to the cube of the mean distance of the planet from the sun. This statement is inaccurate. Since the variation of r is nonlinear, a is *not* the mean value of r.

In the special case where the orbit of body A is a circle of radius r, the quantity a becomes the radius. Then

$$T^2 = \frac{4\pi^2}{GM} r^3 \tag{12.129a}$$

12.15.7 Determination of Instantaneous Velocity

The instantaneous velocity of body A under the gravitational attraction of body B can be expressed in terms of the instantaneous value of either θ or r. We shall first compute the radial and transverse components of the velocity. The expressions for these components appear in Art. 12.14.12.

Differentiating Eq. (12.121) with respect to θ and then applying Eq. (12.119), we obtain the following:

$$\frac{dr}{d\theta} = \frac{h^2}{GM} \frac{e \sin(\theta + \alpha)}{[1 + e \cos(\theta + \alpha)]^2}$$

$$v_r = \frac{dr}{dt} = \frac{dr}{d\theta}\frac{d\theta}{dt} = \frac{dr}{d\theta}\frac{h}{r^2} = \frac{GM}{h} e \sin(\theta + \alpha)$$

$$v_{tr} = r\frac{d\theta}{dt} = r\frac{h}{r^2} = \frac{h}{r} = \frac{GM[1 + e\cos(\theta + \alpha)]}{h}$$

Setting $v^2 = v_r^2 + v_{tr}^2$ and simplifying, we obtain

$$v^2 = \left(\frac{GM}{h}\right)^2 [1 + 2e \cos(\theta + \alpha) + e^2] \tag{12.130}$$

This equation can be transformed to one that contains r rather than θ by solving Eq. (12.121) for $e \cos(\theta + \alpha)$ and substituting the resulting expression. The following equation emerges:

$$v^2 = \left(\frac{GM}{h}\right)^2 (e^2 - 1) + \frac{2GM}{r} \tag{12.131}$$

The second equation reveals that the velocity of body A is maximum when that body is at its perihelion. If the orbit is elliptic, the velocity is minimum when body A is at its aphelion. If the orbit is hyperbolic, the square of the velocity of body A approaches $(GM/h)^2(e^2 - 1)$ as a limit as the body recedes indefinitely from body B. Moreover, if the orbit is hyperbolic, it approaches a straight line, as shown in Fig. 3.21.

In the special case where the orbit of body A is a circle of radius r, $e = 0$, and Eq. (12.123*c*) yields

$$v = \sqrt{\frac{GM}{r}} \tag{12.131a}$$

EXAMPLE 12.92 With reference to Example 12.90, what is the velocity of body A when $\theta = 15°$?

SOLUTION The situation in Example 12.90 falls under Case 1, and $\alpha = 0$. From the equation for r, $e = 3.876$.

$$h = r_0 v_0 = 4 \times 1.6 = 6.4 \text{ m}^2/\text{s} \qquad GM = 2.1 \text{ m}^3/\text{s}^2$$

Substitution in Eq. (12.130) yields

$$v^2 = \left(\frac{2.1}{6.4}\right)^2 (1 + 2 \times 3.876 \cos 15° + 3.876^2)$$

$$v = 1.591 \text{ m/s}$$

Since θ is relatively small, the velocity at the given point is close to the initial velocity of 1.6 m/s.

12.15.8 Motion of Rockets

We shall now investigate the motion of rockets launched from the earth. In the present context, body A is the rocket and body B is the earth. The distance r between the bodies becomes the distance from the center of the earth to the rocket. The terms *perihelion* and *aphelion* are now replaced with the terms *perigee* and *apogee*, respectively. Rockets are usually elevated to a specified height above the earth's surface and then launched horizontally. In this situation, angle β in Fig. 12.71*b* is 90°, and the equations of Art. 12.15.5 are applicable.

The notation is as follows:

g = mean gravitational acceleration of a body near the earth's surface

M_e = mass of the earth

R_e = mean radius of the earth

We shall apply the following values:

$$g = 9.807 \text{ m/s}^2 \ (32.18 \text{ ft/s}^2)$$

$$R_e = 6378 \text{ km} = 6.378 \times 10^6 \text{ m} \ (20.926 \times 10^6 \text{ ft; } 3963 \text{ mi})$$

The gravitational attraction of the earth on a body of mass m that is near the earth's surface (in absolute value) is

$$F = \frac{GM_e m}{R_e^2} = mg$$

$$\therefore GM_e = gR_e^2 = 9.807(6.378 \times 10^6)^2 = 398.9 \times 10^{12} \text{ m}^3/\text{s}^2 \ (14.089 \times 10^{15} \text{ ft}^3/\text{s}^2)$$

EXAMPLE 12.93 A satellite circles the earth at an altitude of 1300 km (4.265×10^6 ft). Compute its velocity and period of revolution.

SOLUTION Let v denote the velocity.

$$r = 6378 + 1300 = 7678 \text{ km} = 7.678 \times 10^6 \text{ m} \ (25.191 \times 10^6 \text{ ft})$$

By Eq. (12.131*a*),

$$v = \sqrt{\frac{GM_e}{r}} = \sqrt{\frac{398.9 \times 10^{12}}{7.678 \times 10^6}} = 7208 \text{ m/s} \ (23{,}650 \text{ ft/s})$$

By Eq. (12.129*a*),

$$T^2 = \frac{4\pi^2(7.678 \times 10^6)^3}{398.9 \times 10^{12}} \qquad T = 6693 \text{ s} = 111.6 \text{ min}$$

EXAMPLE 12.94 A rocket will be launched horizontally at an altitude of 1200 km (3.937×10^6 ft). What is the escape velocity at this altitude?

SOLUTION

$$r_0 = 6378 + 1200 = 7578 \text{ km} \ (24.863 \times 10^6 \text{ ft})$$

By Eq. (12.128),

$$v_e = \sqrt{\frac{2GM_e}{r_0}} = \sqrt{\frac{2 \times 398.9 \times 10^{12}}{7.578 \times 10^6}} = 10{,}260 \text{ m/s} \ (33{,}660 \text{ ft/s})$$

EXAMPLE 12.95 A rocket will be launched horizontally at an altitude of 1260 km (4.134×10^6 ft). What must be its launch velocity if it is to orbit the earth and clear the earth's surface by 200 km (0.6562×10^6 ft)? If it is launched at this velocity, what will be the extreme values of its velocity?

SOLUTION Refer to Fig. 12.73. At its launching, the rocket is at its apogee, and it reaches its perigee when $\theta = 180°$. The extreme values of r are as follows:

$$r_{max} = r_0 = 6378 + 1260 = 7638 \text{ km } (25.060 \times 10^6 \text{ ft})$$

$$r_{min} = 6378 + 200 = 6578 \text{ km } (21.582 \times 10^6 \text{ ft})$$

Applying Eq. (12.121*a*) with $\cos\theta = -1$, we obtain

$$r_{min} = \frac{r_0^2 v_0^2}{2GM_e - r_0 v_0^2}$$

Solving,

$$v_0 = \sqrt{\frac{2GM_e r_{min}}{r_0(r_0 + r_{min})}} = \sqrt{\frac{2 \times 398.9 \times 6.578 \times 10^{18}}{7.638 \times 14.216 \times 10^{12}}}$$

$$= 6{,}952 \text{ m/s } (22{,}810 \text{ ft/s})$$

Since the rocket is at its apogee at launching, this calculated value is its minimum velocity. Equation (12.131) yields the following:

$$v_{max}^2 - v_{min}^2 = 2GM_e\left(\frac{1}{r_{min}} - \frac{1}{r_{max}}\right)$$

$$= 2 \times 398.9 \times 10^{12}\left(\frac{1}{6.578} - \frac{1}{7.638}\right)\frac{1}{10^6}$$

$$= 16.8316 \times 10^6$$

$$v_{max}^2 = (6.952 \times 10^3)^2 + 16.8316 \times 10^6$$

$$v_{max} = 8{,}072 \text{ m/s } (26{,}484 \text{ ft/s})$$

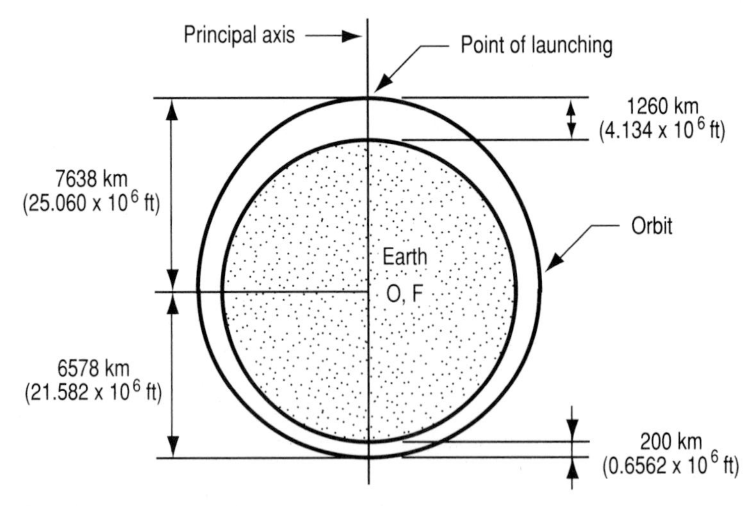

FIGURE 12.73 Orbit of earth satellite.

12.16 REFERENCES

Abramowitz, M., and I. A. Stegun: *Handbook of Mathematical Functions*, National Bureau of Standards (Applied Mathematics Series # 55), Washington, 1964.

Andrews, Larry C.: *Special Functions of Mathematics for Engineers*, 2d ed., McGraw-Hill, New York, 1991.

Gradshteyn, I. S., and I. M. Ryzhik: *Table of Integrals, Series, and Products*, Academic, San Diego, 1980.

Kevorkian, J.: *Partial Differential Equations*, Wadsworth and Brooks/Cole, Pacific Grove, Calif., 1990.

Korn, Granino A., and Theresa M. Korn: *Mathematical Handbook for Scientists and Engineers*, 2d ed., McGraw-Hill, New York, 1968.

Murphy, G. M.: *Ordinary Differential Equations and their Solutions*, Van Nostrand, Princeton, N.J., 1960.

Spanier, Jerome, and Keith B. Oldham: *An Atlas of Functions*, Hemisphere, Washington, 1987.

Tuma, Jan J.: *Engineering Mathematics Handbook*, 3d ed., McGraw-Hill, New York, 1987.

APPENDIX

DERIVATIVES AND INTEGRALS

TABLE A.1 DERIVATIVES

Notation: a, c, and n are constants; u and v are functions of x; e is the base of natural logarithms.

Algebraic Functions

$$\frac{dc}{dx} = 0 \tag{1}$$

$$\frac{d(u+v)}{dx} = \frac{du}{dx} + \frac{dv}{dx} \tag{2}$$

$$\frac{d(uv)}{dx} = u\frac{dv}{dx} + v\frac{du}{dx} \tag{3}$$

$$\frac{d(cu)}{dx} = c\frac{du}{dx} \tag{4}$$

$$\frac{d\left(\dfrac{u}{v}\right)}{dx} = \frac{v\dfrac{du}{dx} - u\dfrac{dv}{dx}}{v^2} \tag{5}$$

$$\frac{d(u^n)}{dx} = nu^{n-1}\frac{du}{dx} \tag{6}$$

Trigonometric Functions

$$\frac{d(\sin u)}{dx} = \cos u\frac{du}{dx} \tag{7}$$

$$\frac{d(\cos u)}{dx} = -\sin u\frac{du}{dx} \tag{8}$$

$$\frac{d(\tan u)}{dx} = \sec^2 u \frac{du}{dx} \tag{9}$$

$$\frac{d(\cot u)}{dx} = -\csc^2 u \frac{du}{dx} \tag{10}$$

$$\frac{d(\sec u)}{dx} = \sec u \tan u \frac{du}{dx} \tag{11}$$

$$\frac{d(\csc u)}{dx} = -\csc u \cot u \frac{du}{dx} \tag{12}$$

Inverse Trigonometric Functions

$$\frac{d(\sin^{-1} u)}{dx} = \frac{1}{\sqrt{1-u^2}} \frac{du}{dx} \tag{13}$$

$$\frac{d(\cos^{-1} u)}{dx} = -\frac{1}{\sqrt{1-u^2}} \frac{du}{dx} \tag{14}$$

$$\frac{d(\tan^{-1} u)}{dx} = \frac{1}{1+u^2} \frac{du}{dx} \tag{15}$$

$$\frac{d(\cot^{-1} u)}{dx} = -\frac{1}{1+u^2} \frac{du}{dx} \tag{16}$$

$$\frac{d(\sec^{-1} u)}{dx} = \frac{1}{u\sqrt{u^2-1}} \frac{du}{dx} \tag{17}$$

$$\frac{d(\csc^{-1} u)}{dx} = -\frac{1}{u\sqrt{u^2-1}} \frac{du}{dx} \tag{18}$$

Logarithmic and Exponential Functions

$$\frac{d(\log_e u)}{dx} = \frac{1}{u} \frac{du}{dx} \tag{19}$$

$$\frac{d(\log_a u)}{dx} = \frac{1}{u} \log_a e \frac{du}{dx} \tag{20}$$

$$\frac{d(e^u)}{dx} = e^u \frac{du}{dx} \tag{21}$$

$$\frac{d(a^u)}{dx} = a^u \log_e a \frac{du}{dx} \tag{22}$$

$$\frac{d(u^v)}{dx} = vu^{v-1} \frac{du}{dx} + u^v \log_e u \frac{dv}{dx} \tag{23}$$

Hyperbolic Functions

$$\frac{d(\sinh u)}{dx} = \cosh u \frac{du}{dx} \tag{24}$$

$$\frac{d(\cosh u)}{dx} = \sinh u \frac{du}{dx} \tag{25}$$

$$\frac{d(\tanh u)}{dx} = \operatorname{sech}^2 u \frac{du}{dx} \tag{26}$$

$$\frac{d(\coth u)}{dx} = -\operatorname{csch}^2 u \frac{du}{dx} \tag{27}$$

$$\frac{d(\operatorname{sech} u)}{dx} = -\operatorname{sech} u \tanh u \frac{du}{dx} \tag{28}$$

$$\frac{d(\operatorname{csch} u)}{dx} = -\operatorname{csch} u \coth u \frac{du}{dx} \tag{29}$$

Inverse Hyperbolic Functions

$$\frac{d(\sinh^{-1} u)}{dx} = \frac{1}{\sqrt{u^2+1}} \frac{du}{dx} \tag{30}$$

$$\frac{d(\cosh^{-1} u)}{dx} = \frac{1}{\sqrt{u^2-1}} \frac{du}{dx} \tag{31}$$

$$\frac{d(\tanh^{-1} u)}{dx} = \frac{1}{1-u^2} \frac{du}{dx} \tag{32}$$

$$\frac{d(\coth^{-1} u)}{dx} = -\frac{1}{u^2-1} \frac{du}{dx} \tag{33}$$

$$\frac{d(\operatorname{sech}^{-1} u)}{dx} = -\frac{1}{u\sqrt{1-u^2}} \frac{du}{dx} \tag{34}$$

$$\frac{d(\operatorname{csch}^{-1} u)}{dx} = -\frac{1}{u\sqrt{u^2+1}} \frac{du}{dx} \tag{35}$$

TABLE A.2 INTEGRALS

The constant of integration is omitted. The natural logarithm is denoted by "ln."

1 General Forms

$$\int u^n \, du = \frac{u^{n+1}}{n+1} \qquad n \neq -1 \tag{1}$$

$$\int u^{-1}\,du = \int \frac{du}{u} = \ln u \tag{2}$$

$$\int u\,dv = uv - \int v\,du \tag{3}$$

2 Algebraic Forms

2A Integrands Containing a + bu

$$\int (a+bu)^n\,du = \frac{(a+bu)^{n+1}}{b(n+1)} \qquad n \neq -1 \tag{4}$$

$$\int \frac{du}{a+bu} = \frac{1}{b}\ln|a+bu| \tag{5}$$

$$\int \frac{u\,du}{a+bu} = \frac{1}{b^2}(a+bu-a\ln|a+bu|) \tag{6}$$

$$\int \frac{u^2\,du}{a+bu} = \frac{1}{b^3}\left[\frac{(a+bu)^2}{2} - 2a(a+bu) + a^2\ln|a+bu|\right] \tag{7}$$

$$\int \frac{du}{u(a+bu)} = -\frac{1}{a}\ln\left|\frac{a+bu}{u}\right| \tag{8}$$

$$\int \frac{du}{u^2(a+bu)} = -\frac{1}{au} + \frac{b}{a^2}\ln\left|\frac{a+bu}{u}\right| \tag{9}$$

$$\int \frac{du}{(a+bu)^2} = -\frac{1}{b(a+bu)} \tag{10}$$

$$\int \frac{du}{(a+bu)^3} = -\frac{1}{2b(a+bu)^2} \tag{11}$$

$$\int \frac{u\,du}{(a+bu)^2} = \frac{1}{b^2}\left(\ln|a+bu| + \frac{a}{a+bu}\right) \tag{12}$$

$$\int \frac{u\,du}{(a+bu)^3} = \frac{1}{b^2}\left[-\frac{1}{a+bu} + \frac{a}{2(a+bu)^2}\right] \tag{13}$$

$$\int \frac{u^2\,du}{(a+bu)^2} = \frac{1}{b^3}\left(a+bu-2a\ln|a+bu| - \frac{a^2}{a+bu}\right) \tag{14}$$

$$\int \frac{u^2\,du}{(a+bu)^3} = \frac{1}{b^3}\left(\ln|a+bu| + \frac{2a}{a+bu} - \frac{a^2}{2(a+bu)^2}\right) \tag{15}$$

$$\int \frac{du}{u(a+bu)^2} = \frac{1}{a(a+bu)} - \frac{1}{a^2} \ln \left| \frac{a+bu}{u} \right| \tag{16}$$

$$\int \frac{du}{u^2(a+bu)^2} = -\frac{a+2bu}{a^2u(a+bu)} + \frac{2b}{a^3} \ln \left| \frac{a+bu}{u} \right| \tag{17}$$

2B *Integrands Containing* $a^2 \pm u^2$ *or* $u^2 - a^2$

$$\int \frac{du}{a^2+u^2} = \frac{1}{a} \tan^{-1} \frac{u}{a} \tag{18}$$

$$\int \frac{du}{a^2-u^2} = \frac{1}{2a} \ln \left| \frac{a+u}{a-u} \right| \tag{19}$$

$$\int \frac{du}{u^2-a^2} = \frac{1}{2a} \ln \left| \frac{u-a}{u+a} \right| \tag{20}$$

2C *Integrands Containing* $a + bu^2$

$$\int \frac{du}{a+bu^2} = \frac{1}{\sqrt{ab}} \tan^{-1} u \sqrt{\frac{b}{a}} \qquad \text{when } a > 0, b > 0 \tag{21}$$

$$\int \frac{du}{a+bu^2} = \frac{1}{2\sqrt{-ab}} \ln \left| \frac{\sqrt{a}+u\sqrt{-b}}{\sqrt{a}-u\sqrt{-b}} \right| \qquad \text{when } a > 0, b < 0 \tag{22}$$

$$\int \frac{u\,du}{a+bu^2} = \frac{1}{2b} \ln \left| u^2 + \frac{a}{b} \right| \tag{23}$$

$$\int \frac{u^2\,du}{a+bu^2} = \frac{u}{b} - \frac{a}{b} \int \frac{du}{a+bu^2} \tag{24}$$

$$\int \frac{du}{u(a+bu^2)} = \frac{1}{2a} \ln \left| \frac{u^2}{a+bu^2} \right| \tag{25}$$

$$\int \frac{du}{u^2(a+bu^2)} = -\frac{1}{au} - \frac{b}{a} \int \frac{du}{a+bu^2} \tag{26}$$

$$\int \frac{du}{(a+bu^2)^2} = \frac{u}{2a(a+bu^2)} + \frac{1}{2a} \int \frac{du}{a+bu^2} \tag{27}$$

2D *Integrands Containing* $u^m(a + bu^n)^p$.

The following integrals are alternatives.

$$\int u^m(a+bu^n)^p\,du = \frac{u^{m-n+1}(a+bu^n)^{p+1}}{(np+m+1)b} - \frac{(m-n+1)a}{(np+m+1)b} \int u^{m-n}(a+bu^n)^p\,du \tag{28}$$

$$\int u^m(a+bu^n)^p\,du = \frac{u^{m+1}(a+bu^n)^p}{np+m+1} + \frac{anp}{np+m+1} \int u^m(a+bu^n)^{p-1}\,du \tag{29}$$

$$\int u^m(a+bu^n)^p\,du = \frac{u^{m+1}(a+bu^n)^{p+1}}{(m+1)a} - \frac{(np+m+n+1)b}{(m+1)a}\int u^{m+n}(a+bu^n)^p\,du \tag{30}$$

$$\int u^m(a+bu^n)^p\,du = -\frac{u^{m+1}(a+bu^n)^{p+1}}{n(p+1)a} + \frac{np+m+n+1}{n(p+1)a}\int u^m(a+bu^n)^{p+1}\,du \tag{31}$$

2E Integrands Containing $\sqrt{a+bu}$

$$\int \sqrt{a+bu}\,du = \frac{2\sqrt{(a+bu)^3}}{3b} \tag{32}$$

$$\int u\sqrt{a+bu}\,du = -\frac{2(2a-3bu)\sqrt{(a+bu)^3}}{15b^2} \tag{33}$$

$$\int u^2\sqrt{a+bu}\,du = \frac{2(8a^2-12abu+15b^2u^2)\sqrt{(a+bu)^3}}{105b^3} \tag{34}$$

$$\int \frac{du}{\sqrt{a+bu}} = \frac{2\sqrt{a+bu}}{b} \tag{35}$$

$$\int \frac{u\,du}{\sqrt{a+bu}} = -\frac{2(2a-bu)\sqrt{a+bu}}{3b^2} \tag{36}$$

$$\int \frac{u^2\,du}{\sqrt{a+bu}} = \frac{2(8a^2-4abu+3b^2u^2)\sqrt{a+bu}}{15b^3} \tag{37}$$

$$\int \frac{du}{u\sqrt{a+bu}} = \frac{1}{\sqrt{a}}\ln\left|\frac{\sqrt{a+bu}-\sqrt{a}}{\sqrt{a+bu}+\sqrt{a}}\right| \quad \text{when } a>0 \tag{38}$$

$$\int \frac{du}{u\sqrt{a+bu}} = \frac{2}{\sqrt{-a}}\tan^{-1}\sqrt{\frac{a+bu}{-a}} \quad \text{when } a<0 \tag{39}$$

$$\int \frac{du}{u^2\sqrt{a+bu}} = -\frac{\sqrt{a+bu}}{au} - \frac{b}{2a}\int \frac{du}{u\sqrt{a+bu}} \tag{40}$$

$$\int \frac{\sqrt{a+bu}\,du}{u} = 2\sqrt{a+bu} + a\int \frac{du}{u\sqrt{a+bu}} \tag{41}$$

2F Integrands Containing $\sqrt{a^2+u^2}$

$$\int \sqrt{a^2+u^2}\,du = \frac{u\sqrt{a^2+u^2}+a^2\ln|u+\sqrt{a^2+u^2}|}{2} \tag{42}$$

$$\int \sqrt{(a^2+u^2)^3}\,du = \frac{u(2u^2+5a^2)\sqrt{a^2+u^2}}{8} + \frac{3a^4\ln|u+\sqrt{a^2+u^2}|}{8} \tag{43}$$

$$\int \sqrt{(a^2+u^2)^n}\,du = \frac{u\sqrt{(a^2+u^2)^n}}{n+1} + \frac{a^2 n}{n+1}\int \sqrt{(a^2+u^2)^{n-2}}\,du \tag{44}$$

$$\int u\sqrt{(a^2+u^2)^n}\,du = \frac{\sqrt{(a^2+u^2)^{n+2}}}{n+2} \tag{45}$$

$$\int u^2\sqrt{a^2+u^2}\,du = \frac{u(2u^2+a^2)\sqrt{a^2+u^2}}{8} - \frac{a^4 \ln |u+\sqrt{a^2+u^2}|}{8} \tag{46}$$

$$\int \frac{du}{\sqrt{a^2+u^2}} = \ln |u+\sqrt{a^2+u^2}| \tag{47}$$

$$\int \frac{du}{\sqrt{(a^2+u^2)^3}} = \frac{u}{a^2\sqrt{a^2+u^2}} \tag{48}$$

$$\int \frac{u\,du}{\sqrt{a^2+u^2}} = \sqrt{a^2+u^2} \tag{49}$$

$$\int \frac{u\,du}{\sqrt{(a^2+u^2)^3}} = -\frac{1}{\sqrt{a^2+u^2}} \tag{50}$$

$$\int \frac{u^2\,du}{\sqrt{a^2+u^2}} = \frac{u\sqrt{a^2+u^2} - a^2 \ln |u+\sqrt{a^2+u^2}|}{2} \tag{51}$$

$$\int \frac{u^2\,du}{\sqrt{(a^2+u^2)^3}} = -\frac{u}{\sqrt{a^2+u^2}} + \ln |u+\sqrt{a^2+u^2}| \tag{52}$$

$$\int \frac{du}{u\sqrt{a^2+u^2}} = \frac{1}{a} \ln \left| \frac{u}{a+\sqrt{a^2+u^2}} \right| \tag{53}$$

$$\int \frac{du}{u^2\sqrt{a^2+u^2}} = -\frac{\sqrt{a^2+u^2}}{a^2 u} \tag{54}$$

$$\int \frac{du}{u^3\sqrt{a^2+u^2}} = -\frac{\sqrt{a^2+u^2}}{2a^2u^2} + \frac{1}{2a^3} \ln \left| \frac{a+\sqrt{a^2+u^2}}{u} \right| \tag{55}$$

$$\int \frac{\sqrt{a^2+u^2}\,du}{u} = \sqrt{a^2+u^2} - a \ln \left| \frac{a+\sqrt{a^2+u^2}}{u} \right| \tag{56}$$

$$\int \frac{\sqrt{a^2+u^2}\,du}{u^2} = -\frac{\sqrt{a^2+u^2}}{u} + \ln |u+\sqrt{a^2+u^2}| \tag{57}$$

2G Integrands Containing $\sqrt{a^2 - u^2}$

$$\int \sqrt{a^2-u^2}\,du = \frac{1}{2}\left(u\sqrt{a^2-u^2} + a^2 \sin^{-1}\frac{u}{a}\right) \tag{58}$$

$$\int \sqrt{(a^2-u^2)^3}\,du = \frac{u}{8}(5a^2-2u^2)\sqrt{a^2-u^2} + \frac{3a^4}{8}\sin^{-1}\frac{u}{a} \tag{59}$$

$$\int \sqrt{(a^2-u^2)^n}\,du = \frac{u\sqrt{(a^2-u^2)^n}}{n+1} + \frac{a^2 n}{n+1}\int \sqrt{(a^2-u^2)^{n-2}}\,du \tag{60}$$

$$\int u\sqrt{(a^2-u^2)^n}\,du = -\frac{\sqrt{(a^2-u^2)^{n+2}}}{n+2} \tag{61}$$

$$\int u^2\sqrt{a^2-u^2}\,du = \frac{u(2u^2-a^2)\sqrt{a^2-u^2}}{8} + \frac{a^4}{8}\sin^{-1}\frac{u}{a} \tag{62}$$

$$\int \frac{du}{\sqrt{a^2-u^2}} = \sin^{-1}\frac{u}{a} \tag{63}$$

$$\int \frac{du}{\sqrt{(a^2-u^2)^3}} = \frac{u}{a^2\sqrt{a^2-u^2}} \tag{64}$$

$$\int \frac{u\,du}{\sqrt{a^2-u^2}} = -\sqrt{a^2-u^2} \tag{65}$$

$$\int \frac{u\,du}{\sqrt{(a^2-u^2)^3}} = \frac{1}{\sqrt{a^2-u^2}} \tag{66}$$

$$\int \frac{u^2\,du}{\sqrt{a^2-u^2}} = -\frac{u\sqrt{a^2-u^2}}{2} + \frac{a^2}{2}\sin^{-1}\frac{u}{a} \tag{67}$$

$$\int \frac{u^2\,du}{\sqrt{(a^2-u^2)^3}} = \frac{u}{\sqrt{a^2-u^2}} - \sin^{-1}\frac{u}{a} \tag{68}$$

$$\int \frac{du}{u\sqrt{a^2-u^2}} = \frac{1}{a}\ln\left|\frac{u}{a+\sqrt{a^2-u^2}}\right| \tag{69}$$

$$\int \frac{du}{u^2\sqrt{a^2-u^2}} = -\frac{\sqrt{a^2-u^2}}{a^2u} \tag{70}$$

$$\int \frac{du}{u^3\sqrt{a^2-u^2}} = -\frac{\sqrt{a^2-u^2}}{2a^2u^2} + \frac{1}{2a^3}\ln\left|\frac{u}{a+\sqrt{a^2-u^2}}\right| \tag{71}$$

$$\int \frac{\sqrt{a^2-u^2}\,du}{u} = \sqrt{a^2-u^2} - a\ln\left|\frac{a+\sqrt{a^2-u^2}}{u}\right| \tag{72}$$

$$\int \frac{\sqrt{a^2-u^2}\,du}{u^2} = -\frac{\sqrt{a^2-u^2}}{u} - \sin^{-1}\frac{u}{a} \tag{73}$$

2H Integrands Containing $\sqrt{u^2-a^2}$

$$\int \sqrt{u^2-a^2}\,du = \frac{u\sqrt{u^2-a^2}}{2} - a^2\ln\left|u+\sqrt{u^2-a^2}\right| \tag{74}$$

$$\int \sqrt{(u^2-a^2)^3}\,du = \frac{u(2u^2-5a^2)\sqrt{u^2-a^2}}{8} + \frac{3a^4}{8}\ln\left|u+\sqrt{u^2-a^2}\right| \tag{75}$$

$$\int \sqrt{(u^2 - a^2)^n}\, du = \frac{u\sqrt{(u^2 - a^2)^n}}{n+1} - \frac{a^2 n}{n+1} \int \sqrt{(u^2 - a^2)^{n-2}}\, du \tag{76}$$

$$\int u\sqrt{(u^2 - a^2)^n}\, du = \frac{\sqrt{(u^2 - a^2)^{n+2}}}{n+2} \tag{77}$$

$$\int u^2\sqrt{u^2 - a^2}\, du = \frac{u(2u^2 - a^2)\sqrt{u^2 - a^2}}{8} - \frac{a^4}{8} \ln |u + \sqrt{u^2 - a^2}| \tag{78}$$

$$\int \frac{du}{\sqrt{u^2 - a^2}} = \ln |u + \sqrt{u^2 - a^2}| \tag{79}$$

$$\int \frac{du}{\sqrt{(u^2 - a^2)^3}} = -\frac{u}{a^2\sqrt{u^2 - a^2}} \tag{80}$$

$$\int \frac{u\, du}{\sqrt{u^2 - a^2}} = \sqrt{u^2 - a^2} \tag{81}$$

$$\int \frac{u\, du}{\sqrt{(u^2 - a^2)^3}} = -\frac{1}{\sqrt{u^2 - a^2}} \tag{82}$$

$$\int \frac{u^2\, du}{\sqrt{u^2 - a^2}} = \frac{u\sqrt{u^2 - a^2}}{2} + \frac{a^2}{2} \ln |u + \sqrt{u^2 - a^2}| \tag{83}$$

$$\int \frac{u^2\, du}{\sqrt{(u^2 - a^2)^3}} = -\frac{u}{\sqrt{u^2 - a^2}} + \ln |u + \sqrt{u^2 - a^2}| \tag{84}$$

$$\int \frac{du}{u\sqrt{u^2 - a^2}} = \frac{1}{a} \cos^{-1} \frac{a}{u} \tag{85}$$

$$\int \frac{du}{u^2\sqrt{u^2 - a^2}} = \frac{\sqrt{u^2 - a^2}}{a^2 u} \tag{86}$$

$$\int \frac{du}{u^3\sqrt{u^2 - a^2}} = \frac{\sqrt{u^2 - a^2}}{2a^2u^2} + \frac{1}{2a^3} \cos^{-1} \frac{a}{u} \tag{87}$$

$$\int \frac{\sqrt{u^2 - a^2}\, du}{u} = \sqrt{u^2 - a^2} - a \cos^{-1} \frac{a}{u} \tag{88}$$

$$\int \frac{\sqrt{u^2 - a^2}\, du}{u^2} = -\frac{\sqrt{u^2 - a^2}}{u} + \ln |u + \sqrt{u^2 - a^2}| \tag{89}$$

2I Integrands Containing $\sqrt{2au \pm u^2}$

$$\int \sqrt{2au - u^2}\, du = \frac{(u - a)\sqrt{2au - u^2}}{2} + \frac{a^2}{2} \sin^{-1} \frac{u - a}{a} \tag{90}$$

$$\int \sqrt{2au + u^2}\, du = \frac{(u + a)\sqrt{2au + u^2}}{2} - \frac{a^2}{2} \ln |u + a + \sqrt{2au + u^2}| \tag{91}$$

$$\int u\sqrt{2au-u^2}\,du = -\frac{(3a^2+au-2u^2)\sqrt{2au-u^2}}{6} + \frac{a^3}{2}\sin^{-1}\frac{u-a}{a} \tag{92}$$

$$\int u\sqrt{2au+u^2}\,du = \frac{\sqrt{(2au+u^2)^3}}{3} - a\int\sqrt{2au+u^2}\,du \tag{93}$$

$$\int u^n\sqrt{2au-u^2}\,du = -\frac{u^{n-1}\sqrt{(2au-u^2)^3}}{n+2} + \frac{(2n+1)a}{n+2}\int u^{n-1}\sqrt{2au-u^2}\,du \tag{94}$$

$$\int \frac{du}{\sqrt{2au-u^2}} = \sin^{-1}\frac{u-a}{a} \tag{95}$$

$$\int \frac{du}{\sqrt{2au+u^2}} = \ln\left|u+a+\sqrt{2au+u^2}\right| \tag{96}$$

$$\int \frac{du}{\sqrt{(2au-u^2)^3}} = \frac{u-a}{a^2\sqrt{2au-u^2}} \tag{97}$$

$$\int \frac{du}{\sqrt{(2au+u^2)^3}} = -\frac{u+a}{a^2\sqrt{2au+u^2}} \tag{98}$$

$$\int \frac{u\,du}{\sqrt{2au-u^2}} = -\sqrt{2au-u^2} + a\sin^{-1}\frac{u-a}{a} \tag{99}$$

$$\int \frac{u\,du}{\sqrt{2au+u^2}} = \sqrt{2au+u^2} - a\ln\left|u+a+\sqrt{2au+u^2}\right| \tag{100}$$

$$\int \frac{u\,du}{\sqrt{(2au\pm u^2)^3}} = \frac{u}{a\sqrt{2au\pm u^2}} \tag{101}$$

$$\int \frac{u^2\,du}{\sqrt{2au-u^2}} = -\frac{(u+3a)\sqrt{2au-u^2}}{2} + \frac{3a^2}{2}\sin^{-1}\frac{u-a}{a} \tag{102}$$

$$\int \frac{u^n\,du}{\sqrt{2au-u^2}} = -\frac{u^{n-1}\sqrt{2au-u^2}}{n} - \frac{a(1-2n)}{n}\int\frac{u^{n-1}\,du}{\sqrt{2au-u^2}} \tag{103}$$

$$\int \frac{du}{u\sqrt{2au\pm u^2}} = -\frac{\sqrt{2au\pm u^2}}{au} \tag{104}$$

$$\int \frac{du}{u^n\sqrt{2au-u^2}} = \frac{\sqrt{2au-u^2}}{a(1-2n)u^n} + \frac{n-1}{(2n-1)a}\int\frac{du}{u^{n-1}\sqrt{2au-u^2}} \tag{105}$$

$$\int \frac{\sqrt{2au-u^2}\,du}{u} = \sqrt{2au-u^2} + a\sin^{-1}\frac{u-a}{a} \tag{106}$$

$$\int \frac{\sqrt{2au+u^2}\,du}{u} = \sqrt{2au+u^2} + a\ln\left|u+a+\sqrt{2au+u^2}\right| \tag{107}$$

$$\int \frac{\sqrt{2au-u^2}\,du}{u^2} = -\frac{2\sqrt{2au-u^2}}{u} - \sin^{-1}\frac{u-a}{a} \tag{108}$$

$$\int \frac{\sqrt{2au+u^2}\,du}{u^2} = -\frac{2\sqrt{2au+u^2}}{u} + \ln\left|u+a+\sqrt{2au+u^2}\right| \tag{109}$$

$$\int \frac{\sqrt{2au-u^2}\,du}{u^n} = \frac{\sqrt{(2au-u^2)^3}}{(3-2n)au^n} + \frac{n-3}{(2n-3)a}\int \frac{\sqrt{2au-u^2}\,du}{u^{n-1}} \tag{110}$$

2J *Integrands Containing* $au^2 + bu + c$

$$\int \frac{du}{au^2+bu+c} = \frac{2}{\sqrt{4ac-b^2}}\tan^{-1}\frac{2au+b}{\sqrt{4ac-b^2}} \qquad \text{when } 4ac > b^2 \tag{111}$$

$$\int \frac{du}{au^2+bu+c} = \frac{1}{\sqrt{b^2-4ac}}\ln\left|\frac{2au+b-\sqrt{b^2-4ac}}{2au+b+\sqrt{b^2-4ac}}\right| \qquad \text{when } 4ac < b^2 \tag{112}$$

$$\int \frac{u\,du}{au^2+bu+c} = \frac{1}{2a}\ln\left|au^2+bu+c\right| - \frac{b}{2a}\int \frac{du}{au^2+bu+c} \tag{113}$$

$$\int \frac{u^2\,du}{au^2+bu+c} = \frac{u}{a} - \frac{b}{2a^2}\ln\left|au^2+bu+c\right| + \frac{b^2-4ac}{2a^2}\int \frac{du}{au^2+bu+c} \tag{114}$$

$$\int \frac{du}{u(au^2+bu+c)} = \frac{1}{2c}\ln\left|\frac{u^2}{au^2+bu+c}\right| - \frac{b}{2c}\int \frac{du}{au^2+bu+c} \tag{115}$$

$$\int \frac{du}{u^2(au^2+bu+c)} = \frac{b}{2c^2}\ln\left|\frac{au^2+bu+c}{u^2}\right| - \frac{1}{cu} + \left(\frac{b^2}{2c^2} - \frac{a}{c}\right)\int \frac{du}{au^2+bu+c} \tag{116}$$

2K *Integrands Containing* $\sqrt{au^2 + bu + c}$

$$\int \sqrt{au^2+bu+c}\,du = \frac{2au+b}{4a}\sqrt{au^2+bu+c} - \frac{b^2-4ac}{8a}\int \frac{du}{\sqrt{au^2+bu+c}} \tag{117}$$

$$\int \frac{du}{\sqrt{au^2+bu+c}} = \frac{1}{\sqrt{a}}\ln\left|2au+b+2\sqrt{a}\sqrt{au^2+bu+c}\right| \qquad \text{when } a > 0 \tag{118}$$

$$\int \frac{du}{\sqrt{au^2+bu+c}} = \frac{1}{\sqrt{-a}}\sin^{-1}\frac{-2au-b}{\sqrt{b^2-4ac}} \qquad \text{when } a < 0 \tag{119}$$

$$\int \frac{u\,du}{\sqrt{au^2+bu+c}} = \frac{1}{a}\sqrt{au^2+bu+c} - \frac{b}{2a}\int \frac{du}{\sqrt{au^2+bu+c}} \tag{120}$$

$$\int \frac{u^2\,du}{\sqrt{au^2+bu+c}} = \left(\frac{u}{2a} - \frac{3b}{4a^2}\right)\sqrt{au^2+bu+c} + \frac{3b^2-4ac}{8a^2}\int \frac{du}{\sqrt{au^2+bu+c}} \qquad (121)$$

$$\int \frac{du}{u\sqrt{au^2+bu+c}} = -\frac{1}{\sqrt{c}} \ln \left| \frac{\sqrt{au^2+bu+c}+\sqrt{c}}{u} + \frac{b}{2\sqrt{c}} \right| \quad \text{when } c > 0 \qquad (122)$$

$$\int \frac{du}{u\sqrt{au^2+bu+c}} = \frac{1}{\sqrt{-c}} \sin^{-1} \frac{bu+2c}{u\sqrt{b^2-4ac}} \quad \text{when } c < 0 \qquad (123)$$

$$\int \frac{du}{u^2\sqrt{au^2+bu+c}} = -\frac{\sqrt{au^2+bu+c}}{cu} - \frac{b}{2c}\int \frac{du}{u\sqrt{au^2+bu+c}} \qquad (124)$$

2L Miscellaneous Algebraic Forms

$$\int \sqrt{\frac{a+u}{b+u}}\,du = \sqrt{(a+u)(b+u)} + (a-b)\ln\left(\sqrt{a+u}+\sqrt{b+u}\right) \qquad (125)$$

$$\int \sqrt{\frac{a+u}{b-u}}\,du = -\sqrt{(a+u)(b-u)} - (a+b)\sin^{-1}\sqrt{\frac{b-u}{a+b}} \qquad (126)$$

$$\int \sqrt{\frac{a-u}{b+u}}\,du = \sqrt{(a-u)(b+u)} + (a+b)\sin^{-1}\sqrt{\frac{b+u}{a+b}} \qquad (127)$$

$$\int \frac{du}{u\sqrt{u^n-a^2}} = \frac{2}{an}\sec^{-1}\frac{\sqrt{u^n}}{a} \qquad (128)$$

3 Exponential and Logarithmic Forms

$$\int e^u\,du = e^u \qquad (129)$$

$$\int a^u\,du = \frac{a^u}{\ln a} \qquad (130)$$

$$\int e^{au}\,du = \frac{e^{au}}{a} \qquad (131)$$

$$\int a^{bu}\,du = \frac{a^{bu}}{b\ln a} \qquad (132)$$

$$\int \ln u\,du = u\ln u - u \qquad (133)$$

$$\int \frac{du}{\ln u} = \ln|\ln u| + \ln u + \frac{(\ln u)^2}{2\cdot 2!} + \frac{(\ln u)^3}{3\cdot 3!} + \cdots \qquad (134)$$

$$\int \frac{du}{u \ln u} = \ln |\ln u| \tag{135}$$

$$\int ue^{au}\, du = \frac{e^{au}}{a^2}(au - 1) \tag{136}$$

$$\int u^n e^{au}\, du = \frac{u^n e^{au}}{a} - \frac{n}{a}\int u^{n-1} e^{au}\, du \tag{137}$$

$$\int \frac{e^{au}\, du}{u} = \ln u + au + \frac{a^2u^2}{2 \cdot 2!} + \frac{a^3u^3}{3 \cdot 3!} + \cdots \tag{138}$$

$$\int \frac{e^{au}\, du}{u^n} = \frac{1}{n-1}\left(-\frac{e^{au}}{u^{n-1}} + a\int \frac{e^{au}\, du}{u^{n-1}}\right) \tag{139}$$

$$\int ua^u\, du = \frac{a^u u}{\ln a} - \frac{a^u}{(\ln a)^2} \tag{140}$$

$$\int \frac{a^u\, du}{u} = \ln u + u \ln a + \frac{(u \ln a)^2}{2 \cdot 2!} + \frac{(u \ln a)^3}{3 \cdot 3!} + \cdots \tag{141}$$

$$\int u \ln u\, du = \frac{u^2}{2}\left(\ln u - \frac{1}{2}\right) \tag{142}$$

$$\int u^2 \ln u\, du = \frac{u^3}{3}\left(\ln u - \frac{1}{3}\right) \tag{143}$$

$$\int u^n \ln u\, du = \frac{u^{n+1}}{n+1}\left(\ln u - \frac{1}{n+1}\right) \tag{144}$$

$$\int e^{au} \ln u\, du = \frac{e^{au} \ln u}{a} - \frac{1}{a}\int \frac{e^{au}}{u}\, du \tag{145}$$

$$\int \frac{du}{1 + e^u} = \ln\left(\frac{e^u}{1 + e^u}\right) \tag{146}$$

$$\int \frac{du}{a + be^{nu}} = \frac{1}{an}(nu - \ln |a + be^{nu}|) \tag{147}$$

$$\int \frac{du}{ae^{nu} + be^{-nu}} = \frac{1}{n\sqrt{ab}} \tan^{-1}\left(e^{nu}\sqrt{\frac{a}{b}}\right) \tag{148}$$

$$\int \ln (a^2 + u^2)\, du = u \ln (a^2 + u^2) - 2u + 2a \tan^{-1}\frac{u}{a} \tag{149}$$

4 Trigonometric Forms

$$\int \sin u\, du = -\cos u \tag{150}$$

$$\int \sin^2 u\, du = \frac{u}{2} - \frac{1}{4}\sin 2u \tag{151}$$

$$\int \sin^n u\, du = -\frac{\sin^{n-1} u \cos u}{n} + \frac{n-1}{n}\int \sin^{n-2} u\, du \tag{152}$$

$$\int \cos u\, du = \sin u \tag{153}$$

$$\int \cos^2 u\, du = \frac{u}{2} + \frac{1}{4}\sin 2u \tag{154}$$

$$\int \cos^n u\, du = \frac{\cos^{n-1} u \sin u}{n} + \frac{n-1}{n}\int \cos^{n-2} u\, du \tag{155}$$

$$\int \tan u\, du = -\ln|\cos u| \tag{156}$$

$$\int \tan^2 u\, du = \tan u - u \tag{157}$$

$$\int \tan^n u\, du = \frac{\tan^{n-1} u}{n-1} - \int \tan^{n-2} u\, du \tag{158}$$

$$\int \cot u\, du = \ln|\sin u| \tag{159}$$

$$\int \cot^2 u\, du = -\cot u - u \tag{160}$$

$$\int \cot^n u\, du = -\frac{\cot^{n-1} u}{n-1} - \int \cot^{n-2} u\, du \tag{161}$$

$$\int \sec u\, du = \frac{1}{2}\ln\frac{1+\sin u}{1-\sin u} = \ln\left|\tan\left(\frac{\pi}{4}+\frac{u}{2}\right)\right| \tag{162}$$

$$\int \sec^2 u\, du = \tan u \tag{163}$$

$$\int \sec^n u\, du = \frac{\sin u \sec^{n-1} u}{n-1} + \frac{n-2}{n-1}\int \sec^{n-2} u\, du \tag{164}$$

$$\int \csc u\, du = \ln\left|\tan\frac{u}{2}\right| \tag{165}$$

$$\int \csc^2 u\, du = -\cot u \tag{166}$$

$$\int \csc^n u\, du = -\frac{\cos u \csc^{n-1} u}{n-1} + \frac{n-2}{n-1}\int \csc^{n-2} u\, du \tag{167}$$

$$\int \sin u \cos u \, du = \frac{1}{2} \sin^2 u = -\frac{1}{2} \cos^2 u \tag{168}$$

$$\int \sin^n u \cos u \, du = \frac{\sin^{n+1} u}{n+1} \tag{169}$$

$$\int \cos^n u \sin u \, du = -\frac{\cos^{n+1} u}{n+1} \tag{170}$$

$$\int \sin^2 u \cos^2 u \, du = \frac{u}{8} - \frac{1}{32} \sin 4u \tag{171}$$

$$\begin{aligned} \int \sin^n u \cos^m u \, du &= -\frac{\sin^{n-1} u \cos^{m+1} u}{m+n} + \frac{n-1}{m+n} \int \sin^{n-2} u \cos^m u \, du \\ &= \frac{\sin^{n+1} u \cos^{m-1} u}{m+n} + \frac{m-1}{m+n} \int \sin^n u \cos^{m-2} u \, du \end{aligned} \tag{172}$$

$$\begin{aligned} \int \frac{\sin^n u \, du}{\cos^m u} &= \frac{1}{n-m} \left[-\frac{\sin^{n-1} u}{\cos^{m-1} u} + (n-1) \int \frac{\sin^{n-2} u \, du}{\cos^m u} \right] \\ &= \frac{1}{m-1} \left[\frac{\sin^{n+1} u}{\cos^{m-1} u} - (n-m+2) \int \frac{\sin^m u \, du}{\cos^{m-2} u} \right] \end{aligned} \tag{173}$$

$$\begin{aligned} \int \frac{\cos^n u \, du}{\sin^m u} &= -\frac{1}{m-1} \left[\frac{\cos^{n+1} u}{\sin^{m-1} u} + (n-m+2) \int \frac{\cos^n u \, du}{\sin^{m-2} u} \right] \\ &= \frac{1}{n-m} \left[\frac{\cos^{n-1} u}{\sin^{m-1} u} + (n-1) \int \frac{\cos^{n-2} u \, du}{\sin^m u} \right] \end{aligned} \tag{174}$$

$$\int \frac{du}{1 + \sin u} = -\tan \left(\frac{\pi}{4} - \frac{u}{2} \right) \tag{175}$$

$$\int \frac{du}{1 - \sin u} = \tan \left(\frac{\pi}{4} + \frac{u}{2} \right) \tag{176}$$

$$\int \frac{du}{1 + \cos u} = \tan \frac{u}{2} \tag{177}$$

$$\int \frac{du}{1 - \cos u} = -\cot \frac{u}{2} \tag{178}$$

$$\int \frac{du}{a + b \sin u} = \frac{2}{\sqrt{a^2 - b^2}} \tan^{-1} \left[\frac{a \tan (u/2) + b}{\sqrt{a^2 - b^2}} \right] \qquad \text{when } a^2 > b^2 \tag{179}$$

$$\int \frac{du}{a + b \sin u} = \frac{1}{\sqrt{b^2 - a^2}} \ln \left| \frac{a \tan (u/2) + b - \sqrt{b^2 - a^2}}{a \tan (u/2) + b + \sqrt{b^2 - a^2}} \right| \qquad \text{when } a^2 < b^2 \tag{180}$$

$$\int \frac{du}{a + b\cos u} = \frac{2}{\sqrt{a^2 - b^2}} \tan^{-1}\left(\sqrt{\frac{a-b}{a+b}} \tan\frac{u}{2}\right) \quad \text{when } a^2 > b^2 \tag{181}$$

$$\int \frac{du}{a + b\cos u} = \frac{1}{\sqrt{b^2 - a^2}} \ln\left|\frac{\sqrt{b-a}\tan(u/2) + \sqrt{b+a}}{\sqrt{b-a}\tan(u/2) - \sqrt{b+a}}\right| \quad \text{when } a^2 < b^2 \tag{182}$$

$$\int \sec u \tan u \, du = \sec u \tag{183}$$

$$\int \csc u \cot u \, du = -\csc u \tag{184}$$

$$\int \frac{du}{a^2\cos^2 u + b^2 \sin^2 u} = \frac{1}{ab} \tan^{-1}\left(\frac{b\tan u}{a}\right) \tag{185}$$

$$\int \sin mu \sin nu \, du = \frac{\sin(m-n)u}{2(m-n)} - \frac{\sin(m+n)u}{2(m+n)} \tag{186}$$

$$\int \cos mu \cos nu \, du = \frac{\sin(m-n)u}{2(m-n)} + \frac{\sin(m+n)u}{2(m+n)} \tag{187}$$

$$\int \sin mu \cos nu \, du = -\frac{\cos(m-n)u}{2(m-n)} - \frac{\cos(m+n)u}{2(m+n)} \tag{188}$$

$$\int \sin^{-1} u \, du = u \sin^{-1} u + \sqrt{1-u^2} \tag{189}$$

$$\int \cos^{-1} u \, du = u \cos^{-1} u - \sqrt{1-u^2} \tag{190}$$

$$\int \tan^{-1} u \, du = u \tan^{-1} u - \frac{1}{2} \ln(1+u^2) \tag{191}$$

$$\int \cot^{-1} u \, du = u \cot^{-1} u + \frac{1}{2} \ln(1+u^2) \tag{192}$$

$$\int \sec^{-1} u \, du = u \sec^{-1} u - \ln\left|u + \sqrt{u^2-1}\right| \tag{193}$$

$$\int \csc^{-1} u \, du = u \csc^{-1} u + \ln\left|u + \sqrt{u^2-1}\right| \tag{194}$$

5 Algebraic-Trigonometric Forms

$$\int u \sin u \, du = \sin u - u\cos u \tag{195}$$

$$\int u^2 \sin u \, du = 2u \sin u - (u^2 - 2) \cos u \tag{196}$$

$$\int u^n \sin u \, du = -u^n \cos u + n \int u^{n-1} \cos u \, du \tag{197}$$

$$\int u \cos u \, du = \cos u + u \sin u \tag{198}$$

$$\int u^2 \cos u \, du = 2u \cos u + (u^2 - 2) \sin u \tag{199}$$

$$\int u^n \cos u \, du = u^n \sin u - n \int u^{n-1} \sin u \, du \tag{200}$$

$$\int \frac{\sin u}{u} \, du = u - \frac{u^3}{3 \cdot 3!} + \frac{u^5}{5 \cdot 5!} - \frac{u^7}{7 \cdot 7!} + \cdots \tag{201}$$

$$\int \frac{\cos u}{u} \, du = \ln u - \frac{u^2}{2 \cdot 2!} + \frac{u^4}{4 \cdot 4!} - \frac{u^6}{6 \cdot 6!} + \cdots \tag{202}$$

$$\int \frac{u \, du}{1 + \sin u} = 2 \ln \left| \cos \left(\frac{\pi}{4} - \frac{u}{2} \right) \right| - u \tan \left(\frac{\pi}{4} - \frac{u}{2} \right) \tag{203}$$

$$\int \frac{u \, du}{1 - \sin u} = 2 \ln \left| \sin \left(\frac{\pi}{4} - \frac{u}{2} \right) \right| + u \cot \left(\frac{\pi}{4} - \frac{u}{2} \right) \tag{204}$$

$$\int \frac{u \, du}{1 + \cos u} = 2 \ln \left| \cos \frac{u}{2} \right| + u \tan \frac{u}{2} \tag{205}$$

$$\int \frac{u \, du}{1 - \cos u} = 2 \ln \left| \sin \frac{u}{2} \right| - u \cot \frac{u}{2} \tag{206}$$

6 Exponential-Trigonometric Forms

$$\int e^{au} \sin u \, du = \frac{e^{au}(a \sin u - \cos u)}{1 + a^2} \tag{207}$$

$$\int e^{au} \cos u \, du = \frac{e^{au}(\sin u + a \cos u)}{1 + a^2} \tag{208}$$

$$\int e^{au} \sin nu \, du = \frac{e^{au}(a \sin nu - n \cos nu)}{a^2 + n^2} \tag{209}$$

$$\int e^{au} \cos nu \, du = \frac{e^{au}(n \sin nu + a \cos nu)}{a^2 + n^2} \tag{210}$$

INDEX

ABOUT THE AUTHOR

Max Kurtz is a consulting engineer and educator, and a Life Member of the National Society of Professional Engineers. He has more than 30 years' experience training engineering professionals. His previous books include the *Handbook of Engineering Economics* and the *Comprehensive Structural Design Guide*. He is a biographee in *Who's Who in the World.*